Lipo Wang (Ed.)

Support Vector Machines: Theory and Applications

Studies in Fuzziness and Soft Computing, Volume 177

Editor-in-chief
Prof. Janusz Kacprzyk
Systems Research Institute
Polish Academy of Sciences
ul. Newelska 6
01-447 Warsaw
Poland
E-mail: kacprzyk@ibspan.waw.pl

Further volumes of this series can be found on our homepage: springeronline.com

Vol. 162. R. Khosla, N. Ichalkaranje, L.C. Jain
Design of Intelligent Multi-Agent Systems, 2005
ISBN 3-540-22913-2

Vol. 163. A. Ghosh, L.C. Jain (Eds.)
Evolutionary Computation in Data Mining, 2005
ISBN 3-540-22370-3

Vol. 164. M. Nikravesh, L.A. Zadeh, J. Kacprzyk (Eds.)
Soft Computing for Information Prodessing and Analysis, 2005
ISBN 3-540-22930-2

Vol. 165. A.F. Rocha, E. Massad, A. Pereira Jr.
The Brain: From Fuzzy Arithmetic to Quantum Computing, 2005
ISBN 3-540-21858-0

Vol. 166. W.E. Hart, N. Krasnogor, J.E. Smith (Eds.)
Recent Advances in Memetic Algorithms, 2005
ISBN 3-540-22904-3

Vol. 167. Y. Jin (Ed.)
Knowledge Incorporation in Evolutionary Computation, 2005
ISBN 3-540-22902-7

Vol. 168. Yap P. Tan, Kim H. Yap, Lipo Wang (Eds.)
Intelligent Multimedia Processing with Soft Computing, 2005
ISBN 3-540-22902-7

Vol. 169. C.R. Bector, Suresh Chandra
Fuzzy Mathematical Programming and Fuzzy Matrix Games, 2005
ISBN 3-540-23729-1

Vol. 170. Martin Pelikan
Hierarchical Bayesian Optimization Algorithm, 2005
ISBN 3-540-23774-7

Vol. 171. James J. Buckley
Simulating Fuzzy Systems, 2005
ISBN 3-540-24116-7

Vol. 172. Patricia Melin, Oscar Castillo
Hybrid Intelligent Systems for Pattern Recognition Using Soft Computing, 2005
ISBN 3-540-24121-3

Vol. 173. Bogdan Gabrys, Kauko Leiviskä, Jens Strackeljan (Eds.)
Do Smart Adaptive Systems Exist?, 2005
ISBN 3-540-24077-2

Vol. 174. Mircea Negoita, Daniel Neagu, Vasile Palade
Computational Intelligence: Engineering of Hybrid Systems, 2005
ISBN 3-540-23219-2

Vol. 175. Anna Maria Gil-Lafuente
Fuzzy Logic in Financial Analysis, 2005
ISBN 3-540-23213-3

Vol. 176. Udo Seiffert, Lakhmi C. Jain, Patric Schweizer (Eds.)
Bioinformatics Using Computational Intelligence Paradigms, 2005
ISBN 3-540-22901-9

Vol. 177. Lipo Wang (Ed.)
Support Vector Machines: Theory and Applications, 2005
ISBN 3-540-24388-7

Lipo Wang (Ed.)

Support Vector Machines: Theory and Applications

Springer

Professor Lipo Wang
Nanyang Technological University
School of Electrial & Electronic Engineering
Nanyang Avenue
Singapore 639798
Singapore
E-mail: elpwang@ntu.edu.sg

ISSN print edition: 1434-9922
ISSN electronic edition: 1860-0808

ISBN-13 978-3-642-06368-8 e-ISBN-13 978-3-540-32384-6

Springer is a part of Springer Science+Business Media
springeronline.com

Printed in The Netherlands

Cover design: E. Kirchner, Springer Heidelberg

Preface

The support vector machine (SVM) is a supervised learning method that generates input-output mapping functions from a set of labeled training data. The mapping function can be either a classification function, i.e., the category of the input data, or a regression function. For classification, nonlinear kernel functions are often used to transform input data to a high-dimensional feature space in which the input data become more separable compared to the original input space. Maximum-margin hyperplanes are then created. The model thus produced depends on only a subset of the training data near the class boundaries. Similarly, the model produced by Support Vector Regression ignores any training data that is sufficiently close to the model prediction. SVMs are also said to belong to "kernel methods".

In addition to its solid mathematical foundation in statistical learning theory, SVMs have demonstrated highly competitive performance in numerous real-world applications, such as bioinformatics, text mining, face recognition, and image processing, which has established SVMs as one of the state-of-the-art tools for machine learning and data mining, along with other soft computing techniques, e.g., neural networks and fuzzy systems.

This volume is composed of 20 chapters selected from the recent myriad of novel SVM applications, powerful SVM algorithms, as well as enlightening theoretical analysis. Written by experts in their respective fields, the first 12 chapters concentrate on SVM theory, whereas the subsequent 8 chapters emphasize practical applications, although the "decision boundary" separating these two categories is rather "fuzzy".

Kecman first presents an introduction on the SVM, explaining the basic theory and implementation aspects. In the chapter contributed by Ma and Cherkassky, a novel approach to nonlinear classification using a collection of several simple (linear) classifiers is proposed based on a new formulation of the learning problem called multiple model estimation. Pelckmans, Goethals, De Brabanter, Suykens, and De Moor describe componentwise Least Squares Support Vector Machines (LS-SVMs) for the estimation of additive models consisting of a sum of nonlinear components.

Motivated by the statistical query model, Mitra, Murthy and Pal study an active learning strategy to solve the large quadratic programming problem of SVM design in data mining applications. Kaizhu Huang, Haiqin Yang, King, and Lyu propose a unifying theory of the Maxi-Min Margin Machine (M4) that subsumes the SVM, the minimax probability machine, and the linear discriminant analysis. Vogt and Kecman present an active-set algorithm for quadratic programming problems in SVMs, as an alternative to working-set (decomposition) techniques, especially when the data set is not too large, the problem is ill-conditioned, or when high precision is needed.

Being aware of the abundance of methods for SVM model selection, Anguita, Boni, Ridella, Rivieccio, and Sterpi carefully analyze the most well-known methods and test some of them on standard benchmarks to evaluate their effectiveness. In an attempt to minimize bias, Peng, Heisterkamp, and Dai propose locally adaptive nearest neighbor classification methods by using locally linear SVMs and quasiconformal transformed kernels. Williams, Wu, and Feng discuss two geometric methods to improve SVM performance, i.e., (1) adapting kernels by magnifying the Riemannian metric in the neighborhood of the boundary, thereby increasing class separation, and (2) optimally locating the separating boundary, given that the distributions of data on either side may have different scales.

Song, Hu, and Xulei Yang derive a Kuhn-Tucker condition and a decomposition algorithm for robust SVMs to deal with overfitting in the presence of outliers. Lin and Sheng-de Wang design a fuzzy SVM with automatic determination of the membership functions. Kecman, Te-Ming Huang, and Vogt present the latest developments and results of the Iterative Single Data Algorithm for solving large-scale problems.

Exploiting regularization and subspace decomposition techniques, Lu, Plataniotis, and Venetsanopoulos introduce a new kernel discriminant learning method and apply the method to face recognition. Kwang In Kim, Jung, and Hang Joon Kim employ SVMs and neural networks for automobile license plate localization, by classifying each pixel in the image into the object of interest or the background based on localized color texture patterns. Mattera discusses SVM applications in signal processing, especially the problem of digital channel equalization. Chu, Jin, and Lipo Wang use SVMs to solve two important problems in bioinformatics, i.e., cancer diagnosis based on microarray gene expression data and protein secondary structure prediction.

Emulating the natural nose, Brezmes, Llobet, Al-Khalifa, Maldonado, and Gardner describe how SVMs are being evaluated in the gas sensor community to discriminate different blends of coffee, different types of vapors and nerve agents. Zhan presents an application of the SVM in inverse problems in ocean color remote sensing. Liang uses SVMs for non-invasive diagnosis of delayed gastric emptying from the cutaneous electrogastrograms (EGGs). Rojo-Álvarez, García-Alberola, Artés-Rodríguez, and Arenal-Maíz apply SVMs, together with bootstrap resampling and principal component analysis, to tachycardia discrimination in implantable cardioverter defibrillators.

I would like to express my sincere appreciation to all authors and reviewers who have spent their precious time and efforts in making this book a reality. I wish to especially thank Professor Vojislav Kecman, who graciously took on the enormous task of writing a comprehensive introductory chapter, in addition to his other great contributions to this book. My gratitude also goes to Professor Janusz Kacprzyk and Dr. Thomas Ditzinger for their kindest support and help with this book.

Singapore
January 2005

Lipo Wang

Contents

Support Vector Machines – An Introduction

V. Kecman

The University of Auckland, School of Engineering,
Auckland, New Zealand

This is a book about learning from empirical data (i.e., examples, samples, measurements, records, patterns or observations) by applying support vector machines (SVMs) a.k.a. kernel machines. The basic aim of this introduction[1] is to give, as far as possible, a condensed (but systematic) presentation of a novel learning paradigm embodied in SVMs. Our focus will be on the constructive learning algorithms for both the classification (pattern recognition) and regression (function approximation) problems. Consequently, we will not go into all the subtleties and details of the statistical learning theory (SLT) and structural risk minimization (SRM) which are theoretical foundations for the learning algorithms presented below. Instead, a quadratic programming based learning leading to parsimonious SVMs will be presented in a gentle way – starting with linear separable problems, through the classification tasks having overlapped classes but still a linear separation boundary, beyond the linearity assumptions to the nonlinear separation boundary, and finally to the linear and nonlinear regression problems. The adjective "parsimonious" denotes an SVM with a small number of support vectors. The scarcity of the model results from a sophisticated learning that matches the model capacity to the data complexity ensuring a good performance on the future, previously unseen, data.

Same as the neural networks or similarly to them, SVMs possess the well-known ability of being universal approximators of any multivariate function to any desired degree of accuracy. Consequently, they are of particular interest for modeling the unknown, or partially known, highly nonlinear, complex systems, plants or processes. Also, at the very beginning, and just to be sure what the whole book is about, we should state clearly when there is no need for an application of SVMs' model-building techniques. In short, whenever there exists a good and reliable analytical closed-form model (or it is possible to

[1]This introduction strictly follows the School of Engineering of The University of Auckland Report 616. The right to use the material from this report is received with gratitude.

V. Kecman: *Support Vector Machines – An Introduction*, StudFuzz **177**, 1–47 (2005)
www.springerlink.com

devise one) there is no need to resort to learning from empirical data by SVMs (or by any other type of a learning machine).

1 Basics of Learning from Data

SVMs have been developed in the reverse order to the development of neural networks (NNs). SVMs evolved from the sound theory to the implementation and experiments, while the NNs followed more heuristic path, from applications and extensive experimentation to the theory. It is interesting to note that the very strong theoretical background of SVMs did not make them widely appreciated at the beginning. The publication of the first papers by Vapnik, Chervonenkis and co-workers in 1964/65 went largely unnoticed till 1992. This was due to a widespread belief in the statistical and/or machine learning community that, despite being theoretically appealing, SVMs are neither suitable nor relevant for practical applications. They were taken seriously only when excellent results on practical learning benchmarks were achieved in digit recognition, computer vision and text categorization. Today, SVMs show better results than (or comparable outcomes to) NNs and other statistical models, on the most popular benchmark problems (see some results and comparisons in [3, 4, 12, 20]).

The learning problem setting for SVMs is as follows: there is some unknown and nonlinear dependency (mapping, function) $y = f(\mathbf{x})$ between some high-dimensional input vector $\mathbf{x}$ and scalar output y (or the vector output $\mathbf{y}$ as in the case of multiclass SVMs). There is no information about the underlying joint probability functions. Thus, one must perform a *distribution-free learning.* The only information available is a training data set $D = \{(\mathbf{x}_i, y_i) \in X \times Y\}, i = 1, l$, where l stands for the number of the training data pairs and is therefore equal to the size of the training data set D. Often, y_i is denoted as d_i, where d stands for a desired (target) value. Hence, SVMs belong to the supervised learning techniques.

Note that this problem is similar to the classic statistical inference. However, there are several very important differences between the approaches and assumptions in training SVMs and the ones in classic statistics and/or NNs modeling. Classic statistical inference is based on the following three fundamental assumptions:

1. Data can be modeled by a set of linear in parameter functions; this is a foundation of a parametric paradigm in learning from experimental data.
2. In the most of real-life problems, a stochastic component of data is the normal probability distribution law, that is, the underlying joint probability distribution is a Gaussian distribution.
3. Because of the second assumption, the induction paradigm for parameter estimation is the maximum likelihood method, which is reduced to the minimization of the sum-of-errors-squares cost function in most engineering applications.

All three assumptions on which the classic statistical paradigm relied turned out to be inappropriate for many contemporary real-life problems [35] because of the following facts:

1. Modern problems are high-dimensional, and if the underlying mapping is not very smooth the linear paradigm needs an exponentially increasing number of terms with an increasing dimensionality of the input space X (an increasing number of independent variables). This is known as "the curse of dimensionality".
2. The underlying real-life data generation laws may typically be very far from the normal distribution and a model-builder must consider this difference in order to construct an effective learning algorithm.
3. From the first two points it follows that the maximum likelihood estimator (and consequently the sum-of-error-squares cost function) should be replaced by a new induction paradigm that is uniformly better, in order to model non-Gaussian distributions.

In addition to the three basic objectives above, the novel SVMs' problem setting and inductive principle have been developed for standard contemporary data sets which are typically high-dimensional and sparse (meaning, the data sets contain small number of the training data pairs).

SVMs are the so-called "nonparametric" models. "Nonparametric" does not mean that the SVMs' models do not have parameters at all. On the contrary, their "learning" (selection, identification, estimation, training or tuning) is the crucial issue here. However, unlike in classic statistical inference, the parameters are not predefined and their number depends on the training data used. In other words, parameters that define the capacity of the model are data-driven in such a way as to match the model capacity to data complexity. This is a basic paradigm of the structural risk minimization (SRM) introduced by Vapnik and Chervonenkis and their coworkers that led to the new learning algorithm. Namely, there are two basic constructive approaches possible in designing a model that will have a good generalization property [33, 35]:

1. choose an appropriate structure of the model (order of polynomials, number of HL neurons, number of rules in the fuzzy logic model) and, keeping the estimation error (a.k.a. confidence interval, a.k.a. variance of the model) fixed in this way, minimize the training error (i.e., empirical risk), or
2. keep the value of the training error (a.k.a. an approximation error, a.k.a. an empirical risk) fixed (equal to zero or equal to some acceptable level), and minimize the confidence interval.

Classic NNs implement the first approach (or some of its sophisticated variants) and SVMs implement the second strategy. In both cases the resulting model should resolve the trade-off between under-fitting and over-fitting the training data. The final model structure (its order) should ideally *match the learning machines capacity with training data complexity.* This important difference in two learning approaches comes from the minimization of different

Table 1. Basic Models and Their Error (Risk) Functionals

Multilayer Perceptron (NN)	Regularization Network (Radial Basis Functions Network)	Support Vector Machine
$R = \underbrace{\sum_{i=1}^{l} (d_i - f(\mathbf{x}_i, \mathbf{w}))^2}_{\text{Closeness to data}}$	$R = \underbrace{\sum_{i=1}^{l} (d_i - f(\mathbf{x}_i, \mathbf{w}))^2}_{\text{Closeness to data}} + \lambda \underbrace{\lVert \mathbf{P}f \rVert^2}_{\text{Smoothness}}$	$R = \sum_{i=1}^{l} \underbrace{L\varepsilon}_{\text{Closeness to data}} + \underbrace{\Omega(l, h)}_{\text{capacity of a machine}}$

Closeness to data = training error, a.k.a. empirical risk

cost (error, loss) functionals. Table 1 tabulates the basic risk functionals applied in developing the three contemporary statistical models.

d_i stands for desired values, $\mathbf{w}$ is the weight vector subject to training, λ is a regularization parameter, $\mathbf{P}$ is a smoothness operator, L_ε is a loss function of SVMs', h is a VC dimension and Ω is a function bounding the capacity of the learning machine. In classification problems L_ε is typically 0–1 loss function, and in regression problems L_ε is the so-called Vapnik's ε-insensitivity loss (error) function

$$L_\varepsilon = |y - f(\mathbf{x},\ \mathbf{w})|_\varepsilon = \begin{cases} 0, & \text{if } |y - f(\mathbf{x},\ \mathbf{w})| \leq \varepsilon \\ |y - f(\mathbf{x}, \mathbf{w})| - \varepsilon\,, & \text{otherwise}\,. \end{cases} \tag{1}$$

where ε is a radius of a tube within which the regression function must lie, after the successful learning. (Note that for $\varepsilon = 0$, the interpolation of training data will be performed). It is interesting to note that [11] has shown that under some constraints the SV machine can also be derived from the framework of regularization theory rather than SLT and SRM. Thus, *unlike the classic adaptation algorithms (that work in the L_2 norm), SV machines represent novel learning techniques which perform SRM.* In this way, the SV machine creates a model with minimized VC dimension and when the VC dimension of the model is low, the expected probability of error is low as well. This means good performance on previously unseen data, i.e. a good generalization. This property is of particular interest because the model that generalizes well is a good model and not the model that performs well on training data pairs. Too good a performance on training data is also known as an extremely undesirable overfitting.

As it will be shown below, in the "simplest" pattern recognition tasks, support vector machines use a linear separating hyperplane to create a *classifier with a maximal margin.* In order to do that, the learning problem for the SV machine will be cast as a *constrained nonlinear optimization* problem. In this setting the cost function will be quadratic and the constraints linear (i.e., one will have to solve a classic *quadratic programming problem*).

In cases when given classes cannot be linearly separated in the original input space, the SV machine first (non-linearly) transforms the original input space into a higher dimensional *feature space.* This transformation can be achieved by using various nonlinear mappings; polynomial, sigmoidal as in multilayer perceptrons, RBF mappings having as the basis functions radially symmetric functions such as Gaussians, or multiquadrics or different spline functions. After this nonlinear transformation step, the task of an SV machine in finding the linear optimal separating hyperplane in this feature space is "relatively trivial". Namely, the optimization problem to solve in a feature space will be of the same kind as the calculation of a maximal margin separating hyperplane in the original input space for linearly separable classes. How, after the specific nonlinear transformation, nonlinearly separable problems in input space can become linearly separable problems in a feature space will be shown later.

In a probabilistic setting, there are three basic components in all learning-from-data tasks: A *generator* of random inputs $\mathbf{x}$, a *system* whose *training responses* $y(d)$ are used for training the learning machine, and a *learning machine* which, by using inputs $\mathbf{x}_i$ and system's responses y_i, should learn (estimate, model) the unknown dependency between these two sets of variables defined by the weight vector $\mathbf{w}$ (Fig. 1). This figure shows the most common learning setting that some readers may have already seen in various

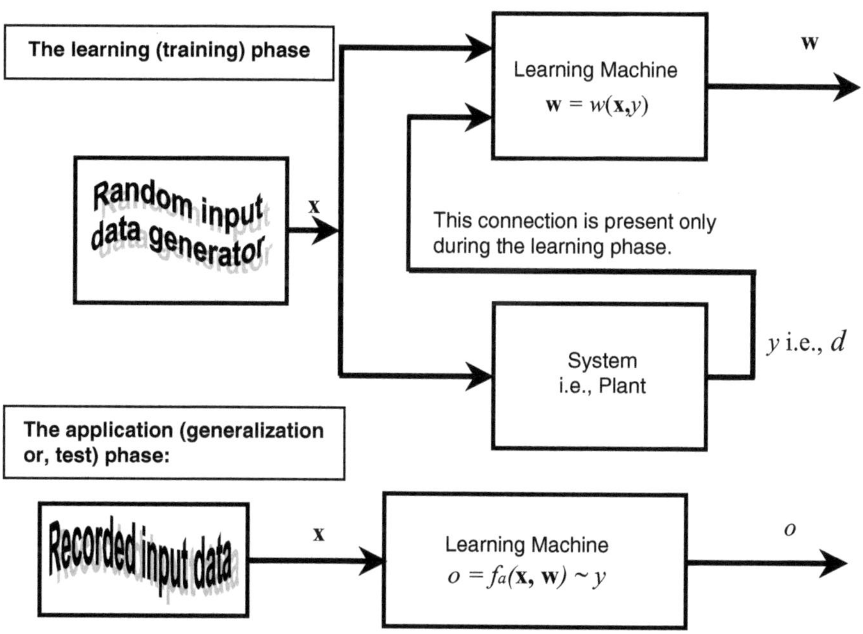

Fig. 1. A model of a learning machine (*top*) $\mathbf{w} = w(\mathbf{x}, y)$ that during *the training phase* (by observing inputs $\mathbf{x}_i$ to, and outputs y_i from, the system) estimates (learns, adjusts, trains, tunes) its parameters (weights) $\mathbf{w}$, and in this way learns mapping $y = f(\mathbf{x}, \mathbf{w})$ performed by the system. The use of $f_a(\mathbf{x}, \mathbf{w}) \sim y$ denotes that *we will rarely* try to *interpolate* training data pairs. We would rather seek an *approximating function* that can generalize well. After the training, at the *generalization* or *test phase*, the output from a machine $o = f_a(\mathbf{x}, \mathbf{w})$ is expected to be "a good" estimate of a system's true response y

other fields – notably in statistics, NNs, control system identification and/or signal processing. During the (successful) training phase a learning machine should be able to find the relationship between an input space X and an output space Y, by using data D in regression tasks (or to find a function that separates data within the input space, in classification ones). The result of a learning process is an "approximating function" $f_a(\mathbf{x}, \mathbf{w})$, which in statistical literature is also known as, a *hypothesis* $f_a(\mathbf{x}, \mathbf{w})$. This function approximates the underlying (or true) dependency between the input and output in the case of regression, and the decision boundary, i.e., separation function, in a classification. The chosen hypothesis $f_a(\mathbf{x}, \mathbf{w})$ belongs to a *hypothesis space of functions* $H(f_a \in H)$, and it is a function that minimizes some *risk functional* $R(\mathbf{w})$.

It may be practical to remind the reader that under the general name "approximating function" we understand any mathematical structure that maps inputs $\mathbf{x}$ into outputs y. Hence, an "approximating function" may be: a multilayer perceptron NN, RBF network, SV machine, fuzzy model, Fourier truncated series or polynomial approximating function. Here we discuss SVMs. A set of parameters $\mathbf{w}$ is the very subject of learning and generally these parameters are called *weights*. These parameters may have different geometrical and/or physical meanings. Depending upon the hypothesis space of functions H we are working with, the parameters $\mathbf{w}$ are usually:

- the hidden and the output layer weights in multilayer perceptrons,
- the rules and the parameters (for the positions and shapes) of fuzzy subsets,
- the coefficients of a polynomial or Fourier series,
- the centers and (co)variances of Gaussian basis functions as well as the output layer weights of this RBF network,
- the support vector weights in SVMs.

There is another important class of functions in learning from examples tasks. A learning machine tries to capture an unknown *target function* $f_0(\mathbf{x})$ that is believed to belong to some target space T, or to a class T, that is also called a *concept class*. Note that we rarely know the target space T and that our learning machine generally does not belong to the same class of functions as an unknown target function $f_0(\mathbf{x})$. Typical examples of target spaces are continuous functions with s continuous derivatives in n variables; Sobolev spaces (comprising square integrable functions in n variables with s square integrable derivatives), band-limited functions, functions with integrable Fourier transforms, Boolean functions, etc. In the following, we will assume that the target space T is a space of differentiable functions. The basic problem we are facing stems from the fact that we know very little about the possible underlying function between the input and the output variables. All we have at our disposal is a training data set of labeled examples drawn by independently sampling an $(X \times Y)$ space according to some unknown probability distribution.

The learning-from-data problem is ill-posed. (This will be shown on Figs. 2 and 3 for a regression and classification examples respectively). The basic source of the ill-posedness of the problem is due to the infinite number of possible solutions to the learning problem. At this point, just for the sake of illustration, it is useful to remember that all functions that interpolate data points will result in a zero value for training error (empirical risk) as shown (in the case of regression) in Fig. 2. The figure shows a simple example of three-out-of-infinitely-many different interpolating functions of training data pairs sampled from a noiseless function $y = \sin(x)$.

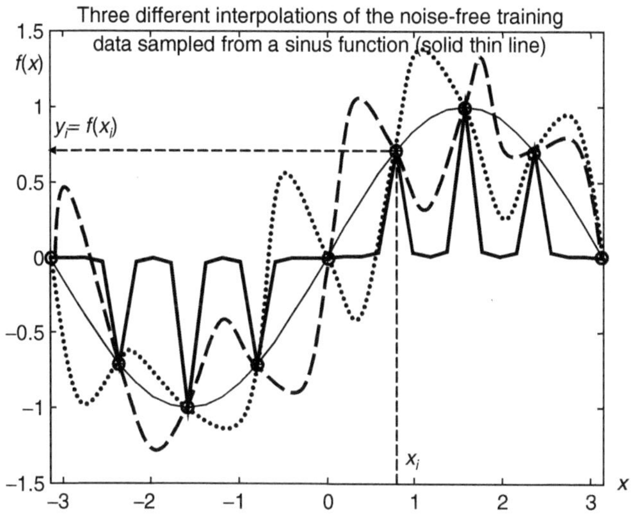

Fig. 2. Three-out-of-infinitely-many interpolating functions resulting in a training error equal to 0. However, a *thick solid*, *dashed* and *dotted lines* are bad models of a true function $y = \sin(x)$ (*thin dashed line*)

In Fig. 2, each interpolant results in a training error equal to zero, but at the same time, each one is a very bad model of the true underlying dependency between x and y, because all three functions perform very poorly outside the training inputs. In other words, none of these three particular interpolants can generalize well. However, not only interpolating functions can mislead. There are many other approximating functions (learning machines) that will minimize the empirical risk (approximation or training error) but not necessarily the generalization error (true, expected or guaranteed risk). This follows from the fact that a learning machine is trained by using some particular sample of the true underlying function and consequently it always produces biased approximating functions. These approximants depend necessarily on the specific training data pairs (i.e., the training sample) used.

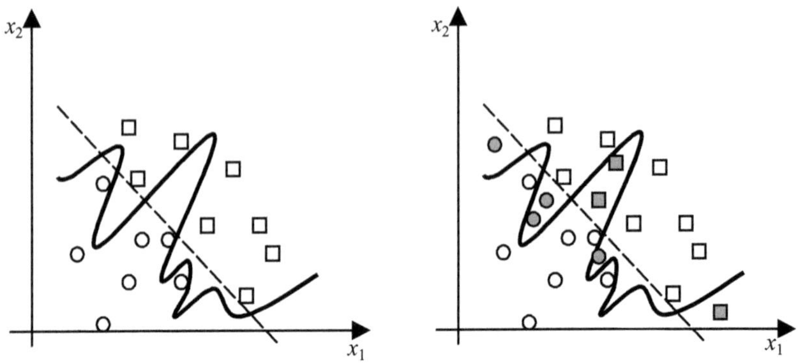

Fig. 3. Overfitting in the case of linearly separable classification problem. *Left:* The perfect classification of the training data (*empty circles* and *squares*) by both low order linear model (*dashed line*) and high order nonlinear one (*solid wiggly curve*). *Right:* Wrong classification of all the test data shown (*filled circles* and *squares*) by a high capacity model, but correct one by the simple linear separation boundary

Figure 3 shows an extremely simple classification example where the classes (represented by the empty training circles and squares) are linearly separable. However, in addition to a linear separation (dashed line) the learning was also performed by using a model of a high capacity (say, the one with Gaussian basis functions, or the one created by a high order polynomial, over the 2-dimensional input space) that produced a perfect separation boundary (empirical risk equals zero) too. However, such a model is overfitting the data and it will definitely perform very badly on, during the training unseen, test examples. Filled circles and squares in the right hand graph are all wrongly classified by the nonlinear model. Note that a simple linear separation boundary correctly classifies both the training and the test data.

A solution to this problem proposed in the framework of the SLT is restricting the hypothesis space H of approximating functions to a set smaller than that of the target function T while simultaneously controlling the flexibility (complexity) of these approximating functions. This is ensured by an introduction of a novel induction principle of the SRM and its algorithmic realization through the SV machine. The Structural Risk Minimization principle [31] tries to minimize an expected risk (the cost function) R comprising two terms as given in Table 1 for the SVMs $R = \Omega(l,h) + \sum_{i=1}^{l} L_\varepsilon = \Omega(l,h) + R_{\mathrm{emp}}$ and it is based on the fact that for the classification learning problem with a probability of at least $1-\eta$ the bound

$$R(\mathbf{w}_n) \leq \Omega\left(\frac{h}{l}, \frac{\ln(\eta)}{l}\right) + R_{\mathrm{emp}}(\mathbf{w}_n) , \tag{2a}$$

holds. The first term on the right hand side is named a VC confidence (confidence term or confidence interval) that is defined as

$$\Omega\left(\frac{h}{l}, \frac{\ln(\eta)}{l}\right) = \sqrt{\frac{h\left[\ln\left(\frac{2l}{h}\right)+1\right] - \ln\left(\frac{\eta}{4}\right)}{l}}\,. \tag{2b}$$

The parameter h is called the VC (Vapnik-Chervonenkis) dimension of a set of functions. It describes the capacity of a set of functions implemented in a learning machine. For binary classification h is the maximal number of points which can be separated (shattered) into two classes in all possible 2^h ways by using the functions of the learning machine.

An SV (learning) machine can be thought of as

- a set of functions implemented in an SVM,
- an induction principle and,
- an algorithmic procedure for implementing the induction principle on the given set of functions.

The notation for risks given above by using $R(\mathbf{w}_n)$ denotes that an expected risk is calculated over a set of functions $f_{an}(\mathbf{x}, \mathbf{w}_n)$ of increasing complexity. Different bounds can also be formulated in terms of other concepts such as *growth function* or *annealed VC entropy*. Bounds also differ for regression tasks. More detail can be found in ([33], as well as in [3]). However, the general characteristics of the dependence of the confidence interval on the number of training data l and on the VC dimension h is similar and given in Fig. 4.

Equations (2a) show that when the number of training data increases, i.e., for $l \to \infty$ (with other parameters fixed), an expected (true) risk $R(\mathbf{w}_n)$ is very close to empirical risk $R_{\text{emp}}(\mathbf{w}_n)$ because $\Omega \to 0$. On the other hand, when the probability $1-\eta$ (also called a confidence level which should not be confused with the confidence term Ω) approaches 1, the generalization bound grows large, because in the case when $\eta \to 0$ (meaning that the confidence level $1-\eta \to 1$), the value of $\Omega \to \infty$. This has an obvious intuitive interpretation [3] in that any learning machine (model, estimates) obtained from a finite number of training data cannot have an arbitrarily high confidence level. There is always a trade-off between the accuracy provided by bounds and the degree of confidence (in these bounds). Figure 4 also shows that the VC confidence interval increases with an increase in a VC dimension h for a fixed number of the training data pairs l.

The SRM is a novel inductive principle for learning from finite training data sets. It proved to be very useful when dealing with *small samples*. The basic idea of the SRM is to choose (from a large number of possibly candidate learning machines), a model of the right capacity to describe the given training data pairs. As mentioned, this can be done by restricting the hypothesis space H of approximating functions and simultaneously by controlling their flexibility (complexity). Thus, learning machines will be those parameterized models that, by increasing the number of parameters (typically called weights w_i here), form a nested structure in the following sense

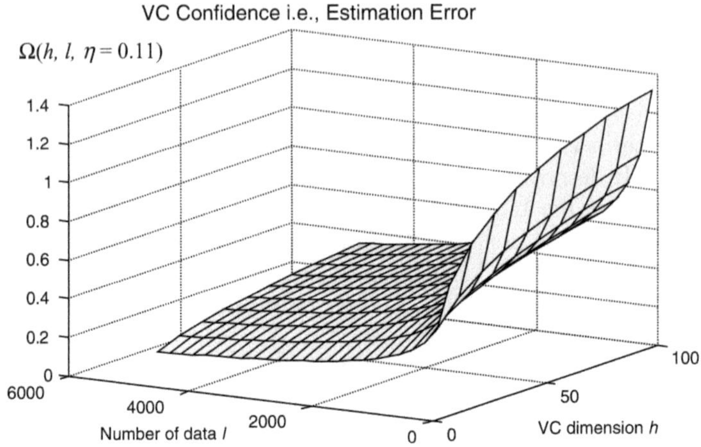

Fig. 4. The dependency of VC confidence $\Omega(h, l, \eta)$ on the number of training data l and the VC dimension $h(h < l)$ for a fixed confidence level $1 - \eta = 1 - 0.11 = 0.89$

$$H_1 \subset H_2 \subset H_3 \subset \ldots H_{n-1} \subset H_n \subset \ldots \subset H \tag{3}$$

In such a nested set of functions, every function always contains a previous, less complex, function. Typically, H_n may be: a set of polynomials in one variable of degree n; fuzzy logic model having n rules; multilayer perceptrons, or RBF network having n HL neurons, SVM structured over n support vectors. The goal of learning is one of a *subset selection* that matches training data complexity with approximating model capacity. In other words, a learning algorithm chooses an optimal polynomial degree or, an optimal number of HL neurons or, an optimal number of FL model rules, for a polynomial model or NN or FL model respectively. For learning machines linear in parameters, this complexity (expressed by the VC dimension) is given by the number of weights, i.e., by the number of "free parameters". For approximating models nonlinear in parameters, the calculation of the VC dimension is often not an easy task. Nevertheless, even for these networks, by using simulation experiments, one can find a model of appropriate complexity.

2 Support Vector Machines in Classification and Regression

Below, we focus on the algorithm for implementing the SRM induction principle on the given set of functions. It implements the strategy mentioned previously – it keeps the training error fixed and minimizes the confidence interval. We first consider a "simple" example of linear decision rules (i.e., the separating functions will be hyperplanes) for binary classification (dichotomization) of linearly separable data. In such a problem, we are able to perfectly classify data pairs, meaning that an empirical risk can be set to zero. It is the easiest classification problem and yet an excellent introduction of all relevant and important ideas underlying the SLT, SRM and SVM.

Our presentation will gradually increase in complexity. It will begin with a *Linear Maximal Margin Classifier for Linearly Separable Data* where there is no sample overlapping. Afterwards, we will allow some degree of overlapping of training data pairs. However, we will still try to separate classes by using linear hyperplanes. This will lead to the *Linear Soft Margin Classifier for Overlapping Classes.* In problems when linear decision hyperplanes are no longer feasible, the mapping of an input space into the so-called feature space (that "corresponds" to the HL in NN models) will take place resulting in the *Nonlinear Classifier.* Finally, in the subsection on *Regression by SV Machines* we introduce same approaches and techniques for solving regression (i.e., function approximation) problems.

2.1 Linear Maximal Margin Classifier for Linearly Separable Data

Consider the problem of binary classification or dichotomization. Training data are given as

$$(\mathbf{x}_1, y_1), (\mathbf{x}_2, y_2), \ldots, (\mathbf{x}_l, y_l), \mathbf{x} \in \Re^n, y \in \{+1, -1\} \tag{4}$$

For reasons of visualization only, we will consider the case of a two-dimensional input space, i.e., $\mathbf{x} \in \Re^2$. Data are linearly separable and there are many different hyperplanes that can perform separation (Fig. 5). (Actually, for $\mathbf{x} \in \Re^2$, the separation is performed by "planes" $w_1x_1 + w_2x_2 + b = 0$. In other words, the decision boundary, i.e., the separation line in input space is defined by the equation $w_1x_1 + w_2x_2 + b = 0$. How to find "the best" one?

The difficult part is that all we have at our disposal are sparse training data. Thus, we want to find the optimal separating function without knowing the underlying probability distribution $P(\mathbf{x}, y)$. There are many functions that can solve given pattern recognition (or functional approximation) tasks. In such a problem setting, the SLT (developed in the early 1960s by Vapnik and Chervonenkis) shows that it is crucial to restrict the class of functions implemented by a learning machine to one with a complexity that is suitable for the amount of available training data.

In the case of a classification of linearly separable data, this idea is transformed into the following approach – among all the hyperplanes that minimize the training error (i.e., empirical risk) find the one with the largest margin. This is an intuitively acceptable approach. Just by looking at Fig. 5 we will find that the dashed separation line shown in the right graph seems to promise good classification while facing previously unseen data (meaning, in the generalization phase). Or, at least, it seems to probably be better in generalization than the dashed decision boundary having smaller margin shown in the left

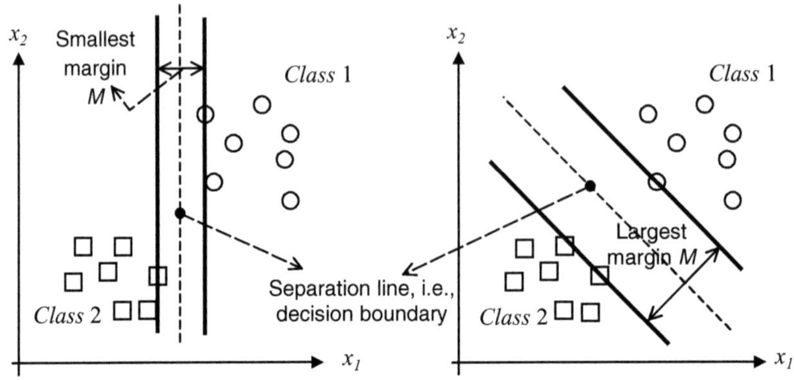

Fig. 5. Two-out-of-many separating lines: a good one with a large margin (*right*) and a less acceptable separating line with a small margin (*left*)

graph. This can also be expressed as that a classifier with smaller margin will have higher expected risk.

By using given training examples, during the learning stage, our machine finds parameters $\mathbf{w} = [w_1 w_2 \dots w_n]^T$ and b of a discriminant or decision function $d(\mathbf{x}, \mathbf{w}, b)$ given as

$$d(\mathbf{x}, \mathbf{w}, b) = \mathbf{w}^T \mathbf{x} + b = \sum_{i=1}^{n} w_i x_i + b \; , \tag{5}$$

where $\mathbf{x}, \mathbf{w} \in \Re^n$, and the scalar b is called *a bias*. (Note that the dashed separation lines in Fig. 5 represent the line that follows from $d(\mathbf{x}, \mathbf{w}, b) = 0$). After the successful training stage, by using the weights obtained, the learning machine, given previously unseen pattern $\mathbf{x}_p$, produces output 0 according to an *indicator function* given as

$$i_F = o = \text{sign}(d(\mathbf{x}_p, \mathbf{w}, b)) \; , \tag{6}$$

where 0 is the standard notation for the *output* from the learning machine. In other words the decision rule is:

$$\text{if}\, d(\mathbf{x}_p, \mathbf{w}, b) > 0, \text{the pattern}\, \mathbf{x}_p \,\text{belongs to a class 1 (i.e.,}\, o = y_1 = +1) \; ,$$

and

$$\text{if,}\, d(\mathbf{x}_p, \mathbf{w}, b) < 0 \,\text{the pattern}\, \mathbf{x}_p \,\text{belongs to a class 2 (i.e.,}\, o = y_2 = -1) \; .$$

The *indicator function* i_F given by (6) is a step-wise (i.e., a stairs-wise) function (see Figs. 6 and 7). At the same time, the decision (or discriminant) function $d(\mathbf{x}, \mathbf{w}, b)$ is a hyperplane. Note also that both a decision hyperplane d and the indicator function i_F live in an $n + 1$-dimensional space or they lie "over" a training pattern's n-dimensional input space. There is one more

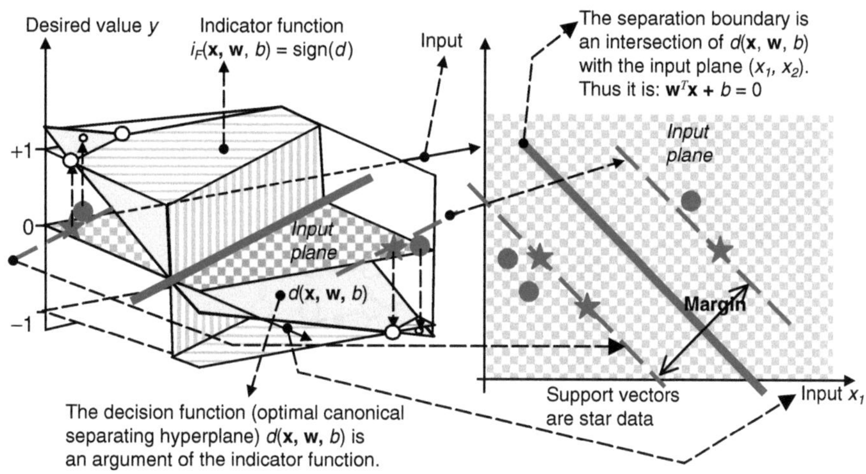

Fig. 6. The definition of a decision (discriminant) function or hyperplane $d(\mathbf{x}, \mathbf{w}, b)$, a decision (separating) boundary $d(\mathbf{x}, \mathbf{w}, b) = 0$ and an indicator function $i_F = \text{sign}(d(\mathbf{x}, \mathbf{w}, b))$ whose value represents a learning, or SV, machine's output 0

mathematical object in classification problems called a *separation boundary* that lives in the same n-dimensional space of input vectors $\mathbf{x}$. Separation boundary separates vectors $\mathbf{x}$ into two classes. Here, in cases of linearly separable data, the boundary is also a (separating) hyperplane but of a lower order than $d(\mathbf{x}, \mathbf{w}, b)$. The decision (separation) boundary is an intersection of a decision function $d(\mathbf{x},\ \mathbf{w}, b)$ and a space of input features. It is given by

$$d(\mathbf{x}, \mathbf{w}, b) = 0\ . \tag{7}$$

All these functions and relationships can be followed, for two-dimensional inputs $\mathbf{x}$, in Fig. 6. In this particular case, the decision boundary i.e., separating (hyper)plane is actually a separating line in an $x_1 - x_2$ plane and a decision function $d(\mathbf{x},\ \mathbf{w},\ b)$ is a plane over the 2-dimensional space of features, i.e., over an $x_1 - x_2$ plane.

In the case of 1-dimensional training patterns x (i.e., for 1-dimensional inputs x to the learning machine), decision function $d(x, \mathbf{w}, b)$ is a straight line in an x-y plane. An intersection of this line with an x-axis defines a point that is a separation boundary between two classes. This can be followed in Fig. 7. Before attempting to find an optimal separating hyperplane having the largest margin, we introduce the concept of the *canonical hyperplane*. We depict this concept with the help of the 1-dimensional example shown in Fig. 7.

Not quite incidentally, the decision plane $d(\mathbf{x}, \mathbf{w}, b)$ shown in Fig. 6 is also a *canonical* plane. Namely, the values of d and of i_F are the same and both are equal to $|1|$ for the support vectors depicted by stars. At the same time, for all other training patterns $|d| > |i_F|$. In order to present a notion

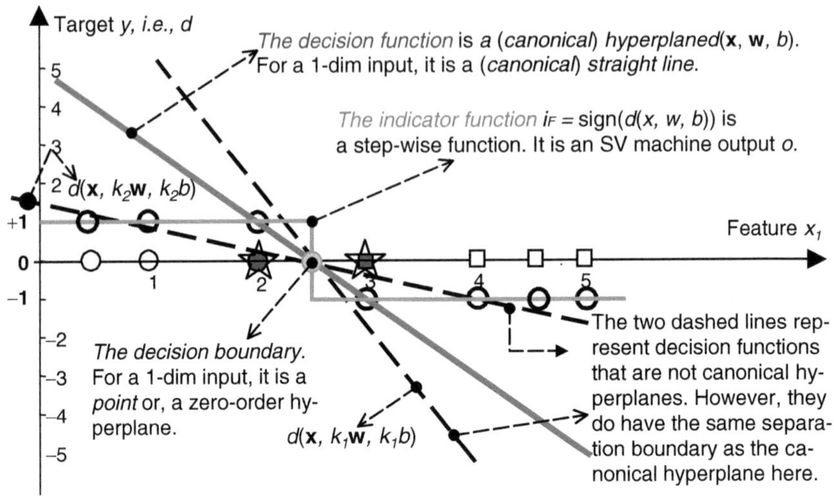

Fig. 7. SV classification for 1-dimensional inputs by the linear decision function. Graphical presentation of a *canonical hyperplane*. For 1-dimensional inputs, it is actually a canonical straight line (depicted as a *thick straight solid line*) that passes through points (+2, +1) and (+3, −1) defined as the support vectors (*stars*). The two *dashed lines* are the two other decision hyperplanes (i.e., *straight lines*). The training input patterns $\{x_1 = 0.5, x_2 = 1, x_3 = 2\} \in$ *Class* 1 have a desired or target value (label) $y_1 = +1$. The inputs $\{x_4 = 3, x_5 = 4, x_6 = 4.5, x_7 = 5\} \in$ *Class* 2 have the label $y_2 = -1$

of this new concept of the canonical plane, first note that there are many hyperplanes that can correctly separate data. In Fig. 7 three different decision functions $d(\mathbf{x}, \mathbf{w}, b)$ are shown. There are infinitely many more. In fact, given $d(\mathbf{x}, \mathbf{w}, b)$, all functions $d(\mathbf{x}, k\mathbf{w}, kb)$, where k is a positive scalar, are correct decision functions too. Because parameters $(\mathbf{w}, b)$ describe the same separation hyperplane as parameters $(k\mathbf{w}, kb)$ there is a need to introduce the notion of a *canonical hyperplane:*

A hyperplane is in the canonical form with respect to training data $\mathbf{x} \in X$ if

$$\underbrace{\min}_{x_i \in X} |\mathbf{w}^T \mathbf{x}_i + b| = 1 \ . \tag{8}$$

The solid line $d(\mathbf{x}, \mathbf{w}, b) = -2x + 5$ in Fig. 7 fulfills (8) because *its minimal absolute value for the given six training patterns* belonging to two classes is 1. It achieves this value for two patterns, chosen as support vectors, namely for $x_3 = 2$, and $x_4 = 3$. For all other patterns, $|d| > 1$.

Note an interesting detail regarding the notion of a canonical hyperplane that is easily checked. There are many different hyperplanes (planes and straight lines for 2-D and 1-D problems in Figs. 6 and 7 respectively) that have the same separation boundary (solid line and a dot in Figs. 6 (right) and 7 respectively). At the same time there are far fewer hyperplanes that can be defined as canonical ones fulfilling (8). In Fig. 7, i.e., for a 1-dimensional input vector x, the canonical hyperplane is unique. This is not the case for training patterns of higher dimension. Depending upon the configuration of class' elements, various canonical hyperplanes are possible.

Therefore, there is a need to define an *optimal* canonical hyperplane (OCSH) as a canonical hyperplane having a *maximal margin*. This search for a separating, maximal margin, canonical hyperplane is the ultimate learning goal in statistical learning theory underlying SV machines. Carefully note the adjectives used in the previous sentence. The hyperplane obtained from a limited training data must have a *maximal margin* because it will *probably* better classify new data. It must be in *canonical* form because this will ease the quest for significant patterns, here called support vectors. The canonical form of the hyperplane will also simplify the calculations. Finally, the resulting hyperplane must ultimately *separate* training patterns.

We avoid the derivation of an expression for the calculation of a distance (margin M) between the closest members from two classes for its simplicity. The curious reader can derive the expression for M as given below, or it can look in [15] or other books. The margin M can be derived by both the geometric and algebraic argument and is given as

$$M = \frac{2}{\|\mathbf{w}\|} \, . \tag{9}$$

This important result will have a great consequence for the constructive (i.e., learning) algorithm in a design of a maximal margin classifier. It will lead to solving a quadratic programming (QP) problem which will be shown shortly. Hence, the "good old" gradient learning in NNs will be replaced by solution of the QP problem here. This is the next important difference between the NNs and SVMs and follows from the implementation of SRM in designing SVMs, instead of a minimization of the sum of error squares, which is a standard cost function for NNs.

Equation (9) is a very interesting result showing that minimization of a norm of a hyperplane normal weight vector $\|\mathbf{w}\| = \sqrt{(\mathbf{w}^T\mathbf{w})} = \sqrt{w_1^2 + w_2^2 + \cdots + w_n^2}$ leads to a maximization of a margin M. Because a minimization of $\sqrt{f}$ is equivalent to the minimization of f, the minimization of a norm $\|\mathbf{w}\|$ equals a minimization of $\mathbf{w}^T\mathbf{w} = \sum_{i=1}^{n} w_i^2 = w_1^2 + w_2^2 + \cdots + w_n^2$, and this leads to a maximization of a margin M. Hence, the learning problem is

$$\text{minimize} \quad \frac{1}{2}\mathbf{w}^T\mathbf{w} \, , \tag{10a}$$

subject to constraints introduced and given in (10b) below. (A multiplication of $\mathbf{w}^T\mathbf{w}$ by 0.5 is for numerical convenience only, and it does not change the solution). Note that in the case of linearly separable classes empirical error equals zero ($R_{\text{emp}} = 0$ in (2a)) and minimization of $\mathbf{w}^T\mathbf{w}$ corresponds to a minimization of a confidence term Ω. The OCSH, i.e., a separating hyperplane with the largest margin defined by $M = 2/\|\mathbf{w}\|$, specifies *support vectors,* i.e., training data points closest to it, which satisfy $y_j[\mathbf{w}^T\mathbf{x}_j + b] \equiv 1,\ j = 1, N_{SV}$. For all the other (non-SVs data points) the OCSH satisfies inequalities $y_i[\mathbf{w}^T\mathbf{x}_i + b] > 1$. In other words, for all the data, OCSH should satisfy the following constraints

$$y_i[\mathbf{w}^T\mathbf{x}_i + b] \geq 1, \quad i = 1, \ldots, l \tag{10b}$$

where l denotes a number of training data points, and N_{SV} stands for a number of SVs. The last equation can be easily checked visually in Figs. 6 and 7 for 2-dimensional and 1-dimensional input vectors $\mathbf{x}$ respectively. Thus, in order to find the optimal separating hyperplane having a maximal margin, a learning machine should minimize $\|\mathbf{w}\|^2$ subject to the inequality constraints (10b). This is a *classic quadratic optimization problem with inequality constraints.* Such an optimization problem is solved by the *saddle point* of the Lagrange functional (Lagrangian)[2]

$$L(\mathbf{w}, b, \alpha) = \frac{1}{2}\mathbf{w}^T\mathbf{w} - \sum_{i=1}^{l} \alpha_i \{y_i[\mathbf{w}^T\mathbf{x}_i + b] - 1\}\ , \tag{11}$$

where the α_i are Lagrange multipliers. The search for an optimal *saddle point* $(\mathbf{w}_0, b_0, \alpha_0)$ is necessary because Lagrangian L must be *minimized* with respect to $\mathbf{w}$ and b, and has to be *maximized* with respect to nonnegative α_i (i.e., $\alpha_i \geq 0$ should be found). This problem can be solved either in a *primal space* (which is the space of parameters $\mathbf{w}$ and b) or in a *dual space* (which is the space of Lagrange multipliers α_i). The second approach gives insightful results and we will consider the solution in a dual space below. In order to do that, we use Karush-Kuhn-Tucker (KKT) conditions for the optimum of a constrained function. In our case, both the objective function (11) and constraints (10b) are *convex* and KKT conditions are *necessary* and *sufficient* conditions for a maximum of (11). These conditions are: at the saddle point $(\mathbf{w}_0,\ b_0,\ \alpha_0)$, derivatives of Lagrangian L with respect to primal variables should vanish which leads to,

$$\frac{\partial L}{\partial \mathbf{w}_0} = 0\ , \text{ i.e.,} \qquad \mathbf{w}_0 = \sum_{i=1}^{l} \alpha_i y_i \mathbf{x}_i\ , \tag{12}$$

$$\frac{\partial L}{\partial b_0} = 0\ , \text{ i.e.,} \qquad \sum_{i=1}^{l} \alpha_i y_i = 0\ , \tag{13}$$

and the KKT complementarity conditions below (stating that at the solution point the products between dual variables and constraints equals zero) must also be satisfied,

[2] In forming the Lagrangian, for constraints of the form $f_i > 0$, the inequality constraints equations are multiplied by *nonnegative* Lagrange multipliers (i.e., $\alpha_i \geq 0$) and *subtracted* from the objective function.

$$\alpha_i\{y_i[\mathbf{w}^T\mathbf{x}_i + b] - 1\} = 0, \quad i = 1,\ l\ . \tag{14}$$

Substituting (12) and (13) into a *primal variables Lagrangian* $L(\mathbf{w}, b, \boldsymbol{\alpha})$ (11), we change to the *dual variables Lagrangian* $L_d(\alpha)$

$$L_d(\alpha) = \sum_{i=1}^{l} \alpha_i - \frac{1}{2}\sum_{i,j=1}^{l} y_i y_j \alpha_i \alpha_j \mathbf{x}_i^T \mathbf{x}_j\ . \tag{15}$$

In order to find the optimal hyperplane, a dual Lagrangian $L_d(\boldsymbol{\alpha})$ has to be *maximized* with respect to nonnegative α_i (i.e., α_i must be in the nonnegative quadrant) and with respect to the equality constraint as follows

$$\alpha_i \geq 0\ , \qquad i = 1,\ l \tag{16a}$$

$$\sum_{i=1}^{l} \alpha_i y_i = 0 \tag{16b}$$

Note that the dual Lagrangian $L_d(\boldsymbol{\alpha})$ is expressed in terms of training data and depends *only* on the *scalar products* of input patterns $(\mathbf{x}_i^T\mathbf{x}_j)$. The dependency of $L_d(\boldsymbol{\alpha})$ on a scalar product of inputs will be very handy later when analyzing nonlinear decision boundaries and for general nonlinear regression. Note also that the number of unknown variables equals the number of training data l. After learning, the number of free parameters is equal to the number of SVs but it does not depend on the dimensionality of input space. Such a *standard quadratic optimization problem* can be expressed in a *matrix notation* and formulated as follows:

Maximize

$$L_d(\boldsymbol{\alpha}) = -0.5\boldsymbol{\alpha}^T\mathbf{H}\boldsymbol{\alpha} + \mathbf{f}^T\boldsymbol{\alpha}\ , \tag{17a}$$

subject to

$$\mathbf{y}^T\boldsymbol{\alpha} = 0\ , \tag{17b}$$

$$\alpha_i \geq 0, \quad i = 1,\ l\ , \tag{17c}$$

where $\boldsymbol{\alpha} = [\alpha_1, \alpha_2, \ldots, \alpha_l]^T$, $\mathbf{H}$ denotes the Hessian matrix ($H_{ij} = y_i y_j(\mathbf{x}_i\mathbf{x}_j) = y_i y_j \mathbf{x}_i^T\mathbf{x}_j$) of this problem, and $\mathbf{f}$ is an $(l,\ 1)$ unit vector $\mathbf{f} = \mathbf{1} = [1\ 1 \ldots 1]^T$. (Note that maximization of (17a) equals a minimization of $L_d(\boldsymbol{\alpha}) = 0.5\boldsymbol{\alpha}^T\mathbf{H}\boldsymbol{\alpha} - \mathbf{f}^T\boldsymbol{\alpha}$, subject to the same constraints). Solutions α_{0i} of the dual optimization problem above determine the parameters $\mathbf{w}_o$ and b_o of the optimal hyperplane according to (12) and (14) as follows

$$\mathbf{w}_o = \sum_{i=1}^{l} \alpha_{0i} y_i \mathbf{x}_i\ , \tag{18a}$$

$$\begin{aligned} b_o &= \frac{1}{N_{SV}}\left(\sum_{s=1}^{N_{SV}}\left(\frac{1}{y_s} - \mathbf{x}_s^T\mathbf{w}_o\right)\right. \\ &= \frac{1}{N_{SV}}\left(\sum_{s=1}^{N_{SV}}(y_s - \mathbf{x}_s^T\mathbf{w}_o)\right.\ , \quad s = 1,\ N_{SV}\ . \end{aligned} \tag{18b}$$

In deriving (18b) we used the fact that y can be either +1 or −1, and $1/y = y$. N_{SV} denotes the number of support vectors. There are two important observations about the calculation of $\mathbf{w}_o$. First, an optimal weight vector $\mathbf{w}_o$, is obtained in (18a) as a linear combination of the training data points and second, $\mathbf{w}_o$ (same as the bias term b_o) is calculated by using only the selected data points called *support vectors* (SVs). The fact that the summations in (18a) goes over all training data patterns (i.e., from 1 to l) is irrelevant because the Lagrange multipliers for all non-support vectors equal zero ($\alpha_{0i} = 0$, $i = N_{SV}+1,\ l$). Finally, having calculated $\mathbf{w}_o$ and b_o we obtain a decision hyperplane $d(\mathbf{x})$ and an indicator function $i_F = 0 = \text{sign}(d(\mathbf{x}))$ as given below

$$D(\mathbf{x}) = \sum_{i=1}^{l} w_{0i}x_i + b_o = \sum_{i=1}^{l} y_i\alpha_i\mathbf{x}_i^T\mathbf{x} + b_o\ , \quad i_F = 0 = \text{sign}(d(\mathbf{x}))\ . \tag{19}$$

Training data patterns having non-zero Lagrange multipliers are called *support vectors*. For linearly separable training data, all support vectors lie on the margin and they are generally just a small portion of all training data (typically, $N_{SV} \ll l$). Figures 6, 7 and 8 show the geometry of standard results for non-overlapping classes.

Before presenting applications of OCSH for both overlapping classes and classes having nonlinear decision boundaries, we will comment only on whether and how SV based linear classifiers actually implement the SRM principle. The more detailed presentation of this important property can be found in [15, 25].

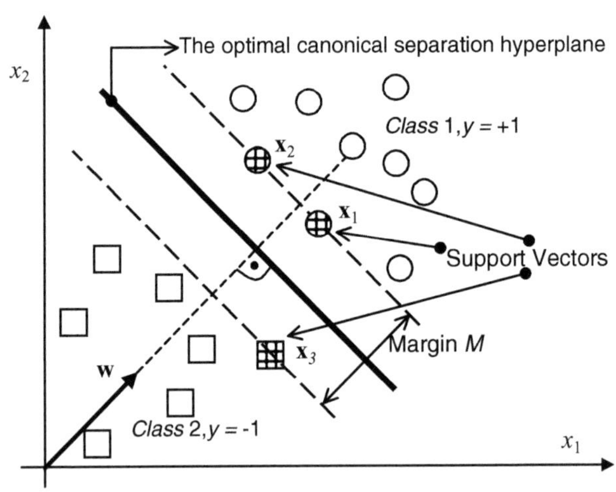

Fig. 8. The optimal canonical separating hyperplane with the largest margin intersects halfway between the two classes. The points closest to it (satisfying $y_j\left|\mathbf{w}^T\mathbf{x}_j + b\right| = 1, j = 1,\ N_{SV}$) are *support vectors* and the OCSH satisfies y_i $(\mathbf{w}^T\mathbf{x}_i + b) \geq 1\ i = 1,\ l$ (where l denotes the number of training data and N_{SV} stands for the number of SV). Three support vectors ($\mathbf{x}_1$ and $\mathbf{x}_2$ from class 1, and $\mathbf{x}_3$ from class 2) are the textured training data

First, it can be shown that an increase in margin reduces the number of points that can be shattered i.e., the increase in margin reduces the VC dimension, and this leads to the decrease of the SVM capacity. In short, by minimizing $\|\mathbf{w}\|$ (i.e., maximizing the margin) the SV machine training actually minimizes the VC dimension and consequently a generalization error (expected risk) at the same time. This is achieved by imposing a structure on the set of canonical hyperplanes and then, during the training, by choosing the one with a minimal VC dimension. A structure on the set of canonical hyperplanes is introduced by considering various hyperplanes having different $\|\mathbf{w}\|$. In other words, we analyze sets S_A such that $\|\mathbf{w}\| \leq A$. Then, if $A_1 \leq A_2 \leq A_3 \leq \ldots \leq A_n$, we introduced a nested set $S_{A1} \subset S_{A2} \subset S_{A3} \subset \ldots \subset S_{An}$. Thus, if we impose the constraint $\|\mathbf{w}\| \leq A$, then the canonical hyperplane cannot be closer than $1/A$ to any of the training points $\mathbf{x}_i$. Vapnik in [33] states that the VC dimension h of a set of canonical hyperplanes in $\Re^n$ such that $\|\mathbf{w}\| \leq A$ is

$$H \leq \min[R^2A^2, n] + 1 , \tag{20}$$

where all the training data points (vectors) are enclosed by a sphere of the smallest radius R. Therefore, a small $\|\mathbf{w}\|$ results in a small h, and minimization of $\|\mathbf{w}\|$ is an implementation of the SRM principle. In other words, a minimization of the canonical hyperplane weight norm $\|\mathbf{w}\|$ minimizes the VC dimension according to (20). See also Fig. 4 that shows how the estimation error, meaning the expected risk (because the empirical risk, due to the linear separability, equals zero) decreases with a decrease of a VC dimension. Finally, there is an interesting, simple and powerful result [33] connecting the generalization ability of learning machines and the number of support vectors. Once the support vectors have been found, we can calculate the bound on the expected probability of committing an error on a test example as follows

$$E_l[P(\text{error})] \leq \frac{E[\text{number of support vectors}]}{l} , \tag{21}$$

where E_l denotes expectation over all training data sets of size l. Note how easy it is to estimate this bound that is independent of the dimensionality of the input space. Therefore, an SV machine having a small number of support vectors will have good generalization ability even in a very high-dimensional space.

2.2 Linear Soft Margin Classifier for Overlapping Classes

The learning procedure presented above is valid for linearly separable data, meaning for training data sets without overlapping. Such problems are rare in practice. At the same time, there are many instances when linear separating hyperplanes can be good solutions even when data are overlapped (e.g., normally distributed classes having the same covariance matrices have a linear separation boundary). However, quadratic programming solutions as

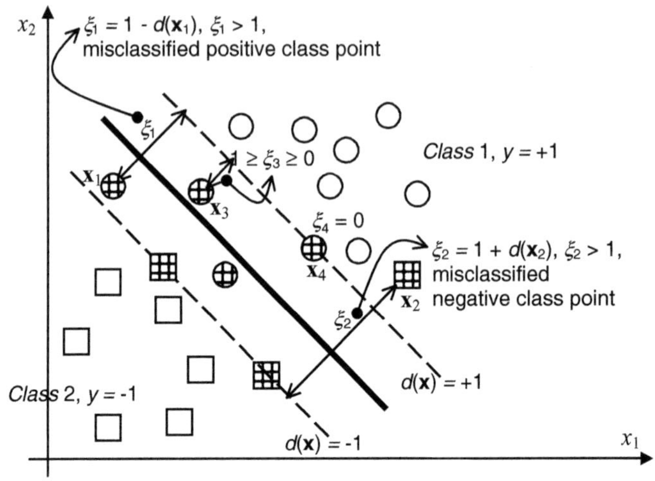

Fig. 9. The soft decision boundary for a dichotomization problem with data overlapping. Separation line (*solid*), margins (*dashed*) and support vectors (textured training data points). 4 SVs in positive class (*circles*) and 3 SVs in negative class (*squares*). 2 misclassifications for positive class and 1 misclassification for negative class

given above cannot be used in the case of overlapping because the constraints $y_i[\mathbf{w}^T\mathbf{x}_i + b] \geq 1,\ i = 1,\ l$ given by (10b) cannot be satisfied. In the case of an overlapping (see Fig. 9), the overlapped data points cannot be correctly classified and for any misclassified training data point $\mathbf{x}_i$, the corresponding α_i will tend to infinity. This particular data point (by increasing the corresponding α_i value) attempts to exert a stronger influence on the decision boundary in order to be classified correctly (see Fig. 9). When the α_i value reaches the maximal bound, it can no longer increase its effect, and the corresponding point will stay misclassified. In such a situation, the algorithm introduced above chooses (almost) all training data points as support vectors. To find a classifier with a maximal margin, the algorithm presented in Sect. 2.1 above, must be changed allowing some data to be unclassified. Better to say, we must leave some data on the "wrong" side of a decision boundary. In practice, we allow a *soft* margin and all data inside this margin (whether on the correct side of the separating line or on the wrong one) are neglected. The width of a soft margin can be controlled by a corresponding penalty parameter C (introduced below) that determines the trade-off between the training error and VC dimension of the model.

The question now is how to measure the degree of misclassification and how to incorporate such a measure into the hard margin learning algorithm given by (10a). The simplest method would be to form the following learning problem

$$\text{minimize} \quad \frac{1}{2}\mathbf{w}^T\mathbf{w} + C \text{ (number of misclassified data)} \,, \tag{22}$$

where C is a penalty parameter, trading off the margin size (defined by $\|\mathbf{w}\|$, i.e., by $\mathbf{w}^T\mathbf{w}$) for the number of misclassified data points. Large C leads to small number of misclassifications, bigger $\mathbf{w}^T\mathbf{w}$ and consequently to the smaller margin and vice versa. Obviously taking $C = \infty$ requires that the number of misclassified data is zero and, in the case of an overlapping this is not possible. Hence, the problem may be feasible only for some value $C < \infty$.

However, the serious problem with (22) is that the error's counting cannot be accommodated within the handy (meaning reliable, well understood and well developed) quadratic programming approach. Also, the counting only cannot distinguish between huge (or disastrous) errors and close misses! The possible solution is to measure the distances ξ_i of the points crossing the margin from the corresponding margin and trade their sum for the margin size as given below

$$\text{minimize} \ \frac{1}{2}\mathbf{w}^T\mathbf{w} + C \text{ (sum of distances of the wrong side points)} \,, \tag{23}$$

In fact this is exactly how the problem of the data overlapping was solved in [5, 6] – by generalizing the optimal "hard" margin algorithm. They introduced the nonnegative *slack variables* ξ_i $(i = 1,\ l)$ in the statement of the optimization problem for the overlapped data points. Now, instead of fulfilling (10a) and (10b), the separating hyperplane must satisfy

$$\text{minimize} \quad \frac{1}{2}\mathbf{w}^T\mathbf{w} + C\sum_{i=1}^{l}\xi_i \,, \tag{24a}$$

subject to

$$y_i[\mathbf{w}^T\mathbf{x}_i + b] \geq 1 - \xi_i,\ i = 1,\ l,\ \xi_i \geq 0 \,, \tag{24b}$$

i.e., subject to

$$\mathbf{w}^T\mathbf{x}_i + b \geq +1 - \xi_i,\ \text{ for } y_i = +1,\ \xi_i \geq 0, \tag{24c}$$
$$\mathbf{w}^T\mathbf{x}_i + b \leq -1 + \xi_i,\ \text{ for } y_i = -1, \xi_i \geq 0 \,. \tag{24d}$$

Hence, for such a *generalized* optimal separating hyperplane, the functional to be minimized comprises an extra term accounting the cost of overlapping errors. In fact the cost function (24a) can be even more general as given below

$$\text{minimize} \quad \frac{1}{2}\mathbf{w}^T\mathbf{w} + C\sum_{i=1}^{l}\xi_i^k \,, \tag{24e}$$

subject to same constraints. This is a convex programming problem that is usually solved only for $k = 1$ or $k = 2$, and such soft margin SVMs are dubbed $L1$ and $L2$ SVMs respectively. By choosing exponent $k = 1$, neither slack variables ξ_i nor their Lagrange multipliers β_i appear in a dual Lagrangian L_d. Same as for a linearly separable problem presented previously, for $L1$ SVMs ($k = 1$) here, the solution to a quadratic programming problem (24), is given by the saddle point of the primal Lagrangian $L_p(\mathbf{w}, b, \xi, \boldsymbol{\alpha}, \boldsymbol{\beta})$ shown below

$$L_p(\mathbf{w}, b, \xi, \boldsymbol{\alpha}, \boldsymbol{\beta}) = \frac{1}{2}\mathbf{w}^T\mathbf{w} + C\left(\sum_{i=1}^{l}\xi_i\right) - \sum_{i=1}^{l}\alpha_i\left\{y_i\left[\mathbf{w}^T\mathbf{x}_i + b\right] - 1 + \xi_i\right\} - \sum_{i=1}^{l}\beta_i\xi_i, \text{ for } L1 \text{ SVM} \tag{25}$$

where α_i and β_i are the Lagrange multipliers. Again, we should find an *optimal* saddle point $(\mathbf{w}_0, b_0, \xi_0, \boldsymbol{\alpha}_0, \boldsymbol{\beta}_0)$ because the Lagrangian L_p has to be *minimized* with respect to $\mathbf{w}$, b and ξ, and *maximized* with respect to nonnegative α_i and β_i. As before, this problem can be solved in either a *primal space* or *dual space* (which is the space of Lagrange multipliers α_i and β_i). Again, we consider a solution in a dual space as given below by using

- standard conditions for an optimum of a constrained function

$$\frac{\partial L}{\partial \mathbf{w}_0} = 0, \text{ i.e., } \mathbf{w}_0 = \sum_{i=1}^{l}\alpha_i y_i \mathbf{x}_i\ , \tag{26}$$

$$\frac{\partial L}{\partial b_0} = 0, \text{ i.e., } \sum_{i=1}^{l}\alpha_i y_i = 0\ , \tag{27}$$

$$\frac{\partial L}{\partial \xi_{i0}} = 0, \text{ i.e., } \alpha_i + \beta_i = C\ , \tag{28}$$

and the KKT complementarity conditions below,

$$\alpha_i\{y_i[\mathbf{w}^T\mathbf{x}_i + b] - 1 + \xi_i\} = 0,\ i = 1,\ l\ , \tag{29a}$$

$$\beta_i\xi_i = (C - \alpha_i)\xi_i = 0,\ i = 1,\ l\ . \tag{29b}$$

At the optimal solution, due to the KKT conditions (29), the last two terms in the primal Lagrangian L_p given by (25) vanish and the *dual variables Lagrangian* $L_d(\alpha)$, for $L1$ SVM, is not a function of β_i. In fact, it is same as the hard margin classifier's L_d given before and repeated here for the soft margin one,

$$L_d(\alpha) = \sum_{i=1}^{l}\alpha_i - \frac{1}{2}\sum_{i,j=1}^{l}y_i y_j \alpha_i \alpha_j \mathbf{x}_i^T\mathbf{x}_j\ . \tag{30}$$

In order to find the optimal hyperplane, a dual Lagrangian $L_d(\alpha)$ has to be *maximized* with respect to nonnegative and (unlike before) smaller than or equal to C, α_i. In other words with

$$C \geq \alpha_i \geq 0, \qquad i = 1,\ l\ , \tag{31a}$$

and under the constraint (27), i.e., under

$$\sum_{i=1}^{l} \alpha_i y_i = 0 \,. \tag{31b}$$

Thus, the final quadratic optimization problem is practically the same as that for the separable case, with the only difference being in the modified bounds of the Lagrange multipliers α_i. The penalty parameter C, which is now the upper bound on α_i, is determined by the user. The selection of a "good" or "proper" C is always done experimentally by using some cross-validation technique. Note that in the previous linearly separable case, without data overlapping, this upper bound $C = \infty$. We can also readily change to the matrix notation of the problem above as in (17a). Most important of all is that the learning problem is expressed only in terms of unknown Lagrange multipliers α_i, and known inputs and outputs. Furthermore, optimization does not solely depend upon inputs $\mathbf{x}_i$ which can be of a very high(inclusive of an infinite) dimension, but it depends upon a scalar product of input vectors $\mathbf{x}_i$. It is this property we will use in the next section where we design SV machines that can create nonlinear separation boundaries. Finally, expressions for both a *decision function* $d(\mathbf{x})$ and an *indicator function* $i_F = \text{sign}(d(\mathbf{x}))$ for a soft margin classifier are same as for linearly separable classes and are also given by (19).

From (29a) follows that there are only three possible solutions for α_i (see Fig. 9):

1. $\alpha_i = 0,\ \xi_i = 0$, data point $\mathbf{x}_i$ is correctly classified,
2. $C > \alpha_i > 0, \rightarrow$ then, the two complementarity conditions must result in $y_i[\mathbf{w}^T\mathbf{x}_i + b] - 1 + \xi_i = 0$, and $\xi_i = 0$. Thus, $y_i[\mathbf{w}^T\mathbf{x}_i + b] = 1$ and $\mathbf{x}_i$ is a support vector. The support vectors with $C \geq \alpha_i \geq 0$ are called *unbounded* or *free support vectors.* They lie on the two margins,
3. $\alpha_i = C, \rightarrow$ then, $y_i[\mathbf{w}^T\mathbf{x}_i + b] - 1 + \xi_i = 0$, and $\xi_i \geq 0$, and $\mathbf{x}_i$ is a support vector. The support vectors with $\alpha_i = C$ are called *bounded support vectors.* They lie on the "wrong" side of the margin. For $1 > \xi_i \geq 0$, $\mathbf{x}_i$ is still correctly classified, and if $\xi_i \geq 1$, $\mathbf{x}_i$ is misclassified.

For $L2$ SVM the second term in the cost function (24e) is quadratic, i.e., $C\sum_{i=1}^{l} \xi_i^2$, and this leads to changes in a dual optimization problem,

$$L_d(\alpha) = \sum_{i=1}^{l} \alpha_i - \frac{1}{2} \sum_{i,j=1}^{l} y_i y_j \alpha_i \alpha_j \left(\mathbf{x}_i^T \mathbf{x}_j + \frac{\delta_{ij}}{C} \right), \tag{32}$$

$$\text{subject to } \alpha_i \geq 0, i = 1,\ l, \text{ and } \sum_{i=1}^{l} \alpha_i y_i = 0 \,. \tag{33}$$

where, $\delta_{ij} = 1$ for $i = j$, and it is zero otherwise. Note the change in Hessian matrix elements given by second terms in (32), as well as that there is no longer an upper bound on α_i. The $1/C$ is added to the diagonal entries of $\mathbf{H}$ ensuring its positive definiteness and stabilizing the solution. The detailed analysis and comparisons of the $L1$ and $L2$ SVMs is presented in [1]. It can be also found in the author's report 616 [17]. We use the most popular $L1$ SVMs here, because they usually produce more sparse solutions, i.e., they create a decision function by using fewer SVs than the $L2$ SVMs.

2.3 The Nonlinear Classifier

The linear classifiers presented in two previous sections are very limited. Mostly, classes are not only overlapped but the genuine separation functions are nonlinear hypersurfaces. A nice and strong characteristic of the approach presented above is that it can be easily (and in a relatively straightforward manner) extended to create nonlinear decision boundaries. The motivation for such an extension is that an SV machine that can create a nonlinear decision hypersurface will be able to classify nonlinearly separable data. This will be achieved by considering a linear classifier in the so-called *feature space* that will be introduced shortly. A very simple example of a need for designing nonlinear models is given in Fig. 10 where the true separation boundary is quadratic. It is obvious that no errorless linear separating hyperplane can be found now. The best linear separation function shown as a dashed straight line would make six misclassifications (textured data points; 4 in the negative class and 2 in the positive one). Yet, if we use the nonlinear separation

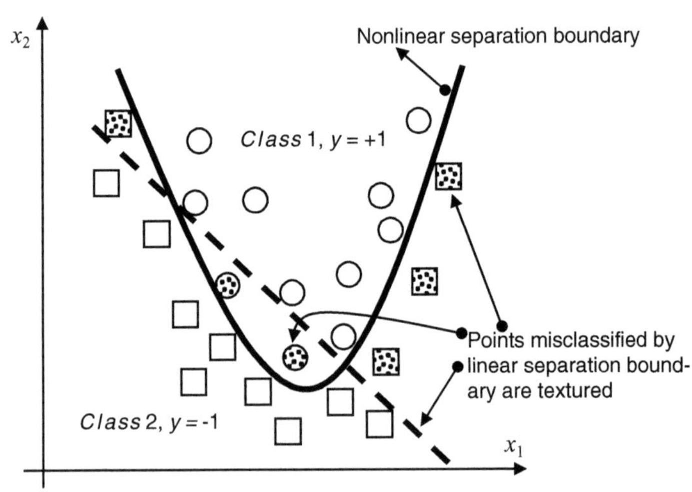

Fig. 10. A nonlinear SVM without data overlapping. A true separation is a quadratic curve. The nonlinear separation line (*solid*), the linear one (*dashed*) and data points misclassified by the linear separation line (the textured training data points) are shown. There are 4 misclassified negative data and 2 misclassified positive ones. SVs are not shown

boundary we are able to separate two classes without any error. Generally, for n-dimensional input patterns, instead of a nonlinear curve, an SV machine will create a nonlinear separating hypersurface.

The basic idea in designing nonlinear SV machines is to map input vectors $\mathbf{x} \in \Re^n$ into vectors $\mathbf{\Phi}(\mathbf{x})$ of a higher dimensional *feature space* F (where $\mathbf{\Phi}$ represents mapping: $\Re^n \rightarrow \Re^f$), and to solve a linear classification problem in this feature space

$$\mathbf{x} \in \Re^n \rightarrow \mathbf{\Phi}(\mathbf{x}) = [\phi_1(\mathbf{x})\ \phi_2(\mathbf{x}), \ldots, \phi_n(\mathbf{x})]^T \in \Re^f \ , \qquad (34)$$

A mapping $\mathbf{\Phi}(\mathbf{x})$ is chosen in advance. i.e., it is a fixed function. Note that an input space ($\mathbf{x}$-space) is spanned by components x_i of an input vector $\mathbf{x}$ and a feature space F ($\mathbf{\Phi}$-space) is spanned by components $\phi_i(\mathbf{x})$ of a vector $\mathbf{\Phi}(\mathbf{x})$. By performing such a mapping, we hope that in a $\mathbf{\Phi}$-space, our learning algorithm will be able to linearly separate images of $\mathbf{x}$ by applying the linear SVM formulation presented above. (In fact, it can be shown that for a whole class of mappings the linear separation in a feature space is always possible. Such mappings will correspond to the positive definite kernels that will be shown shortly). We also expect this approach to again lead to solving a quadratic optimization problem with similar constraints in a $\mathbf{\Phi}$-space. The solution for an indicator function $i_F(\mathbf{x}) = \text{sign}(\mathbf{w}^T\mathbf{\Phi}(\mathbf{x}) + b) = \text{sign}(\sum_{i=1}^{l} y_i\alpha_i\mathbf{\Phi}^T(\mathbf{x}_i)\mathbf{\Phi}(\mathbf{x}) + b)$, which is a linear classifier in a feature space, will create a nonlinear separating hypersurface in the original input space given by (35) below. (Compare this solution with (19) and note the appearances of scalar products in both the original X-space and in the feature space F). The equation for an $i_F(\mathbf{x})$ just given above can be rewritten in a "neural networks" form as follows

$$\begin{aligned} i_F(\mathbf{x}) &= \text{sign}\left(\sum_{i=1}^{l} y_i\alpha_i\mathbf{\Phi}^T(\mathbf{x}_i)\mathbf{\Phi}(\mathbf{x}) + b\right) \\ &= \text{sign}\left(\sum_{i=1}^{l} y_i\alpha_i k(\mathbf{x}_i, \mathbf{x}) + b\right) = \text{sign}\left(\sum_{i=1}^{l} v_i k(\mathbf{x}_i, \mathbf{x}) + b\right) \qquad (35) \end{aligned}$$

where v_i corresponds to the output layer weights of the "SVM's network" and $k(\mathbf{x}_i, \mathbf{x})$ denotes the value of the kernel function that will be introduced shortly. (v_i equals $y_i\alpha_i$ in the classification case presented above and it is equal to $(\alpha_i - \alpha_i^*)$ in the regression problems). Note the difference between the weight vector $\mathbf{w}$ which norm should be minimized and which is the vector of the same dimension as the feature space vector $\mathbf{\Phi}(\mathbf{x})$ and the weightings $v_i = \alpha_i y_i$ that are scalar values composing the weight vector $\mathbf{v}$ which dimension equals the number of training data points l. The $(l - N_{SVs})$ of v_i components are equal to zero, and only N_{SVs} entries of $\mathbf{v}$ are nonzero elements.

A simple example below (Fig. 11) should exemplify the idea of a nonlinear mapping to (usually) higher dimensional space and how it happens that the data become linearly separable in the F-space.

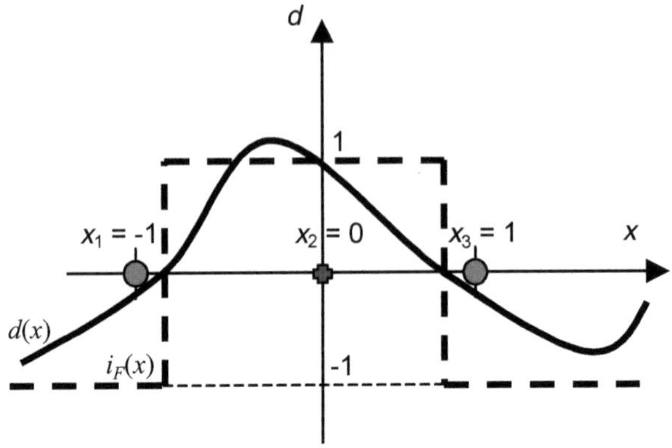

Fig. 11. A nonlinear 1-dimensional classification problem. One possible solution is given by the decision function $d(x)$ (*solid curve*) i.e., by the corresponding indicator function defined as $i_F = \text{sign}(d(x))$ (*dashed* stepwise function)

Consider solving the simplest 1-D classification problem given the input and the output (desired) values as follows: $\mathbf{x} = [-1\ 0\ 1]^T$ and $\mathbf{d} = \mathbf{y} = [-1\ 1\ -1]^T$.

Here we choose the following mapping to the feature space: $\mathbf{\Phi}(\mathbf{x}) = [\phi_1(\mathbf{x})\phi_2(\mathbf{x})\phi_3(\mathbf{x})]^T = [x^2\ \sqrt{2x}\ 1]^T$. The mapping produces the following three points in the feature space (shown as the *rows* of the matrix $\mathbf{F}$ (F standing for features))

$$\mathbf{F} = \begin{bmatrix} 1 & -\sqrt{2} & 1 \\ 0 & 0 & 1 \\ 1 & \sqrt{2} & 1 \end{bmatrix}.$$

These three points are linearly separable by the plane $\phi_3(\mathbf{x}) = 2\phi_1(\mathbf{x})$ in a feature space as shown in Fig. 12. It is easy to show that the mapping obtained by $\mathbf{\Phi}(\mathbf{x}) = [x^2\quad \sqrt{2x}\quad 1]^T$ is a scalar product implementation of a quadratic kernel function $(\mathbf{x}_i^T\mathbf{x}_j + 1)^2 = k(\mathbf{x}_i, \mathbf{x}_j)$. In other words, $\mathbf{\Phi}^T(\mathbf{x}_i)\mathbf{\Phi}(\mathbf{x}_j) = k(\mathbf{x}_i, \mathbf{x}_j)$. This equality will be introduced shortly.

There are two basic problems when mapping an input $\mathbf{x}$-space into higher order F-space:

(i) the choice of mapping $\mathbf{\Phi}(\mathbf{x})$ that should result in a "rich" class of decision hypersurfaces,
(ii) the calculation of the scalar product $\mathbf{\Phi}^T(\mathbf{x})\mathbf{\Phi}(\mathbf{x})$ that can be computationally very discouraging if the number of features f (i.e., dimensionality f of a feature space) is very large.

The second problem is connected with a phenomenon called the "*curse of dimensionality*". For example, to construct a decision surface corresponding to a polynomial of degree *two* in an n-D input space, a dimensionality of a feature space $f = n(n+3)/2$. In other words, a feature space is spanned by f coordinates of the form

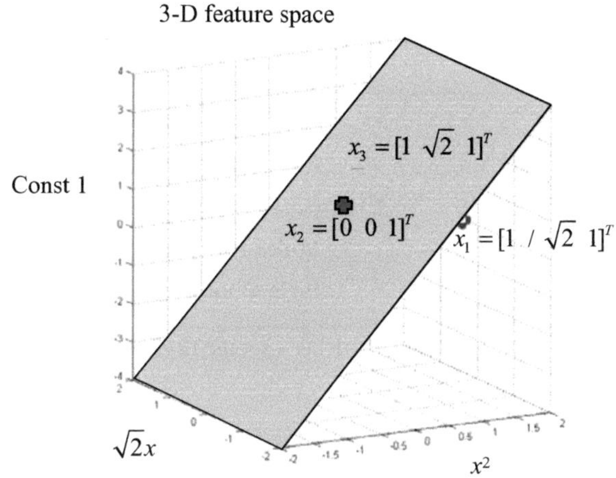

Fig. 12. The three data points of a problem in Fig. 11 are linearly separable in the feature space (obtained by the mapping $\mathbf{\Phi}(\mathbf{x}) = [\phi_1(\mathbf{x})\ \phi_2(\mathbf{x})\ \phi_3(\mathbf{x})]^T = [x^2\ \sqrt{2x}\ 1]^T$). The separation boundary is given as the plane $\phi_3(\mathbf{x}) = 2\phi_1(\mathbf{x})$ shown in the figure

$z_1 = x_1, \ldots, z_n = x_n$ (n coordinates), $z_{n+1} = (x_1)^2, \ldots, z_{2n} = (x_n)^2$ (next n coordinates), $z_{2n+1} = x_1x_2, \ldots, z_f = x_nx_{n-1}$ ($n(n-1)/2$ coordinates),

and the separating hyperplane created in this space, is a second-degree polynomial in the input space [35]. Thus, constructing a polynomial of degree two only, in a 256-dimensional input space, leads to a dimensionality of a feature space $f = 33,152$. Performing a scalar product operation with vectors of such, or higher, dimensions, is not a cheap computational task. The problems become serious (and fortunately only seemingly unsolvable) if we want to construct a polynomial of degree 4 or 5 in the same 256-dimensional space leading to the construction of a decision hyperplane in a billion-dimensional feature space.

This explosion in dimensionality can be avoided by noticing that in the quadratic optimization problem given by (15) and (30), as well as in the final expression for a classifier, *training data only appear in the form of scalar products* $\mathbf{x}_i^T\mathbf{x}_j$. These products will be replaced by scalar products $\mathbf{\Phi}^T(\mathbf{x})\mathbf{\Phi}(\mathbf{x})_i = [\phi_1(\mathbf{x}),\ \phi_2(\mathbf{x}), \ldots, \phi_n(\mathbf{x})]^T\ [\phi_1(\mathbf{x}_i),\ \phi_2(\mathbf{x}_i), \ldots, \phi_n(\mathbf{x}_i)]$ in a feature space F, and the latter can be and will be expressed by using the *kernel function* $K(\mathbf{x}_i,\ \mathbf{x}_j) = \mathbf{\Phi}^T(\mathbf{x}_i)\mathbf{\Phi}(\mathbf{x}_j)$.

Note that a *kernel function* $K(\mathbf{x}_i, \mathbf{x}_j)$ is a function in input space. Thus, the basic advantage in using kernel function $K(\mathbf{x}_i, \mathbf{x}_j)$ is in avoiding performing a mapping $\mathbf{\Phi}(\mathbf{x})$ et all. Instead, the required scalar products in a feature space $\mathbf{\Phi}^T(\mathbf{x}_i)\mathbf{\Phi}(\mathbf{x}_j)$, are calculated directly by computing kernels $K(\mathbf{x}_i, \mathbf{x}_j)$ for given training data vectors in an input space. In this way, we bypass a possibly extremely high dimensionality of a feature space F. Thus, by using

Table 2. Popular Admissible Kernels

Kernel Functions	Type of Classifier
$K(\mathbf{x},\mathbf{x}_i) = (\mathbf{x}^T\mathbf{x}_i)$	Linear, dot product, kernel, CPD
$K(\mathbf{x},\mathbf{x}_i) = \left[(\mathbf{x}^T\mathbf{x}_i)+1\right]^d$	Complete polynomial of degree d, PD
$K(\mathbf{x},\mathbf{x}_i) = e^{\frac{1}{2}\left[(\mathbf{x}-\mathbf{x}_i)^T\Sigma^{-1}(\mathbf{x}-\mathbf{x}_i)\right]}$	Gaussian RBF, PD
$K(\mathbf{x},\mathbf{x}_i) = \tanh\left[(\mathbf{x}^T\mathbf{x}_i)+b\right]^*$	Multilayer perceptron, CPD
$K(\mathbf{x},\mathbf{x}_i) = \frac{1}{\sqrt{\lVert\mathbf{x}-\mathbf{x}_i\rVert^2+\beta}}$	Inverse multiquadric function, PD

* only for certain values of b, (C)PD = (conditionally) positive definite

the chosen kernel $K(\mathbf{x}_i,\mathbf{x}_j)$, we can construct an SVM that operates in an infinite dimensional space (such an kernel function is a Gaussian kernel function given in Table 2 below). In addition, as will be shown below, by applying kernels we do not even have to know what the actual mapping $\mathbf{\Phi}(\mathbf{x})$ is. A kernel is a function K such that

$$K(\mathbf{x}_i,\ \mathbf{x}_j) = \mathbf{\Phi}^T(\mathbf{x}_i)\mathbf{\Phi}(\mathbf{x}_j)\ . \tag{36}$$

There are many possible kernels, and the most popular ones are given in Table 2. All of them should fulfill the so-called Mercer's conditions. The Mercer's kernels belong to a set of *reproducing kernels*. For further details see [2, 15, 19, 26, 35].

The simplest is a linear kernel defined as $K(\mathbf{x}_i,\ \mathbf{x}_j) = \mathbf{x}_i^T\mathbf{x}_j$. Below we show a few more kernels:

Polynomial Kernels

Let $\mathbf{x} \in \Re^2$ i.e., $\mathbf{x} = [x_1\, x_2]^T$, and if we choose $\mathbf{\Phi}(\mathbf{x}) = [x_1^2 \quad \sqrt{2}x_1x_2 \quad x_1^2]^T$ (i.e., there is an $\Re^2 \to \Re^3$ mapping), then the dot product

$$\begin{aligned}
\mathbf{\Phi}^T(\mathbf{x}_i)\mathbf{\Phi}(\mathbf{x}_j) &= \left[x_{i1}^2\ \ \sqrt{2}x_{i1}x_{i2}\ \ x_{i1}^2\right]\left[x_{j1}^2\ \ \sqrt{2}x_{j1}x_{j2}\ \ x_{j1}^2\right]^T \\
&= \left[x_{i1}^2\,x_{j1}^2 + 2x_{i1}x_{i2}\,x_{j1}x_{i2} + x_{i2}^2\,x_{j2}^2\right] \\
&= (\mathbf{x}_i^T\mathbf{x}_j)^2 = K(\mathbf{x}_i,\ \mathbf{x}_j),\ \text{or} \\
&\quad K(\mathbf{x}_i,\ \mathbf{x}_j) = (\mathbf{x}_i^T\mathbf{x}_j)^2 = \mathbf{\Phi}^T(\mathbf{x}_i)\mathbf{\Phi}(\mathbf{x}_j)\ .
\end{aligned}$$

Note that in order to calculate the scalar product in a feature space $\mathbf{\Phi}^T(\mathbf{x}_i)\mathbf{\Phi}(\mathbf{x}_j)$, we do not need to perform the mapping $\mathbf{\Phi}(\mathbf{x}) = [x_1^2\ \sqrt{2}x_1x_2\ x_1^2]^T$ et all. Instead, we calculate this product directly in the input space by computing $(\mathbf{x}_i^T\mathbf{x}_j)^2$. This is very well known under the popular name of *the kernel trick*. Interestingly, note also that other mappings such as an

$\Re^2 \to \Re^3$ mapping given by $\mathbf{\Phi}(\mathbf{x}) = [x_1^2 - x_2^2 \quad 2x_1x_2 \quad x_1^2 + x_2^2]$, or an
$\Re^2 \to \Re^4$ mapping given by $\mathbf{\Phi}(\mathbf{x}) = [x_1^2 \quad x_1x_2 \quad x_1x_2 \quad x_2^2]$,

also accomplish the same task as $(\mathbf{x}_i^T\mathbf{x}_j)^2$.

Now, assume the following mapping

$$\mathbf{\Phi}(\mathbf{x}) = [1 \quad \sqrt{2}x_1 \quad \sqrt{2}x_2 \quad \sqrt{2}x_1x_2 \quad x_1^2 \quad x_2^2] ,$$

i.e., there is an $\Re^2 \to \Re^5$ mapping plus bias term as the constant 6th dimension's value. Then the dot product in a feature space $\mathbf{F}$ is given as

$$\begin{aligned}\mathbf{\Phi}^T(\mathbf{x}_i)\mathbf{\Phi}(\mathbf{x}_j) &= 1 + 2x_{i1}x_{j1} + 2x_{i2}x_{j2} + 2x_{i1}x_{i2}x_{j1}x_{i2} + x_{i1}^2x_{j1}^2 + x_{i2}^2x_{j2}^2 \\ &= 1 + 2(\mathbf{x}_i^T\mathbf{x}_j) + (\mathbf{x}_i^T\mathbf{x}_j)^2 = (\mathbf{x}_i^T\mathbf{x}_j + 1)^2 = K(\mathbf{x}_i, \mathbf{x}_j) \ , \text{ or} \\ & K(\mathbf{x}_i, \ \mathbf{x}_j) = (\mathbf{x}_i^T\mathbf{x}_j + 1)^2 = \mathbf{\Phi}^T(\mathbf{x}_i)\mathbf{\Phi}(\mathbf{x}_j) \ .\end{aligned}$$

Thus, the last mapping leads to the second order *complete* polynomial.

Many candidate functions can be applied to a convolution of an inner product (i.e., for kernel functions) $K(\mathbf{x}, \ \mathbf{x}_i)$ in an SV machine. Each of these functions constructs a different nonlinear decision hypersurface in an input space. In the first three rows, the Table 2 shows the three most popular kernels in SVMs' in use today, and the inverse multiquadric one as an interesting and powerful kernel to be proven yet.

The positive definite (PD) kernels are the kernels which Gramm matrix $\mathbf{G}$ (a.k.a. Grammian) calculated by using all the l training data points is positive definite (meaning all its eigenvalues are strictly positive, i.e., $\lambda_i > 0, \ i = 1, \ l$)

$$\mathbf{G} = \mathbf{K}(\mathbf{x}_i, \ \mathbf{x}_j) = \begin{bmatrix} k(\mathbf{x}_1,\mathbf{x}_1) & k(\mathbf{x}_1,\mathbf{x}_2) & \cdots & k(\mathbf{x}_1,\mathbf{x}_l) \\ k(\mathbf{x}_2,\mathbf{x}_1) & k(\mathbf{x}_2,\mathbf{x}_2) & \vdots & k(\mathbf{x}_2,\mathbf{x}_l) \\ \vdots & \vdots & \vdots & \vdots \\ k(\mathbf{x}_l,\mathbf{x}_1) & k(\mathbf{x}_l,\mathbf{x}_2) & \cdots & k(\mathbf{x}_l,\mathbf{x}_l) \end{bmatrix} . \qquad (37)$$

The kernel matrix $\mathbf{G}$ is a symmetric one. Even more, any symmetric positive definite matrix can be regarded as a kernel matrix, that is – as an inner product matrix in some space.

Finally, we arrive at the point of presenting the learning in nonlinear classifiers (in which we are ultimately interested here). The learning algorithm for a nonlinear SV machine (classifier) follows from the design of an *optimal separating hyperplane* in a *feature space*. This is the same procedure as the construction of a "hard" (15) and "soft" (30) margin classifiers in an $\mathbf{x}$-space previously. In a $\mathbf{\Phi}(\mathbf{x})$-space, the dual Lagrangian, given previously by (15) and (30), is now

$$L_d(\alpha) = \sum_{i=1}^{l} \alpha_i - \frac{1}{2} \sum_{i,j=1}^{l} \alpha_i \alpha_j y_i y_j \mathbf{\Phi}_i^T \mathbf{\Phi}_j \ , \qquad (38)$$

and, according to (36), by using chosen kernels, we should maximize the following dual Lagrangian

$$L_d(\alpha) = \sum_{i=1}^{l} \alpha_i - \frac{1}{2} \sum_{i,j=1}^{l} \alpha_i \alpha_j y_i y_j K(\mathbf{x}_i, \mathbf{x}_j) \qquad (39)$$

subject to

$$\alpha_i \geq 0, \quad i = 1, \, l \quad \text{and} \quad \sum_{i=1}^{l} \alpha_i y_i = 0 \,. \tag{39a}$$

In a more general case, because of a noise or due to generic class' features, there will be an overlapping of training data points. Nothing but constraints for α_i change. Thus, the nonlinear "soft" margin classifier will be the solution of the quadratic optimization problem given by (39) subject to constraints

$$C \geq \alpha_i \geq 0, \quad i = 1, \, l \quad \text{and} \quad \sum_{i=1}^{l} \alpha_i y_i = 0 \,. \tag{39b}$$

Again, the only difference to the separable nonlinear classifier is the upper bound C on the Lagrange multipliers α_i. In this way, we limit the influence of training data points that will remain on the "wrong" side of a separating nonlinear hypersurface. After the dual variables are calculated, the decision hypersurface $d(\mathbf{x})$ is determined by

$$d(\mathbf{x}) = \sum_{i=1}^{l} y_i \alpha_i K(\mathbf{x}, \, \mathbf{x}_i) + b = \sum_{i=1}^{l} v_i K(\mathbf{x}, \, \mathbf{x}_i) + b \tag{40}$$

and the indicator function is $i_F(\mathbf{x}) = \text{sign}[d(\mathbf{x})] = \text{sign}[\sum_{i=1}^{l} v_i K(\mathbf{x}, \mathbf{x}_i) + b]$.

Note that the summation is not actually performed over all training data but rather over the support vectors, because only for them do the Lagrange multipliers differ from zero. The existence and calculation of a bias b is now not a direct procedure as it is for a linear hyperplane. Depending upon the applied kernel, the bias b can be implicitly part of the kernel function. If, for example, Gaussian RBF is chosen as a kernel, it can use a bias term as the $f + 1^{\text{st}}$ feature in F-space with a constant output $= +1$, but not necessarily. In short, all PD kernels do not necessarily need an explicit bias term b, but b can be used. (More on this can be found in the Kecman, Huang, and Vogt's chapter, as well as in the Vogt and Kecman's one in this book). Same as for the linear SVM, (39) can be written in a matrix notation as

maximize

$$L_d(\alpha) = -0.5\alpha^T \mathbf{H} \alpha + \mathbf{f}^T \alpha \,, \tag{41a}$$

subject to

$$\mathbf{y}^T \alpha = 0 \,, \tag{41b}$$

$$C \geq \alpha_i \geq 0 \,, \qquad i = 1, \, l \,, \tag{41c}$$

where $\alpha = [\alpha_1, \alpha_2, \ldots, \alpha_l]^T$, $\mathbf{H}$ denotes the Hessian matrix ($H_{ij} = y_i y_j K(\mathbf{x}_i, \, \mathbf{x}_j)$) of this problem and $\mathbf{f}$ is an $(l, 1)$ unit vector $\mathbf{f} = \mathbf{1} = [1\,1 \ldots 1]^T$. Note that if $K(\mathbf{x}_i, \mathbf{x}_j)$ is the positive definite matrix, so is the matrix $y_i y_j K(\mathbf{x}_i, \mathbf{x}_j)$.

The following 1-D example (just for the sake of graphical presentation) will show the creation of a linear decision function in a feature space and a corresponding nonlinear (quadratic) decision function in an input space.

Suppose we have 4 1-D data points given as $x_1 = 1,\ x_2 = 2,\ x_3 = 5,\ x_4 = 6$, with data at 1, 2, and 6 as class 1 and the data point at 5 as class 2, i.e., $y_1 = -1,\ y_2 = -1,\ y_3 = 1,\ y_4 = -1$. We use the polynomial kernel of degree 2, $K(x, y) = (xy + 1)^2$. C is set to 50, which is of lesser importance because the constraints will be not imposed in this example for maximal value for the dual variables α_i will be smaller than $C = 50$.

Case 1: Working with a bias term b as given in (40).
We first find $\alpha_i (i = 1, \ldots, 4)$ by solving dual problem (41a) having a Hessian matrix

$$\mathbf{H} = \begin{bmatrix} 4 & 9 & -36 & 49 \\ 9 & 25 & -121 & 169 \\ -36 & -121 & 676 & -961 \\ 49 & 169 & -961 & 1369 \end{bmatrix}.$$

Alphas are $\alpha_1 = 0,\ \alpha_2 = 2.499999,\ \alpha_3 = 7.333333,\ \alpha_4 = 4.833333$ and the bias b will be found by using (18b), or by fulfilling the requirements that the values of a decision function at the support vectors should be the given y_i. The model (decision function) is given by

$$\begin{aligned} d(x) &= \sum_{i=1}^{4} y_i \alpha_i K(x, x_i) + b = \sum_{i=1}^{4} v_i (x x_i + 1)^2 + b, \text{ or by} \\ d(x) &= 2.499999(-1)(2x+1)^2 + 7.333333(1)(5x+1)^2 \\ &\quad + 4.833333(-1)(6x+1)^2 + b \\ d(x) &= -0.666667x^2 + 5.333333x + b\,. \end{aligned}$$

Bias b is determined from the requirement that at the SV points 2, 5 and 6, the outputs must be -1, 1 and -1 respectively. Hence, $b = -9$, resulting in the decision function

$$d(x) = -0.666667x^2 + 5.333333x - 9\,.$$

The nonlinear (quadratic) decision function and the indicator one are shown in Fig. 13.

Note that in calculations above 6 decimal places have been used for alpha values. The calculation is numerically very sensitive, and working with fewer decimals can give very approximate or wrong results.

The complete polynomial kernel as used in the case 1, is *positive definite* and there is no need to use an explicit bias term b as presented above. Thus, one can use the same second order polynomial model without the bias term b. Note that in this particular case there is no equality constraint equation that originates from an equalization of the primal Lagrangian derivative in respect

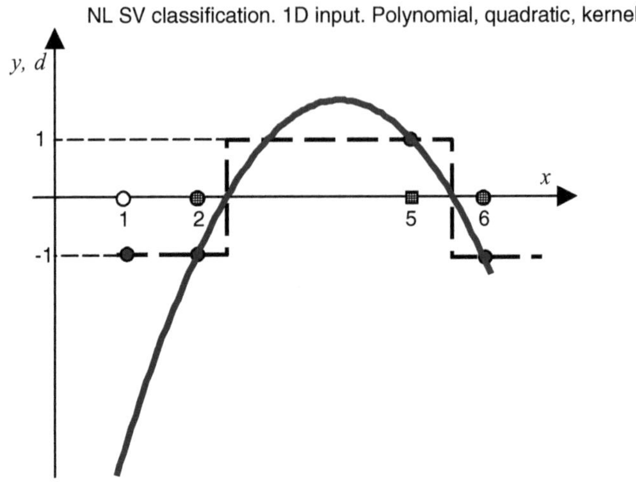

Fig. 13. The nonlinear decision function (*solid*) and the indicator function (*dashed*) for 1-D overlapping data. By using a complete second order polynomial the model with and without a bias term b are same

to the bias term b to zero. Hence, we do not use (41b) while using a positive definite kernel without bias as it will be shown below in the case 2.

Case 2: Working without a bias term b
Because we use the same second order polynomial kernel, the Hessian matrix $\mathbf{H}$ is same as in the case 1. The solution without the equality constraint for alphas is: $\alpha_1 = 0$, $\alpha_2 = 24.999999$, $\alpha_3 = 43.333333$, $\alpha_4 = 27.333333$.

The model (decision function) is given by

$$
\begin{aligned}
d(x) &= \sum_{i=1}^{4} y_i \alpha_i K(x, x_i) = \sum_{i=1}^{4} v_i (x x_i + 1)^2, \text{ or by} \\
d(x) &= 2.499999(-1)(2x+1)^2 + 43.333333(1)(5x+1)^2 \\
&\quad + 27.333333(-1)(6x+1)^2 \\
d(x) &= -0.666667x^2 + 5.333333x - 9 \,.
\end{aligned}
$$

Thus the nonlinear (quadratic) decision function and consequently the indicator function in the two particular cases are equal.

XOR Example

In the next example shown by Figs. 14 and 15 we present all the important mathematical objects of a nonlinear SV classifier by using a classic XOR (*exclusive-or*) problem. The graphs show all the mathematical functions (objects) involved in a nonlinear classification. Namely, the nonlinear decision function $d(\mathbf{x})$, the NL indicator function $i_F(\mathbf{x})$, training data $(\mathbf{x}_i)$, support vectors $(\mathbf{x}_{SV})_i$ and separation boundaries.

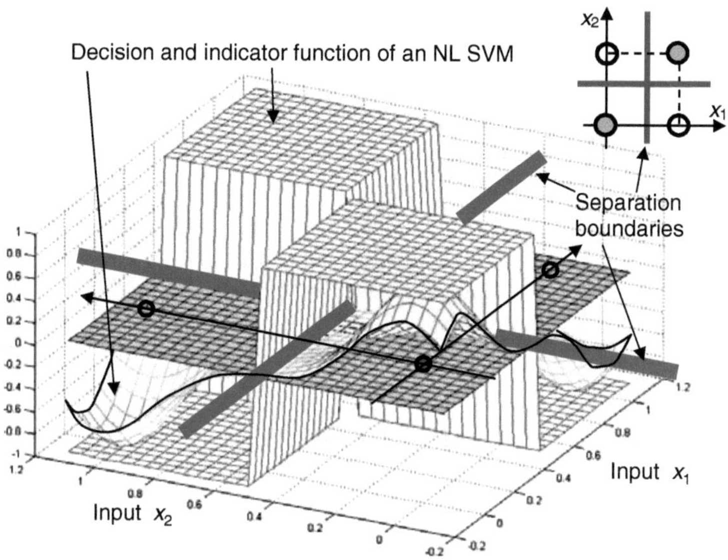

Fig. 14. XOR problem. Kernel functions (2-D Gaussians) are not shown. The nonlinear decision function, the nonlinear indicator function and the separation boundaries are shown. All four data are chosen as support vectors

The same objects will be created in the cases when the input vector $\mathbf{x}$ is of a dimensionality $n > 2$, but the visualization in these cases is not possible. In such cases one talks about the decision hyper-function (hyper-surface) $d(\mathbf{x})$, indicator hyper-function (hyper-surface) $i_F(\mathbf{x})$, training data $(\mathbf{x}_i)$, support vectors $(\mathbf{x}_{SV})_i$ and separation hyper-boundaries (hyper-surfaces).

Note the different character of a $d(\mathbf{x})$, $i_F(\mathbf{x})$ and separation boundaries in the two graphs given below. However, in both graphs all the data are correctly classified. The analytic solution to the Fig. 15 for the *second order polynomial kernel* (i.e., for $(\mathbf{x}_i^T\mathbf{x}_j + 1)^2 = \mathbf{\Phi}^T(\mathbf{x}_i)\mathbf{\Phi}(\mathbf{x}_j)$, where $\mathbf{\Phi}(\mathbf{x}) = [1\ \sqrt{2}x_1\ \sqrt{2}x_2\ \sqrt{2}x_1x_2\ x_1^2\ x_2^2]$, no explicit bias and $C = \infty$) goes as follows. Inputs and desired outputs are, $\mathbf{x} = \left[\begin{smallmatrix} 0 & 1 & 1 & 0 \\ 0 & 1 & 0 & 1 \end{smallmatrix}\right]^T$, $\mathbf{y} = \mathbf{d} = [1\quad 1\quad -1\quad -1]^T$. The dual Lagrangian (39) has the Hessian matrix $\mathbf{H} = \left[\begin{smallmatrix} 1 & 1 & -1 & -1 \\ 1 & 9 & -4 & -4 \\ -1 & -4 & 4 & 1 \\ -1 & -4 & 1 & 4 \end{smallmatrix}\right]$.

The optimal solution can be obtained by taking the derivative of L_d with respect to dual variables $\alpha_i (i = 1, 4)$ and by solving the resulting linear system of equations taking into account the constraints. The solution to

$$\begin{aligned}
\alpha_1 + \alpha_2 - \alpha_3 - \alpha_4 &= 1\,, \\
\alpha_1 + 9\alpha_2 - 4\alpha_3 - 4\alpha_4 &= 1\,, \\
-\alpha_1 - 4\alpha_2 + 4\alpha_3 + \alpha_4 &= 1\,, \\
-\alpha_1 - 4\alpha_2 + \alpha_3 + 4\alpha_4 &= 1\,,
\end{aligned}$$

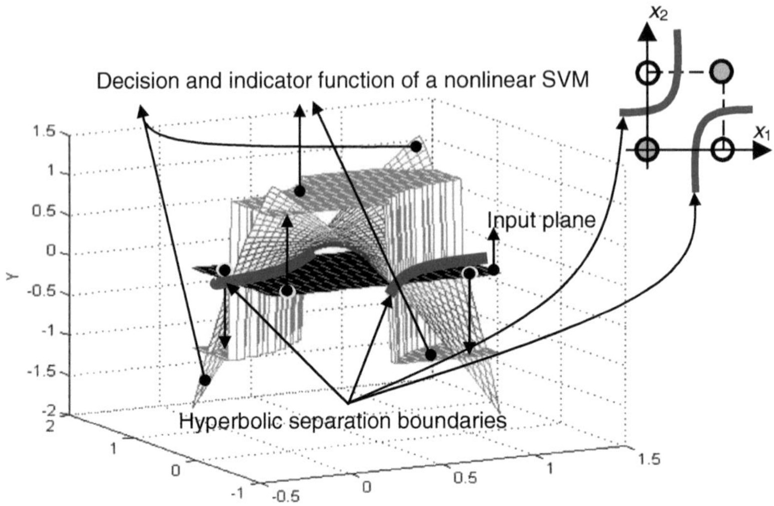

Fig. 15. XOR problem. Kernel function is a 2-D polynomial. The nonlinear decision function, the nonlinear indicator function and the separation boundaries are shown. All four data are support vectors

subject to $\alpha_i > 0$, $(i = 1, 4)$, is $\alpha_1 = 4.3333$, $\alpha_2 = 2.0000$, $\alpha_3 = 2.6667$ and $\alpha_4 = 2.6667$. The decision function in a 3-D space is

$$\begin{aligned} d(\mathbf{x}) &= \sum_{i=1}^{4} y_i \alpha_i \mathbf{\Phi}^T(\mathbf{x}_i)\mathbf{\Phi}(\mathbf{x}) \\ &= \big(4.3333\big[1 \quad 0 \quad 0 \quad 0 \quad 0 \quad 0\big] + 2\big[1 \quad \sqrt{2} \quad \sqrt{2} \quad \sqrt{2} \quad 1 \quad 1\big] \\ &\quad - 2.6667\big[1 \quad \sqrt{2} \quad 0 \quad 0 \quad 1 \quad 0\big] \\ &\quad - 2.6667\big[1 \quad 0 \quad \sqrt{2} \quad 0 \quad 0 \quad 1\big]\big)\,\mathbf{\Phi}(\mathbf{x}) \\ &= [1 \quad -0.9429 \quad -0.9429 \quad 2.8284 \quad -0.6667 \quad -0.6667] \\ &\quad \times \left[1 \quad \sqrt{2}x_1 \quad \sqrt{2}x_2 \quad \sqrt{2}x_1x_2 \quad x_1^2 \quad x_2^2\right]^T, \end{aligned}$$

and finally

$$d(\mathbf{x}) = 1 - 1.3335x_1 - 1.3335x_2 + 4x_1x_2 - 0.6667x_1^2 - 0.6667x_2^2\,.$$

It is easy to check that the values of $d(\mathbf{x})$ for all the training inputs in $\mathbf{x}$ equal the desired values in $\mathbf{d}$. The $d(\mathbf{x})$ is the saddle-like function shown in Fig. 15.

Here we have shown the derivation of an expression for $d(\mathbf{x})$ by using explicitly a mapping $\mathbf{\Phi}$. Again, we do not have to know what mapping $\mathbf{\Phi}$ is at all. By using *kernels* in *input space*, we calculate a *scalar product* required in a (*possibly high dimensional*) *feature space* and we avoid mapping $\mathbf{\Phi}(\mathbf{x})$. This is known as kernel "trick". It can also be useful to remember that the way in which the kernel "trick" was applied in designing an SVM can be utilized in all other algorithms that depend on the scalar product (e.g., in principal component analysis or in the nearest neighbor procedure).

2.4 Regression by Support Vector Machines

In the regression, we estimate the functional dependence of the dependent (output) variable $y \in \Re$ on an n-dimensional input variable $\mathbf{x}$. Thus, unlike in pattern recognition problems (where the desired outputs y_i are discrete values e.g., Boolean) we deal with *real valued* functions and we model an $\Re^n$ to $\Re^1$ mapping here. Same as in the case of classification, this will be achieved by training the SVM model on a training data set first. Interestingly and importantly, a learning stage will end in the same shape of a dual Lagrangian as in classification, only difference being in a dimensionalities of the Hessian matrix and corresponding vectors which are of a double size now e.g., $\mathbf{H}$ is a $(2l, 2l)$ matrix.

Initially developed for solving classification problems, SV techniques can be successfully applied in regression, i.e., for a functional approximation problems [8, 34]. The general regression learning problem is set as follows – the learning machine is given l training data from which it attempts to learn the input-output relationship (dependency, mapping or function) $f(\mathbf{x})$. A training data set $D = \{[\mathbf{x}(i), y(i)] \in \Re^n \times \Re, \ i = 1, \dots, l\}$ consists of l pairs $(\mathbf{x}_1, y_1)$, $(\mathbf{x}_2, y_2), \dots, (\mathbf{x}_l, y_l)$, where the inputs $\mathbf{x}$ are n-dimensional vectors $\mathbf{x} \in \Re^n$ and system responses $y \in \Re$, are continuous values.

We introduce all the relevant and necessary concepts of SVM's regression in a gentle way starting again with a *linear regression hyperplane* $f(\mathbf{x}, \mathbf{w})$ given as

$$f(\mathbf{x}, \mathbf{w}) = \mathbf{w}^T \mathbf{x} + b \,. \tag{41d}$$

In the case of SVM's regression, we measure the *error of approximation* instead of the margin used in classification. The most important difference in respect to classic regression is that we use a novel loss (error) functions here. This is the Vapnik's *linear loss function* with *ε-insensitivity zone* defined as

$$E(\mathbf{x}, y, f) = |y - f(\mathbf{x}, \mathbf{w})|_\varepsilon = \begin{cases} 0 & \text{if } |y - f(\mathbf{x}, \mathbf{w})| \le \varepsilon \\ |y - f(\mathbf{x}, \mathbf{w})| - \varepsilon, & \text{otherwise}\,. \end{cases}, \tag{43a}$$

or as,

$$e(\mathbf{x}, y, f) = \max(0, |y - f(\mathbf{x}, \mathbf{w})| - \varepsilon) \,. \tag{43b}$$

Thus, the loss is equal to 0 if the difference between the predicted $f(\mathbf{x}_i, \mathbf{w})$ and the measured value y_i is less than ε. Vapnik's ε-insensitivity loss function (43) defines an ε tube (Fig. 17). If the predicted value is within the tube the loss (error or cost) is zero. For all other predicted points outside the tube, the loss equals the magnitude of the difference between the predicted value and the radius ε of the tube.

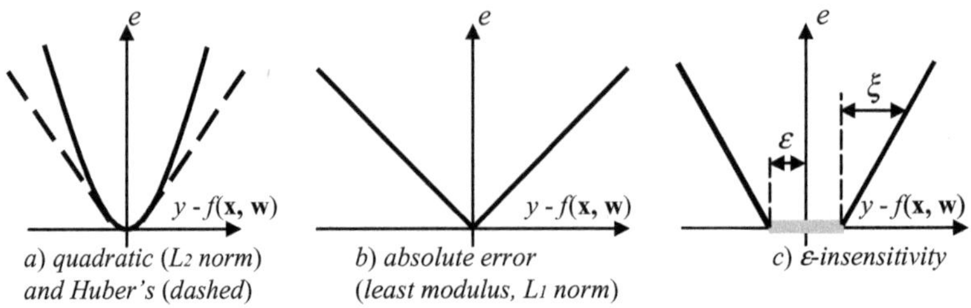

Fig. 16. Loss (error) functions

The two classic error functions are: a square error, i.e., L_2 norm $(y - f)^2$, as well as an absolute error, i.e., L_1 norm, least modulus $|y - f|$ introduced by Yugoslav scientist Rudjer Boskovic in 18th century [9]. The latter error function is related to Huber's error function. An application of Huber's error function results in a *robust regression*. It is the most reliable technique if nothing specific is known about the model of a noise. We do no present Huber's loss function here in analytic form. Instead, we show it by a dashed curve in Fig. 16a. In addition, Fig. 16 shows typical shapes of all mentioned error (loss) functions above.

Note that for $\varepsilon = 0$, Vapnik's loss function equals a least modulus function. Typical graph of a (nonlinear) regression problem as well as all relevant mathematical objects required in learning unknown coefficients w_i are shown in Fig. 17.

We will formulate an SVM regression's algorithm for the linear case first and then, for the sake of an NL model design, we will apply mapping to a feature space, utilize the kernel "trick" and construct a nonlinear regression

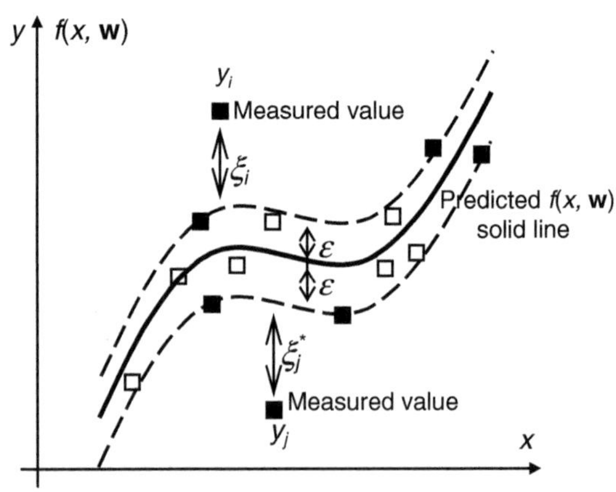

Fig. 17. The parameters used in (1-dimensional) support vector regression Filled ■ are support vectors, and the empty □ ones are not. Hence, SVs can appear only on the tube boundary or outside the tube

hypersurface. This is actually the same order of presentation as in classification tasks. Here, for the regression, we "measure" the empirical error term R_{emp} by Vapnik's ε-insensitivity loss function given by (43) and shown in Fig. 16c (while the minimization of the confidence term Ω will be realized through a minimization of $\mathbf{w}^T\mathbf{w}$ again). The empirical risk is given as

$$R^{\varepsilon}_{\text{emp}}(\mathbf{w}, b) = \frac{1}{l}\sum_{i=1}^{l} \left|y_i - \mathbf{w}^T\mathbf{x}_i - b\right|_{\varepsilon}, \tag{44}$$

Figure 18 shows two linear approximating functions as dashed lines inside an ε-tube having the same empirical risk $R^{\varepsilon}_{\text{emp}}$ as the regression function $f(\mathbf{x}, \mathbf{w})$ on the training data.

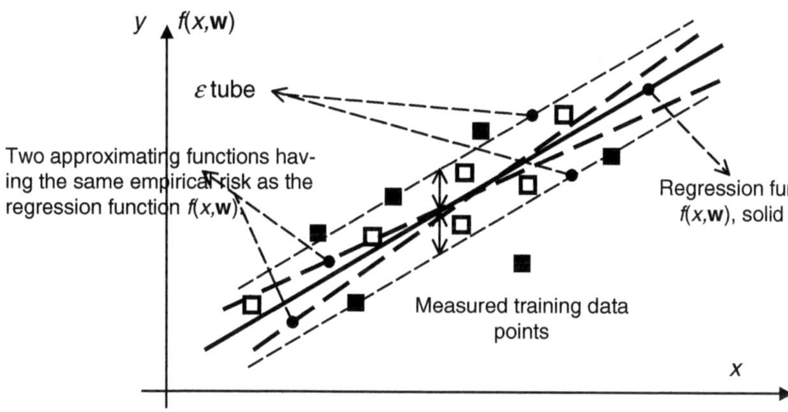

Fig. 18. Two linear approximations inside an ε tube (*dashed lines*) have the same empirical risk $R^{\varepsilon}_{\text{emp}}$ on the training data as the regression function

As in classification, we try to minimize both the empirical risk $R^{\varepsilon}_{\text{emp}}$ and $||\mathbf{w}||^2$ simultaneously. Thus, we construct a linear regression hyperplane $f(\mathbf{x}, \mathbf{w}) = \mathbf{w}^T\mathbf{x} + b$ by minimizing

$$R = \frac{1}{2}||\mathbf{w}||^2 + C\sum_{i=1}^{l} |y_i - f(\mathbf{x}_i, \mathbf{w})|_{\varepsilon}\,. \tag{45}$$

Note that the last expression resembles the ridge regression scheme. However, we use Vapnik's ε-insensitivity loss function instead of a squared error now. From (43a) and Fig. 17 it follows that for all training data outside an ε-tube,

$$|y - f(\mathbf{x}, \mathbf{w})| - \varepsilon = \xi \quad \text{for data "above" an } \varepsilon\text{-tube, or}$$
$$|y - f(\mathbf{x}, \mathbf{w})| - \varepsilon = \xi^* \quad \text{for data "below" an } \varepsilon\text{-tube}\,.$$

Thus, minimizing the risk R above equals the minimization of the following risk

$$R_{\mathbf{w},\xi,\xi^*} = \left[\frac{1}{2}\|\mathbf{w}\|^2 + C\left(\sum_{i=1}^{l}\xi_i + \sum_{i=1}^{l}\xi_i^*\right)\right] , \tag{46}$$

under constraints

$$y_i - \mathbf{w}^T\mathbf{x}_i - b \le \varepsilon + \xi_i, \quad i = 1, l \; , \tag{47a}$$
$$\mathbf{w}^T\mathbf{x}_i + b - y_i \le \varepsilon + \xi_i^*, \quad i = 1, l \; , \tag{47b}$$
$$\xi_i \ge 0, \xi_i^* \ge 0, \quad i = 1, l \; . \tag{47c}$$

where ξ_i and ξ_i^* are slack variables shown in Fig. 17 for measurements "above" and "below" an ε-tube respectively. Both slack variables are positive values. Lagrange multipliers α_i and α_i^* (that will be introduced during the minimization below) related to the first two sets of inequalities above, will be nonzero values for training points "above" and "below" an ε-tube respectively. Because no training data can be on both sides of the tube, either α_i or α_i^* will be nonzero. For data points inside the tube, both multipliers will be equal to zero. Thus $\alpha_i \; \alpha_i^* = 0$.

Note also that the constant C that influences a trade-off between an approximation error and the weight vector norm $\|\mathbf{w}\|$ is a design parameter that is chosen by the user. An increase in C penalizes larger errors i.e., it forces ξ_i and ξ_i^* to be small. This leads to an approximation error decrease which is achieved only by increasing the weight vector norm $\|\mathbf{w}\|$. However, an increase in $\|\mathbf{w}\|$ increases the confidence term Ω and does not guarantee a small generalization performance of a model. Another design parameter which is chosen by the user is the required precision embodied in an ε value that defines the size of an ε-tube. The choice of ε value is easier than the choice of $C$and it is given as either maximally allowed or some given or desired percentage of the output values y_i (say, $\varepsilon = 0.1$ of the mean value of $\mathbf{y}$).

Similar to procedures applied in the SV classifiers' design, we solve the constrained optimization problem above by forming a *primal variables Lagrangian* as follows,

$$\begin{aligned} L_p(\mathbf{w}, b, \xi_i, \xi_i^*, \alpha_i, \alpha_i^*, \beta_i, \beta_i^*) = & \frac{1}{2}\mathbf{w}^T\mathbf{w} + C\sum_{i=1}^{l}(\xi_i + \xi_i^*) - \sum_{i=1}^{l}(\beta_i^*\xi_i^* + \beta_i\xi_i) \\ & - \sum_{i=1}^{l}\alpha_i\left[\mathbf{w}^T\mathbf{x}_i + b - y_i + \xi + \xi_i\right] \\ & - \sum_{i=1}^{l}\alpha_i^*\left[y_i - \mathbf{w}^T\mathbf{x}_i, -b + \xi + \xi_i^*\right] . \end{aligned} \tag{48}$$

A primal variables Lagrangian $L_p(\mathbf{w}, b, \xi_i, \xi_i^*, \alpha_i\alpha_i^*, \beta_i, \beta_i^*)$ has to be *minimized* with respect to primal variables $\mathbf{w}, b, \xi_i$ and ξ_i^* and *maximized* with respect to nonnegative Lagrange multipliers $\alpha, \alpha_i^*, \beta$ and β_i^*. Hence, the function has the saddle point at the optimal solution $(\mathbf{w}_0, b_0, \xi_{i0}, \xi_{i0}^*)$ to the original problem. At the optimal solution the partial derivatives of L_p in respect to primal variables vanishes. Namely,

$$\frac{\partial L_p(\mathbf{w}_0, b_0, \xi_{i0}, \xi_{i0}^*, \alpha_i, \alpha_i^*, \beta_i, \beta_i^*)}{\partial \boldsymbol{w}} = \boldsymbol{w}_0 - \sum_{i=1}^{l} (\alpha_i - \alpha_i^*)\mathbf{x}_i = 0 \; , \tag{49}$$

$$\frac{\partial L_p(\mathbf{w}_0, b_0, \xi_{i0}, \xi_{i0}^*, \alpha_i, \alpha_i^*, \beta_i, \beta_i^*)}{\partial b} = \sum_{i=1}^{l} (\alpha_i - \alpha_i^*) = 0 \; , \tag{50}$$

$$\frac{\partial L_p(\mathbf{w}_0, b_0, \xi_{i0}, \xi_{i0}^*, \alpha_i, \alpha_i^*, \beta_i, \beta_i^*)}{\partial \xi_i} = C - \alpha_i - \beta_i = 0 \; , \tag{51}$$

$$\frac{\partial L_p(\mathbf{w}_0, b_0, \xi_{i0}, \xi_{i0}^*, \alpha_i, \alpha_i^*, \beta_i, \beta_i^*)}{\partial \xi_i^*} = C - \alpha_i^* - \beta_i^* = 0 \; . \tag{52}$$

Substituting the KKT above into the primal L_p given in (48), we arrive at the problem of the *maximization of a dual variables Lagrangian* $L_d(\alpha, \alpha^*)$ below,

$$\begin{aligned} L_d(\alpha_i, \alpha_i^*) &= -\frac{1}{2} \sum_{i,j=1}^{l} (\alpha_i - \alpha_i^*)(\alpha_j - \alpha_j^*)\mathbf{x}_i^T \mathbf{x}_j - \varepsilon \sum_{i=1}^{l} (\alpha_i + \alpha_i^*) + \sum_{i=1}^{l} (\alpha_i - \alpha_i^*) y_i \\ &= -\frac{1}{2} \sum_{i,j=1}^{l} (\alpha_i - \alpha_i^*)(\alpha_j - \alpha_j^*)\mathbf{x}_i^T \mathbf{x}_j - \sum_{i=1}^{l} (\varepsilon - y_i)\alpha_i - \sum_{i=1}^{l} (\varepsilon + y_i)\alpha_i^* \end{aligned} \tag{53}$$

subject to constraints

$$\sum_{i=1}^{l} \alpha_i^* = \sum_{i=1}^{l} \alpha_i \quad \text{or} \quad \sum_{i=1}^{l} (\alpha_i - \alpha_i^*) = 0 \; , \tag{54a}$$

$$0 \le \alpha_i \le C \quad i = 1, l \; , \tag{54b}$$

$$0 \le \alpha_i^* \le C \quad i = 1, l \; . \tag{54c}$$

Note that the dual variables Lagrangian $L_d(\boldsymbol{\alpha}, \boldsymbol{\alpha}^*)$ is expressed in terms of Lagrange multipliers α_i and α_i^* only. However, the size of the problem, with respect to the size of an SV classifier design task, is doubled now. There are $2l$ unknown dual variables ($l\,\alpha_i - s$ and $l\,\alpha_i^* - s$) for a linear regression and the Hessian matrix $\mathbf{H}$ of the quadratic optimization problem in the case of regression is a $(2l, 2l)$ matrix. The *standard quadratic optimization problem* above can be expressed in a *matrix notation* and formulated as follows:

$$\text{minimize } L_d(\boldsymbol{\alpha}) = -0.5\boldsymbol{\alpha}^T \mathbf{H}\boldsymbol{\alpha} + \mathbf{f}^T \boldsymbol{\alpha} \; , \tag{55}$$

subject to (54) where $\boldsymbol{\alpha} = [\alpha_1, \alpha_2, \ldots \alpha_l, \alpha_1^*, \alpha_2^*, \ldots, \alpha_1^*]^T, \mathbf{H} = [\mathbf{G} \;\; -\mathbf{G}; \; -\mathbf{G} \;\; \mathbf{G}], \mathbf{G}$ is an (l, l) matrix with entries $G_{ij} = [\mathbf{x}_i^T \mathbf{x}_j]$ for a linear regression, and $\mathbf{f} = [\varepsilon - y_1, \varepsilon - y_2, \ldots, \varepsilon - y_l, \varepsilon + y_1, \varepsilon + y_2, \ldots, \varepsilon + y_l]^T$. (Note that G_{ij}, as given above, is a badly conditioned matrix and we rather use $G_{ij} = [\mathbf{x}_i^T \mathbf{x}_j + 1]$ instead). Again, (55) is written in a form of some standard optimization routine that typically *minimizes* given objective function subject to same constraints (54).

The learning stage results in l Lagrange multiplier *pairs* (α_i, α_i^*). After the learning, the number of nonzero parameters α_i or α_i^* is equal to the number of SVs. However, this number does not depend on the dimensionality of input space and this is particularly important when working in very high dimensional spaces. Because at least one element of each pair $(\alpha_i, \alpha_i^*), i = 1, l$, is zero, the product of α_i and α_i^* is always zero, i.e., $\alpha_i \alpha_i^* = 0$.

At the optimal solution the following *KKT complementarity conditions* must be fulfilled

$$\alpha_i(\mathbf{w}^T\mathbf{x}_i + b - y_i + \varepsilon + \xi_i) = 0 \; , \tag{56}$$

$$\alpha_i^*(-\mathbf{w}^T\mathbf{x}_i - b + y_i + \varepsilon + \xi_i^*) = 0 \; , \tag{57}$$

$$\beta_i \xi_i = (C - \alpha_i)\xi_i = 0 \; , \tag{58}$$

$$\beta_i^* \xi_i^* = (C - \alpha_i^*)\xi_i^* = 0 \; , \tag{59}$$

(58) states that for $0 < \alpha_i < C, \xi_i = 0$ holds. Similarly, from (59) follows that for $0 < \alpha_i^* < C, \xi^* = 0$ and, for $0 < \alpha_i, \alpha_i^* < C$, from (56) and (57) follows,

$$\mathbf{w}^T\mathbf{x}_i + b - y_i + \varepsilon = 0 \; , \tag{60}$$

$$-\mathbf{w}^T\mathbf{x}_i - b + y_i + \varepsilon = 0 \; . \tag{61}$$

Thus, for all the data points fulfilling $y - f(\mathbf{x}) = +\varepsilon$, dual variables α_i must be between 0 and C, or $0 < \alpha_i < C$, and for the ones satisfying $y - f(\mathbf{x}) = -\varepsilon, \alpha_i^*$ take on values $0 < \alpha_i^* < C$. These data points are called the *free* (or *unbounded*) support vectors. They allow computing the value of the bias term b as given below

$$b = y_i - \mathbf{w}^T\mathbf{x}_i - \varepsilon, \text{ for } 0 < \alpha_i < C \; , \tag{62a}$$

$$b = y_i - \mathbf{w}^T\mathbf{x}_i + \varepsilon, \text{ for } 0 < \alpha_i^* < C \; . \tag{62b}$$

The calculation of a bias term b is numerically very sensitive, and it is better to compute the bias b by averaging over all the *free* support vector data points.

The final observation follows from (58) and (59) and it tells that for all the data points outside the ε-tube, i.e., when $\xi_i > 0$ and $\xi_i^* > 0$, both α_i and α_i^* equal C, i.e., $\alpha_i = C$ for the points above the tube and $\alpha_i^* = C$ for the points below it. These data are the so-called *bounded support vectors*. Also, for all the training data points within the tube, or when $|y - f(\mathbf{x})| < \varepsilon$, both α_i and α_i^* equal zero and they are neither the support vectors nor do they construct the decision function $f(\mathbf{x})$.

After calculation of Lagrange multipliers α_i and α_i^*, using (49) we can find an *optimal* (desired) weight vector of the *regression hyperplane* as

$$\mathbf{w}_0 = \sum_{i=1}^{l} (\alpha_i - \alpha_i^*)\mathbf{x}_i \; . \tag{63}$$

The best regression hyperplane obtained is given by

$$f(\mathbf{x},\mathbf{w}) = \mathbf{w}_0^T\mathbf{x} + b = \sum_{i=1}^{l}(\alpha_i - \alpha_i^*)\mathbf{x}_i^T\mathbf{x} + b\,. \tag{64}$$

More interesting, more common and the most challenging problem is to aim at solving the *nonlinear regression tasks.* A generalization to nonlinear regression is performed in the same way the nonlinear classifier is developed from the linear one, i.e., by carrying the mapping to the feature space, or by using kernel functions instead of performing the complete mapping which is usually of extremely high (possibly of an infinite) dimension. Thus, the nonlinear regression function in an input space will be devised by considering a linear regression hyperplane in the *feature space.*

We use the same basic idea in designing SV machines for creating a *nonlinear regression function.* First, a mapping of input vectors $\mathbf{x} \in \Re^n$ into vectors $\mathbf{\Phi}(\mathbf{x})$ of a higher dimensional *feature space* F (where $\mathbf{\Phi}$ represents mapping: $\Re^n \rightarrow \Re^f$) takes place and then, we solve a linear regression problem in this feature space. A mapping $\mathbf{\Phi}(\mathbf{x})$ is again the chosen in advance, or fixed, function. Note that an input space ($\mathbf{x}$-space) is spanned by components x_i of an input vector $\mathbf{x}$ and a feature space F ($\mathbf{\Phi}$-space) is spanned by components $\phi_i(\mathbf{x})$ of a vector $\mathbf{\Phi}(\mathbf{x})$. By performing such a mapping, we hope that in a $\mathbf{\Phi}$-space, our learning algorithm will be able to perform a linear regression hyperplane by applying the linear regression SVM formulation presented above. We also expect this approach to again lead to solving a quadratic optimization problem with inequality constraints in the feature space. The (linear in a feature space F) solution for the regression hyperplane $f = \mathbf{w}^T\mathbf{\Phi}(\mathbf{x}) + b$, will create a nonlinear regressing hypersurface in the original input space. The most popular kernel functions are *polynomials* and *RBF* with *Gaussian kernels.* Both kernels are given in Table 2.

In the case of the nonlinear regression, the learning problem is again formulated as the maximization of a dual Lagrangian (55) with the Hessian matrix $\mathbf{H}$ structured in the same way as in a linear case, i.e. $\mathbf{H} = [\mathbf{G}\ -\mathbf{G};\ -\mathbf{G}\ \mathbf{G}]$ but with the changed Grammian matrix $\mathbf{G}$ that is now given as

$$\mathbf{G} = \begin{bmatrix} \mathbf{G}_{11} & \cdots & \mathbf{G}_{1l} \\ \vdots & \mathbf{G}_{ii} & \vdots \\ \mathbf{G}_{l1} & \cdots & \mathbf{G}_{ll} \end{bmatrix}, \tag{65}$$

where the entries $G_{ij} = \mathbf{\Phi}^T(\mathbf{x}_i)\mathbf{\Phi}(\mathbf{x}_j) = K(\mathbf{x}_i,\ \mathbf{x}_j),\ i,j = 1,\ l.$

After calculating Lagrange multiplier vectors α and α^*, we can find an optimal weighting vector of the *kernels expansion* as

$$\mathbf{v}_0 = \alpha - \alpha^*\,. \tag{66}$$

Note however the difference in respect to the linear regression where the expansion of a decision function is expressed by using the optimal weight vector

$\mathbf{w}_0$. Here, in an NL SVMs' regression, the optimal weight vector $\mathbf{w}_0$ could often be of infinite dimension (which is the case if the Gaussian kernel is used). Consequently, we neither calculate $\mathbf{w}_0$ nor we have to express it in a closed form. Instead, we create the best nonlinear regression function by using the weighting vector $\mathbf{v}_0$ and the kernel (Grammian) matrix $\mathbf{G}$ as follows,

$$f(\mathbf{x}, \mathbf{w}) = \mathbf{G}\mathbf{v}_0 + b\,. \tag{67}$$

In fact, the last result follows from the very setting of the learning (optimizing) stage in a feature space where, in all the equations above from (47) to (64), we replace $\mathbf{x}_i$ by the corresponding feature vector $\mathbf{\Phi}(\mathbf{x}_i)$. This leads to the following changes:

- instead $G_{ij} = \mathbf{x}_i^T\mathbf{x}_j$ we get $G_{ij} = \mathbf{\Phi}^T(\mathbf{x}_i)\mathbf{\Phi}(\mathbf{x}_j)$ and, by using the kernel function $K(\mathbf{x}_i,\ \mathbf{x}_j) = \mathbf{\Phi}^T(\mathbf{x}_i)\mathbf{\Phi}(\mathbf{x}_j)$, it follows that $G_{ij} = K(\mathbf{x}_i,\ \mathbf{x}_j)$.
- Similarly, (63) and (64) change as follows:

$$\mathbf{w}_0 = \sum_{i=1}^{l} (\alpha_i - \alpha_i^*)\mathbf{\Phi}(\mathbf{x}_i), \tag{68}$$

$$\begin{aligned} f(\mathbf{x},\ \mathbf{w}) &= \mathbf{w}_0^T\mathbf{\Phi}(\mathbf{x}) + b = \sum_{i=1}^{l} (\alpha_i - \alpha_i^*)\mathbf{\Phi}^T(\mathbf{x}_i)\mathbf{\Phi}(\mathbf{x}) + b, \\ &= \sum_{i=1}^{l}(\alpha_i - \alpha_i^*)K(\mathbf{x}_i, \mathbf{x}) + b\,. \end{aligned} \tag{69}$$

If the bias term b is explicitly used as in (67) then, for an NL SVMs' regression, it can be calculated from the upper SVs as,

$$\left.\begin{aligned} b &= y_i - \sum_{j=1}^{N\, free upper\, SVs} (\alpha_j - \alpha_j^*)\mathbf{\Phi}^T(\mathbf{x}_j)\mathbf{\Phi}(\mathbf{x}_i) - \varepsilon \\ &= y_i - \sum_{j=1}^{N\, free upper\, SVs} (\alpha_j - \alpha_j^*)K(\mathbf{x}_i, \mathbf{x}_j) - \varepsilon \end{aligned}\right., \quad \text{for } 0 < \alpha_i < C\,, \tag{70a}$$

or from the lower ones as,

$$\left.\begin{aligned} b &= y_i - \sum_{j=1}^{N\, free lower\, SVs} (\alpha_j - \alpha_j^*)\mathbf{\Phi}^T(\mathbf{x}_j)\mathbf{\Phi}(\mathbf{x}_i) + \varepsilon \\ &= y_i - \sum_{j=1}^{N\, free lower\, SVs} (\alpha_j - \alpha_j^*)K(\mathbf{x}_i, \mathbf{x}_j) + \varepsilon \end{aligned}\right., \quad \text{for } 0 < \alpha_i^* < C\,. \tag{70b}$$

Note that $\alpha_j^* = 0$ in (70a) and so is $\alpha_j = 0$ in (70b). Again, it is much better to calculate the bias term b by an averaging *over all* the *free* support vector data points.

There are a few learning parameters in constructing SV machines for regression. The three most relevant are the insensitivity zone ε, the penalty parameter C (that determines the trade-off between the training error and VC dimension of the model), and the shape parameters of the kernel function (variances of a Gaussian kernel, order of the polynomial, or the shape parameters of the inverse multiquadrics kernel function). All three parameters' sets should be selected by the user. To this end, the most popular method is a cross-validation. Unlike in a classification, for not too noisy data (primarily without huge outliers), the penalty parameter C could be set to infinity and the modeling can be controlled by changing the insensitivity zone ε and shape parameters only. The *example* below shows how an increase in an insensitivity zone ε has smoothing effects on modeling highly noise polluted data. Increase in ε means a reduction in requirements on the accuracy of approximation. It decreases the number of SVs leading to higher data compression too. This can be readily followed in the lines and Fig. 19 below.

Example: The task here is to construct an SV machine for modeling measured data pairs. The underlying function (known to us but, not to the SVM) is a sinus function multiplied by the square one (i.e., $f(x) = x^2 \sin(x)$) and it is corrupted by 25% of normally distributed noise with a zero mean. Analyze the influence of an insensitivity zone ε on modeling quality and on a compression of data, meaning on the number of SVs.

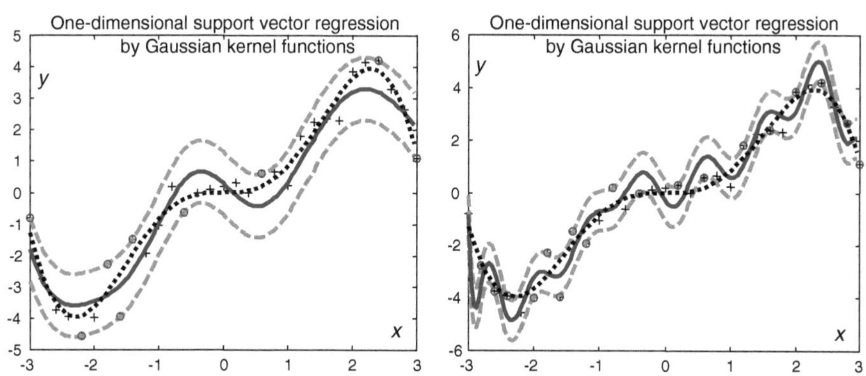

Fig. 19. The influence of an insensitivity zone ε on the model performance. A nonlinear SVM creates a regression function f with Gaussian kernels and models a highly polluted (25% noise) function $x^2 \sin(x)$ (*dotted*). 31 training data points (plus signs) are used. *Left*: $\varepsilon = 1$; 9 SVs are chosen (*encircled plus signs*). *Right*: $\varepsilon = 0.75$; the 18 chosen SVs produced a better approximation to noisy data and, consequently, there is the tendency of overfitting

3 Implementation Issues

In both the classification and the regression the learning problem boils down to solving the QP problem subject to the so-called box-constraints and to the equality constraint in the case that a model with a bias term b is used. The SV training works almost perfectly for not too large data basis. However, when the number of data points is large (say $l > 2{,}000$) the QP problem becomes extremely difficult to solve with standard QP solvers and methods. For example, a classification training set of 50,000 examples amounts to a Hessian matrix $\mathbf{H}$ with $2.5 * 10^9$ (2.5 billion) elements. Using an 8-byte floating-point representation we need 20,000 Megabytes = 20 Gigabytes of memory [21]. This cannot be easily fit into memory of present standard computers, and this is the single basic disadvantage of the SVM method. There are three approaches that resolve the QP for large data sets. Vapnik in [33] proposed the *chunking method* that is the decomposition approach. Another *decomposition* approach is suggested in [21]. The sequential minimal optimization (Platt, 1997) algorithm is of different character and it seems to be an "error back propagation" for an SVM learning. A systematic exposition of these various techniques is not given here, as all three would require a lot of space. However, the interested reader can find a description and discussion about the algorithms mentioned above in two chapters here. The Vogt and Kecman's chapter discusses the application of an *active set* algorithm in solving small to medium sized QP problems. For such data sets and when the high precision is required the active set approach in solving QP problems seems to be superior to other approaches (notably the interior point methods and SMO algorithm). The Kecman, Huang, and Vogt's chapter introduces the efficient *iterative single data algorithm* (*ISDA*) for solving huge data sets (say more than 100,000 or 500,000 or over 1 million training data pairs). It seems that ISDA is the fastest algorithm at the moment for such large data sets still ensuring the convergence to the global optimal solution for dual variables (see the comparisons with SMO in the mentioned chapter). This means that the ISDA provides the exact, and not the approximate, solution to original dual problem.

Let us conclude the presentation of SVMs part by summarizing the basic constructive steps that lead to the SV machine.

A training and design of a support vector machine is an *iterative* algorithm and it involves the following steps:

(a) define your problem as the classification or as the regression one,
(b) preprocess your input data: select the most relevant features, scale the data between $[-1, 1]$, or to the ones having zero mean and variances equal to one, check for possible outliers (strange data points),
(c) select the kernel function that determines the hypothesis space of the decision and regression function in the classification and regression problems respectively,

(d) select the "shape', i.e., "smoothing" parameter of the kernel function (for example, polynomial degree for polynomials and variances of the Gaussian RBF kernels respectively),
(e) choose the penalty factor C and, in the regression, select the desired accuracy by defining the insensitivity zone ε too,
(f) solve the QP problem in l and $2l$ variables in the case of classification and regression problems respectively,
(g) validate the model obtained on some previously, during the training, unseen test data, and if not pleased iterate between steps (d) (or, eventually c) and (g).

The optimizing part f) is for large or huge training data sets computationally extremely demanding. Luckily enough there are many sites for downloading the reliable, fast and free QP solvers. A simple search on the internet will reveal many of them. Particularly, in addition to the classic ones such as MINOS or LOQO for example, there are many more free QP solvers designed specially for the SVMs. The most popular ones are – the LIBSVM, SVMlight, SVM Torch, mySVM and SVM Fu. There are matlab based ones too. Good educational SVMs' software designed in matlab and named LEARNSC, with very good graphic presentations of all relevant objects in an SVM modeling, can be downloaded from the author's book site www.support-vector.ws too.

Finally, it should be mentioned that there are many alternative formulations and approaches to the QP based SVMs described above. Notably, they are the linear programming SVMs [10, 13, 14, 15, 16, 18, 27], υ-SVMs [25] and least squares support vector machines [29]. Their description is far beyond this introduction and the curious readers are referred to references given above.

References

1. Abe, S., Support Vector Machines for Pattern Classification (in print), Springer-Verlag, London, 2004
2. Aizerman, M.A., E.M. Braverman, and L.I. Rozonoer, 1964. Theoretical foundations of the potential function method in pattern recognition learning, Automation and Remote Control 25, 821–837
3. Cherkassky, V., F. Mulier, 1998. Learning From Data: Concepts, Theory and Methods, John Wiley & Sons, New York, NY
4. Chu F., L. Wang, 2003, Gene expression data analysis using support vector machines, Proceedings of the 2003 IEEE International Joint Conference on Neural Networks (Portland, USA, July 20–24, 2003), pp. 2268–2271
5. Cortes, C., 1995. Prediction of Generalization Ability in Learning Machines. PhD Thesis, Department of Computer Science, University of Rochester, NY
6. Cortes, C., Vapnik, V. 1995. Support Vector Networks. Machine Learning 20:273–297
7. Cristianini, N., Shawe-Taylor, J., 2000, An introduction to Support Vector Machines and other kernel-based learning methods, Cambridge University Press, Cambridge, UK

8. Drucker, H., C.J.C. Burges, L. Kaufman, A. Smola, V. Vapnik. 1997. Support vector regression machines, Advances in Neural Information Processing Systems 9, 155–161, MIT Press, Cambridge, MA
9. Eisenhart, C., 1962. Roger Joseph Boscovich and the Combination of Observationes, Actes International Symposium on R. J. Boskovic, pp. 19–25, Belgrade – Zagreb – Ljubljana, YU
10. Frieß, T., R. F. Harrison, 1998, Linear programming support vectors machines for pattern classification and regression estimation and the set reduction algorithm, TR RR-706, University of Sheffield, Sheffield, UK
11. Girosi, F., 1997. An Equivalence Between Sparse Approximation and Support Vector Machines, AI Memo 1606, MIT
12. Graepel, T., R. Herbrich, B. Schölkopf, A. Smola, P. Bartlett, K.–R. Müller, K. Obermayer, R. Williamson, 1999, Classification on proximity data with LP–machines, Proc. of the 9th Intl. Conf. on Artificial NN, ICANN 99, Edinburgh, 7–10 Sept.
13. Hadzic, I., V. Kecman, 1999, Learning from Data by Linear Programming, NZ Postgraduate Conference Proceedings, Auckland, Dec. 15–16
14. Kecman V., Arthanari T., Hadzic I., 2001, LP and QP Based Learning From Empirical Data, IEEE Proceedings of IJCNN 2001, Vol 4., pp. 2451–2455, Washington, DC
15. Kecman, V., 2001, Learning and Soft Computing, Support Verctor machines, Neural Networks and Fuzzy Logic Models, The MIT Press, Cambridge, MA, The book's web site is: http://www.support-vector.ws
16. Kecman V., Hadzic I., 2000, Support Vectors Selection by Linear Programming, Proceedings of the International Joint Conference on Neural Networks (IJCNN 2000), Vol. 5, pp. 193–198, Como, Italy
17. Kecman, V., 2004, Support Vector Machines Basics, Report 616, School of Engineering, The University of Auckland, Auckland, NZ
18. Mangasarian, O.L., 1965, Linear and Nonlinear Separation of Patterns by Linear Programming, Operations Research 13, pp. 444–452
19. Mercer, J., 1909. Functions of positive and negative type and their connection with the theory of integral equations. Philos. Trans. Roy. Soc. London, A 209:415{446}
20. Meyer D., F. Leisch, K. Hornik, 2003, The support vector machine under test, Neurocomputing 55, pp. 169–186
21. Osuna, E., R. Freund, F. Girosi. 1997. Support vector machines: Training and applications. AI Memo 1602, Massachusetts Institute of Technology, Cambridge, MA
22. Platt, J.C. 1998. Sequential minimal optimization: A fast algorithm for training support vector machines. Technical Report MSR-TR-98-14, Microsoft Research, 1998xs
23. Poggio, T., S. Mukherjee, R. Rifkin, A. Rakhlin, A. Verri, b, CBCL Paper #198/AI Memo# 2001-011, Massachusetts Institute of Technology, Cambridge, MA, 2001
24. Schölkopf B., C. Burges, A. Smola, (Editors), 1999. Advances in Kernel Methods – Support Vector Learning, MIT Press, Cambridge, MA
25. Schölkopf B., A. Smola, Learning with Kernels – Support Vector Machines, Optimization, and Beyond, The MIT Press, Cambridge, MA, 2002

26. Smola, A., B. Schölkopf, 1997. On a Kernel-based Method for Pattern Recognition, Regression, Approximation and Operator Inversion. GMD Technical Report No. 1064, Berlin
27. Smola, A., T.T. Friess, B. Schölkopf, 1998, Semiparametric Support Vector and Linear Programming Machines, NeuroCOLT2 Technical Report Series, NC2-TR-1998-024, also in In Advances in Neural Information Processing Systems 11, 1998
28. Support Vector Machines Web Site: http://www.kernel-machines.org/
29. Suykens, J.A.K., T. Van Gestel, J. De Brabanter, B. De Moor, J. Vandewalle, 2002, Least Squares Support Vector Machines, World Scientific Pub. Co., Singapore
30. Vapnik, V.N., A.Y. Chervonenkis, 1968. On the uniform convergence of relative frequencies of events to their probabilities. (In Russian), Doklady Akademii Nauk USSR, 181, (4)
31. Vapnik, V. 1979. Estimation of Dependences Based on Empirical Data [in Russian]. Nauka, Moscow. (English translation: 1982, Springer Verlag, New York)
32. Vapnik, V.N., A.Y. Chervonenkis, 1989. The necessary and sufficient condititons for the consistency of the method of empirical minimization [in Russian], yearbook of the Academy of Sciences of the USSR on Recognition, Classification, and Forecasting, 2, 217–249, Moscow, Nauka, (English transl.: The necessary and sufficient condititons for the consistency of the method of empirical minimization. Pattern Recognitio and Image Analysis, 1, 284–305, 1991)
33. Vapnik, V.N., 1995. The Nature of Statistical Learning Theory, Springer Verlag Inc, New York, NY
34. Vapnik, V., S. Golowich, A. Smola. 1997. Support vector method for function approximation, regression estimation, and signal processing, In Advances in Neural Information Processing Systems 9, MIT Press, Cambridge, MA
35. Vapnik, V.N., 1998. Statistical Learning Theory, J. Wiley & Sons, Inc., New York, NY

Multiple Model Estimation for Nonlinear Classification

Y. Ma[1] and V. Cherkassky[2]

[1] Honeywell Labs, Honeywell International Inc. 3660 Technology Drive, MN 55418, USA
yunqian.ma@honeywell.com
[2] Department of Electrical and Computer Engineering, University of Minnesota, Minneapolis, MN 55455, USA
cherkass@ece.umn.edu

Abstract. This chapter describes a new method for nonlinear classification using a collection of several simple (linear) classifiers. The approach is based on a new formulation of the learning problem called Multiple Model Estimation. Whereas standard supervised-learning learning formulations (such as regression and classification) seek to describe a given (training) data set using a single (albeit complex) model, under multiple model formulation the goal is to describe the data using several models, where each (simple) component model describes a subset of the data. We describe practical implementation of the multiple model estimation approach for classification. Several empirical comparisons indicate that the proposed multiple model classification (MMC) method (using linear component models) may yield comparable (or better) prediction accuracy than standard nonlinear SVM classifiers. In addition, the proposed approach has improved interpretation capabilities, and is more robust since it avoids the problem of SVM kernel selection.

Key words: Support Vector Machines, Multiple Model Estimation, Robust classification

1 Introduction and Motivation

Let us consider standard (binary) classification formulation under the general setting for predictive learning [1, 2, 3, 4]. Given finite sample data $(\mathbf{x}_i, y_i), (i = 1, \ldots, n)$, where $\mathbf{x}$ is an input (feature) vector $\mathbf{x} \in R^d$ and $y \in (+1, -1)$ is a class label, the goal is to estimate the mapping (indicator function) $f : \mathbf{x} \to y$ in order to classify future samples. Learning (model estimation) is a procedure for selecting the best indicator function $f(\mathbf{x}, \omega^*)$ from a set of possible models $f(\mathbf{x}, \omega)$ parameterized by a set of parameters $\omega \in \Omega$. For example, the linear discriminant function is $f(\mathbf{x}, \boldsymbol{\omega}) = \text{sign}(g(\mathbf{x}, \omega))$ where

$$g(\mathbf{x}, \omega) = (\mathbf{x} \cdot \mathbf{w}) + b \tag{1}$$

Y. Ma and V. Cherkassky: *Multiple Model Estimation for Nonlinear Classification*, StudFuzz **177**, 49–76 (2005)
www.springerlink.com

Parameters (aka weights) are $\mathbf{w}$ and b (bias parameter). Decision boundary $g(\mathbf{x}, \omega) = 0$ corresponds to $d - 1$ dimensional hyperplane in a d-dimensional input space.

Various statistical and neural network classification methods have been developed for estimating flexible (nonlinear) decision boundaries with finite data [2, 3, 5, 6]. The main assumption underlying all existing classification methods (both linear and nonlinear) is that all available (training) data can be classified by a single decision boundary no matter how complex the classifier is. This leads to the well-known trade-off between the model complexity and its ability to fit (separate) finite training data. Statistical Learning Theory (aka VC theory) provides theoretical analysis of this trade-off, and offers constructive methodology (called Support Vector Machines) for generating flexible (nonlinear) classifiers with controlled complexity [1, 7].

Our approach to forming (complex) decision boundary is based on relaxing the main assumption that all available (training) data can be described by a single classifier. Instead, we assume that training data can be described by several (simple) classifiers. For example, consider the data set shown in Fig. 1a. All training samples in this data set can be separated well using a complex (nonlinear) decision boundary. An alternative approach is to describe this data using a linear model (decision boundary) as shown in Fig. 1a. Then data points far away from this linear decision boundary (shown as bold circles) can be viewed as outliers (i.e., ignored during training). Such a linear model would classify correctly the *majority* of the data. The remaining (minor) portion of the data (marked in bold) can be classified by another linear model. Hence the data in Fig. 1a can be well described by two simple models (i.e., two linear decision boundaries). Similarly, for the data set shown in Fig. 1b one can form a linear decision boundary separating the majority of the data (as

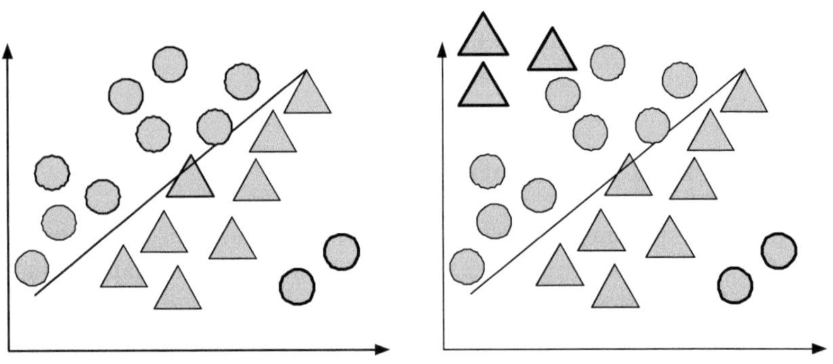

(a) (b)

Fig. 1. Separating training data using linear decision boundaries. The linear model (shown) explains well the majority of training data. (**a**) all training data can be explained well by two linear decision boundaries (**b**) all training data can be modelled by three linear decision boundaries shown in Fig. 1b). Then the remaining samples (misclassified by this major model) can be explained by two additional linear decision boundaries (not shown in Fig. 1b). Hence, the data shown in Fig. 1b can be modeled by three linear models.

Examples in Fig. 1 illustrate the main idea behind the multiple model estimation approach. That is, different subsets of available (training) data can be described well using several (simple) models. However, the approach itself is based on a new formulation of the learning problem, as explained next. Standard formulations (i.e., classification, regression, density estimation) assume that all training data can be described by a single model [1, 2], whereas the multiple model estimation (MME) approach assumes that the data can be described well by several models [8, 9, 10]. However, the number of models and the partitioning of available data (into different models) are both unknown. So the goal of learning (under MME) is to partition available training data into several subsets *and* to estimate corresponding models (for each respective subset). Standard inductive learning formulations (i.e., *single* model estimation) represent a special case of MME.

In the remainder of this section, we clarify the difference between the proposed approach and various existing *multiple learner* systems [11], such as Classification and Regression Trees (CART), Multivariate Adaptive Regression Splines (MARS), mixture of experts etc. [3, 12, 13, 14]; Let us consider the setting of supervised learning, i.e., classification or regression formulation, where the goal is to estimate a mapping $\mathbf{x} \rightarrow y$ in order to classify future samples. The multiple model estimation setting assumes that all training data can be described well by *several* models (mappings) $\mathbf{x} \rightarrow y$. The difference between the traditional (single-model) and multiple model formulations is shown in Fig. 2. Note that many multiple learner systems (for example, modular

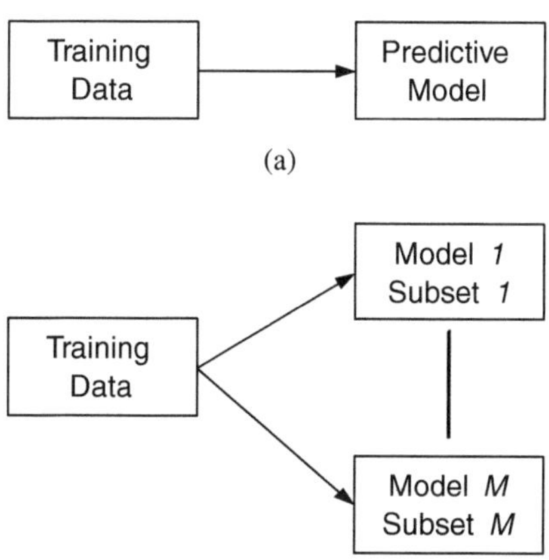

Fig. 2. Two distinct approaches to model estimation: (**a**) traditional single model estimation (**b**) multiple model estimation approach

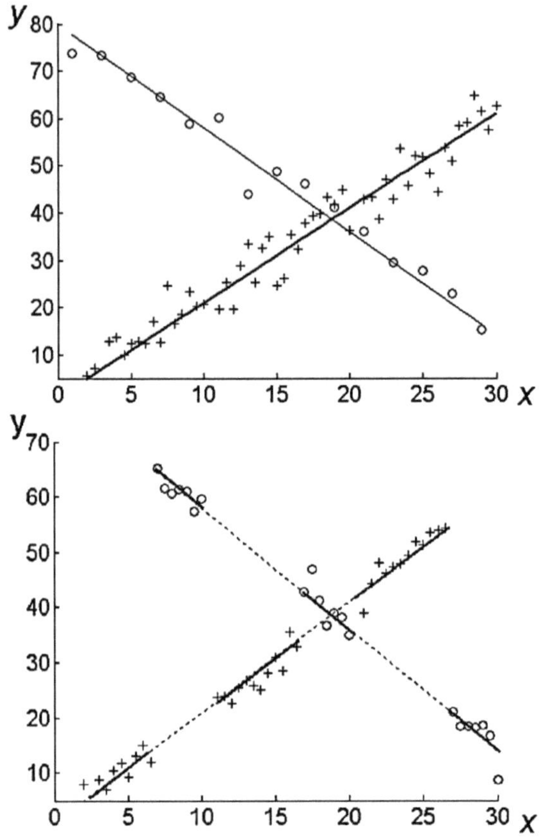

Fig. 3. Two example data sets suitable for multiple model regression formulation. (**a**) two regression models (**b**) single complex regression model

networks or mixture of experts) effectively represent single (complex) model as a collection of several (simple) models, which may appear similar to Fig. 2b. However, this similarity is superficial, since all existing multiple learner approaches still assume standard *single-model* setting for supervised learning. To make this distinction clear, consider example data set in Fig. 3 for (univariate) regression estimation. The first data set in Fig. 3a shows noisy data sampled from two regression models defined on the same **x**-domain. Clearly, existing regression methods (based on a single model formulation) would not be able to estimate accurately both regression models. The second data set (in Fig. 3b) can be interpreted as a single complex (discontinuous, piecewise-linear) regression model. This data set can be estimated, in principle, using the mixture of experts or CART approach. Such approaches are based on a single-model formulation, and hence they need to partition an input space into many regions. However, under multiple model regression approach, this data set can be modeled as only two linear regression models. Hence, the multiple model approach may be better, since it requires estimating just two (linear) models vs six component models using CART or mixture of experts. Two data sets shown in Fig. 3 suggest two general settings in which the multiple model estimation framework may be useful:

- The goal of learning is to *estimate several models*. For example, the data set in Fig. 3a is generated using two different target functions. The goal of learning is to estimate both target functions from (noisy) samples when the correspondence between data samples and target functions is unknown. Methods based on the multiple model estimation formulation can provide accurate regression estimates for both models see [8, 10, 15] for details;
- The goal of learning is to *estimate a single complex model* that can be approximated well by a few simple models, as in the example shown in Fig. 3b. Likewise, under classification formulation, estimating a complex model (i.e., nonlinear decision boundary) can be achieved by learning several simple models (i.e., linear classifiers), as shown in Fig. 1. In this setting, multiple model estimation approach is effectively applied to standard (single model) formulation of the learning problem. Proposed multiple model classification belongs to this setting.

There is a vast body of literature on multiple learner approaches for standard (single model) learning formulation. Following [11], all such methods can be divided into 3 groups:

- *Partitioning Methods* (or Modular Networks) that represent a single complex model by a (small) number of simple models specializing in different regions of the input space. Examples include CART, MARS, mixture of experts etc. [3, 12, 13, 14]. Hence the main goal of learning is to develop an effective strategy for partitioning the input space into a small number of regions.
- *Combining Methods* that use weighted linear combination of several independent predictive models (obtained using the same training data), in order obtain better generalizations. Examples include stacking, committee of networks approach etc.
- *Boosting Methods*, where individual models (classifiers) are trained on the weighted versions of the original data set, and then combined to produce the final model [16].

The proposed multiple model estimation approach may be related to partitioning methods in general, and to mixture models in particular. The mixture modeling approach (for density estimation) assumes that available data originates from several simple density models. The model memberships are modeled as hidden (latent) variables, so that the problem can be transformed into single model density estimation. The parameters of component models and mixing coefficients are estimated via Expectation-Maximization (EM)-type algorithms [17]. For example, for data set in Fig. 3a one can *apply first* various clustering and density estimation techniques to partition the data into (two) structured clusters and *second*, use standard regression methods to each subset of data. This approach is akin to density modeling/estimation, and it generally does not work well for sparse data sets. Instead, one should approach finite sample estimation problems *directly*, [1, 7]. Under the proposed multiple model estimation framework, the goal is to partition available data and to estimate respective models for each subset of the training data, at the same time. Conceptually, multiple model estimation can be viewed as estimating (learning) several simple structures that describe well available (training) data, where each component model is defined in terms of a particular type of the learning problem. For example, for multiple regression formulation each structure is a (single) regression model, whereas for multiple classification each structure is a decision boundary.

Practical implementations of multiple model estimation need to address two fundamental problems, i.e. *model selection and robust estimation*, at the same time. However, all existing constructive learning methods based on the classical statistical framework treat these issues separately. That is, model complexity control is addressed under a single model estimation setting; whereas robust methods typically attempt to describe the majority of the data under a parametric setting (i.e., using a model with known parametric form). This problem is well recognized in the Computer Vision (CV) literature as "the problem of scale". According to [18]: "prior knowledge of scale is often not available, and scale estimates are a function of both noise and modeling error, which are hard to discriminate". Recent work in CV attempts to combine robust estimation with model selection, in order to address this problem [19, 20, 21]. However, these methods still represent an extension of conventional probabilistic (density estimation) approaches. Under the VC-theoretical approach, the goal of learning is to find a model providing good generalization (rather than to estimate a "true" model), so both issues (robustness and model complexity) can be addressed together. In particular, the SVM methodology is based on the concept of margin (aka ϵ-insensitive zone for regression problems). Current SVM research is concerned with effective complexity control (for generalization) under single-model formulation. In contrast, the multiple model estimation learning algorithms [8, 9, 10] employ the concept of margin for controlling *both the model complexity and robustness*. In this chapter, we capitalize on the role of SVM margin, in order to develop new algorithms for multiple model classification.

This chapter is organized as follows. Section 2 presents general multiple model classification procedure. Section 3 describes an SVM-based implementation of the MMC procedure. Section 4 presents empirical comparisons between multiple model classification and standard SVM classificatiers. Finally, discussion and conclusions are given in Sect. 5.

2 Multiple Model Classification

Under MME formulation, the goal is to partition available data and to estimate respective models for each subset of the training data. Hence, multiple model estimation can be viewed as estimating (learning) several "simple structures" that describe well available data. For example, for multiple regression formulation each component is a (single) regression model, whereas for multiple classification each component model is a decision boundary. Since the problem of estimating several models from a single finite data set is inherently complex, we introduce a few assumptions to make it tractable. First, the number of component models is small; second, it is assumed that the majority of the data can be "explained" by a single model. The latter assumption is essential for robust model-free estimation. Assuming that we have a magic robust method that can always accurately estimate a good model for the majority of available data, we suggest an iterative procedure for multiple model estimation shown in Table 1. Note that the generic procedure in Table 1 is rather straightforward, and similar approaches have been used elsewhere, i.e. the "dominant motion" for multiple motion estimation in Computer Vision [22, 23]. The problem, however, lies in specifying a robust method for estimating the major model (in Step 1). Here "robustness" refers to the capability of estimating a major model when available data may be generated by several other "structures". This notion of robustness is different from the traditional robust techniques, because such methods are still based on a single-model formulation, where the goal is resistance (of estimates) with respect to heavy-tailed noise, rather than "structured outliers" [19].

Next we show a simple example to illustrate desirable properties of "robust estimators" for classification. Consider classification data set shown in Fig. 4a, where the decision boundary is formed by two linear models, using a robust method (shown in Fig. 4a), and a traditional (non-robust) CART method (in Fig. 4b). If the "minor portion" of the data (i.e., "△" samples in the upper-right corner of Fig. 4a) varies as shown in Fig. 4c, this does not affect the major

Table 1. General procedure for multiple model estimation

Initialization:	"Available data" = all training samples.
Step 1:	**Estimate major model**, i.e. apply robust estimation method to available data, resulting in a dominant model $M1$ (describing the majority of available data).
Step 2:	**Partition available data** into two subsets, i.e. samples generated by $M1$ and samples generated by other models (the remaining data). This partitioning is performed by analyzing available data samples ordered according to their distance (residuals) to major model $M1$.
Step 3:	**Remove subset of data** generated by major model from available data.
Iterate:	Apply Steps 1∼3 to available data until some stopping criterion is met.

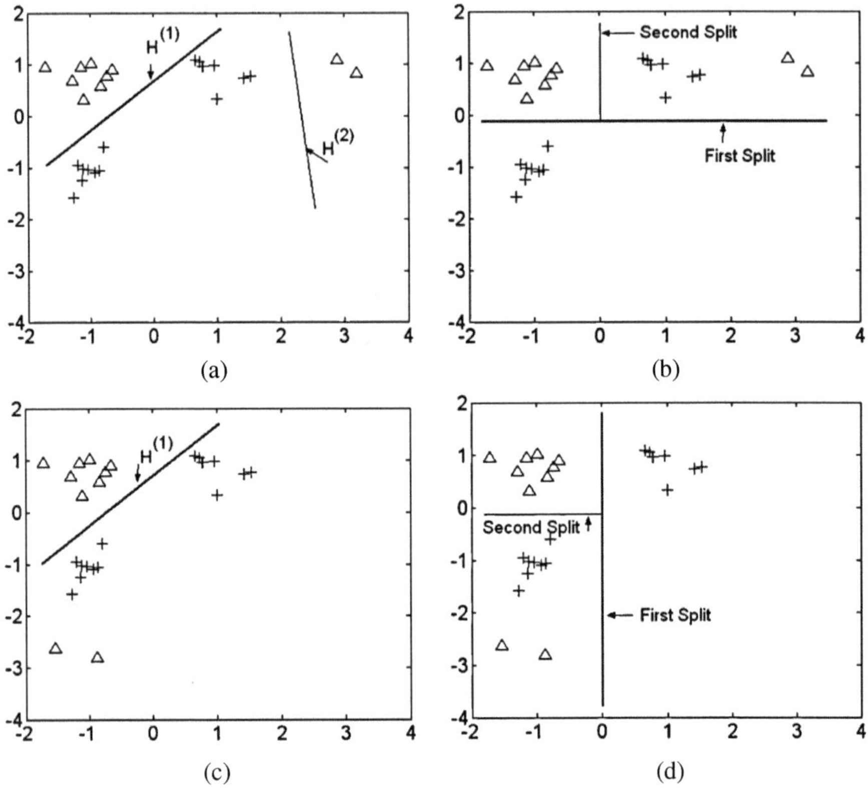

Fig. 4. Comparison of decision boundaries formed by robust method and CART. (**a**) two linear models formed by robust method; (**b**) two linear splits formed by CART for the data set shown in (**a**); (**c**) the first (major) component model formed by a robust method for another data set; (**d**) two linear splits formed by CART for the data set shown in (**c**)

model as shown in Fig. 4c, but this variation in the data will totally change the first split of the CART method (see Fig. 4d). Example in Fig. 4 clearly shows the difference between traditional methods (such as CART) that seek to minimize some loss function for all "available" data during each iteration, and multiple model estimation that seeks to explain the majority of available data during each iteration.

In this chapter, we use a *linear* classifier as the *basic classifier* for each component model. This assumption (about the linearity) is not critical and can be later relaxed. However, it is useful to explain the main ideas underlying the proposed approach. Hence we assume the existence of a hyperplane separating the majority of the data (from one class) from the other class data. More precisely, the *main assumption* is that the majority of the data (of one class) can be described by a single dominant model (i.e., linear decision boundary).

Hence, the remaining samples (that do not fit the dominant model) appear as outliers (with respect to this dominant model). Further, we may distinguish between the two possibilities:

- "Outliers" appear only on one side of the dominant decision boundary;
- "Outliers" appear on both sides of the dominant decision boundary.

Both cases are shown in Fig. 1a and Fig. 1b, respectively. Note that in the first case (shown in Fig. 1a) all available data from one class (labeled as "$\triangle$") can be unambiguously described by the dominant model. That is, all data samples (from this class) lie on the same side of a linear decision boundary, or close to decision boundary if the data is non-separable (as in Fig. 1a). However, in the second case (shown in Fig. 1b) the situation is less clear in the sense that each of the three (linear) decision boundaries can be interpreted as a dominant model, even though we assumed that the middle one is dominant. The situation shown in Fig. 1b leads to several (ambiguous) solutions/interpretations of multiple model classification. Hence, the multiple model classification setting assumes that the dominant model describes well the majority of available data, i.e. that *both conditions* hold:

1. All data samples from one class lie on the same side of a linear decision boundary (or close to decision boundary when the data is non-separable). Stated more generally (for nonlinear component models), this condition implies that all data from one class (say, the first class) belongs to a single convex region, while the data from another class has no such constraint;
2. The majority of the data from another class (the second class) can be explained by the dominant model.

These conditions guarantee that the majority of training data can be "explained" by a (linear) component model during Step 1 of the general procedure in Table 1. For example, the data set shown in Fig. 1a satisfies these conditions, whereas the data set in Fig. 1b does not.

Based on the above assumption, a constructive learning procedure for classification can be described as follows. Given the training data $(\mathbf{x}_i, y_i)$, $(i = 1, \ldots, n)$, $y \in (+1, -1)$, apply an iterative algorithm for Multiple Model Classification (MMC), as shown in Table 2.

The notion of *robust method* in the above procedure is critical for understanding of the proposed approach, as explained next. Robust method is an estimation method that:

- describes well the majority of available data;
- is robust with respect to arbitrary variations of the remaining data.

The *major model* is a decision boundary that classifies correctly the majority of training data (i.e. classifies correctly all samples from one class, and the majority of samples from the second class). We discuss an SVM-based implementation of such a robust classification (implementing Step 1) later in Sect. 3.

Table 2. Multiple Model Classification Algorithm

Initialization:	"Available data" = all training samples.
Step 1:	**Estimate the major classifier (model)**, i.e. apply *robust* classification method to *available data*, resulting in a decision boundary explaining well the majority of the data.
Step 2:	**Partition available data** (from the second class) into two subsets: samples unambiguously classified (explained) by a major model and samples explained by other models (the remaining data).
Step 3:	**Remove the subset of data** (from the second class) classified (explained) by the major model.
Iterate:	Apply Steps 1–3 to available data until some stopping criterion is met.

Next we explain the data partitioning (Step 2). The decision boundary $g(\mathbf{x}) = 0$ formed by robust classification describes (classifies) the majority of available data if the following condition holds for the majority of training samples (from each class):

$$y_j g(\mathbf{x}_j) \geq \delta \text{ where } \delta \leq 0, j = 1, \ldots, l \tag{2}$$

where $(\mathbf{x}_j, y_j)$ denotes training samples (from one class) and l is the number of samples (greater than, say, 70% of samples) from that class. The quantity $y_j g(\mathbf{x}_j)$ describes the distance between a training sample and the decision boundary.

The value $\delta = 0$ corresponds to the case when the majority of data (from one class) is linearly separable from the rest of training data. Small negative parameter value $\delta < 0$ indicates there are some points on the wrong side of decision boundary, i.e. the majority of data from one class is separated from the rest of training data with some overlap. The value of δ quantifies the amount of overlap.

Recall the major model classifies (explains) correctly all samples from one class, and the majority of samples from another class. Let us consider the distribution of residuals $y_j g(\mathbf{x}_j)$ for data samples from the second class, as illustrated in Fig. 5. As evident from Fig. 5 and expression (2), data samples correctly explained (classified) by the major model are either on the correct side of decision boundary, or close to decision boundary (within δ margin). In contrast, the minor portion of the data (from the same class) is far away from decision boundary (see Fig. 5). Therefore, the data explained by the major model (decision boundary) can be easily identified and removed from the original training set (i.e., Step 2 in the procedure outlined above). The proper value of δ can be user-defined, i.e. from the visual inspection of residuals in Fig. 5. Alternatively, this value can be determined automatically, as discussed next. This value should be proportional to the "level of noise" in the data, i.e. to the amount of overlap between the two classes. In particular, when the

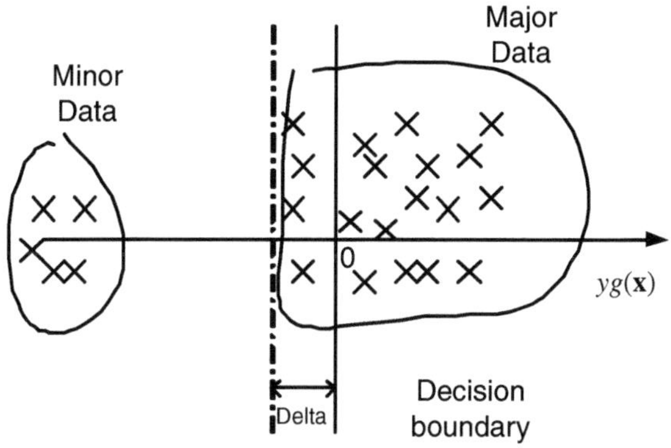

Fig. 5. Distribution of training data relative to a major model (decision boundary)

major model is derived using SVM (as described in Sect. 3), the value of δ should be proportional to the size of margin.

Stopping criterion in the above MMC algorithm can be based on the maximum number of models (decision boundaries) allowed, or on the maximum number of samples allowed in the minor portion of the data during the last iteration (i.e., 3 or 4 samples). In either case, the number of component models (decision boundaries) is not given a priori, but is determined in a data-driven manner.

Next we illustrate the operation of the proposed approach using the following two-dimensional data set:

- Positive class data: two Gaussian clusters centered at $(1, 1)$ and $(-1, -1)$ with variance 0.01. There are 8 samples in each cluster;
- Negative class: two Gaussian clusters centered at $(-1, 1)$ and $(1, -1)$ with variance 0.01. There are 8 samples in the cluster centered at $(-1, 1)$, and 2 samples in the cluster centered at $(1, -1)$.

This training data are shown in Fig. 6a, where the positive class data is labeled as "+", and negative class data is labeled as "$\triangle$". Operation of the proposed multiple model classification approach assumes the existence of robust classification method that can reliably estimate the decision boundary classifying correctly the majority of the data (i.e., major model). Detailed implementation of such a method is given later in Sect. 3. Referring to the iterative procedure for multiple model estimation (given in Table 2), application of robust classification method to available training data (Step 1) results in a major model (denoted as hyperplane $H^{(1)}$) as shown in Fig. 6b. Note the major model $H^{(1)}$ can correctly classify all positive-class data, and it can classify correctly the majority of the negative-class samples. Hence, in Step 2, we remove the majority of the negative-class data (explained by $H^{(1)}$) from available training data. Then we apply robust classification method to the remaining data (during second iteration of an iterative procedure) yielding

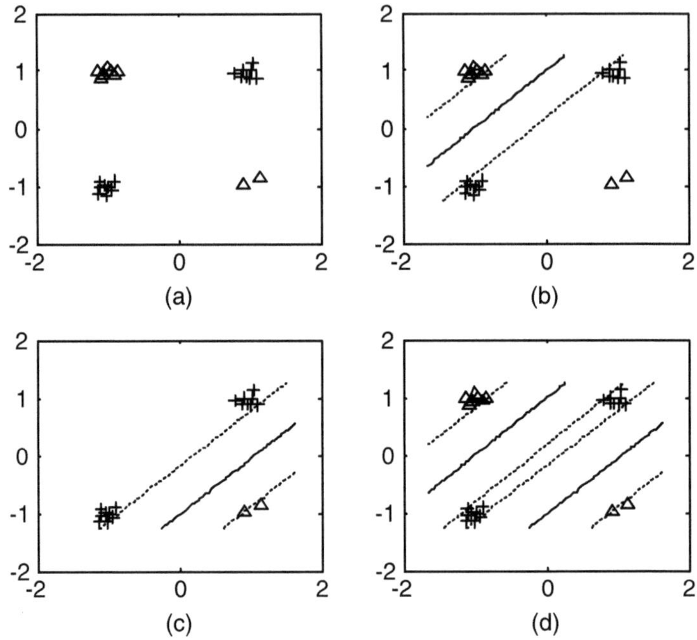

Fig. 6. Application of Multiple Model Classification method to Exclusive-OR like data set. (**a**) Training data; (**b**) Major model: Hyperplane 1; (**c**) Minor model: Hyperplane 2; (**d**) Combination of both component models

the second model (hyperplane) $H^{(2)}$, as shown in Fig. 6c. Two resulting hyperplanes $H^{(1)}$ and $H^{(2)}$ are shown in Fig. 6d.

During the test (or operation) phase, we need to classify given (test) input $\mathbf{x}$ using multiple-model classifier estimated from training data. Recall that multiple model classification yields several models, i.e. hyperplanes $\{H^{(i)}\}$. In order to classify test input $\mathbf{x}$, it is applied first to the major model $H^{(1)}$. There may be two possibilities:

- Input $\mathbf{x}$ can be classified by $H^{(1)}$ unambiguously. In this case, the input is assigned a proper class label and the process is stopped;
- Input $\mathbf{x}$ can not be classified by $H^{(1)}$ unambiguously. In this case, the next model $H^{(2)}$ is used to classify this input on the remaining half of the input space.

Similarly, there may be two possibilities in classifying input $\mathbf{x}$ with respect to $H^{(2)}$, so the above comparison process continues until all models $\{H^{(i)}\}$ are exhausted. Both the learning (model estimation) and operation (test) stages of the proposed multiple model classification approach are shown in Fig. 7.

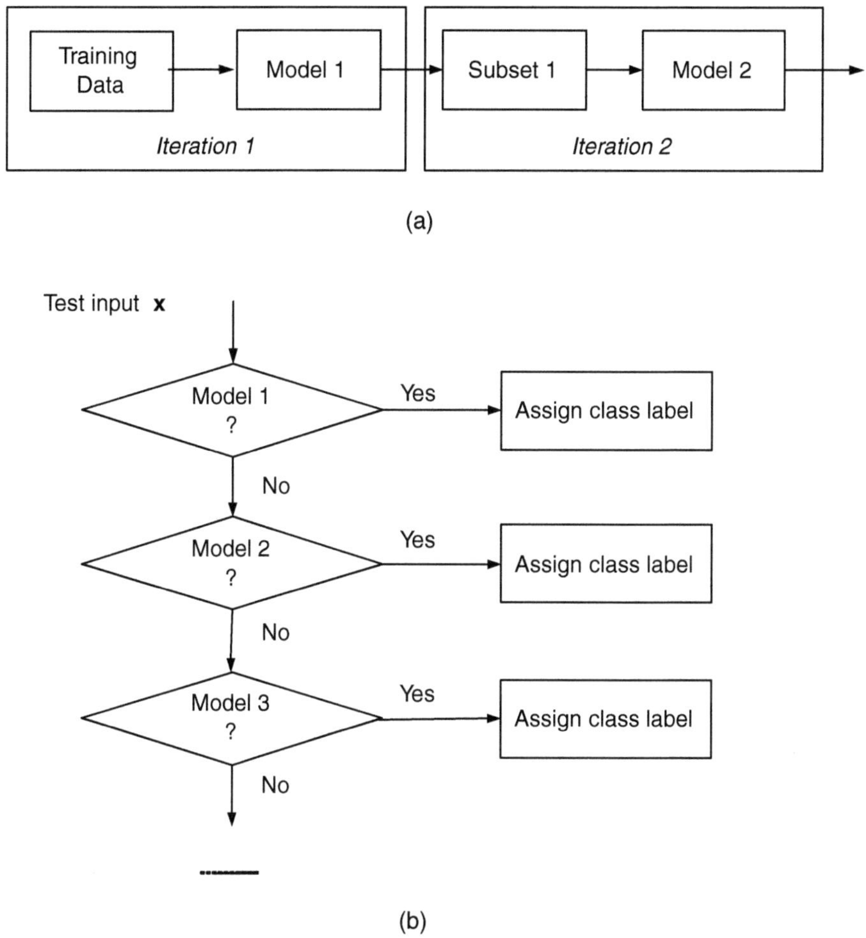

Fig. 7. Multiple model classification approach. (**a**) Training (model estimation) stage; (**b**) operation (test) stage

3 SVM-Based Implementation of Robust Classification

In this section, we discuss implementation of robust classification method used in the procedure for multiple model classification (i.e., *Step 1* of the MMC algorithm given in Table 2). This step insures robust separation of the majority of the training data. As discussed in Sect. 2, "robustness" here refers to the requirement that a robust model (decision boundary) should be insensitive to variations of the minor portion of the data. An implementation described in this section is SVM-based; however, it should be noted that the iterative MMC procedure can use any other robust classification method for implementing Step 1.

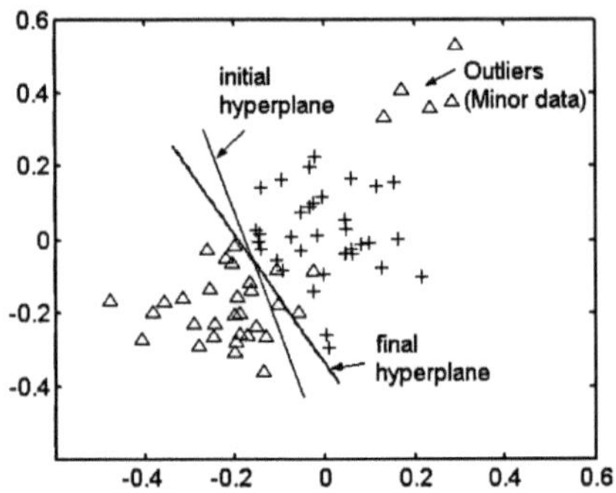

(a)

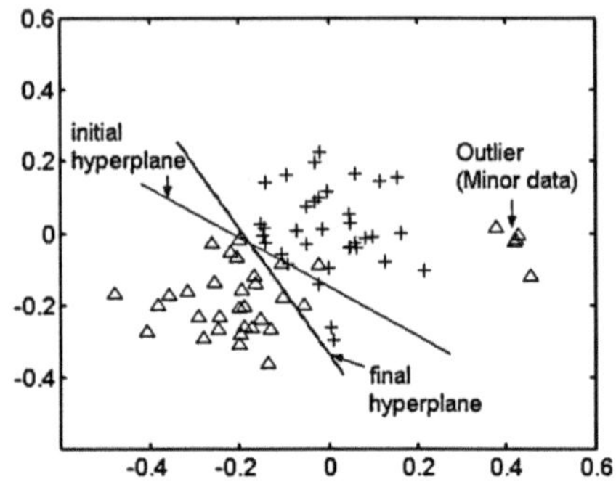

(b)

Fig. 8. Example of robust classification method (based on double application of SVM). (**a**) Major model (final hyperplane) estimated for the first data set; (**b**) Major model (final hyperplane) estimated for the second data set

The main idea of the proposed approach is described next. Recall the assumption (in Sect. 2) that the majority of the data can be described by a single model. Hence, application of a (linear) SVM classifier to all training data should result in a good separation (classification) of the *majority* of training data. The remaining (minor) portion of the data appears as outliers with respect to the SVM model. This minor portion can be identified by analyzing the "slack variables" in SVM solution. This initial SVM model is not robust since the minor portion of the data (outliers) may act as support vectors. However, these outliers can be removed from the training data, and then SVM can be *applied again* to the remaining data. The final SVM model will be robust with respect to any variations of the minor portion of the original training data. Such a "double application of SVM" for robust estimation of the major model (decision boundary) is illustrated in Fig. 8. Note that two data sets shown in Fig. 8 differ only in the minor portion of the data.

In order to describe this method in more technical terms, let us first review linear soft-margin SVM formulation [1, 2, 4, 7]. Given the training data, the (primal) optimization problem is:

$$\text{minimize} \quad \frac{1}{2}||\mathbf{w}||^2 + C\sum_{i=1}^{n}\xi_i \tag{3}$$

$$\text{subject to } y_i(\mathbf{w}\cdot\mathbf{x}_i + b) \geq 1 - \xi_i, i = 1, \ldots, n$$

Solution of constrained optimization problem (3) results in a linear SVM decision boundary $g(\mathbf{x}) = (\mathbf{x}\cdot\mathbf{w}^*) + b^*$. The value of regularization parameter C controls the margin $1/||\mathbf{w}^*||$, i.e. larger C-values result in SVM models with a smaller margin. Each training sample can be characterized by its distance from the margin $\xi_i \geq 0, i = 1, \ldots, n$, aka slack variables [1]. Formally, these slack variables can be expressed as

$$\xi_i = [1 - y_i g(\mathbf{x}_i)]_+ \tag{4}$$

where "+" indicates the positive part of the expression (corresponding to training samples on the wrong side of the margin). Note that correctly classified samples (on the right side of the margin) have zero values of slack variables.

Assuming that the majority of the data can be described by single (linear) SVM model, *initial application* of standard SVM to all training data would result in partitioning training samples into 3 groups:

- Samples correctly classified by SVM model (these samples have *zero slack variables*);
- Samples on the wrong side but close to the margin (these samples have *small slack variables*);
- Samples on the wrong side and far away from the margin (these have *large slack variables*).

Then samples with large slack variables are removed from the data and SVM is *applied the second time* to the remaining data. The final SVM model will be robust with respect to any variations of the minor portion of the original training data. Such a "double application of SVM' for estimating robust model (decision boundary) is summarized in Table 3.

Table 3. Robust classification method: Double Application of SVM (Implementing Step 1 of MMC algorithm in Table 2)

Step (1a):	Apply (linear) SVM classifier to all available data, producing initial hyperplane $g^{(init)}(\mathbf{x}) = 0$
Step (1b):	Calculate the slack variables of training samples with respect to initial SVM hyperplane $g^{(init)}(\mathbf{x}) = 0$, i.e. $\xi_i = [1 - y_i g^{(init)}(\mathbf{x}_i)]_+$. Then order slack variables $\xi_1^* \leq \xi_2^* \leq \dots \leq \xi_n^*$, and remove samples with large slack variables (that is, larger than *threshold* θ) from the training data.
Step (1c):	Apply SVM second time to the remaining data samples (with slack variables smaller than θ). The resulting hyperplane $g(\mathbf{x}) = 0$ represents robust partitioning of the original training data. Note that the final model (hyperplane) is robust with respect to variations of the minor portion of the data by design (since all such samples are removed after initial application of SVM).

Practical application of this algorithm (for robust classification) requires proper setting of parameters C and threshold θ, as explained next.

For the initial application of SVM in *step (1a)*, the goal is to generate linear SVM decision boundary separating (explaining) the major portion of available data. This can be achieved by using sufficiently large C-values, since it is known that the value of SVM margin (for linear SVM) is not sensitive to the

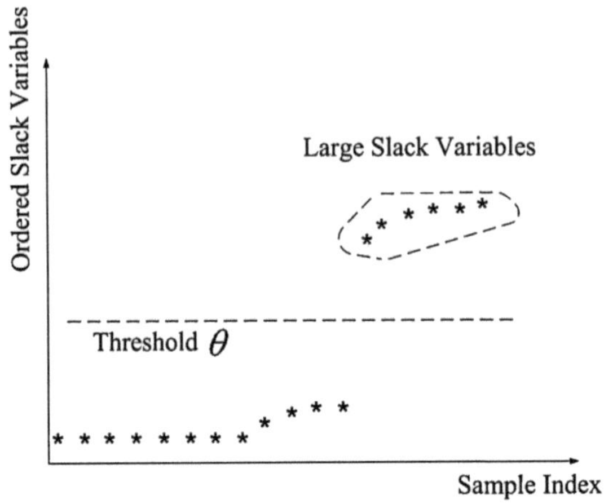

Fig. 9. Distribution of ordered slack variables following initial application of SVM to available data

choice of (large enough) C-values [3]. (The value of regularization parameter C controls the margin, i.e. larger C-values result in SVM models with smaller margin.)

The value of a threshold (see Fig. 9) for separating samples generated by minor model(s) can be set in the range $\theta = (2 \sim 4) * margin$. In all experimental results reported in this chapter, the threshold was set as $\theta = 3 * margin$. Here *margin* denotes the size of margin of the initial SVM hyperplane (in step 1a). During the second application of SVM in *step (1c)*, the goal is to estimate (linear) decision boundary, under standard classification setting. In this case, the value of parameter C can be selected using resampling methods.

Application of the "double SVM" method to available data constitutes the first step of MMC algorithm (shown in Table 3), and it results in a (linear) SVM component model with a certain margin (determined by parameter C selected by resampling). This margin is used to select the value of threshold δ in the second step of MMC algorithm. We used the value of δ equal to three times the margin (of SVM decision boundary for the major model) in all empirical comparisons shown later in this chapter.

The proposed multiple model classification approach partitions the training data into several subsets, so that each subset can be described by a simple (linear) decision boundary. Since this approach results in a hierarchical partitioning of the input space using a combination of linear decision boundaries, it can be directly compared with traditional partitioning methods (such as CART) that construct nonlinear decision boundaries (under single model formulation) via hierarchical partitioning of the input space. Such a comparison with CART is shown in Fig. 10 using a two-dimensional data set. Three initial splits made by the CART algorithm (standard MATLAB implementation using Gini loss function) are shown in Fig. 10a and 10b. The proposed MMC

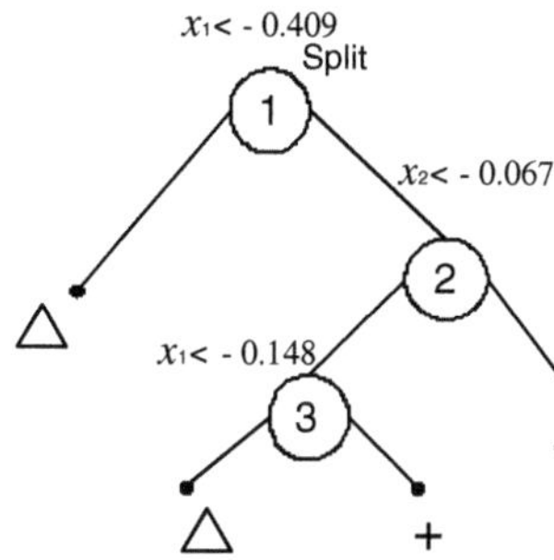

(a)

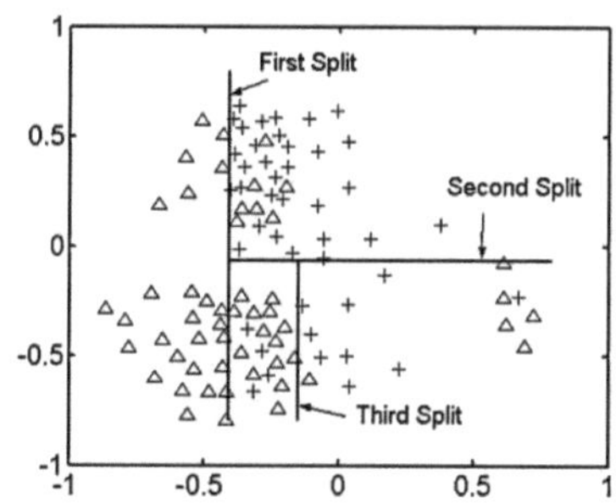

(b)

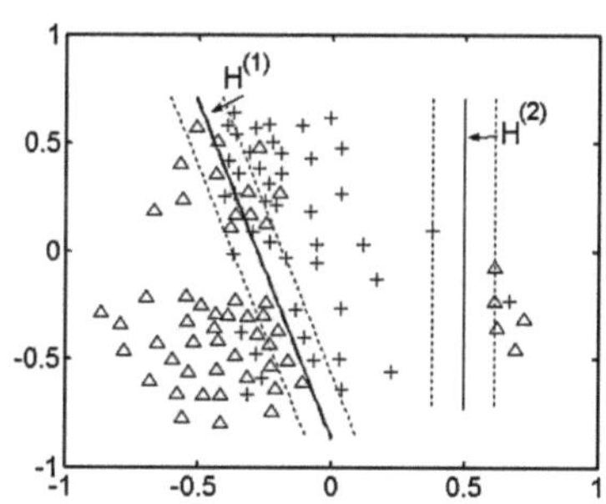

(c)

Fig. 10. Comparison of hierarchical partitioning obtained by CART and MMC. (**a**) Three initial splits obtained by CART; (**b**) Decision boundaries obtained by CART; (**c**) Decision boundaries obtained by using multiple model classification

classification algorithm (in Table 2) using "double SVM" method for robust classification (in Table 3) results in two component models shown in Fig. 10c. Clearly, for this data set, the MMC approach results in a better (more robust and simpler) partitioning of the input space than CART. This example also shows that the CART method may have difficulty producing robust decision boundaries for data sets with large amount of overlap between classes.

4 Experimental Results

This section presents empirical comparisons between the proposed multiple model classification approach and standard nonlinear SVM classification (based on a single model formulation) for several data sets. The proposed method is implemented using linear SVM for estimating each component model, whereas standard approach uses nonlinear SVM model (with kernels). Hence, comparisons intend to show the trade-offs between using several (simple) models under the proposed approach versus using a single complex (nonlinear) model under standard classification formulation. We used radial basis function (RBF) kernels $k(\mathbf{x}_i, \mathbf{x}) = \exp\left(-||\mathbf{x}_i - \mathbf{x}||^2/2p^2\right)$ and polynomial kernels $k(\mathbf{x}_i, \mathbf{x}) = [(\mathbf{x}_i \cdot \mathbf{x}) + 1]^d$ to implement standard (single model) SVM approach [1, 2]. In order to make comparisons "fair', we selected good (near optimal) values of hyper-parameters of nonlinear SVM methods (such as the polynomial degree d, and the width of RBF kernel p) using empirical tuning. Note that the proposed multiple model classification (MMC) does not require empirical tuning of kernel parameters (when it employs linear SVMs).

Experiment 1: The training and test data have the same prior probability 0.5 for both classes. The data is generated as follows:

- Class 1 data (labeled as "+" in Fig. 11) is generated using Gaussian distribution with a center at $(-0.3, 0.7)$ and variance 0.03;
- Class 2 data (labeled as "△" in Fig. 11) is a mixture of two Gaussians centered at $(-0.7, 0.3)$ and $(0.4, 0.7)$, both having the same variance 0.03. The probability that class 2 data is generated from the first cluster is 10/11 (major model) and the probability that class 2 data is generated from the second cluster is 1/11 (minor model).

The training data set (shown in Fig. 11) has 55 samples from Class 1, and 55 samples from Class 2. Note that Class 2 data is generated from 2 clusters, so that 50 samples (major model) are generated by Gaussian cluster centered at (−0.7, 0.3), and 5 samples are generated by a cluster centered at (0.4, 0.7), corresponding to the minor model. A test set of 1100 samples is used to estimate the prediction risk (error rate) of several classification methods under comparison.

Table 4 shows comparisons between MMC (using linear SVM) and the standard nonlinear SVM methods (with polynomial kernels and RBF kernels).

Table 4. Comparison Prediction accuracy for data set 1

	Error Rate (%SV)
RBF ($C = 1, p = 0.2$)	0.0582 (25.5%)
Poly ($C = 1, d = 3$)	0.0673 (26.4%)
MMC ($C = 100$)	0.0555 (14.5%)

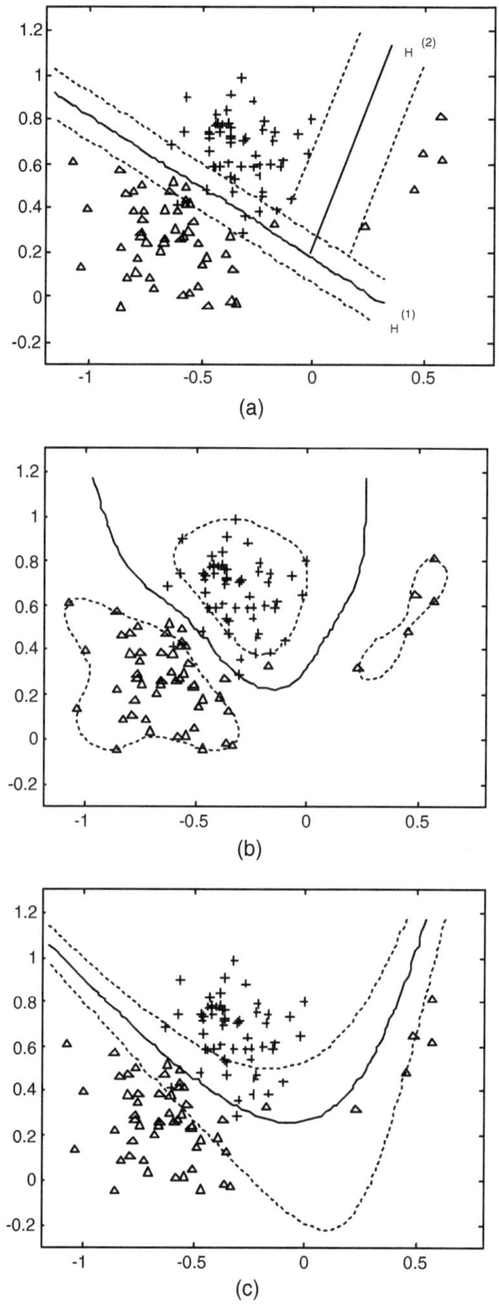

Fig. 11. Comparison of (best) decision boundaries obtained for data set 1. (**a**) MMC (**b**) SVM with RBF kernel, width $p = 0.2$ (**c**) SVM with polynomial kernel, degree $d = 3$

Kernel parameters (polynomial degree d and RBF width parameter p) have been optimally tuned for this data set. Results in Table 4 show the prediction risk (or error rate observed on the test data) and the percentage of support vectors selected by each SVM method with optimal values of hyper-parameters. For the MMC approach, the same (large) C-value was used for all iterations. Actual decision boundaries formed by each method (with optimal parameter values) are shown in Fig. 11. These results indicate that for this data set, the proposed multiple model classification approach is very competitive against standard SVM classifiers. In fact, it provides better prediction accuracy than standard SVM with RBF kernel that is (arguably) optimal for this data set (a mixture of Gaussians). More importantly, the proposed method does not require kernel parameter tuning.

Experiment 2: The training and test data have the same prior probability 5/11 for class 1 data and 6/11 for class 2 data. The data are generated as follows:

- Class 1 data (labeled as "+" in Fig. 12) is uniformly distributed inside the triangle with vertices $(0, 0)$, $(1, 0)$ and $(0, 1)$;
- Class 2 data (labeled as "$\triangle$" in Fig. 12) are a mixture of 3 distributions: a uniform distribution inside the triangular region with vertices at $(1, 0)$, $(0, 1)$ and $(1, 1)$, and two Gaussians centered at $(-0.3, 0.6)$ and $(0.6, -0.3)$. The prior probability that Class 2 data is generated in the triangular region is 50/60 and the prior probabilities that the data is generated by each Gaussian is 8/60 and 2/60, respectively. Two Gaussians centered at $(-0.3, 0.6)$ and $(0.6, -0.3)$ have the same variance 0.01.

The training data set of 110 samples (shown in Fig. 12) has 50 samples from Class 1, and 60 samples from Class 2. Note that Class 2 data is a mixture of 3 distributions, so that 50 samples (major model) are generated inside the triangular region, 8 samples are generated by a Gaussian cluster centered at $(-0.3, 0.6)$, and 2 samples are generated by a Gaussian centered at $(0.6, -0.3)$. A test set of 1100 samples is used to estimate the prediction risk (error rate) of several classification methods under comparison.

Table 5 shows comparisons between the proposed multiple model classification (using linear SVM) and best results for two standard nonlinear SVM methods (with polynomial and RBF kernels). For the MMC approach, the same (large) C-value was used for all iterations. Actual decision boundaries formed by each method (with optimal parameter values) are shown in Fig. 12. For this data set, the proposed multiple model classification approach provides better results than standard SVM classifiers. We also observed that the error rate of standard SVM classifiers varies wildly depending on different values of SVM kernel parameters, whereas the proposed method is more robust as it does not require kernel parameter tuning. This conclusion is consistent with experimental results in Table 4. So the main practical advantage of the proposed method is its robustness with respect to tuning parameters.

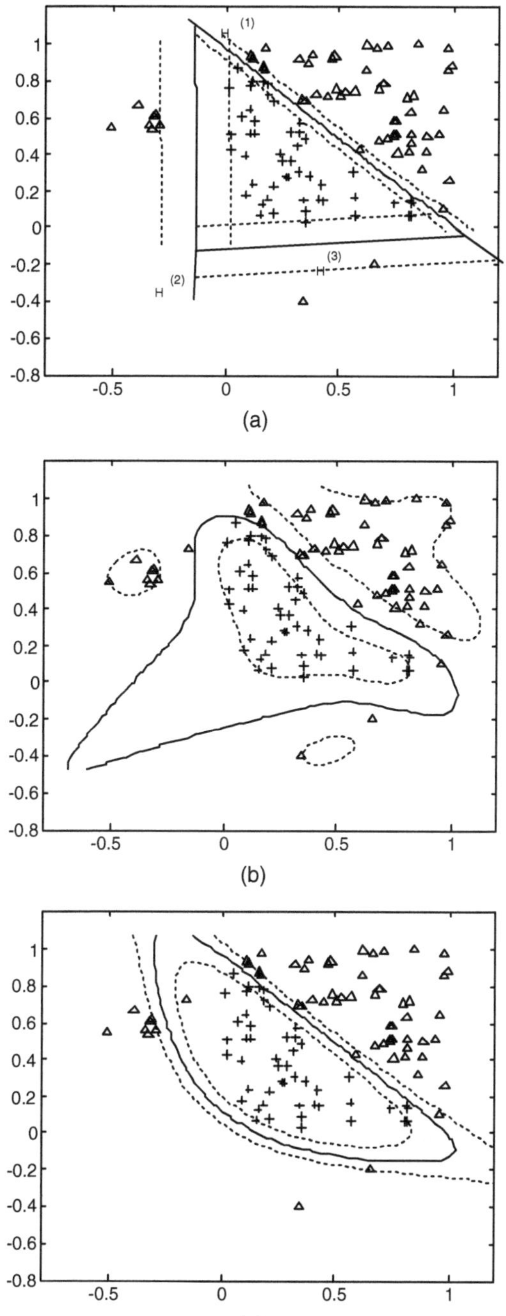

Fig. 12. Comparison of (best) decision boundaries obtained for data set 2. (**a**) MMC (**b**) SVM with RBF kernel, width $p = 0.2$ (**c**) SVM with polynomial kernel, degree $d = 3$

Table 5. Comparison Prediction accuracy for data set 2

	Error Rate (%SV)
RBF ($C = 1, p = 0.2$)	0.0291 (37.3%)
Poly ($C = 100, d = 3$)	0.0473 (16.4%)
MMC ($C = 100$)	0.0155 (15.5%)

Recall that application of the proposed MMC approach requires that the majority of available data can be separated by a single (linear) decision boundary. Of course, this includes (as a special case) the situation when all available data can be modeled by single (linear) SVM classifier. In some applications, however, the above assumption does not hold. For example, multi-category classification problems are frequently modeled as several binary classification problems. In other words, a multi-category classification problem is mapped onto standard binary classification formulation (i.e., all training data is divided into samples from a particular class vs samples from other classes). For example, consider a 3-class problem where each class data contains (roughly) the same number of samples, and the corresponding binary classification problem of estimating (learning) the decision boundary between *Class 1* and the rest of the data (comprising *Class 2* and *Class 3*). In this case, it seems reasonable to apply MMC approach, so that the decision boundary is generated by two models, i.e., *Class 1* vs *Class 2*, and *Class 1* vs *Class 3*. See Fig. 13a. This may suggest an application of the MMC approach to all available data (using a binary classification formulation). However, such a straightforward application of MMC would not work if Class 2 and Class 3 data have a similar number of samples (this violates the assumption that the majority of available data is described by a single model). In order to apply MMC in this setting, one can first modify the available data set by removing (randomly) a portion of Class 3 data (say, 50% of samples are removed), and then applying Step 1 of the "double SVM' algorithm to the remaining data in order to identify the major model (i.e., decision boundary separating Class 1 and Class 2 data). During the second iteration of MMC algorithm (i.e., when estimating the minor model, or decision boundary separating Class 1 and Class 3) we include the removed samples from Class 3. See Fig. 13b and Fig. 13c for illustration of this procedure. Such a modification of the MMC approach (for multiclass problems) has been applied to the well-known IRIS data set.

Experiment 3: The IRIS dataset [24] contains 150 samples describing 3 species (classes): iris setosa, iris versicolor, and iris virginica (50 samples per each class). Two input variables, petal length and petal width, are used for forming classification decision boundaries. Even though the original IRIS dataset has four input variables (sepal length, sepal width, petal length and petal width), it is widely known that the two variables (petal length and width) contain most class discriminating information [25]. Available data (150 samples) are divided into training set (75 samples) and test set (75 samples).

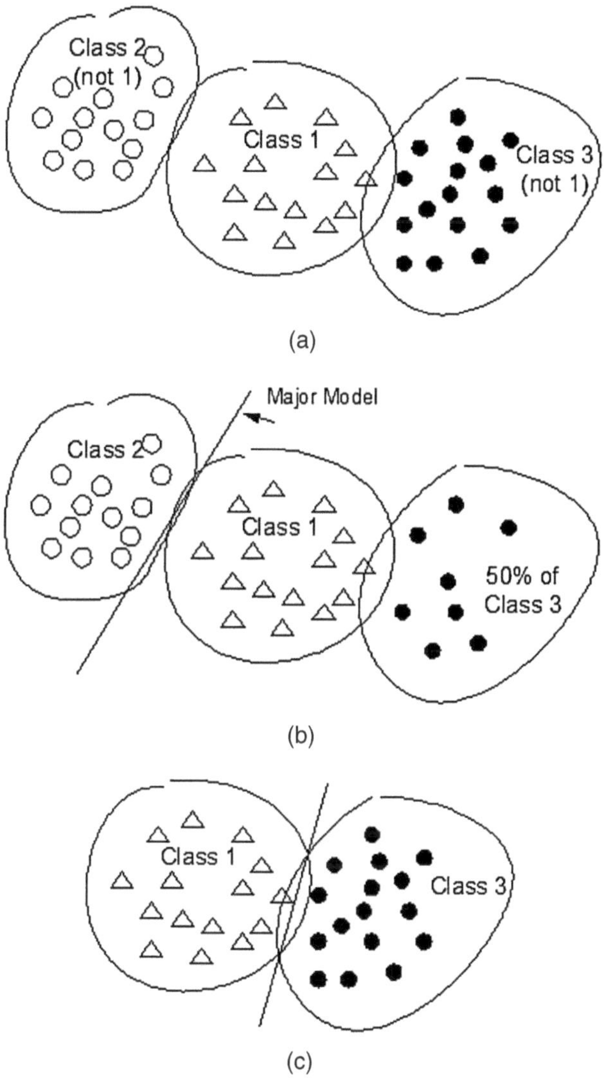

Fig. 13. Application of MMC approach to multi-category classification problems. (**a**) Binary classification formulation for multi-category classification problems; (**b**) Initial application of MMC to a modified data set (to estimate major model); (**c**) Second iteration of MMC algorithm to estimate second (minor) model

The training data is used to estimate the decision boundary, and test data is used to evaluate the classification accuracy. Let us consider the following binary classification problem:

- Positive class data: 25 training samples labeled as *iris versicolor*;

- Negative class data: 50 training samples labeled as *not iris versicolor*. This class includes iris setosa and iris virginica (25 samples each).

This training data (75 samples) is shown in Fig. 14a where samples labeled as iris versicolor are marked as "+", and samples labeled as not iris versicolor are marked as "△". In order to apply MMC to this data set, we first remove (randomly) 50% of training samples labeles as iris virginica, and then apply MMC procedure in order to identify the major model $H^{(1)}$ (i.e., decision boundary between *iris versicolor* and *iris setosa*). Then during the second iteration of MMC algorithm we add the removed samples in order to estimate the minor model $H^{(2)}$ (i.e., decision boundary between *iris versicolor* and *iris virginica*). Both (major and minor) models are shown in Fig. 14a.

For comparison, we also applied two standard nonlinear SVM methods (with polynomial kernels and RBF kernels) to the same IRIS data set (assuming binary classification formulation). Figures 14b and 14c show actual decision boundaries formed by each nonlinear SVM method (with optimal parameter values). Table 6 shows comparison results, in terms of classification accuracy for an independent test set. Notice that all three methods yield the same (low) test error rate, corresponding to a single misclassified test sample. However, the proposed MMC method uses the smallest number of support vectors and hence is arguably more robust that standard nonlinear SVM. Comparison results in Table 6 are consistent with experimental results in Tables 4 and 5. Even though all three methods (shown in Table 6 and Fig. 14) provide the same prediction accuracy, their extrapolation capability (far away from the $\mathbf{x}$-values of training samples) is quite different. For example, the test sample (marked in bold) in Fig. 14 will be classified differently by each method. Specifically, this sample will be classified as *iris versicolor* by the MMC method and by the RBF SVM classifier (see Figs. 14a and 14b), but it will be classified as *not iris versicolor* by the polynomial SVM classifier (see Fig. 14c). Moreover, in the case of SVM classifier with RBF kernel the confidence of prediction will be very low (since this sample lies inside the margin), whereas the confidence level of prediction by MMC method will be very high.

Application of the proposed MMC method to real-life data may result in two distinct outcomes. *First*, the proposed method may yield multiple component models, possibly with improved generalization vs standard (nonlinear)

Table 6. Comparison of Prediction accuracy for IRIS data set

	Error Rate (%SV)
RBF ($C = 1, p = 1$)	0.0267 (25.3%)
Poly ($C = 5, d = 2$)	0.0267 (29.3%)
MMC ($C = 2$)	0.0267 (18.7%)

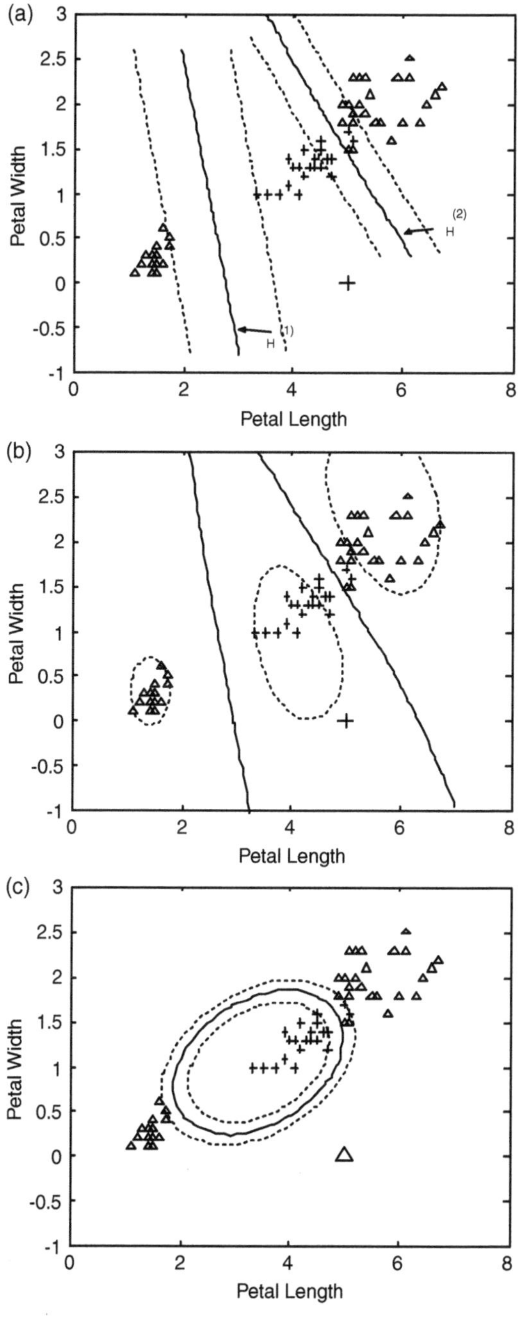

Fig. 14. Comparison of (best) decision boundaries obtained for Iris dataset. (**a**) MMC (**b**) SVM with RBF kernel, width $p = 1$ (**c**) SVM with polynomial kernel, degree $d = 2$

SVM classifier. *Second*, the proposed method may produce a single component model. In this case, MMC is reduced to standard (single-model) linear SVM, as illustrated next.

Experiment 4: The Wine Recognition data set from UCI learning depository contains the results of chemical analysis of 3 different types of wines (grown in the same region in Italy but derived from three different cultivars). The analysis provides the values of 13 descriptors for each of the three types of wines. The goal is to classify each type of wine based on values of these descriptors, and it can be modeled as classification problem (with 3 classes). Class 1 has 59 samples, class 2 has 71 samples, and class 3 has 48 samples. We mapped this problem onto 3 separate binary classification problems, and used 3/4 of the available data as training data, and 1/4 as test data, following [26]. Then the MMC approach was applied to each of three binary classifiers, and produced (in each case) a single linear decision boundary. For this data set, multiple model classification is reduced to standard linear SVM. Moreover, this data (both training and test) is found to be linearly separable, consistent with previous studies [26].

5 Summary and Discussion

We presented a new method for nonlinear classification, based on the Multiple Model Estimation methodology, and described practical implementations of this approach using an iterative application of the modified SVM algorithm. Empirical comparisons for several toy data sets indicate that the proposed multiple model classification (MMC) method (with linear component models) yields prediction accuracy better than /comparable to standard nonlinear SVM classifiers. It may be worth noting that in all empirical comparisons shown in this chapter, the number of support vectors selected under MMC approach is significantly lower than the number of support vectors in standard (nonlinear) SVM classifiers (see Tables 4–6). This observation suggests that MMC approach tends to be more robust, at least for these data sets. However, more empirical comparisons are needed to verify good generalization performance of the proposed MMC method. There are two additional advantages of the MMC method. *First*, the proposed implementation (with linear component models) does not require heuristic tuning of nonlinear SVM kernel parameters in order to achieve good classification accuracy. *Second*, the resulting decision boundary is piecewise-linear and therefore highly interpretable, unlike nonlinear decision boundary obtained using standard SVM approach.

Based on our preliminary comparisons, the MMC approach appears to be competitive vs standard nonlinear SVM classifiers, especially when there are reasons to believe that the data can be explained (modeled) by several simple models. In addition, multiple model estimation approach offers a totally new perspective on the development of classification algorithms and on the interpretation of SVM-based classification models.

In conclusion, we discuss several practical issues important for application of MMC, including its limitations and possible extensions. The main limitation of the iterative procedure for multiple model estimation (in Table 1) is the assumption that (during each iteration) the *majority* of available data can be "explained" well by a *single* component model. If this assumption holds, then the proposed method results in several "simple" models with good generalization; otherwise the proposed multiple model estimation algorithm falls apart. Clearly, this assumption may not hold if the algorithm uses pre-specified SVM parameterizations (i.e., linear SVM). The problem can be overcome by allowing *component models of increasing complexity* during each iteration of the MMC procedure. For example, under multiple model classification setting, the linear model (decision boundary) is tried first. If the majority of available data can not be explained by a linear model, then a more complex (say, quadratic) decision boundary is used during this iteration, etc. Future research may be concerned with practical implementations of such an *adaptive approach* to multiple model estimation, where (during each iteration) the component model complexity can be adapted to ensure that the component model explains well the majority of available data. This approach opens up a number of challenging research issues related to the trade-off between representing (modeling) the data using one complex model vs modeling the same data using several component models (of lower complexity). This requires careful specification of practical strategies for tuning SVM hyper-parameters during each iteration of the MMC algorithm.

Acknowledgement

This work was supported, in part, by NSF grant ECS-0099906.

References

1. Vapnik, V. (1995). The Nature of Statistical Learning Theory, Springer, New York.
2. Cherkassky, V. & Mulier, F. (1998). Learning from Data: Concepts, Theory, and Methods. John Wiley & Sons.
3. Hastie, T., Tibshirani, R., & Friedman, J. (2001). The Elements of Statistical Learning: Data Mining, Inference and Prediction, Springer.
4. Schölkopf, B. & Smola, A. (2002). Learning with Kernels: Support Vector Machines, Regularization, Optimization and Beyond, MIT Press, Cambridge, MA.
5. Bishop, C. (1995). Neural Networks for Pattern Recognition, Oxford: Oxford University Press.
6. Duda, R., Hart, P. & Stork, D. (2000). Pattern Classification, second edition, Wiley, New York.

7. Vapnik, V. (1998). Statistical Learning Theory, Wiley, New York.
8. Cherkassky, V. & Ma, Y. (2005). Multiple Model Regression Estimation, IEEE Trans. on Neural Networks, July, 2005 (To Appear).
9. Ma, Y. & Cherkassky, V. (2003). Multiple Model Classification Using SVM-based Approach, Proc. IJCNN 2003, pp. 1581–1586.
10. Cherkassky, V., Ma, Y. & Wechsler, H. (2004). Multiple Regression Estimation for Motion Analysis and Segmentation, Proc. IJCNN 2004.
11. Ghosh, J. (2002). Multiclassifier Systems: Back to the Future, in Multiple Classifier Systems (MCS2002), J. Kittler and F. Roli (Eds.), LNCS Vol. 2364, pp. 1–15, Springer.
12. Breiman, L., Friedman, J., Olshen, R. & Stone, C. (1984). Classification and Regression Trees. Belmont CA: Wadsworth.
13. Friedman, J. (1991). Multivariate adaptive regression splines (with discussion, Ann. Statist. Vol. 19, pp. 1–141.
14. Jordan, M. & Jacobs, R. (1994). Hierarchical mixtures of experts and the EM algorithm, Neural Computation 6: pp. 181–214.
15. Ma, Y., Multiple Model Estimation using SVM-based learning, PhD thesis, University of Minnesota, 2003.
16. Freund, Y. & Schapire, R. (1997). A decision-theoretic generalization of on-line learning and an application to boosting. J. Comput. System Sci. 55 119–139.
17. Dempster, A., Laird, N. & Rubin, D. (1977). Maximum likelihood from incomplete data via the EM algorithm (with discussion), J. Roy. Stat. Soc., B39, 1–38.
18. Meer, P., Steward, C. & Typer, D. (2000). Robust computer vision: An interdisciplinary challenge, Computer Vision and Image Understanding, 78, 1–7.
19. Chen, H., Meer, P. & Tyler, D. (2001). Robust Regression for Data with Multiple Structures, in Proc. IEEE Conf. on Computer Vision and Pattern Recognition CVPR 2001, pp. 1069–1075.
20. Torr, P., Dick, A. & Cipolla, R. (2000). Layer extraction with a Bayesian model of shapes, In European Conference on Computer Vision, pages II: pp. 273–289.
21. Torr, P. (1999). Model Selection for Two View Geometry: A Review, Shape, Contour and Grouping in Computer Vision, pp. 277–301.
22. Bergen, J., Burt, P., Hingorani, R. & Peleg, S. (1992). A Three-frame algorithm for estimating two-component image motion, IEEE Trans PAMI, 14: 886–895.
23. Irani, M., Rousso, B., & Peleg, S. (1994). Computing Occluding and Transparent Motions, Int. J. Computer Vision, Vol. 12, No. 1, pp. 5–16.
24. Andrews, D. & Herzberg, A. (1985). Data: A collection of problems from Many Fields for the Student and Research Worker, Springer.
25. Gunn, S. (1998). Support Vector Machines for Classification and Regression. Technical Report, Image Speech and Intelligent Systems Research Group, University of Southampton.
26. Roberts, S., Holmes, C. & Denison, D. (2001). Minimum-Entropy Data Clustering Using Reversible Jump Markov Chain Monte Carlo, ICANN 2001, LNCS, pp. 103–110.

Componentwise Least Squares Support Vector Machines

K. Pelckmans[1], I. Goethals, J. De Brabanter[1,2], J.A.K. Suykens[1], and B. De Moor[1]

[1] KULeuven – ESAT – SCD/SISTA,
Kasteelpark Arenberg 10,
3001 Leuven (Heverlee), Belgium
{Kristiaan.Pelckmans,Johan.Suykens}@esat.kuleuven.ac.be
[2] Hogeschool KaHo Sint-Lieven (Associatie KULeuven), Departement Industrieel Ingenieur

Summary. This chapter describes componentwise Least Squares Support Vector Machines (LS-SVMs) for the estimation of additive models consisting of a sum of nonlinear components. The primal-dual derivations characterizing LS-SVMs for the estimation of the additive model result in a single set of linear equations with size growing in the number of data-points. The derivation is elaborated for the classification as well as the regression case. Furthermore, different techniques are proposed to discover structure in the data by looking for sparse components in the model based on dedicated regularization schemes on the one hand and fusion of the componentwise LS-SVMs training with a validation criterion on the other hand.

Key words: LS-SVMs, additive models, regularization, structure detection

1 Introduction

Non-linear classification and function approximation is an important topic of interest with continuously growing research areas. Estimation techniques based on regularization and kernel methods play an important role. We mention in this context smoothing splines [32], regularization networks [22], Gaussian processes [18], Support Vector Machines (SVMs) [6, 23, 31] and many more, see e.g. [16]. SVMs and related methods have been introduced within the context of statistical learning theory and structural risk minimization. In the methods one solves convex optimization problems, typically quadratic programs. Least Squares Support Vector Machines (LS-SVMs)[1] [26, 27] are reformulations to standard SVMs which lead to solving linear

[1]http://www.esat.kuleuven.ac.be/sista/lssvmlab

K. Pelckmans et al.: *Componentwise Least Squares Support Vector Machines*, StudFuzz **177**, 77–98 (2005)
www.springerlink.com

KKT systems for classification tasks as well as regression. In [27] LS-SVMs have been proposed as a class of kernel machines with primal-dual formulations in relation to kernel Fisher Discriminant Analysis (FDA), Ridge Regression (RR), Partial Least Squares (PLS), Principal Component Analysis (PCA), Canonical Correlation Analysis (CCA), recurrent networks and control. The dual problems for the static regression without bias term are closely related to Gaussian processes [18], regularization networks [22] and Kriging [5], while LS-SVMs rather take an optimization approach with primal-dual formulations which have been exploited towards large scale problems and in developing robust versions.

Direct estimation of high dimensional nonlinear functions using a non-parametric technique without imposing restrictions faces the problem of the curse of dimensionality. Several attempts were made to overcome this obstacle, including projection pursuit regression [11] and kernel methods for dimensionality reduction (KDR) [13]. Additive models are very useful for approximating high dimensional nonlinear functions [15, 25]. These methods and their extensions have become one of the widely used nonparametric techniques as they offer a compromise between the somewhat conflicting requirements of flexibility, dimensionality and interpretability. Traditionally, splines are a common modeling technique [32] for additive models as e.g. in MARS (see e.g. [16]) or in combination with ANOVA [19]. Additive models were brought further to the attention of the machine learning community by e.g. [14, 31]. Estimation of the nonlinear components of an additive model is usually performed by the iterative *backfitting algorithm* [15] or a two-stage *marginal integration based* estimator [17]. Although consistency of both is shown under certain conditions, important practical problems (number of iteration steps in the former) and more theoretical problems (the pilot estimator needed for the latter procedure is a too generally posed problem) are still left.

In this chapter we show how the primal-dual derivations characterizing LS-SVMs can be employed to formulate a straightforward solution to the estimation problem of additive models using convex optimization techniques for classification as well as regression problems. Apart from this one-shot optimal training algorithm, the chapter approaches the problem of structure detection in additive models [14, 16] by considering an appropriate regularization scheme leading to sparse components. The additive regularization (AReg) framework [21] is adopted to emulate effectively these schemes based on 2-norms, 1-norms and specialized penalization terms [2]. Furthermore, a validation criterion is considered to select relevant components. Classically, exhaustive search methods (or stepwise procedures) are used which can be written as a combinatorial optimization problem. This chapter proposes a convex relaxation to the component selection problem.

This chapter is organized as follows. Section 2 presents componentwise LS-SVM regressors and classifiers for efficient estimation of additive models and relates the result with ANOVA kernels and classical estimation procedures. Section 3 introduces the additive regularization in this context and shows how to emulate dedicated regularization schemes in order to obtain sparse components. Section 4 considers the problem of component selection based on a validation criterion. Section 5 presents a number of examples.

2 Componentwise LS-SVMs and Primal-Dual Formulations

2.1 The Additive Model Class

Giving a training set defined as $\mathcal{D}_N = \{(x_k, y_k)\}_{k=1}^N \subset \mathbb{R}^D \times \mathbb{R}$ of size N drawn i.i.d. from an unknown distribution F_{XY} according to $y_k = f(x_k) + e_k$ where $f : \mathbb{R}^D \to \mathbb{R}$ is an unknown real-valued smooth function, $E[y_k|X = x_k] = f(x_k)$ and $e_1, \ldots, e_N$ are uncorrelated random errors with $E[e_k|X = x_k] = 0$, $E[(e_k)^2|X = x_k] = \sigma_e^2 < \infty$. The n data points of the validation set are denoted as $\mathcal{D}_n^{(v)} = \{x_j^{(v)}, y_j^{(v)}\}_{j=1}^n$. The following vector notations are used throughout the text: $X = (x_1, \ldots, x_N) \in \mathbb{R}^{D\times N}$, $Y = (y_1, \ldots, y_N)^T \in \mathbb{R}^N$, $X^{(v)} = (x_1^{(v)}, \ldots, x_n^{(v)}) \in \mathbb{R}^{D\times n}$ and $Y^{(v)} = (y_1^{(v)}, \ldots, y_n^{(v)})^T \in \mathbb{R}^n$. The estimator of a regression function is difficult if the dimension D is large. One way to quantify this is the optimal minimax rate of convergence $N^{-2l/(2l+D)}$ for the estimation of an l times differentiable regression function which converges to zero slowly if D is large compared to l [24]. A possibility to overcome the curse of dimensionality is to impose additional structure on the regression function. Although not needed in the derivation of the optimal solution, the input variables are assumed to be uncorrelated (see also *concurvity* [15]) in the applications.

Let superscript $x^d \in \mathbb{R}$ denote the d-th component of an input vector $x \in \mathbb{R}^D$ for all $d = 1, \ldots, D$. Let for instance each component correspond with a different dimension of the input observations. Assume that the function f can be approximated arbitrarily well by a model having the following structure

$$f(x) = \sum_{d=1}^{D} f^d(x^d) + b \,, \tag{1}$$

where $f^d : \mathbb{R} \to \mathbb{R}$ for all $d = 1, \ldots, D$ are unknown real-valued smooth functions and b is an intercept term. The following vector notation is used: $X^d = (x_1^d, \ldots, x_N^d) \in \mathbb{R}^{1\times N}$ and $X^{(v)d} = (x_1^{(v)d}, \ldots, x_n^{(v)d}) \in \mathbb{R}^{1\times n}$. The optimal rate of convergence for estimators based on this model is $N^{-2l/(2l+1)}$ which is independent of D [25]. Most state-of-the-art estimation techniques for additive models can be divided into two approaches [16]:

- *Iterative approaches* use an iteration where in each step part of the unknown components are fixed while optimizing the remaining components. This is motivated as:

$$\hat{f}^{d_1}\left(x_k^{d_1}\right) = y_k - e_k - \sum_{d_2 \neq d_1} \hat{f}^{d_2}\left(x_k^{d_2}\right) , \tag{2}$$

for all $k = 1, \ldots, N$ and $d_1 = 1, \ldots, D$. Once the $N-1$ components of the second term are known, it becomes easy to estimate the lefthandside. For a large class of linear smoothers, such so-called backfitting algorithms are equivalent to a Gauss-Seidel algorithm for solving a big $(ND \times ND)$ set of linear equations [16]. The backfitting algorithm [15] is theoretically and practically well motivated.

- *Two-stages marginalization approaches* construct in the first stage a general black-box pilot estimator (as e.g. a Nadaraya-Watson kernel estimator) and finally estimate the additive components by marginalizing (integrating out) for each component the variation of the remaining components.

2.2 Componentwise LS-SVMs for Regression

At first, a primal-dual formulation is derived for componentwise LS-SVM regressors. The global model takes the form as in (1) for any $x_* \in \mathbb{R}^D$

$$f(x_*; w_d, b) = \sum_{d=1}^{D} f^d(x_*^d; w_d) + b = \sum_{d=1}^{D} {w_d}^T \varphi_d(x_*^d) + b \, . \tag{3}$$

The individual components of an additive model based on LS-SVMs are written as $f^d(x^d; w_d) = w_d^T \varphi_d(x^d)$ in the primal space where $\varphi_d : \mathbb{R} \to \mathbb{R}^{n_{\varphi_d}}$ denotes a potentially infinite ($n_{\varphi_d} = \infty$) dimensional feature map. The regularized least squares cost function is given as [27]

$$\min_{w_d, b, e_k} \mathcal{J}_\gamma(w_d, e) = \frac{1}{2} \sum_{d=1}^{D} {w_d}^T w_d + \frac{\gamma}{2} \sum_{k=1}^{N} e_k^2$$

$$\text{s.t.} \sum_{d=1}^{D} {w_d}^T \varphi_d\left(x_k^d\right) + b + e_k = y_k, \quad k = 1, \ldots, N \, . \tag{4}$$

Note that the regularization constant γ appears here as in classical Tikhonov regularization [30]. The Lagrangian of the constraint optimization problem becomes

$$\mathcal{L}_\gamma(w_d, b, e_k; \alpha_k) = \frac{1}{2} \sum_{d=1}^{D} {w_d}^T w_d + \frac{\gamma}{2} \sum_{k=1}^{N} e_k^2 - \sum_{k=1}^{N} \alpha_k \left(\sum_{d=1}^{D} {w_d}^T \varphi_d\left(x_k^d\right) + b + e_k - y_k \right) . \tag{5}$$

By taking the conditions for optimality $\partial\mathcal{L}_\gamma/\partial\alpha_k = 0$, $\partial\mathcal{L}_\gamma/\partial b = 0$, $\partial\mathcal{L}_\gamma/\partial e_k = 0$ and $\partial\mathcal{L}_\gamma/\partial w_d = 0$ and application of the kernel trick $K^d(x_k^d, x_j^d) = \varphi_d(x_k^d)^T$

$\varphi_d(x_j^d)$ with a positive definite (Mercer) kernel $K^d : \mathbb{R} \times \mathbb{R} \to \mathbb{R}$, one gets the following conditions for optimality

$$\begin{cases} y_k = \sum_{d=1}^{D} {w_d}^T \varphi_d(x_k^d) + b + e_k, & k = 1, \ldots, N \ (a) \\ e_k \gamma = \alpha_k & k = 1, \ldots, N \ (b) \\ w_d = \sum_{k=1}^{N} \alpha_k \varphi_d(x_k^d) & d = 1, \ldots, D \ (c) \\ 0 = \sum_{k=1}^{N} \alpha_k. & (d) \end{cases} \tag{6}$$

Note that condition (6.b) states that the elements of the solution vector α should be proportional to the errors. The dual problem is summarized in matrix notation as

$$\left[\begin{array}{c|c} 0 & 1_N^T \\ \hline 1_N & \Omega + I_N/\gamma \end{array}\right] \left[\begin{array}{c} b \\ \hline \alpha \end{array}\right] = \left[\begin{array}{c} 0 \\ \hline Y \end{array}\right], \tag{7}$$

where $\Omega \in \mathbb{R}^{N \times N}$ with $\Omega = \sum_{d=1}^{D} \Omega^d$ and $\Omega_{kl}^d = K^d(x_k^d, x_l^d)$ for all $k, l = 1, \ldots, N$, which is expressed in the dual variables $\hat{\alpha}$ instead of $\hat{w}$. A new point $x_* \in \mathbb{R}^D$ can be evaluated as

$$\hat{y}_* = \hat{f}^d(x_*; \hat{\alpha}, \hat{b}) = \sum_{k=1}^{N} \hat{\alpha}_k \sum_{d=1}^{D} K^d \left(x_k^d, x_*^d\right) + \hat{b}\,, \tag{8}$$

where $\hat{\alpha}$ and $\hat{b}$ is the solution to (7). Simulating a validation datapoint x_j for all $j = 1, \ldots, n$ by the d-th individual component

$$\hat{y}_j^d = \hat{f}^d\left(x_j^d; \hat{\alpha}\right) = \sum_{k=1}^{N} \hat{\alpha}_k K^d \left(x_k^d, x_j^d\right)\,, \tag{9}$$

which can be summarized as follows: $\hat{Y} = (\hat{y}_1, \ldots, \hat{y}_N)^T \in \mathbb{R}^N$, $\hat{Y}^d = (\hat{y}_1^d, \ldots,. .\hat{y}_N^d)^T \in \mathbb{R}^N$, $\hat{Y}^{(v)} = (\hat{y}_1^{(v)}, \ldots, \hat{y}_n^{(v)})^T \in \mathbb{R}^n$ and $\hat{Y}_d^{(v)} = (\hat{y}_1^{(v)d}, \ldots, \hat{y}_n^{(v)d})^T \in \mathbb{R}^n$.

Remarks

- Note that the componentwise LS-SVM regressor can be written as a linear smoothing matrix [27]:

  $$\hat{Y} = S_\gamma Y\,. \tag{10}$$

  For notational convenience, the bias term is omitted from this description. The smoother matrix $S_\gamma \in \mathbb{R}^{N \times N}$ becomes

  $$S_\gamma = \Omega \left(\Omega + I_N \frac{1}{\gamma}\right)^{-1}. \tag{11}$$

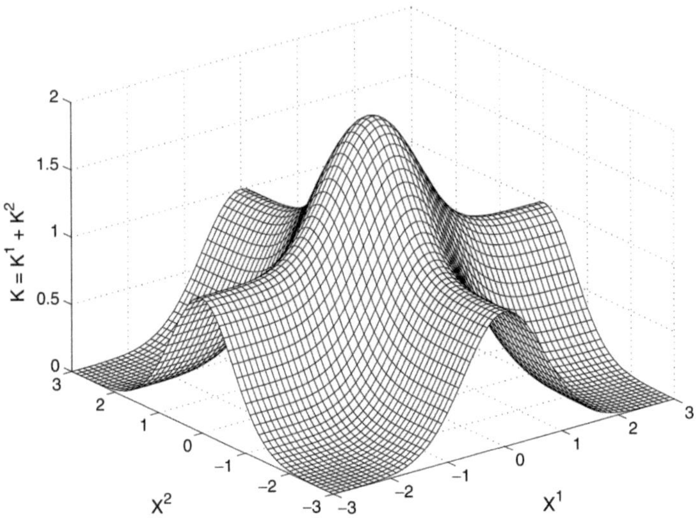

Fig. 1. The two dimensional componentwise Radial Basis Function (RBF) kernel for componentwise LS-SVMs takes the form $K(x_k, x_l) = K^1(x_k^1, x_l^1) + K^2(x_k^2, x_l^2)$ as displayed. The standard RBF kernel takes the form $K(x_k, x_l) = \exp(-\|x_k - x_l\|_2^2/\sigma^2)$ with $\sigma \in \mathbb{R}_0^+$ an appropriately chosen bandwidth

- The set of linear (7) corresponds with a classical LS-SVM regressor where a modified kernel is used

$$K(x_k, x_j) = \sum_{d=1}^{D} K^d\left(x_k^d, x_j^d\right) . \tag{12}$$

Figure 1 shows the modified kernel in case a one dimensional Radial Basis Function (RBF) kernel is used for all D (in the example, $D = 2$) components. This observation implies that componentwise LS-SVMs inherit results obtained for classical LS-SVMs and kernel methods in general. From a practical point of view, the previous kernels (and a fortiori componentwise kernel models) result in the same algorithms as considered in the ANOVA kernel decompositions as in [14, 31].

$$\begin{aligned} K(x_k, x_j) = &\sum_{d=1}^{D} K^d\left(x_k^d, x_j^d\right) \\ &+ \sum_{d_1 \neq d_2} K^{d_1 d_2}\left(\left(x_k^{d_1}, x_k^{d_2}\right)^T, \left(x_j^{d_1}, x_j^{d_2}\right)^T\right) + \ldots, \end{aligned} \tag{13}$$

where the componentwise LS-SVMs only consider the first term in this expansion. The described derivation as such bridges the gap between the estimation of additive models and the use of ANOVA kernels.

2.3 Componentwise LS-SVMs for Classification

In the case of classification, let $y_k, y_j^{(v)} \in \{-1, 1\}$ for all $k = 1, \ldots, N$ and $j = 1, \ldots, n$. The analogous derivation of the componentwise LS-SVM classifier is briefly reviewed. The following model is considered for modeling the data

$$f(x) = \text{sign}\left(\sum_{d=1}^{D} f^d(x^d) + b\right), \tag{14}$$

where again the individual components of the additive model based on LS-SVMs are given as $f^d(x^d) = {w_d}^T \varphi_d(x^d)$ in the primal space where $\varphi_d : \mathbb{R} \rightarrow \mathbb{R}^{n_{\varphi_d}}$ denotes a potentially infinite ($n_{\varphi_d} = \infty$) dimensional feature map. The regularized least squares cost function is given as [26, 27]

$$\begin{aligned} \min_{w_d, b, e_k} \mathcal{J}_\gamma(w_d, e) &= \frac{1}{2}\sum_{d=1}^{D} {w_d}^T w_d + \frac{\gamma}{2}\sum_{k=1}^{N} e_k^2 \\ \text{s.t. } y_k \left(\sum_{d=1}^{D} {w_d}^T \varphi_d\left(x_k^d\right) + b\right) &= 1 - e_k, \quad k = 1, \ldots, N, \end{aligned} \tag{15}$$

where e_k are so-called slack-variables for all $k = 1, \ldots, N$. After construction of the Lagrangian and taking the conditions for optimality, one obtains the following set of linear equations (see e.g. [27]):

$$\left[\begin{array}{c|c} 0 & Y^T \\ \hline Y & \Omega_y + I_N/\gamma \end{array}\right] \left[\begin{array}{c} b \\ \hline \alpha \end{array}\right] = \left[\begin{array}{c} 0 \\ \hline 1_N \end{array}\right], \tag{16}$$

where $\Omega_y \in \mathbb{R}^{N \times N}$ with $\Omega_y = \sum_{d=1}^{D} \Omega_y^d \in \mathbb{R}^{N \times N}$ and $\Omega_{y,kl}^d = y_k y_l K^d(x_k^d, x_l^d)$. New data points $x_* \in \mathbb{R}^D$ can be evaluated as

$$\hat{y}_* = \text{sign}\left(\sum_{k=1}^{N} \hat{\alpha}_k y_k \sum_{d=1}^{D} K^d(x_k^d, x_*^d) + \hat{b}\right). \tag{17}$$

In the remainder of this text, only the regression case is considered. The classification case can be derived straightforwardly along the lines.

3 Sparse Components via Additive Regularization

A regularization method fixes a priori the answer to the ill-conditioned (or ill-defined) nature of the inverse problem. The classical Tikhonov regularization scheme [30] states the answer in terms of the norm of the solution. The formulation of the additive regularization (AReg) framework [21] made it possible to impose alternative answers to the ill-conditioning of the problem at hand. We refer to this AReg level as *substrate LS-SVMs*. An appropriate regularization scheme for additive models is to favor solutions using the smallest number of components to explain the data as much as possible. In this paper, we use the somewhat relaxed condition of *sparse* components to select appropriate components instead of the more general problem of input (or component) selection.

3.1 Level 1: Componentwise LS-SVM Substrate

Using the Additive Regularization (AReg) scheme for componentwise LS-SVM regressors results into the following modified cost function:

$$\min_{w_d, b, e_k} \mathcal{J}_c(w_d, e) = \frac{1}{2}\sum_{d=1}^{D} {w_d}^T w_d + \frac{1}{2}\sum_{k=1}^{N}(e_k - c_k)^2$$

$$\text{s.t. } \sum_{d=1}^{D} {w_d}^T \varphi_d\left(x_k^d\right) + b + e_k = y_k, \quad k = 1, \ldots, N, \tag{18}$$

where $c_k \in \mathbb{R}$ for all $k = 1, \ldots, N$. Let $c = (c_1, \ldots, c_N)^T \in \mathbb{R}^N$. After constructing the Lagrangian and taking the conditions for optimality, one obtains the following set of linear equations, see [21]:

$$\left[\begin{array}{c|c} 0 & 1_N^T \\ \hline 1_N & \Omega + I_N \end{array}\right] \left[\begin{array}{c} b \\ \hline \alpha \end{array}\right] + \left[\begin{array}{c} 0 \\ \hline c \end{array}\right] = \left[\begin{array}{c} 0 \\ \hline Y \end{array}\right] \tag{19}$$

and $e = \alpha + c \in \mathbb{R}^N$. Given a regularization constant vector c, the unique solution follows immediately from this set of linear equations.

However, as this scheme is too general for practical implementation, c should be limited in an appropriate way by imposing for example constraints corresponding with certain model assumptions or a specified cost function. Consider for a moment the conditions for optimality of the componentwise LS-SVM regressor using a regularization term as in ridge regression, one can see that (7) corresponds with (19) if $\gamma^{-1}\alpha = \alpha + c$ for given γ. Once an appropriate c is found which satisfies the constraints, it can be plugged in into the LS-SVM substrate (19). It turns out that one can omit this conceptual second stage in the computations by elimination of the variable c in the constrained optimization problem (see Fig. 2).

Alternatively, a measure corresponding with a (penalized) cost function can be used which fulfills the role of model selection in a broad sense. A variety of such explicit or implicit limitations can be *emulated* based on different criteria (see Fig. 3).

3.2 Level 2: an L_1 Based Component Regularization Scheme (*convex*)

We now study how to obtain sparse components by considering a dedicated regularization scheme. The LS-SVM substrate technique is used to emulate

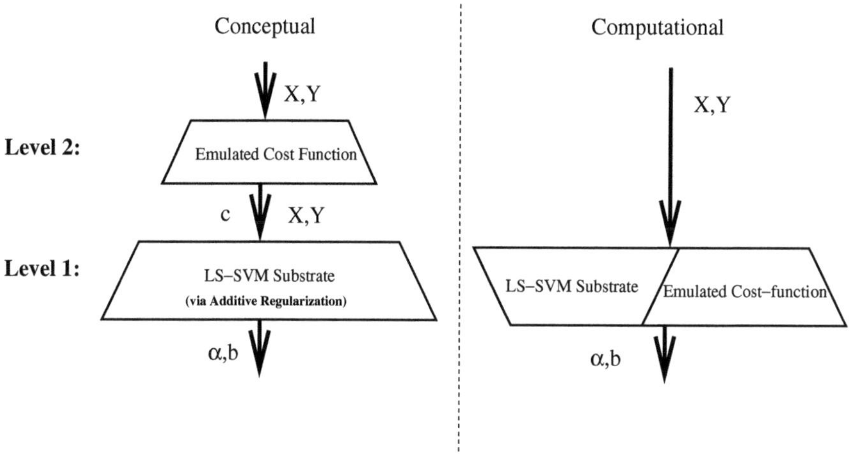

Fig. 2. Graphical representation of the additive regularization framework used for emulating other loss functions and regularization schemes. Conceptually, one differentiates between the newly specified cost function and the LS-SVM substrate, while computationally both are computed simultanously

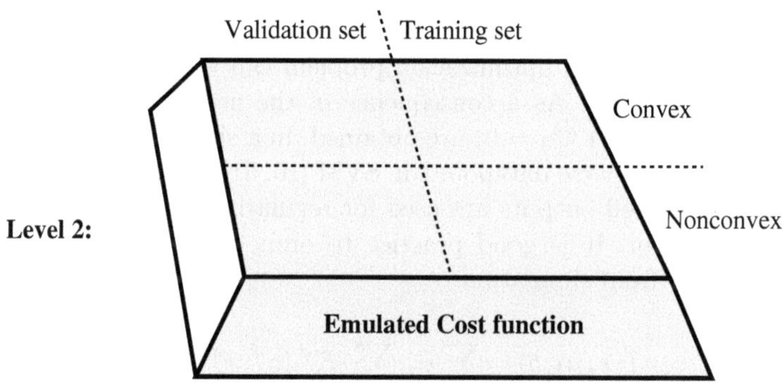

Fig. 3. The level 2 cost functions of Fig. 2 on the conceptual level can take different forms based on validation performance or trainings error. While some will result in convex tuning procedures, other may loose this property depending on the chosen cost function on the second level

the proposed scheme as primal-dual derivations (see e.g. Subsect. 2.2) are not straightforward anymore.

Let $\hat{Y}^d \in \mathbb{R}^N$ denote the estimated training outputs of the d-th submodel f^d as in (9). The component based regularization scheme can be translated as the following constrained optimization problem where the conditions for optimality (18) as summarized in (19) are to be satisfied exactly (after elimination of w)

$$\min_{c,\hat{Y}^d,e_k;\alpha,b} \mathcal{J}_\xi(\hat{Y}^d, e_k) = \frac{1}{2}\sum_{d=1}^{D} \|\hat{Y}^d\|_1 + \frac{\xi}{2}\sum_{k=1}^{N} e_k^2$$
$$\text{s.t.} \begin{cases} 1_N^T\alpha = 0, \\ \Omega\,\alpha + 1_N^T b + \alpha + c = Y, \\ \Omega^d\,\alpha = \hat{Y}^d, \quad \forall d = 1, \ldots, D \\ \alpha + c = e\,, \end{cases} \tag{20}$$

where the use of the robust L_1 norm can be justified as in general no assumptions are imposed on the distribution of the elements of $\hat{Y}^d$. By elimination of c using the equality $e = \alpha + c$, this problem can be written as follows

$$\min_{\hat{Y}^d,e_k;\alpha,b} \mathcal{J}_\xi(\hat{Y}^d, e_k) = \frac{1}{2}\sum_{d=1}^{D} \|\hat{Y}^d\|_1 + \frac{\xi}{2}\sum_{k=1}^{N} e_k^2$$
$$\text{s.t.} \left[\begin{array}{c|c} 0 & 1_N^T \\ \hline 1_N & \Omega \\ \hline 0_N & \Omega^1 \\ \vdots & \vdots \\ 0_N & \Omega^d \end{array}\right] \left[\begin{array}{c} b \\ \hline \alpha \end{array}\right] + \left[\begin{array}{c} 0 \\ \hline e \\ \hline \hat{Y}^1 \\ \vdots \\ \hat{Y}^D \end{array}\right] = \left[\begin{array}{c} 0 \\ \hline Y \\ \hline 0_N \\ \vdots \\ 0_N \end{array}\right]. \tag{21}$$

This convex constrained optimization problem can be solved as a quadratic programming problem. As a consequence of the use of the L_1 norm, often sparse components ($\|\hat{Y}^d\|_1 = 0$) are obtained, in a similar way as sparse variables of LASSO or sparse datapoints in SVM [16, 31]. An important difference is that the estimated outputs are used for regularization purposes instead of the solution vector. It is good practice to omit sparse components on the training dataset from simulation:

$$\hat{f}(x_*; \hat{\alpha}, \hat{b}) = \sum_{i=1}^{N} \hat{\alpha}_i \sum_{d\in\mathcal{S}_D} K^d\left(x_i^d, x_*^d\right) + \hat{b}\,, \tag{22}$$

where $\mathcal{S}_D = \{d|\hat{\alpha}^T\Omega^d\hat{\alpha} \neq 0\}$.

Using the L_2 norm $\sum_{d=1}^{D} \|\hat{Y}^d\|_2^2$ instead leads to a much simpler optimization problem, but additional assumptions (Gaussianity) are needed on the distribution of the elements of $\hat{Y}^d$. Moreover, the component selection has to resort on a significance test instead of the sparsity resulting from (21). A practical algorithm is proposed in Subsect. 5.1 that uses an iteration of L_2 norm based optimizations in order to calculate the optimum of the proposed regularized cost function.

3.3 Level 2 bis: A Smoothly Thresholding Function

This subsection considers extensions to classical formulations towards the use of dedicated regularization schemes for sparsifying components. Consider the componentwise regularized least squares cost function defined as

$$\mathcal{J}_\lambda(w_d, e) = \frac{\lambda}{2}\sum_{d=1}^{D} \ell(w_d) + \frac{1}{2}\sum_{k=1}^{N} e_k^2 , \tag{23}$$

where $\ell(w_d)$ is a penalty function and $\lambda \in \mathbb{R}_0+$ acts as a regularization parameter. We denote $\lambda\ell(\cdot)$ by $\ell_\lambda(\cdot)$, so it may depend on λ. Examples of penalty functions include:

- The L_p penalty function $\ell_\lambda^p(w_d) = \lambda\|w_d\|_p^p$ leads to a bridge regression [10, 12]. It is known that the L_2 penalty function $p = 2$ results in the ridge regression. For the L_1 penalty function the solution is the soft thresholding rule [7]. LASSO, as proposed by [28, 29], is the penalized least squares estimate using the L_1 penalty function (see Fig. 4a).
- Let the indicator function $I_{\{x\in\mathcal{A}\}} = 1$ if $x \in \mathcal{A}$ for a specified set $\mathcal{A}$ and 0 otherwise. When the penalty function is given by $\ell_\lambda(w_d) = \lambda^2 - (\|w_d\|_1 - \lambda)^2 I_{\{\|w_d\|_1<\lambda\}}$ (see Fig. 4b), the solution is a hard-thresholding rule [1].

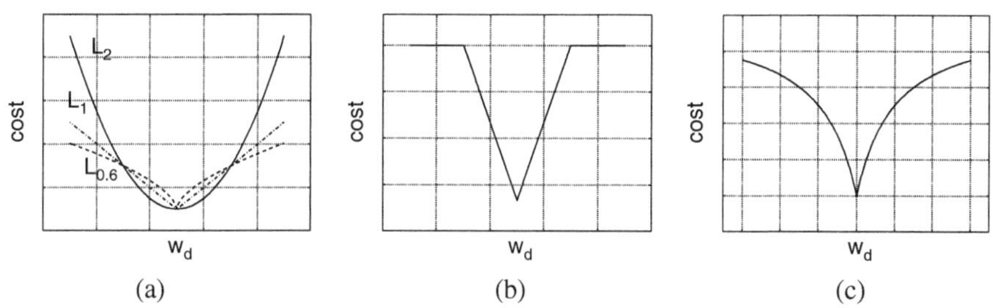

Fig. 4. Typical penalty functions: (**a**) the L_p penalty family for $p = 2, 1$ and 0.6, (**b**) hard thresholding penalty function and (**c**) the transformed L_1 penalty function

The L_p and the hard thresholding penalty functions do not simultaneously satisfy the mathematical conditions for unbiasedness, sparsity and continuity [9]. The hard thresholding has a discontinuous cost surface. The only continuous cost surface (defined as the cost function associated with the solution space) with a thresholding rule in the L_p-family is the L_1 penalty function, but the resulting estimator is shifted by a constant λ. To avoid these drawbacks, [20] suggests the penalty function defined as

$$\ell_\lambda^a(w_d) = \frac{\lambda a\|w_d\|_1}{1 + a\|w_d\|_1} , \tag{24}$$

with $a \in \mathbb{R}_0^+$. This penalty function behaves quite similarly as the Smoothly Clipped Absolute Deviation (SCAD) penalty function as suggested by [8]. The Smoothly Thresholding Penalty (TTP) function (24) improves the properties of the L_1 penalty function and the hard thresholding penalty function (see Fig. 4c), see [2]. The unknowns a and λ act as regularization parameters.

A plausible value for a was derived in [2, 20] as $a = 3.7$. The transformed L_1 penalty function satisfies the oracle inequalities [7]. One can plugin the described semi-norm $\ell_\lambda^a(\cdot)$ to improve the component based regularization scheme (20). Again, the additive regularization scheme is used for the emulation of this scheme

$$\min_{c,\hat{Y}^d,e_k;\alpha,b} \mathcal{J}_\lambda(\hat{Y}^d, e_k) = \frac{1}{2}\sum_{d=1}^{D} \ell_\lambda^a(\hat{Y}^d) + \frac{1}{2}\sum_{k=1}^{N} e_k^2$$
$$\text{s.t.} \begin{cases} 1_N^T\alpha = 0 \ , \\ \Omega\,\alpha + 1_N^T b + \alpha + c = Y \ , \\ \Omega^d\,\alpha = \hat{Y}^d, \quad \forall d = 1,\ldots,D \\ \alpha + c = e \ , \end{cases} \tag{25}$$

which becomes non-convex but can be solved using an iterative scheme as explained later in Subsect. 5.1.

4 Fusion of Componentwise LS-SVMs and Validation

This section investigates how one can tune the componentwise LS-SVMs with respect to a validation criterion in order to improve the generalization performance of the final model. As proposed in [21], fusion of training and validation levels can be investigated from an optimization point of view, while conceptually they are to be considered at different levels.

4.1 Fusion of Componentwise LS-SVMs and Validation

For this purpose, the fusion argument as introduced in [21] is briefly revised in relation to regularization parameter tuning. The estimator of the LS-SVM regressor on the training data for a fixed value γ is given as (4)

$$\textbf{Level 1}: \ (\hat{w},\hat{b}) = \arg\min_{w,b,e} \mathcal{J}_\gamma(w,e) \quad \text{s.t. (4) holds,} \tag{26}$$

which results into solving a linear set of (7) after substitution of w by Lagrange multipliers α. Tuning the regularization parameter by using a validation criterion gives the following estimator

$$\textbf{Level 2}: \ \hat{\gamma} = \arg\min_{\gamma} \sum_{j=1}^{n} (f(x_j;\hat{\alpha},\hat{b}) - y_j)^2 \quad \text{with} \quad (\hat{\alpha},\hat{b}) = \arg\min_{\alpha,b} \mathcal{J}_\gamma \tag{27}$$

satisfying again (4). Using the conditions for optimality (7) and eliminating w and e

$$\textbf{Fusion}: \ (\hat{\gamma}, \hat{\alpha}, \hat{b}) = \arg\min_{\gamma,\alpha,b} \sum_{j=1}^{n} \left(f(x_j; \alpha, b) - y_j\right)^2 \quad \text{s.t.} \quad (7) \text{ holds}, \tag{28}$$

which is referred to as fusion. The resulting optimization problem was noted to be non-convex as the set of optimal solutions w (or dual α's) corresponding with a $\gamma > 0$ is non-convex. To overcome this problem, a re-parameterization of the trade-off was proposed leading to the additive regularization scheme. At the cost of overparameterizing the trade-off, convexity is obtained. To circumvent this drawback, different ways to restrict explicitly or implicitly the (effective) degrees of freedom of the regularization scheme $c \in \mathcal{A} \subset \mathbb{R}^N$ were proposed while retaining convexity [21]. The convex problem resulting from additive regularization is

$$\textbf{Fusion}: \ (\hat{c}, \hat{\alpha}, \hat{b}) = \arg\min_{c\in\mathcal{A},\alpha,b} \sum_{j=1}^{n} \left(f(x_j; \alpha, b) - y_j\right)^2 \quad \text{s.t.} \quad (19) \text{ holds}, \tag{29}$$

and can be solved efficiently as a convex constrained optimization problem if $\mathcal{A}$ is a convex set, resulting immediately in the optimal regularization trade-off and model parameters [4].

4.2 Fusion for Component Selection using Additive Regularization

One possible relaxed version of the component selection problem goes as follows: *Investigate whether it is plausible to drive the components on the validation set to zero without too large modifications on the global training solution.* This is translated as the following cost function much in the spirit of (20). Let $\Omega^{(v)}$ denote $\sum_{d=1}^{D} \Omega^{(v)d} \in \mathbb{R}^{n\times N}$ and $\Omega^{(v)d}_{jk} = K^d(x_j^{(v)d}, x_k^d)$ for all $j = 1, \ldots, n$ and $k = 1, \ldots, N$.

$$(\hat{c}, \hat{Y}^{(v)d}, \hat{w}_d, \hat{e}, \hat{\alpha}, \hat{b}) = \arg\min_{c,\hat{Y}^d,\hat{Y}^{(v)d},e,\alpha,b} \ \frac{1}{2}\sum_{d=1}^{D} \|\hat{Y}^{(v)d}\|_1 + \frac{1}{2}\sum_{d=1}^{D} \|\hat{Y}^d\|_1 + \frac{\xi}{2}\sum_{k=1}^{N} e_k^2$$
$$\text{s.t.} \begin{cases} 1_N^T \alpha = 0 \\ \alpha + c = e \\ \Omega\,\alpha + 1_N b + \alpha + c = Y \\ \Omega^d\,\alpha = \hat{Y}^d\,, \quad \forall d = 1, \ldots, D\,, \\ \Omega^{(v)d}\,\alpha = \hat{Y}^{(v)d}\,, \quad \forall d = 1, \ldots, D\,, \end{cases} \tag{30}$$

where the equality constraints consist of the conditions for optimality of (19) and the evaluation of the validation set on the individual components. Again, this convex problem can be solved as a quadratic programming problem.

4.3 Fusion for Component Selection Using Componentwise LS-SVMs with Additive Regularization

We proceed by considering the following primal cost function for a fixed but strictly positive $\eta = (\eta_1, \ldots, \eta_D)^T \in (\mathbb{R}_0^+)^D$

$$\textbf{Level 1}: \quad \min_{w_d, b, e_k} \mathcal{J}_\eta(w_d, e) = \frac{1}{2} \sum_{d=1}^{D} \frac{{w_d}^T w_d}{\eta_d} + \frac{1}{2} \sum_{k=1}^{N} e_k^2$$
$$\text{s.t.} \sum_{d=1}^{D} {w_d}^T \varphi_d\left(x_k^d\right) + b + e_k = y_k \ ,$$
$$k = 1, \ldots, N \ . \tag{31}$$

Note that the regularization vector appears here similar as in the Tikhonov regularization scheme [30] where each component is regularized individually. The Lagrangian of the constrained optimization problem with multipliers $\alpha^\eta \in \mathbb{R}^N$ becomes

$$\mathcal{L}_\eta(w_d, b, e_k; \alpha_k) = \frac{1}{2} \sum_{d=1}^{D} \frac{{w_d}^T w_d}{\eta_d} + \frac{1}{2} \sum_{k=1}^{N} e_k^2 - \sum_{k=1}^{N} \alpha_k^\eta \left(\sum_{d=1}^{D} {w_d}^T \varphi_d(x_k^d) + b + e_k - y_k \right) . \tag{32}$$

By taking the conditions for optimality $\partial \mathcal{L}_\eta / \partial \alpha_k = 0$, $\partial \mathcal{L}_\eta / \partial b = 0$, $\partial \mathcal{L}_\eta / \partial e_k = 0$ and $\partial \mathcal{L}_\eta / \partial w_d = 0$, one gets the following conditions for optimality

$$\begin{cases} y_k = \sum_{d=1}^{D} {w_d}^T \varphi_d(x_k^d) + b + e_k \ , & k = 1, \ldots, N \ (a) \\ e_k = \alpha_k^\eta \ , & k = 1, \ldots, N \ (b) \\ w_d = \eta_d \sum_{k=1}^{N} \alpha_k^\eta \varphi_d(x_k^d) \ , & d = 1, \ldots, D \ (c) \\ 0 = \sum_{k=1}^{N} \alpha_k^\eta \ . & (d) \end{cases} \tag{33}$$

The dual problem is summarized in matrix notation by application of the kernel trick,

$$\left[\begin{array}{c|c} 0 & 1_N^T \\ \hline 1_N & \Omega^\eta + I_N \end{array} \right] \left[\begin{array}{c} b \\ \hline \alpha^\eta \end{array} \right] = \left[\begin{array}{c} 0 \\ \hline Y \end{array} \right] , \tag{34}$$

where $\Omega^\eta \in \mathbb{R}^{N \times N}$ with $\Omega^\eta = \sum_{d=1}^{D} \eta_d \Omega^d$ and $\Omega_{kl}^d = K^d(x_k^d, x_l^d)$. A new point $x_* \in \mathbb{R}^D$ can be evaluated as

$$\hat{y}_* = \hat{f}(x_*; \hat{\alpha}^\eta, \hat{b}) = \sum_{k=1}^{N} \hat{\alpha}_k^\eta \sum_{d=1}^{D} \eta_d K^d(x_k^d, x_*^d) + \hat{b} \ , \tag{35}$$

where $\hat{\alpha}$ and $\hat{b}$ are the solution to (34). Simulating a training datapoint x_k for all $k = 1, \ldots, N$ by the d-th individual component

$$\hat{y}_k^{\eta,d} = \hat{f}^d\left(x_k^d; \hat{\alpha}^\eta\right) = \eta_d \sum_{l=1}^{N} \hat{\alpha}_l^\eta K^d\left(x_k^d, x_l^d\right) , \tag{36}$$

which can be summarized in a vector $\hat{Y}^{\eta,d} = (\hat{y}_1^d, \ldots, \hat{y}_N^d) \in \mathbb{R}^N$. As in the previous section, the validation performance is used for tuning the regularization parameters

$$\textbf{Level 2}: \ \hat{\eta} = \arg\min_{\eta} \sum_{j=1}^{n} (f(x_j; \hat{\alpha}^\eta, \hat{b}) - y_j)^2 \text{ with } (\hat{\alpha}^\eta, \hat{b}) = \arg\min_{\alpha^\eta, b} \mathcal{J}_\eta , \tag{37}$$

or using the conditions for optimality (34) and eliminating w and e

$$\textbf{Fusion}: \ (\hat{\eta}, \hat{\alpha}^\eta, \hat{b}) = \arg\min_{\eta, \alpha^\eta, b} \sum_{j=1}^{n} \left(f(x_j; \alpha^\eta, b) - y_j\right)^2 \quad \text{s.t.} \quad (34) \text{ holds}, \tag{38}$$

which is a non-convex constrained optimization problem.

Embedding this problem in the additive regularization framework will lead us to a more suitable representation allowing for the use of dedicated algorithms. By relating the conditions (19) to (34), one can view the latter within the additive regularization framework by imposing extra constraints on c. The bias term b is omitted from the remainder of this subsection for notational convenience. The first two constraints reflect training conditions for both schemes. As the solutions α^η and α do not have the same meaning (at least for model evaluation purposes, see (8) and (35)), the appropriate c is determined here by enforcing the same estimation on the training data. In summary:

$$\begin{cases} (\Omega + I_N)\,\alpha + c = Y \\ \left(\left(\sum_{d=1}^{D} \eta_d \Omega^d\right) + I_N\right) \alpha^\eta = Y \\ \left(\sum_{d=1}^{D} \eta_d \Omega^d\right) \alpha^\eta = \Omega\alpha \end{cases} \Rightarrow \begin{cases} (\Omega + I_N)\,\alpha + c = Y \\ \Omega\alpha = \eta^T \otimes I_N \begin{bmatrix} \Omega^1 \\ \cdots \\ \Omega^D \end{bmatrix} (\alpha + c) , \end{cases} \tag{39}$$

where the second set of equations is obtained by eliminating α^η. The last equation of the righthand side represents the set of constraints of the values c for all possible values of η. The product $\otimes$ denotes $\eta^T \otimes I_N = [\eta_1 I_N, \ldots, \eta_D I_N] \in \mathbb{R}^{N \times ND}$. As for the Tikhonov case, it is readily seen that the solution space of c with respect to η is non-convex, however, the constraint on c is recognized as a bilinear form. The fusion problem (38) can be written as

$$\textbf{Fusion}: \ (\hat{\eta}, \hat{\alpha}, \hat{c}) = \arg\min_{\eta, \alpha, c} \left\|\Omega^{(v)}\alpha - Y^{(v)}\right\|_2^2 \quad \text{s.t.} \quad (39) \text{ holds}, \tag{40}$$

where algorithms as alternating least squares can be used.

5 Applications

For practical applications, the following iterative approach is used for solving non-convex cost-functions as (25). It can also be used for the efficient solution of convex optimization problems which become computational heavy in the case of a large number of datapoints as e.g. (21). A number of classification as well as regression problems are employed to illustrate the capabilities of the described approach. In the experiments, hyper-parameters as the kernel parameter (taken to be constants over the components) and the regularization trade-off parameter γ or ξ were tuned using 10-fold cross-validation.

5.1 Weighted Graduated Non-Convexity Algorithm

An iterative scheme was developed based on the graduated non-convexity algorithm as proposed in [2, 3, 20] for the optimization of non-convex cost functions. Instead of using a local gradient (or Newton) step which can be quite involved, an adaptive weighting scheme is proposed: in every step, the relaxed cost function is optimized by using a weighted 2-norm where the weighting terms are chosen based on an initial guess for the global solution. For every symmetric loss function $\ell(|e|) : \mathbb{R}^+ \to \mathbb{R}^+$ which is monotonically increasing, there exists a bijective transformation $t : \mathbb{R} \to \mathbb{R}$ such that for every $e = y - f(x;\theta) \in \mathbb{R}$

$$\ell(e) = (t(e))^2 \ . \tag{41}$$

The proposed algorithm for computing the solution for semi-norms employs iteratively convex relaxations of the prescribed non-convex norm. It is somewhat inspired by the simulated annealing optimization technique for optimizing global optimization problems. The weighted version is based on the following derivation

$$\ell(e_k) = (\nu_k e_k)^2 \Leftrightarrow \nu_k = \sqrt{\frac{\ell(e_k)}{e_k^2}} \ , \tag{42}$$

where the e_k for all $k = 1, \ldots, N$ are the residuals corresponding with the solutions to $\theta = \arg\min_\theta \ell\left(y_k - f(x_k;\theta)\right)$ This is equal to the solution of the convex optimization problem $e_k = \arg\min_\theta \left(\nu_k(y_k - f(x_k;\theta))\right)^2$ for a set of ν_k satisfying (42). For more stable results, the gradient of the penalty function ℓ and the quadratic approximation can be takne equal as follows by using an intercept parameter $\mu_k \in \mathbb{R}$ for all $k = 1, \ldots, N$:

$$\begin{cases} \ell(e_k) = (\nu_k e_k)^2 + \mu_k \\ \ell'(e_k) = 2\nu_k^2 e_k \end{cases} \Leftrightarrow \left[\begin{array}{c|c} e_k^2 & 1 \\ \hline 2e_k & 0 \end{array}\right] \begin{bmatrix} \nu_k^2 \\ \mu_k \end{bmatrix} = \begin{bmatrix} \ell(e_k) \\ \ell'(e_k) \end{bmatrix} , \tag{43}$$

where $\ell'(e_k)$ denotes the derivative of ℓ evaluated in e_k such that a minimum of $\mathcal{J}_\ell$ also minimizes the weighted equivalent (the derivatives are equal). Note that the constant intercepts μ_k are not relevant in the weighted optimization problem. Under the assumption that the two consecutive relaxations $\ell^{(t)}$ and $\ell^{(t+1)}$ do not have too different global solutions, the following algorithm is a plausible practical tool:

Algorithm 1 (Weighted Graduated Non-Convexity Algorithm) For the optimization of semi-norms ($\ell(\cdot)$), a practical approach is based on deforming gradually a 2-norm into the specific loss function of interest. Let ζ be a strictly decreasing series $1, \zeta^{(1)}, \zeta^{(2)}, \ldots, 0$. A plausible choice for the initial convex cost function is the least squares cost function $\mathcal{J}_{\mathrm{LS}}(e) = \|e\|_2^2$.

1. Compute the solution $\theta^{(0)}$ for L_2 norm $J_{\mathrm{LS}}(e) = \|e\|_2^2$ with residuals $e_k^{(0)}$;
2. $t = 0$ and $\nu^{(0)} = 1_N$;
3. Consider the following relaxed cost function $\mathcal{J}^{(t)}(e) = (1 - \zeta_t)\ell(e) + \zeta_t \mathcal{J}_{\mathrm{LS}}(e)$;
4. Estimate the solution $\theta^{(t+1)}$ and corresponding residuals $e_k^{(t+1)}$ of the cost function $\mathcal{J}^{(t)}$ using the weighted approximation $\mathcal{J}_{\mathrm{approx}} = (\nu_k^{(t)} e_k)^2$ of $J^{(t)}(e_k)$
5. Reweight the residuals using weighted approximative squares norms as derived in (43):
6. $t := t + 1$ and iterate step (3, 4, 5, 6) until convergence.

When iterating this scheme, most ν_k will be smaller than 1 as the least squares cost function penalizes higher residuals (typically outliers). However, a number of residuals will have increasing weight as the least squares loss function is much lower for small residuals.

5.2 Regression Examples

To illustrate the additive model estimation method, a classical example was constructed as in [15, 31]. The data were generated according to $y_k = 10\,\mathrm{sinc}(x_k^1) + 20\,(x_k^2 - 0.5)^2 + 10\,x_k^3 + 5\,x_k^4 + e_k$ were $e_k \sim \mathcal{N}(0,1)$, $N = 100$ and the input data X are randomly chosen from the interval $[0,1]^{10}$. Because of the Gaussian nature of the noise model, only results from least squares methods are reported. The described techniques were applied on this training dataset and tested on an independent test set generated using the saem rules. Table 1 reports whether the algorithm recovered the structure in the data (if so, the measure is 100%). The experiment using the smoothly tresholding penalized (STP) cost function was designed as follows: for every 10 components, a version was provided for the algorithm for the use of a linear kernel and another for the use of a RBF kernel (resulting in 20 new components). The regularization scheme was able to select the components with the appropriate kernel (a nonlinear RBF kernel for X^1 and X^2 and linear ones for X^3 and X^4), except for one spurious component (A RBF kernel was selected for the fifth component). Figure 5 provide a schamatical illustration of the algorithm.

Table 1. Results on test data of numerical experiments on the Vapnik regression dataset. The sparseness is expressed in the rate of components which is selected only if the input is relevant (100% means the original structure was perfectly recovered)

Method	Test Set Performance L_2	L_1	L_∞	Sparse Components % recovered
LS-SVMs	0.1110	0.2582	0.8743	0%
componentwise LS-SVMs (7)	0.0603	0.1923	0.6249	0%
L_1 regularization (21)	0.0624	0.1987	0.6601	100%
STP with RBF (25)	0.0608	0.1966	0.6854	100%
STP with RBF and lin (25)	0.0521	0.1817	0.5729	95%
Fusion with AReg (30)	0.0614	0.1994	0.6634	100%
Fusion with comp. reg. (40)	0.0601	0.1953	0.6791	100%

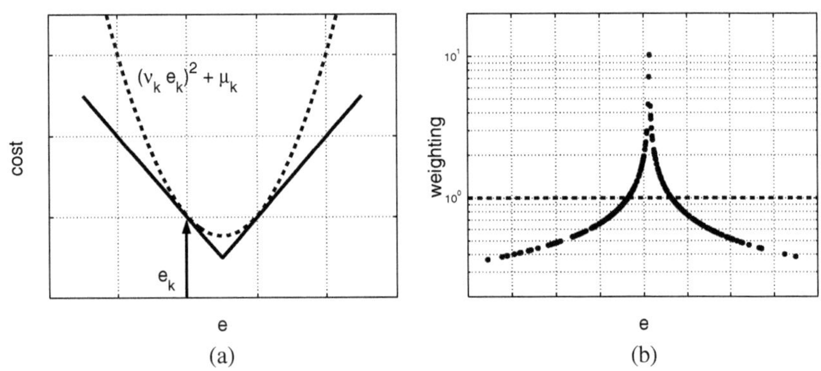

Fig. 5. (**a**) Weighted L_2-norm (*dashed*) approximation $(\nu_k e_k)^2 + \mu_k$ of the L_1-norm (*solid*) $\ell(e) = |e|_1$ which follows from the linear set of (43) once the optimal e_k are known; (**b**) the weighting terms ν_k for a sequence of e_k and $k = 1, \ldots, N$ such that $(\nu_k e_k)^2 + \mu_k = |e_k|_1$ and $2\nu_k^2 e_k = l'(e_k) = \text{sign}(e_k)$ for an appropriate μ_k

5.3 Classification Example

An additive model was estimated by an LS-SVM classifier based on the `spam` data as provided on the UCI benchmark repository, see e.g. [16]. The data consists of word frequencies from 4601 email messages, in a study to screen email for spam. A test set of size 1536 was drawn randomly from the data leaving 3065 to training purposes. The inputs were preprocessed using following transformation $p(x) = \log(1 + x)$ and standardized to unit variance. Figure 7 gives the indicator functions as found using a regularization based technique to detect structure as described in Subsect. 3.3. The structure detection algorithm selected only 6 out of the 56 provided indicators. Moreover, the componentwise approach describes the form of the contribution of each indicator, resulting in an highly interpretable model.

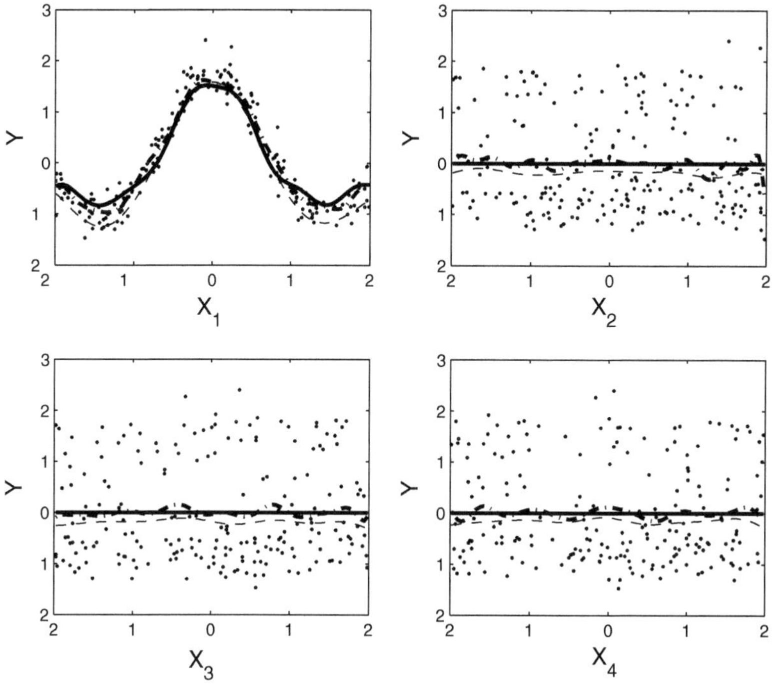

Fig. 6. Example of a toy dataset consisting of four input components X^1, X^2, X^3 and X^4 where only the first one is relevant to predict the output $f(x) = \text{sinc}(x^1)$. A componentwise LS-SVM regressor (*dashed line*) has good prediction performance, while the L_1 penalized cost function of Subsect. 3.2 also recovers the structure in the data as the estimated components correspnding with X^2, X^3 and X^4 are sparse

6 Conclusions

This chapter describes nonlinear additive models based on LS-SVMs which are capable of handling higher dimensional data for regression as well as classification tasks. The estimation stage results from solving a set of linear equations with a size approximatively equal to the number of training datapoints. Furthermore, the additive regularization framework is employed for formulating dedicated regularization schemes leading to structure detection. Finally, a fusion argument for component selection and structure detection based on training componentwise LS-SVMs and validation performance is introduced to improve the generalization abilities of the method. Advantages of using componentwise LS-SVMs include the efficient estimation of additive models with respect to classical practice, interpretability of the estimated model, opportunities towards structure detection and the connection with existing statistical techniques.

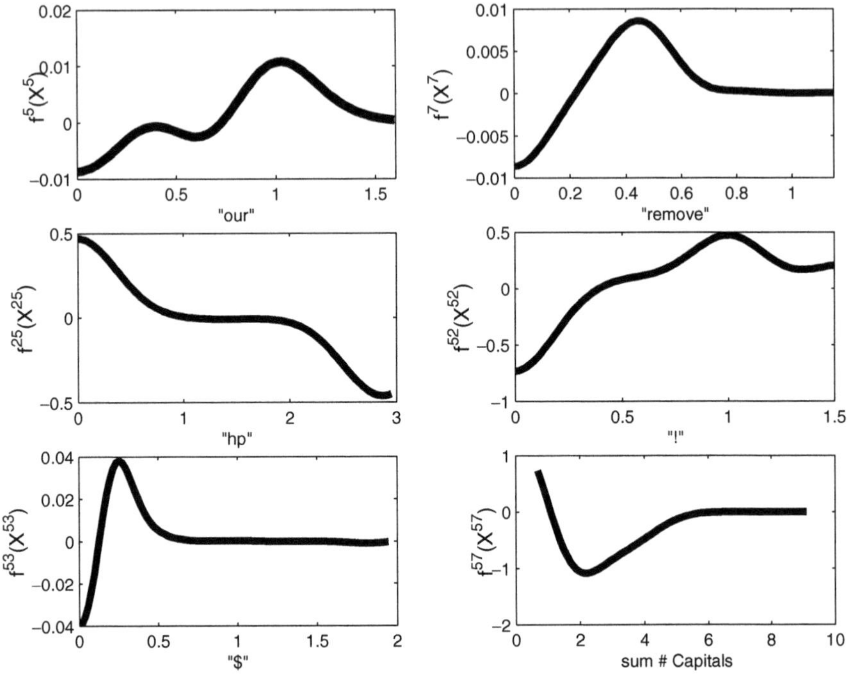

Fig. 7. Results of the spam dataset. The non-sparse components as found by application of Subsect. 3.3 are shown suggesting a number of usefull indicator variables for classifing a mail message as spam or non-spam. The final classifier takes the form $f(X) = f^5(X^5) + f^7(X^7) + f^{25}(X^{25}) + f^{52}(X^{52}) + f^{53}(X^{53}) + f^{56}(X^{56})$ where 6 relevant components were selected out of the 56 provided indicators

Acknowledgements

This research work was carried out at the ESAT laboratory of the Katholieke Universiteit Leuven. Research Council KUL: GOA-Mefisto 666, GOA AMBioRICS, several PhD/postdoc & fellow grants; Flemish Government: FWO: PhD/postdoc grants, projects, G.0240.99 (multilinear algebra), G.0407.02 (support vector machines), G.0197.02 (power islands), G.0141.03 (Identification and cryptography), G.0491.03 (control for intensive care glycemia), G.0120.03 (QIT), G.0452.04 (new quantum algorithms), G.0499.04 (Robust SVM), G.0499.04 (Statistics) research communities (ICCoS, ANMMM, MLDM); AWI: Bil. Int. Collaboration Hungary/Poland; IWT: PhD Grants, GBOU (McKnow) Belgian Federal Science Policy Office: IUAP P5/22 ('Dynamical Systems and Control: Computation, Identification and Modelling', 2002-2006) ; PODO-II (CP/40: TMS and Sustainability); EU: FP5-Quprodis; ERNSI; Eureka 2063-IMPACT; Eureka 2419-FliTE; Contract Research/ agreements: ISMC/IPCOS, Data4s, TML, Elia, LMS, Mastercard is supported by grants from several funding agencies and sources. GOA-Ambiorics, IUAP

V, FWO project G.0407.02 (support vector machines) FWO project G.0499.04 (robust statistics) FWO project G.0211.05 (nonlinear identification) FWO project G.0080.01 (collective behaviour) JS is an associate professor and BDM is a full professor at K.U.Leuven Belgium, respectively.

References

1. Antoniadis, A. (1997). Wavelets in statistics: A review. *Journal of the Italian Statistical Association* (6), 97–144.
2. Antoniadis, A. and J. Fan (2001). Regularized wavelet approximations (with discussion). *Journal of the American Statistical Association* **96**, 939–967.
3. Blake, A. (1989). Comparison of the efficiency of deterministic and stochastic algorithms for visual reconstruction. *IEEE Transactions on Image Processing* **11**, 2–12.
4. Boyd, S. and L. Vandenberghe (2004). *Convex Optimization.* Cambridge University Press.
5. Cressie, N.A.C. (1993). *Statistics for spatial data.* Wiley.
6. Cristianini, N. and J. Shawe-Taylor (2000). *An Introduction to Support Vector Machines.* Cambridge University Press.
7. Donoho, D.L. and I.M. Johnstone (1994). Ideal spatial adaption by wavelet shrinkage. *Biometrika* **81**, 425–455.
8. Fan, J. (1997). Comments on wavelets in statistics: A review. *Journal of the Italian Statistical Association* (6), 131–138.
9. Fan, J. and R. Li (2001). Variable selection via nonconvex penalized likelihood and its oracle properties. *Journal of the American Statistical Association* **96**(456), 1348–1360.
10. Frank, L.E. and J.H. Friedman (1993). A statistical view of some chemometric regression tools. *Technometrics* (35), 109–148.
11. Friedmann, J.H. and W. Stuetzle (1981). Projection pursuit regression. *Journal of the American Statistical Association* **76**, 817–823.
12. Fu, W.J. (1998). Penalized regression: the bridge versus the LASSO. *Journal of Computational and Graphical Statistics* (7), 397–416.
13. Fukumizu, K., F.R. Bach and M.I. Jordan (2004). Dimensionality reduction for supervised learning with reproducing kernel Hilbert spaces. *Journal of Machine Learning Reasearch* (5), 73–99.
14. Gunn, S.R. and J.S. Kandola (2002). Structural modelling with sparse kernels. *Machine Learning* **48**(1), 137–163.
15. Hastie, T. and R. Tibshirani (1990). *Generalized addidive models.* London: Chapman and Hall.
16. Hastie, T., R. Tibshirani and J. Friedman (2001). *The Elements of Statistical Learning.* Springer-Verlag. Heidelberg.
17. Linton, O.B. and J.P. Nielsen (1995). A kernel method for estimating structured nonparameteric regression based on marginal integration. *Biometrika* **82**, 93–100.
18. MacKay, D.J.C. (1992). The evidence framework applied to classification networks. *Neural Computation* **4**, 698–714.
19. Neter, J., W. Wasserman and M.H. Kutner (1974). *Applied Linear Statistical Models.* Irwin.

20. Nikolova, M. (1999). Local strong homogeneity of a regularized estimator. *SIAM Journal on Applied Mathematics* **61**, 633–658.
21. Pelckmans, K., J.A.K. Suykens and B. De Moor (2003). Additive regularization: Fusion of training and validation levels in kernel methods. *(Submitted for Publication) Internal Report 03-184, ESAT-SISTA, K.U.Leuven (Leuven, Belgium).*
22. Poggio, T. and F. Girosi (1990). Networks for approximation and learning. In: *Proceedings of the IEEE.* Vol. 78. Proceedings of the IEEE. pp. 1481–1497.
23. Schoelkopf, B. and A. Smola (2002). *Learning with Kernels.* MIT Press.
24. Stone, C.J. (1982). Optimal global rates of convergence for nonparametric regression. *Annals of Statistics* **13**, 1040–1053.
25. Stone, C.J. (1985). Additive regression and other nonparameteric models. *Annals of Statistics* **13**, 685–705.
26. Suykens, J.A.K. and J. Vandewalle (1999). Least squares support vector machine classifiers. *Neural Processing Letters* **9**(3), 293–300.
27. Suykens, J.A.K., T. Van Gestel, J. De Brabanter, B. De Moor and J. Vandewalle (2002). *Least Squares Support Vector Machines.* World Scientific, Singapore.
28. Tibshirani, R.J. (1996). Regression shrinkage and selection via the LASSO. *Journal of the Royal Statistical Society* (58), 267–288.
29. Tibshirani, R.J. (1997). The LASSO method for variable selection in the cox model. *Statistics in Medicine* (16), 385–395.
30. Tikhonov, A.N. and V.Y. Arsenin (1977). *Solution of Ill-Posed Problems.* Winston. Washington DC.
31. Vapnik, V.N. (1998). *Statistical Learning Theory.* John Wiley and Sons.
32. Wahba, G. (1990). *Spline models for observational data.* SIAM.

Active Support Vector Learning with Statistical Queries

P. Mitra, C.A. Murthy, and S.K. Pal

Machine Intelligence Unit,
Indian Statistical Institute,
Calcutta 700 108, India.
{pabitra_r,murthy,sankar}@isical.ac.in

Abstract. The article describes an active learning strategy to solve the large quadratic programming (QP) problem of support vector machine (SVM) design in data mining applications. The learning strategy is motivated by the statistical query model. While most existing methods of active SVM learning query for points based on their proximity to the current separating hyperplane, the proposed method queries for a set of points according to a distribution as determined by the current separating hyperplane and a newly defined concept of an adaptive confidence factor. This enables the algorithm to have more robust and efficient learning capabilities. The confidence factor is estimated from local information using the k nearest neighbor principle. Effectiveness of the method is demonstrated on real life data sets both in terms of generalization performance and training time.

Key words: Data mining, query learning, incremental learning, statistical queries

1 Introduction

The support vector machine (SVM) [17] has been successful as a high performance classifier in several domains including pattern recognition, data mining and bioinformatics. It has strong theoretical foundations and good generalization capability. A limitation of the SVM design algorithm, particularly for large data sets, is the need to solve a quadratic programming (QP) problem involving a dense $n \times n$ matrix, where n is the number of points in the data set. Since QP routines have high complexity, SVM design requires huge memory and computational time for large data applications. Several approaches exist for circumventing the above shortcomings. These include simpler optimization criterion for SVM design, e.g., the linear SVM and the kernel adatron, specialized QP algorithms like the cojugate gradient method, decomposition techniques which break down the large QP problem into a series of

P. Mitra, C.A. Murthy, and S.K. Pal: *Active Support Vector Learning with Statistical Queries*, StudFuzz **177**, 99–111 (2005)
www.springerlink.com

smaller QP sub-problems, sequential minimal optimization (SMO) algorithm and its various extensions, Nystrom approximations [18] and greedy Bayesian methods [15]. Many of these approaches are discussed in [13]. A simple method to solve the SVM QP problem has been described by Vapnik, which is known as "chunking" [2]. The chunking algorithm uses the fact that the solution of the SVM problem remains the same if one removes the points that correspond to zero Lagrange multipliers of the QP problem (the non-SV points). The large QP problem can thus be broken down into a series of smaller QP problems, whose ultimate goal is to identify all of the non-zero Lagrange multipliers (SVs) while discarding the zero Lagrange multipliers (non-SVs). At every step, chunking solves a QP problem that consists of the non-zero Lagrange multiplier points from the previous step, and a chunk of p other points. At the final step, the entire set of non-zero Lagrange multipliers has been identified; thereby solving the large QP problem. Several variations of chunking algorithm exist depending upon the method of forming the chunks [5]. Chunking greatly reduces the training time compared to batch learning of SVMs. However, it may not handle large-scale training problems due to slow convergence of the chunking steps when p new points are chosen randomly.

Recently, active learning has become a popular paradigm for reducing the sample complexity of large scale learning tasks [4, 7]. It is also useful in situations where unlabeled data is plentiful but labeling is expensive. In active learning, instead of learning from "random samples', the learner has the ability to select its own training data. This is done iteratively, and the output of a step is used to select the examples for the next step. In the context of support vector machine active learning can be used to speed up chunking algorithms. In [3], a query learning strategy for large margin classifiers is presented which iteratively requests the label of the data point closest to the current separating hyperplane. This accelerates the learning drastically compared to random sampling. An active learning strategy based on version space splitting is presented in [16]. The algorithm attempts to select the points which split the current version space into two halves having equal volumes at each step, as they are likely to be the actual support vectors. Three heuristics for approximating the above criterion are described, the simplest among them selects the point closest to the current hyperplane as in [3]. A greedy optimal strategy for active SV learning is described in [12]. Here, logistic regression is used to compute the class probabilities, which is further used to estimate the expected error after adding an example. The example that minimizes this error is selected as a candidate SV. Note that the method was developed only for querying single point, but the result reported in [12] used batches of different sizes, in addition to single point.

Although most of these active learning strategies query only for a single point at each step, several studies have noted that the gain in computational time can be obtained by querying multiple instances at a time. This motivates the formulation of active learning strategies which query for multiple points. Error driven methods for incremental support vector learning with multiple points are described in [9]. In [9] a chunk of p new points having a fixed ratio of correctly classified and misclassified points are used to update the current SV set. However, no guideline is provided for choosing the above ratio. Another major limitation of all the above strategies is that they are essentially greedy methods where the selection of a new point is influenced only by the current hypothesis (separating hyperplane) available. The greedy margin based methods are weak because focusing purely on the boundary points produces a kind of non-robustness, with the algorithm never asking itself whether a large number of examples far from the current boundary do in fact have the correct implied labels. In the above setup, learning may be severely hampered in two situations: a "bad" example is queried which drastically worsens the current hypothesis, and the current hypothesis itself is far from the optimal hypothesis (e.g., in the initial phase of learning). As a result, the examples queried are less likely to be the actual support vectors.

The present article describes an active support vector learning algorithm which is a probabilistic generalization of purely margin based methods. The methodology is motivated by the model of *learning from statistical queries* [6] which captures the natural notion of learning algorithms that construct a hypothesis based on statistical properties of large samples rather than the idiosyncrasies of a particular example. A similar probabilistic active learning strategy is presented in [14]. The present algorithm involves estimating the likelihood that a new example belongs to the actual support vector set and selecting a set of p new points according to the above likelihood, which are then used along with the current SVs to obtain the new SVs. The likelihood of an example being a SV is estimated using a combination of two factors: the margin of the particular example with respect to the current hyperplane, and the degree of confidence that the current set of SVs provides the actual SVs. The degree of confidence is quantified by a measure which is based on the local properties of each of the current support vectors and is computed using the nearest neighbor estimates.

The aforesaid strategy for active support vector learning has several advantages. It allows for querying multiple instances and hence is computationally more efficient than those that are querying for a single example at a time. It not only queries for the error points or points close to the separating hyperplane but also a number of other points which are far from the separating hyperplane and also correctly classified ones. Thus, even if a current hypothesis is erroneous there is scope for it being corrected owing to the later points. If only error points were selected the hypothesis might actually become worse. The ratio of selected points lying close to the separating hyperplane (and misclassified points) to those far from the hyperplane is decided by the confidence factor, which varies adaptively with iteration. If the current SV set is close to the optimal one, the algorithm focuses only on the low margin points and ignores the redundant points that lie far from the hyperplane. On the other hand, if the confidence factor is low (say, in the initial learning phase) it explores a higher number of interior points. Thus, the trade-off between efficiency and robustness of performance is adequately handled in this framework. This results in a reduction in the total number of labeled points queried by the algorithm, in addition to speed up in training; thereby making the algorithm suitable for applications where labeled data is scarce.

Experiments are performed on four real life classification problems. The size of the data ranges from 684 to 495, 141, dimension from 9 to 294. Our algorithm is found to provide superior performance and faster convergence compared to several related algorithms for incremental and active SV learning.

2 Support Vector Machine

Support vector machines are a general class of learning architecture inspired from statistical learning theory that performs *structural risk minimization* on a nested set structure of separating hyperplanes [17]. Given training data, the SVM training algorithm obtains the optimal separating hyperplane in terms of generalization error. Though SVMs may also be used for regression and multiclass classification, in this article we concentrate only on two-class classification problem.

Algorithm: Suppose we are given a set of examples $(\mathbf{x}_1, y_1), \ldots, (\mathbf{x}_l, y_l), \mathbf{x} \in R^N, y_i \in \{-1, +1\}$. We consider decision functions of the form $sgn((\mathbf{w} \cdot \mathbf{x}) + b)$, where $(\mathbf{w} \cdot \mathbf{x})$ represents the inner product of $\mathbf{w}$ and $\mathbf{x}$. We would like to find a decision function $f_{\mathbf{w},b}$ with the properties

$$y_i((\mathbf{w} \cdot \mathbf{x}_i) + b) \geq 1, \quad i = 1, \ldots, l\,. \tag{1}$$

In many practical situations, a separating hyperplane does not exist. To allow for possibilities of violating (1), slack variables are introduced like

$$\xi_i \geq 0, \quad i = 1, \ldots, l \tag{2}$$

to get

$$y_i((\mathbf{w} \cdot \mathbf{x}_i) + b) \geq 1 - \xi_i, \quad i = 1, \ldots, l\,. \tag{3}$$

The support vector approach for minimizing the generalization error consists of the following:

$$\textbf{Minimize}: \quad \Phi(\mathbf{w}, \xi) = (\mathbf{w} \cdot \mathbf{w}) + C \sum_{i=1}^{l} \xi_i \tag{4}$$

subject to the constraints (2) and (3).

It can be shown that minimizing the first term in (4), amounts to minimizing a bound on the VC-dimension, and minimizing the second term corresponds to minimizing the misclassification error [17]. The above minimization problem can be posed as a constrained quadratic programming (QP) problem.

The solution gives rise to a decision function of the form:

$$f(\mathbf{x}) = sgn\left[\sum_{i=1}^{l} y_i\alpha_i(\mathbf{x}\cdot\mathbf{x_i}) + b\right] .$$

Only a small fraction of the α_i coefficients are non-zero. The corresponding pairs of $\mathbf{x_i}$ entries are known as *support vectors* and they fully define the decision function.

3 Probabilistic Active Support Vector Learning Algorithm

In the context of support vector machines, the target of the learning algorithm is to learn the set of support vectors. This is done by incrementally training a SVM on a set of examples consisting of the previous SVs and a new set of points. In the proposed algorithm, the new set of points, instead of being randomly generated, is generated according to a probability $\mathbf{Pr}_{\chi(\mathbf{x},f(\mathbf{x}))}$. $\chi(\mathbf{x},f(\mathbf{x}))$ denotes the event that the example $\mathbf{x}$ is a SV. $f(\mathbf{x})$ is the optimal separating hyperplane. The methodology is motivated by the statistical query model of learning [6], where the oracle instead of providing actual class labels, provides an (approximate) answer to the statistical query "what is the probability that an example belongs to a particular class?".

We define the probability $\mathbf{Pr}_{\chi(\mathbf{x},f(\mathbf{x}))}$ as follows. Let $\langle\mathbf{w},b\rangle$ be the current separating hyperplane available to the learner.

$$\begin{aligned} P_{\chi(\mathbf{x},f(\mathbf{x}))} &= c \qquad \text{if } y(\mathbf{w}\cdot\mathbf{x}+b)\le 1 \\ &= 1-c \quad \text{otherwise} . \end{aligned} \tag{5}$$

Here c is a *confidence parameter* which denotes how close the current hyperplane $\langle\mathbf{w},b\rangle$ is to the optimal one. y is the label of $\mathbf{x}$.

The significance of $P_{\chi(\mathbf{x},f(\mathbf{x}))}$ is as follows: if c is high, which signifies that the current hyperplane is close to the optimal one, points having margin value less than unity are highly likely to be the actual SVs. Hence, the probability $P_{\chi(\mathbf{x},f(\mathbf{x}))}$ returned to the corresponding query is set to a high value c. When the value c is low, the probability of selecting a point lying within the margin decreases, and a high probability value $(1-c)$ is then assigned to a point having high margin. Let us now describe a method for estimating the confidence factor c.

3.1 Estimating the Confidence Factor for a SV Set

Let the current set of support vectors be denoted by $S = \{\mathbf{s}_1,\mathbf{s}_2,\ldots,\mathbf{s}_l\}$. Also, consider a test set $T = \{\mathbf{x}'_1,\mathbf{x}'_2,\ldots,\mathbf{x}'_m\}$ and an integer k (say, $k=\sqrt{l}$). For every $\mathbf{s}_i \in S$ compute the set of k nearest points in T. Among the k

nearest neighbors let k_i^+ and k_i^- denote the number of points having labels +1 and −1 respectively. The confidence factor c is then defined as

$$c = \frac{2}{lk} \sum_{i=1}^{l} \min(k_i^+, k_i^-) \,. \tag{6}$$

Note that the maximum value of the confidence factor c is unity when $k_i^+ = k_i^- \; \forall i = 1, \ldots, l$, and the minimum value is zero when $\min(k_i^+, k_i^-) = 0$ $\forall i = 1, \ldots, l$. The first case implies that all the support vectors lie near the class boundaries and the set $S = \{\mathbf{s}_1, \mathbf{s}_2, \ldots, \mathbf{s}_l\}$ is close to the actual support vector set. The second case, on the other hand, denotes that the set S consists only of interior points and is far from the actual support vector set. Thus, the confidence factor c measures the degree of closeness of S to the actual support vector set. The higher the value of c is, the closer is the current SV set to the actual SV set.

3.2 Algorithm

The active support vector learning algorithm, which uses the probability $\mathbf{Pr}_{\chi(\mathbf{x}, f(\mathbf{x}))}$, estimated above, is presented below.

Let $A = \{\mathbf{x}_1, \mathbf{x}_2, \ldots, \mathbf{x}_n\}$ denote the entire training set used for SVM design. $SV(B)$ denotes the set of support vectors of the set B obtained using the methodology described in Sect. 2. $S_t = \{\mathbf{s}_1, \mathbf{s}_2, \ldots, \mathbf{s}_l\}$ is the support vector set obtained after tth iteration, and $\langle \mathbf{w}_t, b_t \rangle$ is the corresponding separating hyperplane. $Q_t = \{\mathbf{q}_1, \mathbf{q}_2, \ldots, \mathbf{q}_p\}$ is the set of p points actively queried for at step t. c is the confidence factor obtained using (6). The learning steps involved are given below:

Initialize: Randomly select an initial starting set Q_0 of p instances from the training set A. Set $t = 0$ and $S_0 = SV(Q_0)$. Let the parameters of the corresponding hyperplane be $\langle \mathbf{w}_0, b_0 \rangle$.

While *Stopping Criterion* is not satisfied:
  $Q_t = \emptyset$.
  While Cardinality$(Q_t) \leq p$:
    Randomly select an instance $\mathbf{x} \in A$.
    Let y be the label of $\mathbf{x}$.
    If $y(\mathbf{w}_t \cdot \mathbf{x} + b) \leq 1$:
      Select $\mathbf{x}$ with probability c. Set $Q_t = Q_t \cup \mathbf{x}$.
    Else:
      Select $\mathbf{x}$ with probability $1 - c$. Set $Q_t = Q_t \cup \mathbf{x}$.
    End If
  End While
  $S_t = SV(S_t \cup Q_t)$.
  $t = t + 1$.
End While

The set S_T, where T is the iteration at which the algorithm terminates, contains the final SV set.

Stopping Criterion: Among the p points actively queried at each step t, let p_{nm} points have margin greater than unity ($y(\mathbf{w}_t \cdot \mathbf{x} + b) > 1$). Learning is stopped if the quantity $\frac{c \cdot p_{nm}}{p}$ exceeds a threshold Th (say, $= 0.9$).

The stopping criterion may be interpreted as follows. A high value of the quantity $\frac{p_{nm}}{p}$ implies that the query set contains a small number of points with margin less than unity. No further gain can be thus achieved by the learning process. The value of p_{nm} may also be large when the value of c is low in the initial phase of learning. However, if both c and p_{nm} have high values, the current SV set is close to the actual one (i.e., a good classifier is obtained) and also the margin band is empty (i.e., the learning process is saturated); hence, the learning may be terminated.

4 Experimental Results and Comparison

Organization of the experimental results is as follows. First, the characteristics of the four datasets, used, are discussed briefly. Next, the performance of the proposed algorithm in terms of generalization capability, training time and some related quantities, is compared with two other incremental support vector learning algorithms as well as the batch SVM. Linear SVMs are used in all the cases. The effectiveness of the confidence factor c, used for active querying, is then studied.

4.1 Data Sets

Six public domain datasets are used, two of which are large and three relatively smaller. All the data sets have two overlapping classes. Their characteristics are described below. The data sets are available in the UCI machine learning and KDD repositories [1].

Wisconsin Cancer: The popular Wisconsin breast cancer data set contains 9 features, 684 instances and 2 classes.

Twonorm: This is an artificial data set, having dimension 20, 2 classes and 20,000 points. Each class is drawn from a multivariate normal distribution with unit covariance matrix. Class 1 has mean $(a, a, \ldots, a)$ and class 2 has mean $(-a, -a, \ldots, -a)$. $a = \frac{2}{20^{\frac{1}{2}}}$.

Forest Cover Type: This is a GIS data set representing the forest covertype of a region. There are 54 attributes out of which we select 10 numeric valued attributes. The original data contains $581,012$ instances and 8 classes, out of which only $495,141$ points, belonging to classes 1 and 2, are considered here.

Microsoft Web Data: There are 36818 examples with 294 binary attributes. The task is to predict whether an user visits a particular site.

Table 1. Comparison of performance of SVM design algorithms

Data	Algorithm	a_{test}(%) Mean	SD	$\mathcal{D}$	n_{query}	t_{cpu} (sec)
Cancer	BatchSVM	96.32	0.22	–	–	1291
	IncrSVM	86.10	0.72	10.92	0.83	302
	QuerySVM	96.21	0.27	9.91	0.52	262
	StatQSVM	96.43	0.25	7.82	0.41	171
	SMO	96.41	0.23	–	–	91
Twonorm	BatchSVM	97.46	0.72	–	–	8.01×10^4
	IncrSVM	92.01	1.10	12.70	0.24	770
	QuerySVM	93.04	1.15	12.75	0.07	410
	StatQSVM	96.01	1.52	12.01	0.02	390
	SMO	97.02	0.81	–	–	82
Covertype	IncrSVM	57.90	0.74	–	0.04	4.70×10^4
	QuerySVM	65.77	0.72	–	0.008	3.20×10^4
	StatQSVM	74.83	0.77	–	0.004	2.01×10^4
	SMO	74.22	0.41	–	–	0.82×10^4
Microsoft Web	IncrSVM	52.10	0.22	–	0.10	2.54×10^4
	QuerySVM	52.77	0.78	–	0.04	1.97×10^4
	StatQSVM	63.83	0.41	–	0.01	0.02×10^4
	SMO	65.43	0.17	–	–	0.22×10^4

4.2 Classification Accuracy and Training Time

The algorithm for active SV learning with statistical queries (StatQSVM) is compared with two other techniques for incremental SV learning as well as the actual batch SVM algorithm. Only for the Forest Covertype data set, batch SVM could not be obtained due to its large size. The sequential minimal optimization (SMO) algorithm [10] is also compared for all the data sets. The following incremental algorithms are considered.

(i) Incremental SV learning with random chunk selection [11]. (Denoted by IncrSVM in Table 1.)
(ii) SV learning by querying the point closest to the current separating hyperplane [3]. (Denoted by QuerySVM in Table 1.) This is also the "simple margin" strategy in [16].

Comparison is made on the basis of the following quantities. Results are presented in Table 1.

1. Classification accuracy on test set (a_{test}). The test set has size 10% of that of the entire data set, and contains points which do not belong to the (90%) training set. Means and standard deviations (SDs) over 10 independent runs are reported.

2. Closeness of the SV set: We measure closeness of the SV set ($\tilde{S}$), obtained by an algorithm, with the corresponding actual one (S). These are measured by the distance $\mathcal{D}$ defined as follows [8]:

$$\mathcal{D} = \frac{1}{n_{\tilde{S}}} \sum_{x \in \tilde{S}} \delta(x, S) + \frac{1}{n_S} \sum_{y \in S} \delta(y, \tilde{S}) + Dist(\tilde{S}, S) , \qquad (7)$$

where

$$\delta(x, S) = \min_{y \in S} d(x, y), \ \ \delta(y, \tilde{S}) = \min_{x \in \tilde{S}} d(x, y) ,$$

and $Dist(\tilde{S}, S) = \max\{\max_{x \in \tilde{S}} \delta(x, S), \max_{y \in S} \delta(y, \tilde{S})\}$. $n_{\tilde{S}}$ and n_S are the number of points in $\tilde{S}$ and S respectively. $d(x, y)$ is the usual Euclidean distance between points x and y. The distance measure has been used for quantifying the errors of set approximation algorithms [8], and is related to the ϵ cover of a set.

3. Fraction of training samples queried (n_{query}) by the algorithms.
4. CPU time (t_{cpu}) on a Sun UltraSparc 350MHz workstation.

It is observed from the results shown in Table 1 that all the three incremental learning algorithms require several order less training time as compared to batch SVM design, while providing comparable classification accuracies. Among them the proposed one achieves highest or second highest classification score in least time and number of queries for all the data sets. The superiority becomes more apparent for the Forest Covertype data set, where it significantly outperforms both QuerySVM and IncrSVM. The QuerySVM algorithm performs better than IncrSVM for Cancer, Twonorm and the Forest Covertype data sets.

It can be seen from the values of n_{query} in Table 1, that the total number labeled points queried by StatQSVM is the least among all the methods including QuerySVM. This is inspite of the fact that, StatQSVM needs the label of the randomly chosen points even if they wind up not being used for training, as opposed to QuerySVM, which just takes the point closest to the hyperplane (and so does not require knowing its label until one decides to actually train on it). The overall reduction in n_{query} for StatQSVM is probably achieved by its efficient handling of the exploration – exploitation trade-off in active learning.

The SMO algorithm requires substantially less time compared to the incremental ones. However, SMO is not suitable to applications where labeled data is scarce. Also, SMO may be used along with the incremental algorithms for further reduction in design time.

The nature of convergence of the classification accuracy on test set a_{test} is shown in Fig. 1 for all the data sets. It is be observed that the convergence curve for the proposed algorithm dominates those of QuerySVM and IncrSVM. Since the IncrSVM algorithm selects the chunks randomly, the corresponding curve is smooth and almost monotonic, although its convergence rate is much slower compared to the other two algorithms. On the other hand,

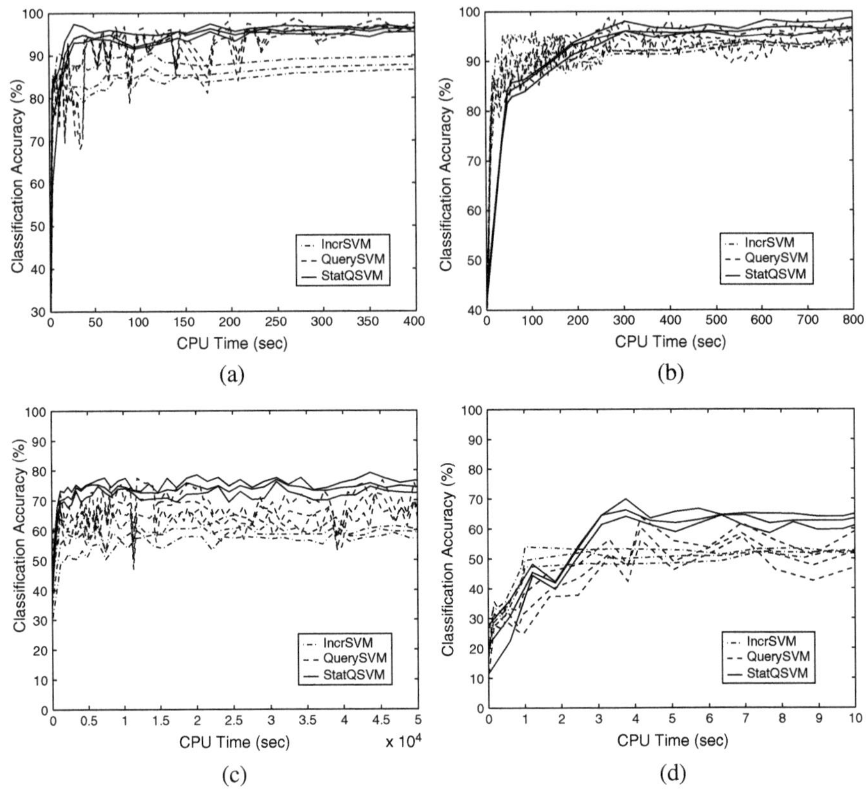

Fig. 1. Variation of a_{test} (maximum, minimum and average over ten runs) with CPU time for (**a**) Cancer, (**b**) Twonorm, (**c**) Forest covertype, (**d**) Microsoft web data

the QuerySVM algorithm selects only the point closest to the current separating hyperplane and achieves a high classification accuracy in few iterations. However, its convergence curve is oscillatory and the classification accuracy falls significantly after certain iterations. This is expected as querying for points close to the current separating hyperplane may often result in gain in performance if the current hyperplane is close to the optimal one. While querying for interior points reduces the risk of performance degradation, it also achieves poor convergence rate. Our strategy for active support vector learning with statistical queries selects a combination of low margin and interior points, and hence maintains a fast convergence rate without oscillatory performance degradation.

In a part of the experiment, the margin distribution of the samples was studied as a measure of generalization performance of the SVM. The distribution in which a larger number of examples have high positive margin values leads to a better generalization performance. It was observed that, although the proposed active learning algorithm terminated before all the actual SVs were identified, the SVM obtained by it produced a better margin distribution than the batch SVM designed using the entire data set. This strengthens the observation of [12] and [3] that active learning along with early stopping improves the generalization performance.

4.3 Effectiveness of the Confidence Factor c

Figure 2 shows the variation of the confidence factor c for the SV sets with distance $\mathcal{D}$. It is observed that for all the data sets c is linearly correlated with $\mathcal{D}$. As the current SV set converges closer to the optimal one, the value of $\mathcal{D}$ decreases and the value of confidence factor c increases. Hence, c is an effective measure of the closeness of the SV set with the actual one.

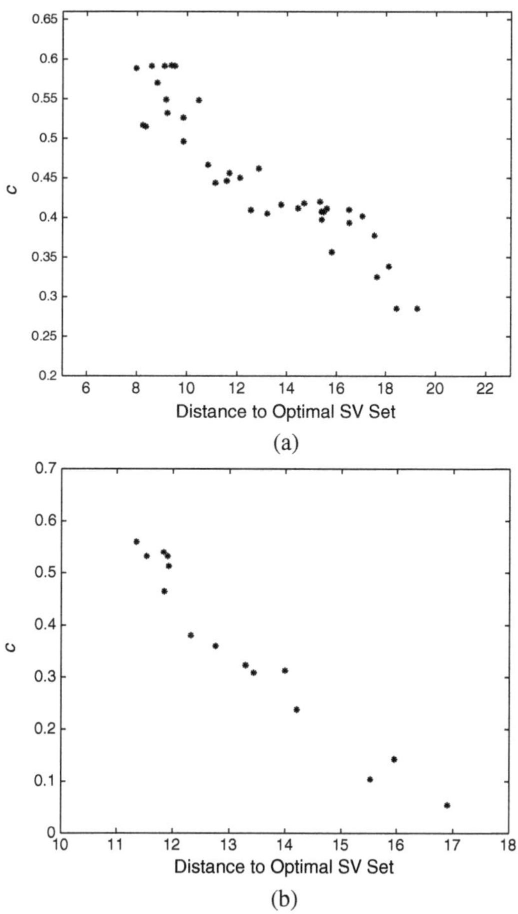

Fig. 2. Variation of confidence factor c and distance $\mathcal{D}$ for (**a**) Cancer, and (**b**) Twonorm data

5 Conclusions and Discussion

A method for probabilistic active SVM learning is presented. Existing algorithms for incremental SV learning either query for points close to the current separating hyperplane or select random chunks consisting mostly of interior points. Both these strategies represent extreme cases; the former one is fast but unstable, while the later one is robust but slowly converging. The former strategy is useful in the final phase of learning, while the later one is more suitable in the initial phase. The proposed active learning algorithm uses an adaptive confidence factor to handle the above trade-off. It more robust than purely margin based methods and potentially faster than random chunk selection because it can, to some extent, avoid calculating margins for non-support vector examples. The superiority of our algorithm is experimentally demonstrated for some real life data sets in terms of both training time and number of queries. The strength of the proposed StatQSVM algorithm lies in the reduction of the number of labeled points queried, rather than just speed up in training. This makes it suitable for environments where labeled data is scarce.

The selection probability (P_χ, (5)), used for active learning, is a two level function of the margin ($y(\mathbf{w} \cdot \mathbf{x} + b)$) of a point $\mathbf{x}$. Continuous functions of margin of $\mathbf{x}$ may also be used. Also, the confidence factor c may be estimated using a kernel based relative class likelihood for more general kernel structures. Logistic framework and probabilistic methods [14] may also be employed for estimating the confidence factor.

References

1. Blake, C. L., Merz , C. J. (1998) UCI Repository of machine learning databases. University of California, Irvine, Dept. of Information and Computer Sciences, http://www.ics.uci.edu/~mlearn/ MLRepository.html
2. Burges, C. J. C. (1998) A tutorial on support vector machines for pattern recognition. Data Mining and Knowledge Discovery, 2, 1–47
3. Campbell, C., Cristianini, N., Smola., A. (2000) Query learning with large margin classifiers. Proc. 17th Intl. Conf. Machine Learning, Stanford, CA, Morgan Kaufman, 111–118
4. Cohn, D., Ghahramani, Z., Jordan, M. (1996) Active learning with statistical models. Journal of AI Research, 4, 129–145
5. Kaufman, L. (1998) Solving the quadratic programming problem arising in support vector classification. Advances in Kernel Methods – Support Vector Learning, MIT Press, 147–168
6. Kearns, M. J. (1993) Efficient noise-tolerant learning from statistical queries. In Proc. 25th ACM Symposium on Theory of Computing, San Diego, CA, 392–401
7. MacKay, D. (1992) information based objective function for active data selection. Neural Computation, 4, 590–604
8. Mandal, D. P., Murthy, C. A., Pal, S. K. (1992) Determining the shape of a pattern class from sampled points in R^2. Intl. J. General Systems, 20, 307–339

9. Mitra, P., Murthy, C. A., Pal, S. K. (2000) Data condensation in large data bases by incremental learning with support vector machines. Proc. 15th Intl. Conf. Pattern Recognition, Barcelona, Spain, 712–715
10. Platt, J. C. (1998) Fast training of support vector machines using sequential minimal optimisation. Advances in Kernel Methods – Support Vector Learning, MIT Press, 185–208
11. Sayeed, N. A., Liu, H., Sung, K. K. (1999) A sudy of support vectors on model independent example selection. Proc. 1st Intl. Conf. Knowledge Discovery and Data Mining, San Diego, CA, 272–276
12. Schohn, G., Cohn, D. (2000) Less is more: Active learning with support vector machines. Proc. 17th Intl. Conf. Machine Learning, Stanford, CA, Morgan Kaufman, 839–846
13. Scholkopf, B., Burges, C. J. C., Smola, A. J. (1998) Advances in Kernel Methods – Support Vector Learning. MIT Press, 1998
14. Seo, S., Wallat, M., Graepel, T., Obermayer, K. (2000) Gaussian process regression: Active data selection and test point rejection. Proc. Int. Joint Conf. Neural Networks, 3, 241–246
15. Tipping, M. E., Faul, A. (2003) Fast marginal likelihood maximization for sparse Bayesian models. Intl. Workshop on AI and Statistics (AISTAT 2003), Key West, FL, Society for AI and Statistics
16. Tong, S., Koller, D. (2001) Support vector machine active learning with application to text classification. Journal of Machine Learning Research, 2, 45–66
17. Vapnik, V. (1998) Statistical Learning Theory. Wiley, New York
18. Williams, C. K. I., Seeger, M. (2001) Using the Nystrom method to speed up kernel machines. Advances in Neural Information Processing System 14 (NIPS'2001), Vancouver, Canada, MIT Press

Local Learning vs. Global Learning: An Introduction to Maxi-Min Margin Machine

K. Huang, H. Yang, I. King, and M.R. Lyu

Department of Computer Science and Engineering,
The Chinese University of Hong Kong,
Shatin, N.T., Hong Kong
{kzhuang, hqyang, king, lyu}@cse.cuhk.edu.hk

Abstract. We present a unifying theory of the Maxi-Min Margin Machine (M^4) that subsumes the Support Vector Machine (SVM), the Minimax Probability Machine (MPM), and the Linear Discriminant Analysis (LDA). As a unified approach, M^4 combines some merits from these three models. While LDA and MPM focus on building the decision plane using global information and SVM focuses on constructing the decision plane in a local manner, M^4 incorporates these two seemingly different yet complementary characteristics in an integrative framework that achieves good classification accuracy. We give some historical perspectives on the three models leading up to the development of M^4. We then outline the M^4 framework and perform investigations on various aspects including the mathematical definition, the geometrical interpretation, the time complexity, and its relationship with other existing models.

Key words: Classification, Local Learning, Global Learning, Hybrid Learning, M^4, Unified Framework

1 Introduction

When constructing a classifier, there is a dichotomy in choosing whether to use local vs. global characteristics of the input data. The framework of using global characteristics of the data, which we refer to as global learning, enjoys a long and distinguished history. When studying real-world phenomena, scientists try to discover the fundamental laws or underlying mathematics that govern these complex phenomena. Furthermore, in practice, due to incomplete information, these phenomena are usually described by using probabilistic or statistical models on sampled data. A common methodology found in these models is to fit a density on the observed data. With the learned density, people can easily perform prediction, inference, and marginalization tasks.

K. Huang et al.: *Local Learning vs. Global Learning: An Introduction to Maxi-Min Margin Machine*, StudFuzz **177**, 113–131 (2005)
www.springerlink.com

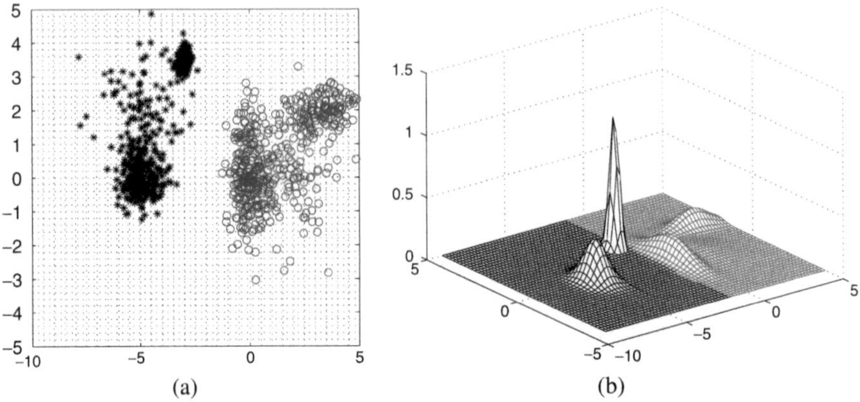

Fig. 1. An illustration of distribution-based classification (also known as the Bayes optimal decision theory). Two Gaussian mixtures are engaged to model the distribution of the two classes of data respectively. The distribution can then be used to construct the decision plane

One type of global learning is generative learning. By assuming a specific model on the observed data, e.g., a Gaussian distribution or a mixture of Gaussian, the phenomena can therefore be described or re-generated. Figure 1(a) illustrates such an example. In this figure, two classes of data are plotted as $*$'s for the first class and $\circ$'s for the other class. The data can thus be modelled as two different mixtures of Gaussian distributions. By knowing only the parameters of these distributions, one can then summarize the phenomena. Furthermore, as illustrated in Figure 1(b), one can clearly employ learned densities to distinguish one class of data from the other class (or simply know how to separate these two classes). This is the well-known Bayes optimal decision problem [1, 2].

One of the main difficulties found in global learning methodologies is the *model selection* problem. More precisely one still needs to select a suitable and appropriate model and its parameters in order to represent the observed data. This is still an open and on-going research topic. Some researchers have argued that it is difficult if not impossible to obtain a general and accurate global learning. Hence, local learning has recently attracted much interests.

Local learning [3, 4, 5] focuses on capturing only useful local information from the observed data. Furthermore, recent research progress and empirical studies demonstrate that the local learning paradigm is superior to global learning in many classification domains.

Local learning is more task-oriented since it omits an intermediate density modelling step in classification tasks. It does not aim to estimate a density from data as in global learning. In fact, it even does not intend to build an accurate model to fit the observed data globally. Therefore, local learning is more direct which results in more accurate and efficient performance. For example, local learning used in learning classifiers from data, tries to employ a subset of input points around the separating hyperplane, while global learning tries to describe the overall phenomena utilizing all input points. Figure 2(a) illustrates the local learning. In this figure, the decision boundary is constructed only based on those filled points, while other points make no contributions to the classification plane (the decision planes are given based on the Gabriel Graph method [6, 7, 8], one of the local learning methods).

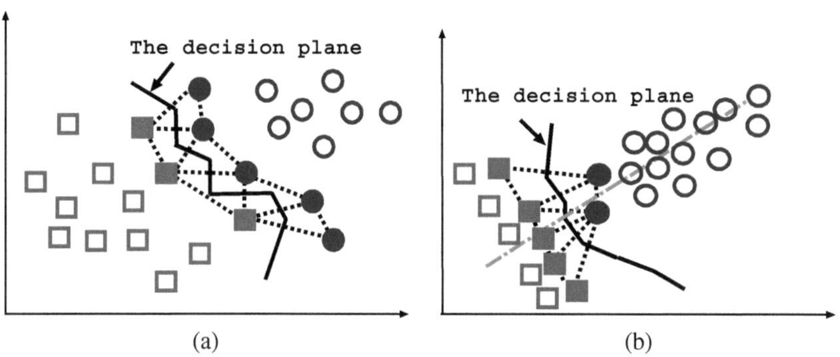

Fig. 2. (**a**) An illustration of local learning (also known as the Gabriel Graph classification). The decision boundary is just determined by some local points indicated as *filled points.* (**b**) An illustration on that local learning cannot grasp data trend

Although local learning appears to contain promising performance, it positions itself at the opposite extreme end to global learning. Employing only local information may lose the overall view of data. Local learning does not grasp the structure of the data, which may prove to be critical for guaranteeing better performance. This can be seen in the example as illustrated in Fig. 2(b). In this figure, the decision boundary (also constructed by the Gabriel Graph classification) is still determined by some local points indicated as filled points. Clearly, this boundary is myopic in nature and does not take into account the overall structure of the data. More specifically, the class associated with ○'s is obviously more likely to scatter than the class associated with □'s in the axis indicated as dashed line. Therefore, instead of simply locating itself in the middle of the filled points, a more promising decision boundary should lie closer to the filled □'s than the filled ○'s. A similar example can also be seen in Sect. 2 on a more principled local learning model, i.e., the current state-of-art-classifier, Support Vector Machines (SVM) [9]. Targeting at unifying this dichotomy, a hybrid learning is introduced in this chapter.

In summary, there are complementary advantages for both local learning and global learning. Global learning summarizes the data and provides the practitioners with knowledge on the structure of data, since with the precise modeling of phenomena, the observations can be accurately regenerated and therefore thoroughly studied or analyzed. However, this also presents

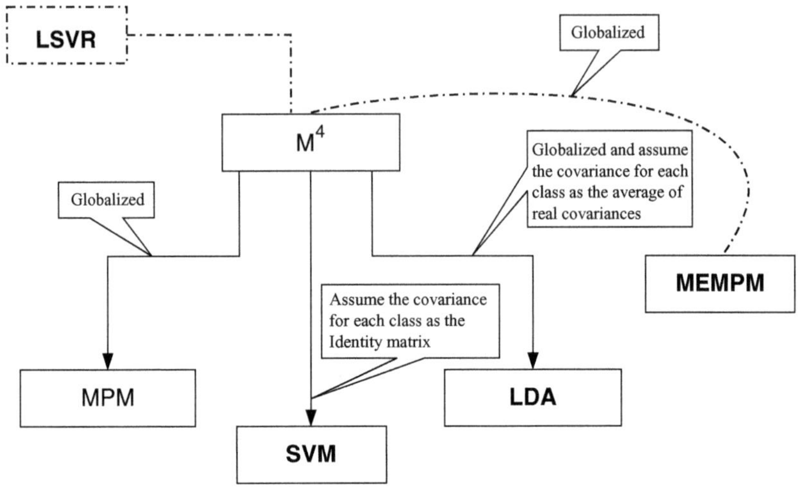

Fig. 3. The relationship between M^4 and other related models

difficulties in how to choose a valid model to describe all the information. In comparison, local learning directly employs part of information critical for the specifically oriented tasks and does not assume a model for the data. Although demonstrated to be superior to global learning in various machine learning tasks, it misses some critical global information. The question here is thus, can reliable global information, independent of specific model assumptions, be combined into local learning from data? This question clearly motivates the development of hybrid learning for which Maxi-Min Margin Machine (M^4) is proposed.

As will be shown later in this chapter, M^4 has built various connections with both global learning and local learning models. As an overview, Fig. 3 briefly illustrates the relationship between M^4 and other models. When it is globalized, M^4 can change into a global learning model, the Minimum Probability Machine (MPM) model [10]. When some assumptions are made on the covariance matrices of data, it becomes another global learning model, the Linear Discriminant Analysis (LDA) [11]. The Support Vector Machine, one of the local learning models, is also its special case when certain conditions are satisfied. Moreover, when compared with a recently-proposed general global learning model, the Minimum Error Minimax Probability Machine (MEMPM) [12], M^4 can derive a very similar special case. Furthermore, a novel regression model, the Local Support Vector Regression (LSVR) [13] can also be connected with M^4.

The rest of this chapter is organized as follows. In the next section, we review the background of both global learning and local learning. In particular, we will provide some historical perspectives on three models, i.e., the Linear Discriminant Analysis, the Minimax Probability Machine, and the Support

Vector Machine. In Sect. 3, we introduce M^4 in details including its model definition, the geometrical interpretation, and its links with other models. Finally, we present remarks to conclude this chapter.

2 Background

In this section, we first review the background of global learning, then followed by the local learning models with emphasis on the current state-of-the-art classifier SVM. We then motivate the hybrid learning model, the Maxi-Min Margin Machine.

2.1 Global Learning

Traditional global learning methods with specific assumptions, i.e., generative learning, may not always coincide with data. Within the context of global learning, researchers begin to investigate approaches with no distributional assumptions. Following this trend, there are non-parametric methods, LDA, and a recently-proposed competitive model MPM. We will review them one by one.

In contrast with generative learning, non-parametric learning does not assume any specific global models before learning. Therefore, no risk will be taken on possible wrong assumptions on the data. Consequently, non-parametric learning appears to set a more valid foundation than generative learning models. One typical non-parametric learning model in the context of classification is the so-called Parzen Window estimation [14].

The Parzen Window estimation also attempts to estimate a density for the observed data. However it employs a different way from generative learning. Parzen window first defines an n-dimensional cell hypercube region R_N over each observation. By defining a window function,

$$w(\mathbf{u}) = \begin{cases} 1 & |\mathbf{u}_j| \leq 1/2 \quad j = 1, 2, \ldots, n \\ 0 & \text{otherwise} , \end{cases} \tag{1}$$

the density is then estimated as

$$p_N(\mathbf{z}) = \frac{1}{N} \sum_{i=1}^{N} \frac{1}{h_N} w\left(\frac{\mathbf{z} - \mathbf{z}^i}{h_N}\right) , \tag{2}$$

where n is the data dimensionality, $\mathbf{z}_i$ for $1 \leq i \leq N$ represents N input data points, and h_N is defined as the length of the edge of R_N.

From the above, one can observe that Parzen Window puts a local density over each observation. The final density is then the statistical result of averaging all the local densities. In practice, the window function can actually be general functions including the most commonly-used Gaussian function.

These non-parametric methods make no underlying assumptions on data and appear to be more general in real cases. Using no parameters actually means using many "parameters" so that each parameter would not dominate other parameters (in the discussed models, the data points can be in fact considered as the "parameters"). In this way, if one parameter fails to work, it will not influence the whole system globally and statistically. However, using many "parameters" also results in serious problems. One of the main problems is that the density is overwhelmingly dependent on the training samples. Therefore, to generate an accurate density, the number of samples needs to be very large (much larger than would be required if we perform the estimation by generative learning approaches). Furthermore the number of data unfortunately increases exponentially with the dimension of data. Hence, it is usually hard to apply non-parametric learning in tasks with high-dimensional data. Another disadvantage caused is its severe requirement for the storage, since all the samples need to be saved beforehand in order to predict new data.

Instead of estimating an accurate distribution over data, an alternative approach is using some robust global information. The Linear Discriminant Analysis builds up a linear classifier by trying to minimize the intra-class distance while maximizing the inter-class distance. In this process, only up to the second order moments which are more robust with respect to the distribution, are adopted. Moreover, recent research has extended this linear classifier into nonlinear classification by using kernelization techniques [15]. In addition, a more recently-proposed model, the Minimax Probability Machine, goes further in this direction. Rather than constructing the decision boundary by estimating specific distributions, this approach exploits the worst-case distribution, which is distribution-free and more robust. With no assumptions on data, this model appears to be more valid in practice and is demonstrated to be competitive with the Support Vector Machine. Furthermore, Huang et al. [12] develop a superset of MPM, called Minimum Error Minimax Probability Machine, which achieves a worst-case distribution-free Bayes Optimal Classifier.

However, the problems for these models are that the robust estimation, e.g., the first and second order moments, may also be inaccurate. Considering specific data points, namely the local characteristic, seems to be necessary in this sense.

2.2 Local Learning

Local learning adopts a largely different way to construct classifiers. This type of learning is task-oriented. In the context of classification, only the final mapping function from the features $\mathbf{z}$ to the class variable c is crucial. Therefore, describing global information from data or explicitly summarizing a distribution, is an intermediate step. Hence the global learning scheme may be deemed wasteful or imprecise especially when the global information cannot be estimated accurately.

Alternatively, recent progress has suggested the local learning methodology. The family of approaches directly pin-points the most critical quantities for classification, while all other information less irrelevant to this purpose is simply omitted. Compared to global learning, this scheme assumes no model and also engages no explicit global information. Among this school of methods are neural networks [16, 17, 18, 19, 20, 21], Gabriel Graph methods [6, 7, 8], and large margin classifiers [5, 22, 23, 24] including Support Vector Machine, a state-of-the-art classifier which achieves superior performance in various pattern recognition tasks. In the following, we will focus on introducing SVM in details.

The Support Vector Machine is established based on minimizing the expected classification risk defined as follows:

$$\mathcal{R}(\boldsymbol{\Theta}) = \int_{\mathbf{z},\mathbf{c}} p(\mathbf{z}, c) l(\mathbf{z}, c, \boldsymbol{\Theta}) , \tag{3}$$

where $\boldsymbol{\Theta}$ represents the chosen model and the associate parameters, which are assumed to be the linear hyperplane in this chapter, and $l(\mathbf{z}, c, \boldsymbol{\Theta})$ is the loss function. Generally $p(\mathbf{z}, c)$ is unknown. Therefore, in practice, the above expected risk is often approximated by the so-called empirical risk:

$$\mathcal{R}_{\mathrm{emp}}(\boldsymbol{\Theta}) = \frac{1}{N} \sum_{j=1}^{N} l(\mathbf{z}^j, c^j, \boldsymbol{\Theta}) . \tag{4}$$

The above loss function describes the extent on how close the estimated class disagrees with the real class for the training data. Various metrics can be used for defining this loss function, including the 0–1 loss and the quadratic loss [25].

However, considering only the training data may lead to the over-fitting problem. In SVM, one big step in dealing with the over-fitting problem has been made, i.e., the margin between two classes should be pulled away in order to reduce the over-fitting risk. Figure 4 illustrates the idea of SVM. Two classes of data, depicted as circles and solid dots are presented in this figure. Intuitively observed, there are many decision hyperplanes, which can be adopted for separating these two classes of data. However, the one plotted in this figure is selected as the favorable separating plane, because it contains the maximum margin between the two classes. Therefore, in the objective function of SVM, a regularization term representing the margin shows up. Moreover, as seen in this figure, only those filled points, called *support vectors*, mainly determine the separating plane, while other points do not contribute to the margin at all. In other word, only several local points are critical for the classification purpose in the framework of SVM and thus should be extracted.

Actually, a more formal explanation and theoretical foundation can be obtained from the Structure Risk Minimization criterion [26, 27]. Therein,

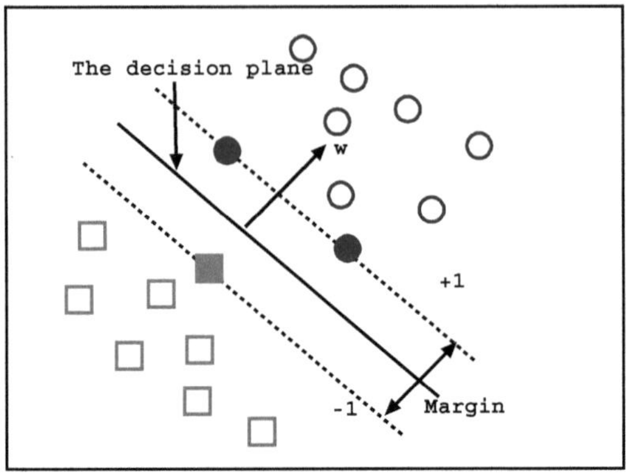

Fig. 4. An illustration of the Support Vector Machine

maximizing the margin between different classes of data is minimizing an upper bound of the expected risk, i.e., the VC dimension bound [27]. However, since the topic is out of the scope of this chapter, interested readers can refer to [9, 27].

2.3 Hybrid Learning

Local learning including SVM has demonstrated its advantages, such as its state-of-the-art performance (the lower generalization error), the optimal and unique solution, and the mathematical tractability [27]. However, it does discard many useful information from data, e.g., the structure information from data. An illustrative example has been seen in Fig. 2 in Sect. 1. In the current state-of-the-art classifier, i.e., SVM, similar problems also occur. This can be seen in Fig. 5. In this figure, the purpose is to separate two catergories of data $\mathbf{x}$ and $\mathbf{y}$, and the classification boundary is intuitively observed to be mainly determined by the dotted axis, i.e., the long axis of the $\mathbf{y}$ data (represented by □'s) or the short axis of the $\mathbf{x}$ data (represented by ○'s). Moreover, along this axis, the $\mathbf{y}$ data are more likely to scatter than the $\mathbf{x}$ data, since $\mathbf{y}$ contains a relatively larger variance in this direction. Noting this "global" fact, a good decision hyperplane seems reasonable to lie closer to the $\mathbf{x}$ side (see the dash-dot line). However, SVM ignores this kind of "global" information, i.e., the statistical trend of data occurrence: The derived SVM decision hyperplane (the solid line) lies unbiasedly right in the middle of two "local" points (the support vectors). The above considerations directly motivate the formulation of the Maxi-Min Margin Machine [28, 29].

3 Maxi-Min Margin Machine

In the following, we first present the scope and the notations. We then for the purpose of clarity, divide M^4 into *separable* and *nonseparable* categories,

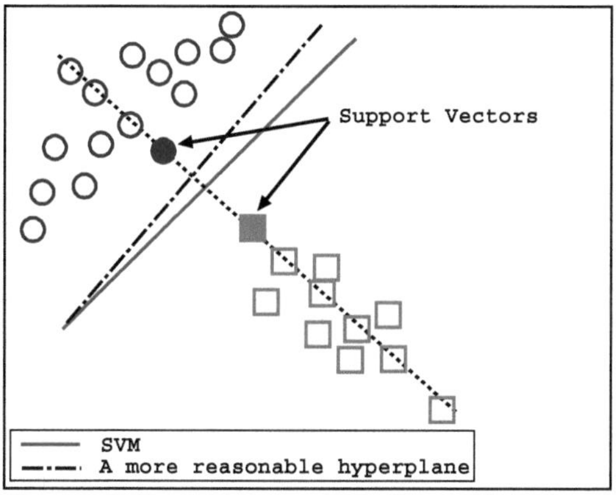

Fig. 5. A decision hyperplane with considerations of both local and global information

and introduce the corresponding hard-margin M^4 (linearly separable) and soft-margin M^4 (linearly non-separable) sequentially. Connections of the M^4 model with other models including SVM, MPM, LDA, and MEMPM will be provided in this section as well.

3.1 Scope and Notations

We only consider two-category classification tasks. Let a training data set contain two classes of samples, represented by $\mathbf{x}_i \in \mathbb{R}^n$ and $\mathbf{y}_j \in \mathbb{R}^n$ respectively, where $i = 1, 2, \ldots, N_\mathbf{x}$, $j = 1, 2, \ldots, N_\mathbf{y}$. The basic task here can be informally described as to find a suitable hyperplane $f(\mathbf{z}) = \mathbf{w}^T\mathbf{z} + b$ separating two classes of data as robustly as possible ($\mathbf{w} \in \mathbb{R}^n \backslash \{\mathbf{0}\}$, $b \in \mathbb{R}$, and $\mathbf{w}^T$ is the transpose of $\mathbf{w}$). Future data points $\mathbf{z}$ for which $f(\mathbf{z}) \geq 0$ are then classified as the class $\mathbf{x}$; otherwise, they are classified as the class $\mathbf{y}$. Throughout this chapter, unless we provide statements explicitly, bold typeface will indicate a vector or matrix, while normal typeface will refer to a scale variable or the component of the vectors.

3.2 Hard Margin Maxi-Min Margin Machine

Assuming the classification samples are separable, we first introduce the model definition and the geometrical interpretation. We then transform the model optimization problem into a sequential Second Order Cone Programming (SOCP) problem and discuss the optimization method.

The formulation for M^4 can be written as:

$$\max_{\rho,\mathbf{w}\neq\mathbf{0},b} \quad \rho \quad s.t. \tag{5}$$

$$\frac{(\mathbf{w}^T\mathbf{x}_i + b)}{\sqrt{\mathbf{w}^T\Sigma_{\mathbf{x}}\mathbf{w}}} \geq \rho, \qquad i = 1, 2, \ldots, N_{\mathbf{x}} \ , \tag{6}$$

$$\frac{-(\mathbf{w}^T\mathbf{y}_j + b)}{\sqrt{\mathbf{w}^T\Sigma_{\mathbf{y}}\mathbf{w}}} \geq \rho, \quad j = 1, 2, \ldots, N_{\mathbf{y}} \ , \tag{7}$$

where Σ_x and Σ_y refer to the covariance matrices of the $\mathbf{x}$ and the $\mathbf{y}$ data, respectively.[1]

This model tries to *maximize* the margin defined as the *minimum* Mahalanobis distance for all training samples,[2] while simultaneously classifying all the data correctly. Compared to SVM, M^4 incorporates the data information in a global way; namely, the covariance information of data or the statistical trend of data occurrence is considered, while SVMs, including l_1-SVM [30] and l_2-SVM [5, 9],[3] simply discard this information or consider the same covariance for each class. Although the above decision plane is presented in a linear form, it has been demonstrated that the standard kernelization trick can be used to extend into the nonlinear decision boundary [12, 29]. Since the focus of this chapter lies at the introduction of M^4, we simply omit the elaboration of the kernelization.

A geometrical interpretation of M^4 can be seen in Fig. 6. In this figure, the $\mathbf{x}$ data are represented by the inner ellipsoid on the left side with its center as $\mathbf{x}_0$, while the $\mathbf{y}$ data are represented by the inner ellipsoid on the right side with its center as $\mathbf{y}_0$. It is observed that these two ellipsoids contain unequal covariances or risks of data occurrence. However, SVM does not consider this global information: Its decision hyperplane (the dotted line) locates unbiasedly in the middle of two support vectors (filled points). In comparison, M^4 defines the margin as a Maxi-Min Mahalanobis distance, which constructs a decision plane (the solid line) with considerations of both the local and global information: the M^4 hyperplane corresponds to the tangent line of two dashed ellipsoids centered at the support vectors (the local information) and shaped by the corresponding covariances (the global information).

3.2.1 Optimization Method

According to [12, 29], the optimization problem for the M^4 model can be cast as a sequential conic programming problem, or more specifically, a sequential SOCP problem. The strategy is based on the "Divide and Conquer" technique. One may note that in the optimization problem of M^4, if ρ is fixed to a

---

[1] For simplicity, we assume Σ_x and Σ_y are always positive definite. In practice, this can be satisfied by adding a small positive amount into their diagonal elements, which is widely used.

[2] This also motivates the name of our model.

[3] l_p-SVM means the "p-norm' distance-based SVM.

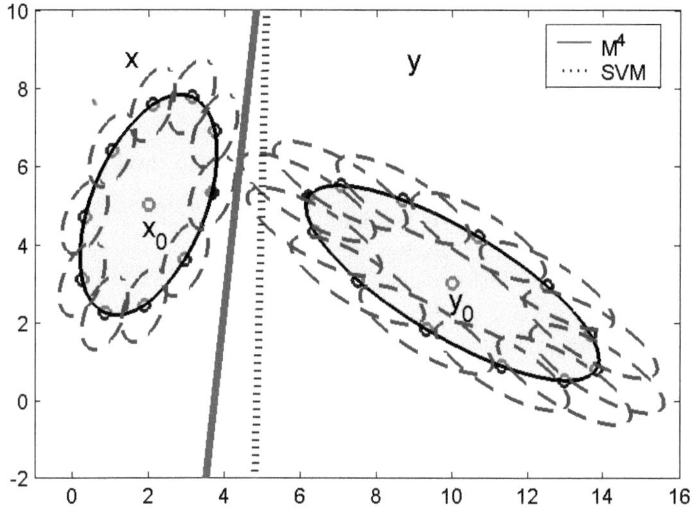

Fig. 6. A geometric interpretation of M^4. The M^4 hyperplane corresponds to the tangent line (*the solid line*) of two small dashed ellipsoids centered at the support vectors (the local information) and shaped by the corresponding covariances (the global information). It is thus more reasonable than the decision boundary calculated by SVM (*the dotted line*)

constant ρ_n, the problem is exactly changed to "conquer" the problem of checking whether the constraints of (6) and (7) can be satisfied. Moreover, as demonstrated by Theorem 1,[4] this "checking" procedure can be stated as an SOCP problem and can be solved in polynomial time. Thus the problem now becomes how ρ is set, which we can use "divide" to handle: If the constraints are satisfied, we can increase ρ_n accordingly; otherwise, we decrease ρ_n.

Theorem 1. *The problem of checking whether there exist* $\mathbf{w}$ *and* b *satisfying the following two sets of constraints (8) and (9) can be transformed as an SOCP problem, which can be solved in polynomial time,*

$$(\mathbf{w}^T\mathbf{x}_i + b) \geq \rho_n\sqrt{\mathbf{w}^T\Sigma_{\mathbf{x}}\mathbf{w}}, \quad i = 1, \ldots, N_{\mathbf{x}} \,, \tag{8}$$

$$-(\mathbf{w}^T\mathbf{y}_j + b) \geq \rho_n\sqrt{\mathbf{w}^T\Sigma_{\mathbf{y}}\mathbf{w}}, \; j = 1, \ldots, N_{\mathbf{y}} \,. \tag{9}$$

Algorithm 3.2.1 lists the detailed step of the optimization procedure, which is also illustrated in Fig. 7.

In Algorithm 3.2.1, if a ρ satisfies the constraints of (6) and (7), we call it a feasible ρ; otherwise, we call it an infeasible ρ.

In practice, many SOCP programs, e.g., Sedumi [31], provide schemes to directly handle the above checking procedure.

[4]Detailed proof can be seen in [12, 29].

Get	$\bar{\mathbf{x}}$, $\bar{\mathbf{y}}$, $\boldsymbol{\Sigma}_{\mathbf{x}}$, $\boldsymbol{\Sigma}_{\mathbf{y}}$;
Initialize	ϵ, ρ_0, $\rho_{\max}$, where ρ_0 is a feasible ρ, and $\rho_{\max}$ is an infeasible ρ with $\rho_0 \le \rho_{\max}$;

Repeat	$\rho_n = \frac{\rho_0+\rho_{\max}}{2}$;
	Call the checking procedure to check whether ρ_n is feasible;
	If ρ_n is feasible
	    $\rho_0 = \rho_n$
	Else
	    $\rho_{\max} = \rho_n$

until $\lvert\rho_0 - \rho_{\max}\rvert \le \epsilon$;	
Assign	$\rho = \rho_n$;

Algorithm 1: Optimization Algorithm of M^4

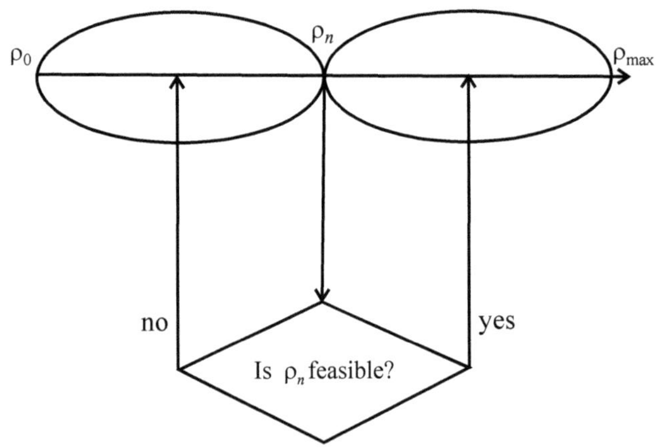

Fig. 7. A graph illustration on the optimization of M4

3.3 Time Complexity

We now analyze the time complexity of M^4. As indicated in [32], if the SOCP is solved based on interior-point methods, it contains a worst-case complexity of $O(n^3)$. If we denote the range of feasible ρ's as $L = \rho_{\max} - \rho_{\min}$ and the required precision as ε, then the number of iterations for M^4 is $\log(L/\varepsilon)$ in the worst case. Adding the cost of forming the system matrix (constraint matrix), which is $O(Nn^3)$ (N represents the number of training points), the total complexity would be $O(n^3 \log(L/\varepsilon) + Nn^3) \approx O(Nn^3)$, which is relatively large but can still be solved in polynomial time.[5]

[5]Note that the system matrix needs to be formed only once.

3.4 Soft Margin Maxi-Min Margin Machine

By introducing slack variables, the M^4 model can be extended to deal with nonseparable cases. This nonseparable version is written as follows:

$$\max_{\rho,\mathbf{w}\neq\mathbf{0},b,\boldsymbol{\xi}} \quad \rho - C\sum_{k=1}^{N_{\mathbf{x}}+N_{\mathbf{y}}} \boldsymbol{\xi}_k \quad s.t. \tag{10}$$

$$(\mathbf{w}^T\mathbf{x}_i + b) \geq \rho\sqrt{\mathbf{w}^T\Sigma_{\mathbf{x}}\mathbf{w}} - \boldsymbol{\xi}_i\ , \tag{11}$$

$$-(\mathbf{w}^T\mathbf{y}_j + b) \geq \rho\sqrt{\mathbf{w}^T\Sigma_{\mathbf{y}}\mathbf{w}} - \boldsymbol{\xi}_{j+N_{\mathbf{x}}}\ , \tag{12}$$

$$\xi_k \geq 0\ ,$$

where $i = 1, \ldots, N_{\mathbf{x}}$, $j = 1, \ldots, N_{\mathbf{y}}$, and $k = 1, \ldots, N_{\mathbf{x}} + N_{\mathbf{y}}$. C is the positive penalty parameter and $\boldsymbol{\xi_k}$ is the slack variable, which can be considered as the extent how the training point $\mathbf{z}_k$ disobeys the ρ margin ($\mathbf{z}_k = \mathbf{x}_k$ when $1 \leq k \leq N_{\mathbf{x}}$; $\mathbf{z}_k = \mathbf{y}_{k-N_{\mathbf{y}}}$ when $N_{\mathbf{x}} + 1 \leq k \leq N_{\mathbf{x}} + N_{\mathbf{y}}$). Thus $\sum_{k=1}^{N_{\mathbf{x}}+N_{\mathbf{y}}} \boldsymbol{\xi}_k$ can be conceptually regarded as the training error or the empirical error. In other words, the above optimization achieves maximizing the minimum margin while minimizing the total training error. The above optimization can further be solved based on a linear search problem [33] combined with the second order cone programming problem [12, 29].

3.5 A Unified Framework

In this section, connections between M^4 and other models are established. More specifically, SVM, MPM, and LDA are actually special cases of M^4 when certain assumptions are made on the data. M^4 therefore represents a unified framework of both global models, e.g., LDA and MPM, and a local model, i.e., SVM.

Corollary 1 M^4 *reduces to the Minimax Probability Machine, when it is "globalized".*

This can be easily seen by expanding and adding the constraints of (6) together. One can immediately obtain the following:

$$\mathbf{w}^T\sum_{i=1}^{N_{\mathbf{x}}}\mathbf{x}_i + N_{\mathbf{x}}b \geq N_{\mathbf{x}}\rho\sqrt{\mathbf{w}^T\Sigma_{\mathbf{x}}\mathbf{w}}\ ,$$

$$\Rightarrow \mathbf{w}^T\overline{\mathbf{x}} + b \geq \rho\sqrt{\mathbf{w}^T\Sigma_{\mathbf{x}}\mathbf{w}}\ , \tag{13}$$

where $\overline{\mathbf{x}}$ denotes the mean of the $\mathbf{x}$ training data.

Similarly, from (7) one can obtain:

$$
\begin{aligned}
& -\left(\mathbf{w}^T \sum_{j=1}^{N_{\mathbf{y}}} \mathbf{y}_j + N_{\mathbf{y}} b\right) \geq N_{\mathbf{y}} \rho \sqrt{\mathbf{w}^T \Sigma_{\mathbf{y}} \mathbf{w}} \,, \\
& \Rightarrow -(\mathbf{w}^T \overline{\mathbf{y}} + b) \geq \rho \sqrt{\mathbf{w}^T \Sigma_{\mathbf{y}} \mathbf{w}} \,,
\end{aligned}
\tag{14}
$$

where $\overline{\mathbf{y}}$ denotes the mean of the $\mathbf{y}$ training data.

Adding (13) and (14), one can obtain:

$$
\begin{aligned}
& \max_{\rho, \mathbf{w} \neq \mathbf{0}} \quad \rho \quad \text{s.t.} \\
& \mathbf{w}^T (\overline{\mathbf{x}} - \overline{\mathbf{y}}) \geq \rho \left(\sqrt{\mathbf{w}^T \Sigma_{\mathbf{x}} \mathbf{w}} + \sqrt{\mathbf{w}^T \Sigma_{\mathbf{y}} \mathbf{w}} \right) .
\end{aligned}
\tag{15}
$$

The above optimization is exactly the MPM optimization [10]. Note, however, that the above procedure is irreversible. This means the MPM is a special case of M^4. In MPM, since the decision is completely determined by the global information, i.e., the mean and covariance matrices [10], the estimates of mean and covariance matrices need to be reliable to assure an accurate performance. However, it cannot always be the case in real-world tasks. On the other hand, M^4 solves this problem in a natural way, because the impact caused by inaccurately estimated mean and covariance matrices can be neutralized by utilizing the local information, namely by satisfying those constraints of (6) and (7) for each local data point.

Corollary 2 M^4 *reduces to the Minimax Probability Machine, when* $\Sigma_{\mathbf{x}} = \Sigma_{\mathbf{y}} = \Sigma = \mathbf{I}$.

Intuitively, as two covariance matrices are assumed to be equal, the Mahalanobis distance changes to the Euclidean distance as used in standard SVM. The M^4 model will naturally reduce to the SVM model (refer to [12, 29] for a detailed proof). From the above, we can consider that two assumptions are implicitly made by SVM: One is the assumption on data "orientation" or data shape, i.e., $\Sigma_{\mathbf{x}} = \Sigma_{\mathbf{y}} = \Sigma$, and the other is the assumption on data "scattering magnitude" or data compactness, i.e., $\Sigma = \mathbf{I}$. However, these two assumptions are inappropriate. We demonstrate this in Fig. 8(a) and Fig. 8(b). We assume the orientation and the magnitude of each ellipsoid represent the data shape and compactness, respectively, in these figures.

Figure 8(a) plots two types of data with the same data orientations but different data scattering magnitudes. It is obvious that, by ignoring data scattering, SVM is improper to locate itself unbiasedly in the middle of the support vectors (filled points), since $\mathbf{x}$ is more possible to scatter in the horizontal axis. Instead, M^4 is more reasonable (see the solid line in this figure). Furthermore, Fig. 8(b) plots the case with the same data scattering magnitudes but different data orientations. Similarly, SVM does not capture the orientation information. In comparison, M^4 grasps this information and demonstrates a more suitable decision plane: M^4 represents the tangent line between two small

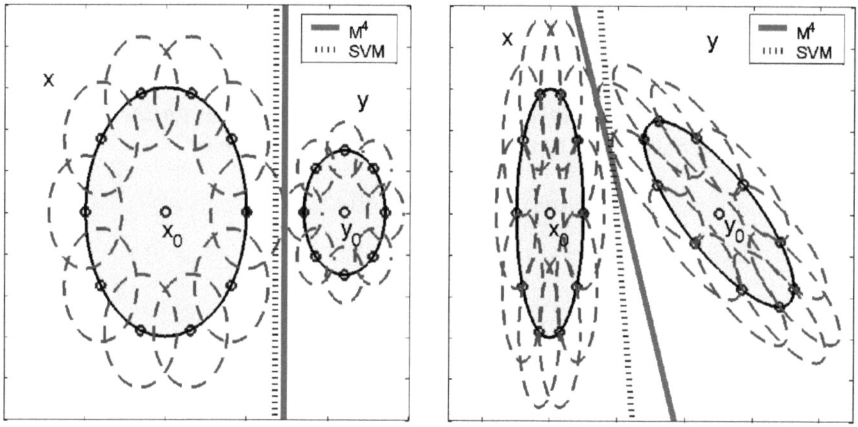

(a) (b)

Fig. 8. An illustration on the connections between SVM, and M^4. (**a**) demonstrates SVM omits the data compactness information. (**b**) demonstrates SVM discards the data orientation information

dashed ellipsoids centered at the support vectors (filled points). Note that SVM and M^4 do not need to generate the same support vectors. In Fig. 8(b), M^4 contains the above two filled points as support vectors, whereas SVM has all the three filled points as support vectors.

Corollary 3 M^4 *reduces to the LDA model, when it is "globalized" and assumes* $\Sigma_{\mathbf{x}} = \Sigma_{\mathbf{y}} = (\widehat{\Sigma_{\mathbf{x}}} + \widehat{\Sigma_{\mathbf{y}}})/2$*, where* $\widehat{\Sigma_{\mathbf{x}}}$ *and* $\widehat{\Sigma_{\mathbf{y}}}$ *are estimates of covariance matrices for the class* $\mathbf{x}$ *and* $\mathbf{y}$ *respectively.*

If we change the denominators in (6) and (7) as $\sqrt{\mathbf{w}^T\widehat{\Sigma_{\mathbf{x}}}\mathbf{w} + \mathbf{w}^T\widehat{\Sigma_{\mathbf{y}}}\mathbf{w}}$, the optimization can be changed as:

$$\max_{\rho,\mathbf{w}\neq\mathbf{0},b} \quad \rho \quad s.t. \tag{16}$$

$$\frac{(\mathbf{w}^T\mathbf{x}_i + b)}{\sqrt{\mathbf{w}^T\widehat{\Sigma_{\mathbf{x}}}\mathbf{w} + \mathbf{w}^T\widehat{\Sigma_{\mathbf{y}}}\mathbf{w}}} \geq \rho\,, \tag{17}$$

$$\frac{-(\mathbf{w}^T\mathbf{y}_j + b)}{\sqrt{\mathbf{w}^T\widehat{\Sigma_{\mathbf{x}}}\mathbf{w} + \mathbf{w}^T\widehat{\Sigma_{\mathbf{y}}}\mathbf{w}}} \geq \rho\,, \tag{18}$$

where $i = 1, \ldots, N_{\mathbf{x}}$ and $j = 1, \ldots, N_{\mathbf{y}}$. If the above two sets of constraints for $\mathbf{x}$ and $\mathbf{y}$ are "globalized" via a procedure similar to that in MPM, the above optimization problem is easily verified to be the following optimization:

$$\max_{\rho,\mathbf{w}\neq\mathbf{0},b} \quad \rho \quad s.t.$$

$$\mathbf{w}^T(\overline{\mathbf{x}} - \overline{\mathbf{y}}) \geq \rho\sqrt{\mathbf{w}^T\Sigma_{\mathbf{x}}\mathbf{w} + \mathbf{w}^T\Sigma_{\mathbf{y}}\mathbf{w}}\,. \tag{19}$$

Note that (19) can be changed as $\rho \leq \frac{|\mathbf{w}^T(\overline{\mathbf{x}}-\overline{\mathbf{y}})|}{\sqrt{\mathbf{w}^T\Sigma_{\mathbf{x}}\mathbf{w}+\mathbf{w}^T\Sigma_{\mathbf{y}}\mathbf{w}}}$, which is exactly the optimization of the LDA.

Corollary 4 *When a "globalized" procedure is performed on the soft margin version,* M[4] *reduces to a large margin classifier as follows:*

$$\max_{\mathbf{w}\neq\mathbf{0},b} \quad \theta t + (1-\theta)s \quad s.t. \tag{20}$$

$$\frac{\mathbf{w}^T\overline{\mathbf{x}}+b}{\sqrt{\mathbf{w}^T\Sigma_{\mathbf{x}}\mathbf{w}}} \geq t\ , \tag{21}$$

$$-\frac{\mathbf{w}^T\overline{\mathbf{y}}+b}{\sqrt{\mathbf{w}^T\Sigma_{\mathbf{y}}\mathbf{w}}} \geq s\ . \tag{22}$$

We can see that the above formula optimize a very similar form as the MEMPM model except that (20) changes to $\min_{\mathbf{w}\neq\mathbf{0},b} \theta\frac{t^2}{1+t^2}+(1-\theta)\frac{s^2}{1+s^2}$ [12]. In MEMPM, $\frac{t^2}{1+t^2}$ $(\frac{s^2}{1+s^2})$ (denoted as α (β)) represents the worst-case accuracy for the classification of future $\mathbf{x}$ ($\mathbf{y}$) data. Thus MEMPM maximizes the weighted accuracy on the future data. In M^4, s and t represent the corresponding margin, which is defined as the distance from the hyperplane to the class center. Therefore, it represents the weighted maximum margin machine in this sense. Moreover, since the conversion function of $g(u) = \frac{u^2}{1+u^2}$ increases monotonically with u, maximizing the above formulae contains a physical meaning similar to the optimization of MEMPM. For the proof, please refer to [12, 29].

3.5.1 Relationship with Local Support Vector Regression

A recently proposed promising model, the Local Support Vector Regression [13], can also be linked with M^4. In regression, the objective is to learn a model from a given data set, $\{(\mathbf{x}_1, y_1), \ldots, (\mathbf{x}_N, y_N)\}$, and then based on the learned model to make accurate predictions of y for future values of $\mathbf{x}$. The LSVR optimization is formulated as follows:

$$\min_{\mathbf{w},b,\xi_i,\xi_i^*} \frac{1}{N}\sum_{i=1}^{N}\sqrt{\mathbf{w}^T\Sigma_i\mathbf{w}} + C\sum_{i=1}^{N}(\xi_i+\xi_i^*)\ , \tag{23}$$

$$\begin{aligned} \text{s.t.}\ \ & y_i - (\mathbf{w}^T\mathbf{x}_i + b) \leq \epsilon\sqrt{\mathbf{w}^T\Sigma_i\mathbf{w}} + \xi_i\ , \\ & (\mathbf{w}^T\mathbf{x}_i + b) - y_i \leq \epsilon\sqrt{\mathbf{w}^T\Sigma_i\mathbf{w}} + \xi_i^*\ , \\ & \xi_i \geq 0, \quad \xi_i^* \geq 0, \quad i = 1, \ldots, N\ , \end{aligned} \tag{24}$$

where ξ_i and ξ_i^* are the corresponding up-side and the down-side errors at the i-th point, respectively. ϵ is a positive constant, which defines the margin width. Σ_i is the covariance matrix formed by the i-th data point and those data points close to it. In the state-of-the-art regression model, namely, the support vector regression [27, 34, 35, 36], the margin width ϵ is fixed. As a comparison in LSVR, this width is adapted automatically and locally with respect to the data volatility. More specifically, suppose $y_i = \mathbf{w}^T\mathbf{x}_i + b$ and $\bar{y}_i = \mathbf{w}^T\bar{\mathbf{x}}_\mathbf{i} + b$. The variance around the i-th data point is written as $\Delta_i = \frac{1}{2k+1}\sum_{j=-k}^{k}(y_{i+j} - \bar{y}_i)^2 = \frac{1}{2k+1}\sum_{j=-k}^{k}(\mathbf{w}^T(\mathbf{x}_{i+j} - \bar{\mathbf{x}}_i))^2 = \mathbf{w}^T\Sigma_i\mathbf{w}$, where $2k$ is the number of data points closest to the i-th data point. Therefore, $\Delta_i = \mathbf{w}^T\Sigma_i\mathbf{w}$ actually captures the volatility in the local region around the i-th data point. LSVR can systematically and automatically vary the tube: If the i-th data point lies in the area with a larger variance of noise, it will contribute to a larger $\epsilon\sqrt{\mathbf{w}^T\Sigma_i\mathbf{w}}$ or a larger local margin. This will result in reducing the impact of the noise around the point; on the other hand, in the case that the i-th data point is in the region with a smaller variance of noise, the local margin (tube), $\epsilon\sqrt{\mathbf{w}^T\Sigma_i\mathbf{w}}$, will be smaller. Therefore, the corresponding point would contribute more in the fitting process.

The LSVR model can be considered as an extension of M^4 into the regression task. Within the framework of classification, M^4 considers different data trends for different classes. Analogously, in the novel LSVR model, we allow different data trends for different regions, which is more suitable for the regression purpose.

4 Conclusion

We present a unifying theory of the Maxi-Min Margin Machine (M^4) that combines two schools of learning thoughts, i.e., local learning and global learning. This hybrid model is shown to subsume both global learning models, i.e., the Linear Discriminant Analysis and the Minimax Probability Machine, and a local learning model, the Support Vector Machine. Moreover, it can be linked with a worst-case distribution-free Bayes optimal classifier, the Minimum Error Minimax Probability Machine and a promising regression model, the Local Support Vector Regression. Historical perspectives, the geometrical interpretation, the detailed optimization algorithm, and various theoretical connections are provided to introduce this novel and promising framework.

Acknowledgements

The work described in this paper was fully supported by two grants from the Research Grants Council of the Hong Kong Special Administrative Region, China (Project No. CUHK4182/03E and Project No. CUHK4235/04E).

References

1. Grzegorzewski, P., Hryniewicz, O., and Gil, M. (2002). Soft methods in probability, statistics and data analysis, Physica-Verlag, Heidelberg; New York.

2. Duda, R. and Hart, P. (1973). Pattern classification and scene analysis: John Wiley & Sons.
3. Girosi, F. (1998). An equivalence between sparse approximation and support vector machines, *Neural Computation* **10(6)**, 1455–1480.
4. Schölkopf, B. and Smola, A. (2002). Learning with Kernels, MIT Press, Cambridge, MA.
5. Smola, A. J., Bartlett, P. L., Scholkopf, B., and Schuurmans, D. (2000). Advances in Large Margin Classifiers, The MIT Press.
6. Barber, C. B., Dobkin, D. P., and Huhanpaa, H. (1996). The Quickhull Algorithm for Convex Hulls, *ACM Transactions on Mathematical Software* **22(4)**, 469–483.
7. J. W. Jaromczyk, G. T. (1992). Relative Neighborhood Graphs And Their Relatives, *Proceedings IEEE* **80(9)**, 1502–1517.
8. Zhang, W. and King, I. (2002). A study of the relationship between support vector machine and Gabriel graph, in Proceedings of IEEE World Congress on Computational Intelligence – International Joint Conference on Neural Networks.
9. Vapnik, V. N. (1998). Statistical Learning Theory, John Wiley & Sons.
10. Lanckriet, G. R. G., Ghaoui, L. E., Bhattacharyya, C., and Jordan, M. I. (2002). A Robust Minimax Approach to Classification, *Journal of Machine Learning Research* **3**, 555–582.
11. Fukunaga, K. (1990). Introduction to Statistical Pattern Recognition, Academic Press, San Diego 2nd edition.
12. Huang, K., Yang, H., King, I., Lyu, M. R., and Chan, L. (2004). The Minimum Error Minimax Probability Machine, Journal of Machine Learning Research, 5:1253–1286, October 2004.
13. Huang, K., Yang, H., King, I., and Lyu, M. R. (2004). Varying the Tube: A Local Support Vector Regression Model, *Technique Report, Dept. of Computer Science and Engineering, The Chinese Univ. of Hong Kong.*
14. Duda, R. O., Hart, P. E., and Stork, D. G. (2000). Pattern Classification, John Wiley & Sons.
15. Mika, S., Ratsch, G., Weston, J., Scholkopf, B., and Muller, K.-R. (1999). Fisher discriminant analysis with kernels, *Neural Networks for Signal Processing IX* pp. 41–48.
16. Anand, R., Mehrotram, G. K., Mohan, K. C., and Ranka, S. (1993). An improved alogrithm for Neural Network Classification of Imbalance Training Sets, *IEEE Transactions on Neural Networks* **4(6)**, 962–969.
17. Fausett, L. (1994). Fundamentals of Neural Networks., New York: Prentice Hall.
18. Haykin, S. (1994). Neural Networks: A Comprehensive Foundation., New York: Macmillan Publishing.
19. Mehra, P. and Wah., B. W. (1992). Artificial neural networks: concepts and theory, Los Alamitos, California: IEEE Computer Society Press.
20. Patterson, D. (1996). Artificial Neural Networks., Singapore: Prentice Hall.
21. Ripley, B. (1996). Pattern Recognition and Neural Networks, Press Syndicate of the University of Cambridge.
22. Cristianini, N. and Shawe-Taylor, J. (2000). An Introduction to Support Vector Machines and Other Kernel-based Learning Methods, Cambridge University Press, Cambridge, U.K.; New York.
23. Schölkopf, B. Burges, C. and Smola, A. (ed.) (1999). Advances in Kernel Methods: Support Vector Learning, MIT Press, Cambridge, Massachusetts.

24. Schölkopf, B. and Smola, A. (ed.) (2002). Learning with kernels: support vector machines, regularization, optimization and beyond, MIT Press, Cambridge, Massachusetts.
25. Trivedi, P. K. (1978). Estimation of a Distributed Lag Model under Quadratic Loss, *Econometrica* **46(5)**, 1181–1192.
26. Burges, C. J. C. (1998). A Tutorial on Support Vector Machines for Pattern Recognition, *Data Mining and Knowledge Discovery* **2(2)**, 121–167.
27. Vapnik, V. N. (1999). The Nature of Statistical Learning Theory, Springer, New York 2nd edition.
28. Huang, K., Yang, H., King, I., and Lyu, M. R. (2004). Learning large margin classifiers locally and globally, in the Twenty-First International Conference on Machine Learning (ICML-2004): pp. 401–408.
29. Huang, K., Yang, H., King, I., and Lyu, M. R. (2004). Maxi-Min Margin Machine: Learning large margin classifiers globally and locally, Journal of Machine Learning, submitted.
30. Zhu, J., Rosset, S., Hastie, T., and Tibshirani, R. (2003). 1-norm Support Vector Machines, In Advances in Neural Information Processing Systems (NIPS 16).
31. Sturm, J. F. (1999). Using SeDuMi 1.02, a MATLAB toolbox for optimization over symmetric cones, *Optimization Methods and Software* **11**, 625–653.
32. Lobo, M., Vandenberghe, L., Boyd, S., and Lebret, H. (1998). Applications of second order cone programming, *Linear Algebra and its Applications* **284**, 193–228.
33. Bertsekas, D. P. (1999). Nonlinear Programming, Athena Scientific, Belmont, Massashusetts 2nd edition.
34. Drucker, H., Burges, C., Kaufman, L., Smola, A., and Vapnik, V. N. (1997). Support Vector Regression Machines, in Michael C. Mozer, Michael I. Jordan, and Thomas Petsche (ed.), Advances in Neural Information Processing Systems, volume **9**, The MIT Press pp. 155–161.
35. Gunn, S. (1998). Support vector machines for classification and regression Technical Report NC2-TR-1998-030 Faculty of Engineering and Applied Science, Department of Electronics and Computer Science, University of Southampton.
36. Smola, A. and Schölkopf, B. (1998). A tutorial on support vector regression Technical Report NC2-TR-1998-030 NeuroCOLT2.

Active-Set Methods for Support Vector Machines

M. Vogt[1] and V. Kecman[2]

[1] Darmstadt University of Technology, Institute of Automatic Control, Landgraf-Georg-Strasse 4, 64283 Darmstadt, Germany
mvogt@iat.tu-darmstadt.de
[2] University of Auckland, School of Engineering, Private Bag 92019 Auckland, New Zealand
v.kecman@auckland.ac.nz

Abstract. This chapter describes an active-set algorithm for quadratic programming problems that arise from the computation of support vector machines (SVMs). Currently, most SVM optimizers implement working-set (decomposition) techniques because of their ability to handle large data sets. Although these show good results in general, active-set methods are a reasonable alternative – in particular if the data set is not too large, if the problem is ill-conditioned, or if high precision is needed. Algorithms are derived for classification and regression with both fixed and variable bias term. The material is completed by acceleration and approximation techniques as well as a comparison with other optimization methods in application examples.

Key words: support vector machines, classification, regression, quadratic programming, active-set algorithm

1 Introduction

Support vector machines (SVMs) have become popular for classification and regression tasks [10, 11] since they can treat large input dimensions and show good generalization behavior. The method has its foundation in classification and has later been extended to regression. SVMs are computed by solving *quadratic programming* (QP) problems

$$\min_{\mathbf{a}} \quad J(\mathbf{a}) = \mathbf{a}^{\mathrm{T}}\mathbf{Q}\mathbf{a} + \mathbf{q}^{\mathrm{T}}\mathbf{a} \tag{1a}$$

$$\text{s.t.} \quad \mathbf{F}\mathbf{a} \geq \mathbf{f} \tag{1b}$$

$$\mathbf{G}\mathbf{a} = \mathbf{g} \tag{1c}$$

the sizes of which are dependent on the number N of training data. The settings for different SVM types will be derived in (10), (18), (29) and (37).

M. Vogt and V. Kecman: *Active-Set Methods for Support Vector Machines*, StudFuzz **177**, 133–158 (2005)
www.springerlink.com

1.1 Optimization Methods

The dependency on the size N of the training data set is the most critical issue of SVM optimization as N may be very large and the memory consumption is roughly $\mathcal{O}(N^2)$ if the whole QP problem (1) needs to be stored in memory. For that, the choice of an optimization method has to consider mainly the problem size and the memory consumption of the algorithm, see Fig. 1.

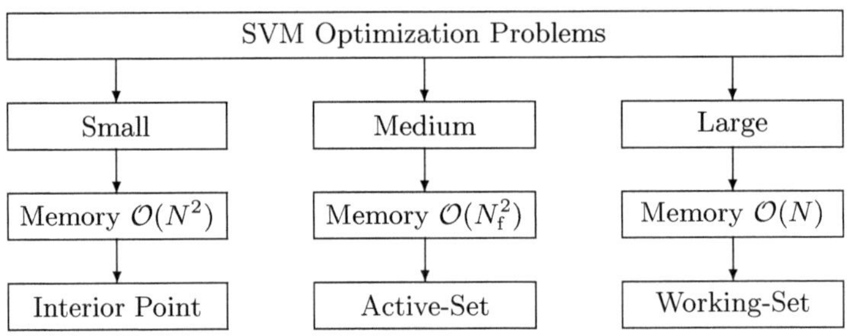

Fig. 1. QP optimization methods for different training data set sizes

If the problem is small enough to be stored completely in memory (on current PC hardware up to approximately 5000 data), *interior point* methods are suitable. They are known to be the most precise QP solvers [7, 10] but have a memory consumption of $\mathcal{O}(N^2)$. For very large data sets on the other hand, there is currently no alternative to *working-set* methods (*decomposition* methods) like SMO [8], ISDA [4] or similar strategies [1]. This class of methods has basically a memory consumption of $\mathcal{O}(N)$ and can therefore cope even with large scale problems. *Active-set* algorithms are appropriate for medium-size problems because they need $\mathcal{O}(N_\mathrm{f}^2 + N)$ memory where N_f is the number of *free* (unbounded) variables. Although N_f is typically much smaller than the number of the data, it dominates the memory consumption for large data sets due to its quadratic dependency.

Common SVM software packages rely on working-set methods because N is often large in practical applications. However, in some situations this is not the optimal approach, e.g., if the problem is ill-conditioned, if the SVM parameters (C and ε) are not chosen carefully, or if high precision is needed. This seems to apply in particular to regression, see Sect. 5. *Active-set* algorithms are the classical solvers for QP problems. They are known to be robust, but they are sometimes slower and (as stated above) require more memory than working-set algorithms. Their robustness is in particular useful for cross-validation techniques where the SVM parameters are varied over a wide range. Only few attempts have been made to utilize this technique for SVMs. E.g., in [5] it is applied to a modified SVM classification problem. An implementation for standard SVM classification can be found in [12], for regression problems in [13]. Also the *Chunking* algorithm [11] is closely related.

1.2 Active-Set Algorithms

The basic idea is to find the active set $\mathcal{A}$, i.e., those inequality constraints that are fulfilled with equality. If $\mathcal{A}$ is known, the Karush-Kuhn-Tucker (KKT) conditions reduce to a simple system of linear equations which yields the solution of the QP problem [7]. Because $\mathcal{A}$ is unknown in the beginning, it is constructed iteratively by adding and removing constraints and testing if the solution remains feasible.

The construction of $\mathcal{A}$ starts with an initial active set $\mathcal{A}^0$ containing the indices of the *bounded* variables (lying on the boundary of the feasible region) whereas those in $\mathcal{F}^0 = \{1, \ldots, N\} \backslash \mathcal{A}^0$ are *free* (lying in the interior of the feasible region). Then the following steps are performed repeatedly for $k = 1, 2, \ldots$:

A1. Solve the KKT system for all variables in $\mathcal{F}^k$.
A2. If the solution is feasible, find the variable in $\mathcal{A}^k$ that violates the KKT conditions most, move it to $\mathcal{F}^k$, then go to A1.
A3. Otherwise find an intermediate value between old and new solution lying on the border of the feasible region, move one bounded variable from $\mathcal{F}^k$ to $\mathcal{A}^k$, then go to A1.

The intermediate solution in step A3 is computed as $\mathbf{a}^k = \eta\bar{\mathbf{a}}^k + (1-\eta)\mathbf{a}^{k-1}$ with maximal $\eta \in [0, 1]$ (*affine scaling*), where $\bar{\mathbf{a}}^k$ is the solution of the linear system in step A1. I.e., the new iterate $\mathbf{a}^k$ lies on the connecting line of $\mathbf{a}^{k-1}$ and $\bar{\mathbf{a}}^k$, see Fig. 2. The optimum is found if during step A2 no violating variable is left in $\mathcal{A}^k$.

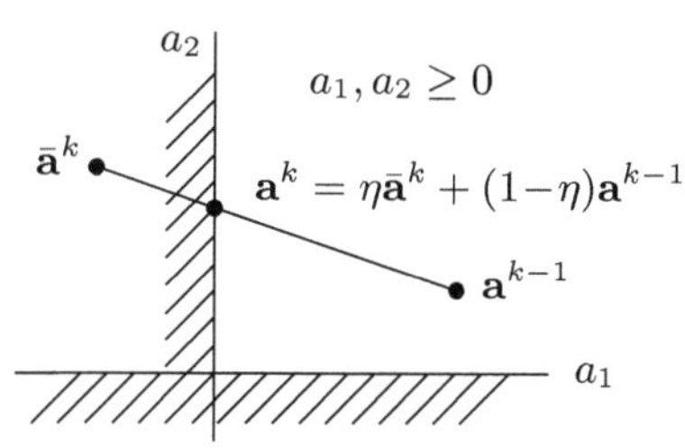

Fig. 2. Affine scaling of the non-feasible solution

This basic algorithm is used for all cases described in the next sections, only the structure of the KKT system in step A1 and the conditions in step A2 are different. Sections 2 and 3 describe how to use the algorithm for SVM classification and regression tasks. In this context the derivation of the dual problems is repeated in order to introduce the distinction between fixed and variable bias term. Section 4 considers the efficient solution of the KKT system, several acceleration techniques and the approximation of the solution with a limited number of support vectors. Application examples for both classification and regression are given in Sect. 5.

2 Support Vector Machine Classification

A two-class classification problem is given by the data set $\{(\mathbf{x}_i, y_i)\}_{i=1}^N$ with the class labels $y_i \in \{-1, 1\}$. Linear classifiers aim to find a decision function $f(\mathbf{x}) = \mathbf{w}^\mathrm{T}\mathbf{x} + b$ so that $f(\mathbf{x}_i) > 0$ for $y_i = 1$ and $f(\mathbf{x}_i) < 0$ for $y_i = -1$. The decision boundary is the intersection of $f(\mathbf{x})$ and the input space, see Fig. 3.

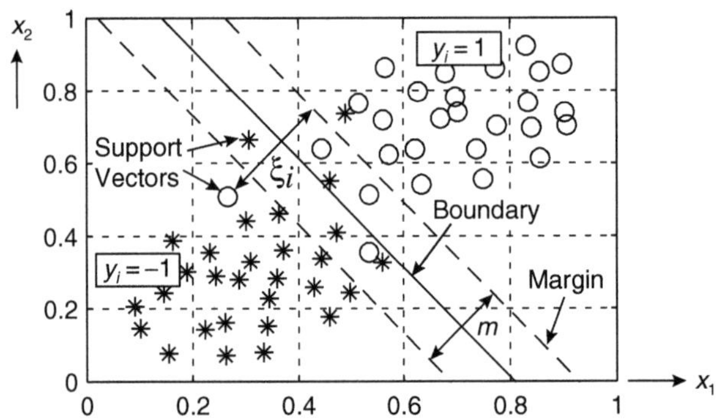

Fig. 3. Separating two overlapping classes with a linear decision function

For separable classes, a SVM classifier computes a decision function having a *maximal margin* m with respect to the two classes, so that all data lie outside the margin, i.e., $y_i f(\mathbf{x}_i) \geq 1$. Since $\mathbf{w}$ is the normal vector of the separating hyperplane in its canonical form [10, 11], the margin can be expressed as $m = 2/\mathbf{w}^\mathrm{T}\mathbf{w}$. In the case of non-separable classes, *slack variables* ξ_i are introduced measuring the distance to the data lying on the wrong side of the margin. They do not only make the constraints feasible but are also penalized by a factor C in the loss function to keep the deviations small [11]. These ideas lead to the *soft margin* classifier:

$$\min_{\mathbf{w},\boldsymbol{\xi}} \quad J_\mathrm{p}(\mathbf{w}, \boldsymbol{\xi}) = \frac{1}{2}\mathbf{w}^\mathrm{T}\mathbf{w} + C\sum_{i=1}^{N}\xi_i \tag{2a}$$

$$\text{s.t.} \quad y_i(\mathbf{w}^\mathrm{T}\mathbf{x}_i + b) \geq 1 - \xi_i \tag{2b}$$

$$\xi_i \geq 0, \quad i = 1, \ldots, N \tag{2c}$$

The parameter C describes the trade-off between maximal margin and correct classification. The *primal* problem (2) is now transformed into its *dual* one by introducing the *Lagrange multipliers* $\boldsymbol{\alpha}$ and $\boldsymbol{\beta}$ of the $2N$ primal constraints. The Lagrangian is given by

$$L_{\mathrm{p}}(\mathbf{w},\boldsymbol{\xi},b,\boldsymbol{\alpha},\boldsymbol{\beta})=\frac{1}{2}\mathbf{w}^{\mathrm{T}}\mathbf{w}+C\sum_{i=1}^{N}\xi_i-\sum_{i=1}^{N}\alpha_i[y_i(\mathbf{w}^{\mathrm{T}}\mathbf{x}_i+b)-1+\xi_i]-\sum_{i=1}^{N}\beta_i\xi_i \quad (3)$$

having a minimum with respect to the primal variables $\mathbf{w}$, $\boldsymbol{\xi}$ and b, and a maximum with respect to the dual variables $\boldsymbol{\alpha}$ and $\boldsymbol{\beta}$ (*saddle point condition*). According to the KKT condition (48a) the minimization is performed with respect to the primal variables in order to find the optimum:

$$\frac{\partial L_{\mathrm{p}}}{\partial \mathbf{w}}=\mathbf{0} \;\Rightarrow\; \mathbf{w}=\sum_{i=1}^{N}y_i\alpha_i\mathbf{x}_i \quad (4a)$$

$$\frac{\partial L_{\mathrm{p}}}{\partial \xi_i}=0 \;\Rightarrow\; \alpha_i+\beta_i=C, \qquad i=1,\ldots,N \quad (4b)$$

Although b is also a primal variable, we defer the minimization with respect to b for a moment. Instead, (4) is used to eliminate $\mathbf{w}$, $\boldsymbol{\xi}$ and $\boldsymbol{\beta}$ from the Lagrangian which leads to

$$L_{\mathrm{p}}^{*}(\boldsymbol{\alpha},b)=-\frac{1}{2}\sum_{i=1}^{N}\sum_{j=1}^{N}y_iy_j\alpha_i\alpha_j\mathbf{x}_i^{\mathrm{T}}\mathbf{x}_j+\sum_{i=1}^{N}\alpha_i-b\sum_{i=1}^{N}y_i\alpha_i\,. \quad (5)$$

To solve *nonlinear* classification problems, the linear SVM is applied to *features* $\boldsymbol{\Phi}(\mathbf{x})$ (instead of the inputs $\mathbf{x}$), where $\boldsymbol{\Phi}$ is a given feature map (see Fig. 4). Since $\mathbf{x}$ occurs in (5) only in scalar products $\mathbf{x}_i^{\mathrm{T}}\mathbf{x}_j$, we define the *kernel function*

$$K(\mathbf{x},\mathbf{x}')=\boldsymbol{\Phi}^{\mathrm{T}}(\mathbf{x})\boldsymbol{\Phi}(\mathbf{x}')\,, \quad (6)$$

and finally (5) becomes

$$L_{\mathrm{p}}^{*}(\boldsymbol{\alpha},b)=-\frac{1}{2}\sum_{i=1}^{N}\sum_{j=1}^{N}y_iy_j\alpha_i\alpha_jK_{ij}+\sum_{i=1}^{N}\alpha_i-b\sum_{i=1}^{N}y_i\alpha_i \quad (7)$$

with the abbreviation $K_{ij}=K(\mathbf{x}_i,\mathbf{x}_j)$. In the following, kernels are always assumed to be symmetric and positive definite. This class of functions includes most of the common kernels [10], e.g.

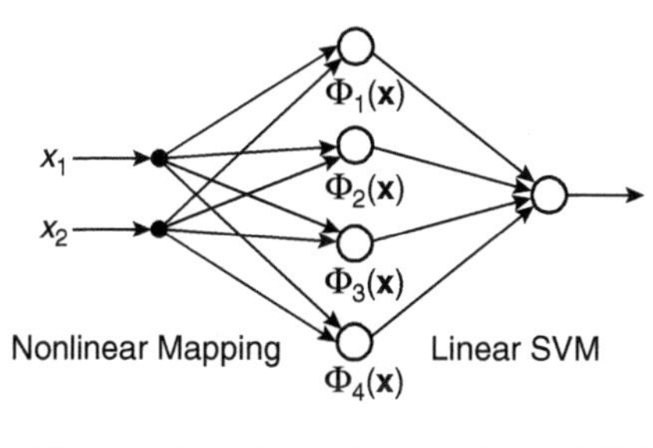

Fig. 4. Structure of a nonlinear SVM

Linear kernel (scalar product):	$K(\mathbf{x},\mathbf{x}') = \mathbf{x}^{\mathrm{T}}\mathbf{x}'$
Inhomogeneous polynomial kernel:	$K(\mathbf{x},\mathbf{x}') = (\mathbf{x}^{\mathrm{T}}\mathbf{x}' + c)^p$
Gaussian (RBF) kernel:	$K(\mathbf{x},\mathbf{x}') = \exp(-\frac{1}{2}\|\mathbf{x}-\mathbf{x}'\|^2/\sigma^2)$
Sigmoidal (MLP) kernel:	$K(\mathbf{x},\mathbf{x}') = \tanh(\mathbf{x}^{\mathrm{T}}\mathbf{x}' + d)$

The conditions (48d) and (4b) yield additional restrictions for the dual variables

$$0 \le \alpha_i \le C\,, \quad i = 1,\ldots,N \tag{8}$$

and from (4a) and (6) follows that

$$f(\mathbf{x}) = \mathbf{w}^{\mathrm{T}}\boldsymbol{\Phi}(\mathbf{x}) + b = \sum_{\alpha_i \neq 0} y_i\alpha_i K(\mathbf{x}_i,\mathbf{x}) + b\,. \tag{9}$$

This shows the strengths of the kernel concept: SVMs can easily handle extremely large feature spaces since the primal variables $\mathbf{w}$ and the feature map $\boldsymbol{\Phi}$ are needed neither for the optimization nor in the decision function. Vectors $\mathbf{x}_i$ with $\alpha_i \neq 0$ are called *support vectors*. Usually only a small fraction of the data set are support vectors, typically about 10%. In Fig. 3, these are the data points lying on the margin ($\xi_i = 0$ and $0 < \alpha_i < C$) or on the "wrong" side of the margin ($\xi_i > 0$ and $\alpha_i = C$).

From the algorithmic point of view, an important decision has to be made at this stage: weather the bias term b is treated as a variable or kept fixed during optimization. The next two sections derive active-set algorithms for both cases.

2.1 Classification with Fixed Bias Term

We first consider the bias term b to be fixed, including the most important case $b = 0$. This is possible if the kernel function provides an implicit bias, e.g., in the case of *positive definite* kernel functions [4, 9, 14]. The only effect is that slightly more support vectors are computed. The main advantage of a fixed bias term is a simpler algorithm since no additional equality constraint needs to be imposed during optimization (like below in (18)):

$$\min_{\boldsymbol{\alpha}} \quad J_{\mathrm{d}}(\boldsymbol{\alpha}) = \frac{1}{2}\sum_{i=1}^{N}\sum_{j=1}^{N} y_i y_j \alpha_i \alpha_j K_{ij} - \sum_{i=1}^{N}\alpha_i + b\sum_{i=1}^{N} y_i\alpha_i \tag{10a}$$

$$\text{s.t.} \quad 0 \le \alpha_i \le C\,, \quad i = 1,\ldots,N \tag{10b}$$

Note that $J_{\mathrm{d}}(\boldsymbol{\alpha})$ equals $-L_{\mathrm{p}}^*(\boldsymbol{\alpha}, b)$ with a given b. For $b = 0$ (the "no-bias SVM") the last term of the objective function (10a) vanishes. The reason for the change of the sign in the objective function is that optimization algorithms usually assume *minimization* rather than *maximization* problems.

If b is kept fixed, the SVM is computed by solving the box-constrained convex QP problem (10), which is one of the most convenient QP cases. To solve it with the active-set method described in Sect. 1, the KKT conditions of this problem must be found. Its Lagrangian is

$$\begin{aligned} L_{\mathrm{d}}(\boldsymbol{\alpha}, \boldsymbol{\lambda}, \boldsymbol{\mu}) = & \frac{1}{2}\sum_{i=1}^{N}\sum_{j=1}^{N} y_i y_j \alpha_i \alpha_j K_{ij} - \sum_{i=1}^{N}\alpha_i + b\sum_{i=1}^{N} y_i \alpha_i \\ & - \sum_{i=1}^{N}\lambda_i \alpha_i - \sum_{i=1}^{N}\mu_i (C - \alpha_i) \end{aligned} \tag{11}$$

where λ_i and μ_i are the Lagrange multipliers of the constraints $\alpha_i \geq 0$ and $\alpha_i \leq C$, respectively. Introducing the prediction errors $E_i = f(\mathbf{x}_i) - y_i$, the KKT conditions can be derived for $i = 1, \ldots, N$ (see App. A):

$$\frac{\partial L_{\mathrm{d}}}{\partial \alpha_i} = y_i E_i - \lambda_i + \mu_i = 0 \tag{12a}$$

$$0 \leq \alpha_i \leq C \tag{12b}$$

$$\lambda_i \geq 0, \quad \mu_i \geq 0 \tag{12c}$$

$$\alpha_i \lambda_i = 0, \quad (C - \alpha_i)\mu_i = 0 \tag{12d}$$

According to α_i, three cases have to be considered:

$$\boxed{0 < \alpha_i < C} \quad (i \in \mathcal{F}) \Rightarrow \lambda_i = \mu_i = 0 \\ \Rightarrow \sum_{j \in \mathcal{F}} y_j \alpha_j K_{ij} = y_i - \sum_{j \in \mathcal{A}_C} y_j \alpha_j K_{ij} - b \tag{13a}$$

$$\boxed{\alpha_i = 0} \quad (i \in \mathcal{A}_0) \quad \Rightarrow \lambda_i = y_i E_i > 0 \\ \Rightarrow \mu_i = 0 \tag{13b}$$

$$\boxed{\alpha_i = C} \quad (i \in \mathcal{A}_C) \quad \Rightarrow \lambda_i = 0 \\ \Rightarrow \mu_i = -y_i E_i > 0 \tag{13c}$$

$\mathcal{A}_0$ denotes the set of lower bounded variables $\alpha_i = 0$, whereas $\mathcal{A}_C$ comprises the upper bounded ones with $\alpha_i = C$. The above conditions are exploited in each iteration step k. Case (13a) establishes the linear system in step A1 for the currently free variables $i \in \mathcal{F}^k$. Cases (13b) and (13c) are the conditions that must be fulfilled for the variables in $\mathcal{A}^k = \mathcal{A}_0^k \cup \mathcal{A}_C^k$ in the optimum, i.e., step A2 of the algorithm searches for the worst violator among these variables. Note that $\mathcal{A}_0^k \cap \mathcal{A}_C^k = \emptyset$ because $\alpha_i^k = 0$ and $\alpha_i^k = C$ cannot be met simultaneously. The variables $\alpha_i = C$ in $\mathcal{A}_C^k$ also occur on the right hand side of the linear system (13a).

The implementation uses the *coefficients* $a_i = y_i \alpha_i$ instead of the Lagrange multipliers α_i. This is done to keep the same formulation for the regression algorithm in Sect. 3, and because it slightly accelerates the computation. With this modification, in step A1 the linear system

$$\mathbf{H}^k \bar{\mathbf{a}}^k = \mathbf{c}^k \tag{14}$$

with

$$\left.\begin{aligned} \bar{a}_i^k &= y_i \bar{\alpha}_i^k \\ h_{ij}^k &= K_{ij} \\ c_i^k &= y_i - \sum_{j \in \mathcal{A}_C^k} a_j^k K_{ij} - b \end{aligned}\right\} \quad \text{for } i, j \in \mathcal{F}^k \tag{15}$$

has to be solved. $\mathbf{H}^k$ is called *reduced* or *projected* Hessian. In the case of box constraints, it results from the complete Hessian $\mathbf{Q}$ in (1) by dropping all rows and columns belonging to constraints that are currently regarded as active. If $\mathcal{F}^k$ contains p free variables, then $\mathbf{H}^k$ is a $p \times p$ matrix. It is positive definite since positive definite kernels are assumed for all algorithms. For that, (14) can be solved by the methods described in Sect. 4. Step A2 computes

$$\lambda_i^k = +y_i E_i^k \quad \text{for } i \in \mathcal{A}_0^k \tag{16a}$$

$$\mu_i^k = -y_i E_i^k \quad \text{for } i \in \mathcal{A}_C^k \tag{16b}$$

and checks if they are positive, i.e., if the KKT conditions are valid for $i \in \mathcal{A}^k = \mathcal{A}_0^k \cup \mathcal{A}_C^k$. Among the negative multipliers, the most negative one is selected and moved to $\mathcal{F}^k$. In practice, the KKT conditions are checked with precision τ, so that a variable α_i is accepted as optimal if $\lambda_i^k > -\tau$ and $\mu_i^k > -\tau$.

2.2 Classification with Variable Bias Term

Most SVM algorithms do not keep the bias term fixed but compute it during optimization. In that case b is a primal variable, and the Lagrangian (3) can be minimized with respect to it:

$$\frac{\partial L_\mathrm{p}}{\partial b} = 0 \Rightarrow \sum_{i=1}^{N} y_i \alpha_i = 0 \tag{17}$$

On the one hand (17) removes the last term from (5), on the other hand it is an additional constraint that must be considered in the optimization problem:

$$\min_{\boldsymbol{\alpha}} \quad J_\mathrm{d}(\boldsymbol{\alpha}) = \frac{1}{2} \sum_{i=1}^{N} \sum_{j=1}^{N} y_i y_j \alpha_i \alpha_j K_{ij} - \sum_{i=1}^{N} \alpha_i \tag{18a}$$

$$\text{s.t.} \quad 0 \le \alpha_i \le C\,, \quad i = 1, \ldots, N \tag{18b}$$

$$\sum_{i=1}^{N} y_i \alpha_i = 0 \tag{18c}$$

This modification changes the Lagrangian (11) to

$$\begin{aligned} L_{\mathrm{d}}(\boldsymbol{\alpha}, \boldsymbol{\lambda}, \boldsymbol{\mu}, \nu) = & \frac{1}{2}\sum_{i=1}^{N}\sum_{j=1}^{N} y_i y_j \alpha_i \alpha_j K_{ij} - \sum_{i=1}^{N} \alpha_i \\ & - \sum_{i=1}^{N} \lambda_i \alpha_i - \sum_{i=1}^{N} \mu_i (C - \alpha_i) - \nu \sum_{i=1}^{N} y_i \alpha_i \end{aligned} \tag{19}$$

and its derivatives to

$$\frac{\partial L_{\mathrm{d}}}{\partial \alpha_i} = y_i \sum_{j=1}^{N} y_j \alpha_j K_{ij} - 1 - \lambda_i + \mu_i - \nu y_i = 0\,, \quad i = 1, \ldots, N \tag{20}$$

where ν is the Lagrange multiplier of the equality constraint (18c). It can be easily seen that $\nu = -b$, i.e., L_{d} is the same as (11) with the important difference that b is not fixed any more. With the additional equality constraint (18c) and again with $a_i = y_i \alpha_i$ the linear system becomes

$$\begin{pmatrix} \mathbf{H}^k & \mathbf{e} \\ \mathbf{e}^{\mathrm{T}} & 0 \end{pmatrix} \begin{pmatrix} \bar{\mathbf{a}}^k \\ b^k \end{pmatrix} = \begin{pmatrix} \mathbf{c}^k \\ d^k \end{pmatrix} \begin{matrix} \} \ p \text{ rows} \\ \} \ 1 \text{ row} \end{matrix} \tag{21}$$

with

$$d^k = - \sum_{j \in \mathcal{A}_C^k} a_j^k \quad \text{and} \quad \mathbf{e} = (1, \ldots, 1)^{\mathrm{T}}\,. \tag{22}$$

One possibility to solve this indefinite system is to use factorization methods for indefinite matrices like the Bunch-Parlett decomposition [3]. But since we retain the assumption that $K(\mathbf{x}_i, \mathbf{x}_j)$ is positive definite, the Cholesky decomposition $\mathbf{H} = \mathbf{R}^{\mathrm{T}}\mathbf{R}$ is available (see Sect. 4), and the system (21) can be solved by exploiting its block structure. For that, a Gauss transformation is applied to the blocks of the matrix, i.e., the first block row is multiplied by $(\mathbf{u}^k)^{\mathrm{T}} := \mathbf{e}^{\mathrm{T}}(\mathbf{H}^k)^{-1}$. Subtracting the second row yields

$$(\mathbf{u}^k)^{\mathrm{T}} \mathbf{e} b^k = (\mathbf{u}^k)^{\mathrm{T}} \mathbf{c}^k - d^k\,. \tag{23}$$

Since this is a scalar equation, it is simply divided by $(\mathbf{u}^k)^{\mathrm{T}}\mathbf{e}$ in order to find b^k. This technique is effective here because only *one* additional row/column has been appended to $\mathbf{H}^k$. The complete solution of the block system is done by the following procedure:

$$\begin{aligned} &\textit{Solve} \quad (\mathbf{R}^k)^{\mathrm{T}} \mathbf{R}^k \mathbf{u}^k = \mathbf{e} \quad \textit{for} \quad \mathbf{u}^k\,. \\ &\textit{Compute} \quad b^k = -\left(\sum_{j \in \mathcal{A}_C^k} a_j^k + \sum_{j \in \mathcal{A}_C^k} u_j^k c_j^k \right) \Big/ \sum_{j \in \mathcal{A}_C^k} u_j^k\,. \\ &\textit{Solve} \quad (\mathbf{R}^k)^{\mathrm{T}} \mathbf{R}^k \bar{\mathbf{a}}^k = \mathbf{c}^k - \mathbf{e} b^k \quad \textit{for} \quad \bar{\mathbf{a}}^k\,. \end{aligned}$$

The computation of λ_i^k and μ_i^k remains the same as in (16) for fixed bias term.

An additional topic has to be considered here: For a variable bias term, the *Linear Independence Constraint Qualification (LICQ)* [7] is violated when for each α_i one inequality constraint is active, e.g., when the algorithm is initialized with $\alpha_i = 0$ for $i = 1, \ldots, N$. Then the gradients of the active inequality constraints and the equality constraint are linear dependent. The algorithm uses *Bland's rule* to avoid *cycling* in these cases.

3 Support Vector Machine Regression

Like in classification, we start from the linear regression problem. The goal is to fit a linear function $f(\mathbf{x}) = \mathbf{w}^{\mathrm{T}}\mathbf{x} + b$ to a given data set $\{(\mathbf{x}_i, y_i)\}_{i=1}^N$. Whereas most other learning methods minimize the sum of squared errors, SVMs try to find a maximal *flat* function, so that all data lie within an *insensitivity zone* of size ε around the function. Outliers are treated by *two* sets of slack variables ξ_i and ξ_i^* measuring the distance above and below the insensitivity zone, respectively, see Fig. 5 (for a nonlinear example) and [10]. This concept results in the following primal problem:

$$\min_{\mathbf{w},\boldsymbol{\xi},\boldsymbol{\xi}^*} \quad J_{\mathrm{p}}(\mathbf{w},\boldsymbol{\xi},\boldsymbol{\xi}^*) = \frac{1}{2}\mathbf{w}^{\mathrm{T}}\mathbf{w} + C\sum_{i=1}^{N}(\xi_i + \xi_i^*) \tag{24a}$$

$$\text{s.t.} \quad y_i - \mathbf{w}^{\mathrm{T}}\mathbf{x}_i - b \leq \varepsilon + \xi_i \tag{24b}$$

$$\mathbf{w}^{\mathrm{T}}\mathbf{x}_i + b - y_i \leq \varepsilon + \xi_i^* \tag{24c}$$

$$\xi_i, \xi_i^* \geq 0\,, \quad i = 1, \ldots, N \tag{24d}$$

To apply the same technique as for classification, the Lagrangian

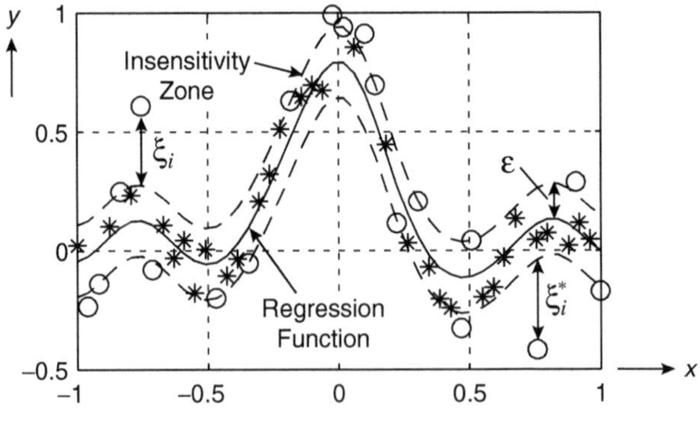

Fig. 5. Nonlinear support vector machine regression

$$L_{\mathrm{p}}(\mathbf{w},b,\boldsymbol{\xi},\boldsymbol{\xi}^*,\boldsymbol{\alpha},\boldsymbol{\alpha}^*,\boldsymbol{\beta},\boldsymbol{\beta}^*) = \frac{1}{2}\mathbf{w}^{\mathrm{T}}\mathbf{w} + C\sum_{i=1}^{N}(\xi_i+\xi_i^*) - \sum_{i=1}^{N}(\beta_i\xi_i + \beta_i^*\xi_i^*)$$
$$- \sum_{i=1}^{N}\alpha_i(\varepsilon + \xi_i - y_i + \mathbf{w}^{\mathrm{T}}\mathbf{x}_i + b) \tag{25}$$
$$- \sum_{i=1}^{N}\alpha_i^*(\varepsilon + \xi_i^* + y_i - \mathbf{w}^{\mathrm{T}}\mathbf{x}_i - b)$$

of the primal problem (24) is needed. $\boldsymbol{\alpha}$, $\boldsymbol{\alpha}^*$, $\boldsymbol{\beta}$ and $\boldsymbol{\beta}^*$ are the dual variables, i.e., the Lagrange multipliers of the primal constraints. As in Sect. 2, the saddle point condition can be exploited to minimize L_{p} with respect to the primal variables $\mathbf{w}$, $\boldsymbol{\xi}$ and $\boldsymbol{\xi}^*$, which results in a function that only contains $\boldsymbol{\alpha}$, $\boldsymbol{\alpha}^*$ and b:

$$L_{\mathrm{p}}^*(\boldsymbol{\alpha},\boldsymbol{\alpha}^*,b) = \frac{1}{2}\sum_{i=1}^{N}\sum_{j=1}^{N}(\alpha_i-\alpha_i^*)(\alpha_j-\alpha_j^*)K_{ij}$$
$$- \sum_{i=1}^{N}(\alpha_i-\alpha_i^*)y_i + \varepsilon\sum_{i=1}^{N}(\alpha_i+\alpha_i^*) + b\sum_{i=1}^{N}(\alpha_i-\alpha_i^*) \tag{26}$$

The scalar product $\mathbf{x}_i^{\mathrm{T}}\mathbf{x}_j$ has already been substituted by the kernel function $K_{ij} = K(\mathbf{x}_i,\mathbf{x}_j)$ to introduce nonlinearity to the SVM, see (6) and Fig. 5. The bias term b is untouched so far because the next sections offer again two possibilities (fixed and variable b) that lead to different algorithms. In both cases, the inequality constraints

$$0 \le \alpha_i^{(*)} \le C\,, \quad i = 1,\ldots,N \tag{27}$$

resulting from (48d) must be fulfilled. Since a data point cannot lie above *and* below the insensitivity zone simultaneously, the dual variables $\boldsymbol{\alpha}$ and $\boldsymbol{\alpha}^*$ are not independent. At least one of the primal constraints (24b) and (24c) must be met with equality for each i. The KKT conditions then imply that $\alpha_i\alpha_i^* = 0$. The output of regression SVMs is computed as

$$f(\mathbf{x}) = \sum_{\alpha_i^{(*)}\neq 0}(\alpha_i-\alpha_i^*)K(\mathbf{x}_i,\mathbf{x}) + b\,. \tag{28}$$

The notation $\alpha_i^{(*)}$ is used as an abbreviation if an (in-) equality is valid for both α_i and α_i^*.

3.1 Regression with Fixed Bias Term

The kernel function is still assumed to be positive definite so that b can be kept fixed or even omitted. The QP problem (10) is similar for regression SVMs. It is built from (26) and (27) by treating the bias term as a fixed parameter:

$$\begin{aligned}\min_{\boldsymbol{\alpha},\,\boldsymbol{\alpha}^*} \quad J_{\mathrm{d}}(\boldsymbol{\alpha},\boldsymbol{\alpha}^*) = &\frac{1}{2}\sum_{i=1}^{N}\sum_{j=1}^{N}(\alpha_i-\alpha_i^*)(\alpha_j-\alpha_j^*)K_{ij} - \sum_{i=1}^{N}(\alpha_i-\alpha_i^*)y_i \\ &+ \varepsilon\sum_{i=1}^{N}(\alpha_i+\alpha_i^*) + b\sum_{i=1}^{N}(\alpha_i-\alpha_i^*)\end{aligned} \tag{29a}$$

$$\text{s.t.} \quad 0 \le \alpha_i^{(*)} \le C\,, \quad i=1,\ldots,N \tag{29b}$$

Again, (29) is formulated as a *minimization* problem by setting $J_{\mathrm{d}}(\boldsymbol{\alpha},\boldsymbol{\alpha}^*) = -L_{\mathrm{p}}^*(\boldsymbol{\alpha},\boldsymbol{\alpha}^*,b)$ with fixed b. For $b=0$ the last term vanishes so that (29) differs from the standard problem (37) only in the absence of the equality constraint (37c). To find the steps A1 and A2 of an active-set algorithm that solves (29) its Lagrangian

$$\begin{aligned}L_{\mathrm{d}}(\boldsymbol{\alpha},\boldsymbol{\alpha}^*,\boldsymbol{\lambda},\boldsymbol{\lambda}^*,\boldsymbol{\mu},\boldsymbol{\mu}^*) = &\frac{1}{2}\sum_{i=1}^{N}\sum_{j=1}^{N}(\alpha_i-\alpha_i^*)(\alpha_j-\alpha_j^*)K_{ij} \\ &- \sum_{i=1}^{N}(\alpha_i-\alpha_i^*)y_i + \varepsilon\sum_{i=1}^{N}(\alpha_i+\alpha_i^*) + b\sum_{i=1}^{N}(\alpha_i-\alpha_i^*) \\ &- \sum_{i=1}^{N}\lambda_i\alpha_i - \sum_{i=1}^{N}\mu_i(C-\alpha_i) - \sum_{i=1}^{N}\lambda_i^*\alpha_i^* - \sum_{i=1}^{N}\mu_i^*(C-\alpha_i^*)\end{aligned} \tag{30}$$

is required. Compared to classification, two additional sets of multipliers λ_i^* (for $\alpha_i^* \ge 0$) and μ_i^* (for $\alpha_i^* \le C$) are needed here. Using the prediction errors $E_i = f(\mathbf{x}_i) - y_i$, the KKT conditions for $i=1,\ldots,N$ are

$$\frac{\partial L_{\mathrm{d}}}{\partial \alpha_i} = \varepsilon + E_i - \lambda_i + \mu_i = 0 \tag{31a}$$

$$\frac{\partial L_{\mathrm{d}}}{\partial \alpha_i^*} = \varepsilon - E_i - \lambda_i^* + \mu_i^* = 0 \tag{31b}$$

$$0 \le \alpha_i^{(*)} \le C \tag{31c}$$

$$\lambda_i^{(*)} \ge 0, \quad \mu_i^{(*)} \ge 0 \tag{31d}$$

$$\alpha_i^{(*)}\lambda_i^{(*)} = 0, \quad (C-\alpha_i^{(*)})\mu_i^{(*)} = 0\,. \tag{31e}$$

According to α_i and α_i^*, five cases have to be considered:

$\boxed{0 < \alpha_i < C,\ \alpha_i^* = 0}$ $\quad (i \in \mathcal{F})$

$$\begin{aligned}&\Rightarrow\ \lambda_i = \mu_i = \mu_i^* = 0,\ \lambda_i^* = 2\varepsilon > 0 \\ &\Rightarrow\ \sum_{j\in\mathcal{F}^{(*)}} a_j K_{ij} = y_i - \varepsilon - \sum_{j\in\mathcal{A}_C^{(*)}} a_j K_{ij}\end{aligned} \tag{32a}$$

$\boxed{0 < \alpha_i^* < C,\ \alpha_i = 0}$ $\quad (i \in \mathcal{F}^*)$

$$\begin{aligned} &\Rightarrow \lambda_i^* = \mu_i = \mu_i^* = 0,\ \lambda_i = 2\varepsilon > 0 \\ &\Rightarrow \sum_{j \in \mathcal{F}^{(*)}} a_j K_{ij} = y_i + \varepsilon - \sum_{j \in \mathcal{A}_C^{(*)}} a_j K_{ij} \end{aligned} \tag{32b}$$

$\boxed{\alpha_i = \alpha_i^* = 0}$ $\quad (i \in \mathcal{A}_0 \cap \mathcal{A}_0^*)$

$$\begin{aligned} &\Rightarrow \lambda_i = \varepsilon + E_i > 0, \quad \lambda_i^* = \varepsilon - E_i > 0 \\ &\Rightarrow \mu_i = 0, \quad \mu_i^* = 0 \end{aligned} \tag{32c}$$

$\boxed{\alpha_i = C,\ \alpha_i^* = 0}$ $\quad (i \in \mathcal{A}_C)$

$$\begin{aligned} &\Rightarrow \lambda_i = 0, \quad \lambda_i^* = \varepsilon - E_i > 0 \\ &\Rightarrow \mu_i = -\varepsilon - E_i > 0, \quad \mu_i^* = 0 \end{aligned} \tag{32d}$$

$\boxed{\alpha_i = 0,\ \alpha_i^* = C}$ $\quad (i \in \mathcal{A}_C^*)$

$$\begin{aligned} &\Rightarrow \lambda_i = \varepsilon + E_i > 0, \quad \lambda_i^* = 0 \\ &\Rightarrow \mu_i = 0, \quad \mu_i^* = -\varepsilon + E_i > 0 \end{aligned} \tag{32e}$$

Obviously, there are more than five cases but only these five can occur due to $\alpha_i \alpha_i^* = 0$: If one of the variables is free ((32a) and (32b)) or equal to C ((32d) and (32e)), the other one must be zero. The structure of the sets $\mathcal{A}_0^*$ and $\mathcal{A}_C^*$ is identical to that of $\mathcal{A}_0$ and $\mathcal{A}_C$, but it considers the variables α_i^* instead of α_i. It follows from the reasoning above that $\mathcal{A}_C \subseteq \mathcal{A}_0^*$, $\mathcal{A}_C^* \subseteq \mathcal{A}_0$ and $\mathcal{A}_C \cap \mathcal{A}_C^* = \emptyset$. Similar to classification, the cases (32a) and (32b) form the linear system for step A1 and the cases (32c) – (32e) are the conditions to be checked in step A2 of the algorithm.

The regression algorithm uses the SVM coefficients $a_i = \alpha_i - \alpha_i^*$. With this abbreviation, the number of variables reduces from $2N$ to N and many similarities to classification can be observed. The linear system is almost the same as (14):

$$\mathbf{H}^k \bar{\mathbf{a}}^k = \mathbf{c}^k \tag{33}$$

with

$$\begin{aligned} &\left.\begin{aligned} \bar{a}_i^k &= \bar{\alpha}_i^k - \bar{\alpha}_i^{*k} \\ h_{ij}^k &= K_{ij} \end{aligned}\right\} \quad \text{for } i \in \mathcal{F}^k \cup \mathcal{F}^{*k} \\ &c_i^k = y_i - \sum_{j \in \mathcal{A}_C^k \cup \mathcal{A}_C^{*k}} a_j^k K_{ij} + \begin{cases} -\varepsilon & \text{for } i \in \mathcal{F}^k \\ +\varepsilon & \text{for } i \in \mathcal{F}^{*k} \end{cases} \end{aligned} \tag{34}$$

only the right hand side has been modified by $\pm\varepsilon$. Step A2 of the algorithm computes

$$\left.\begin{aligned}\lambda_i^k &= \varepsilon + E_i^k \\ \lambda_i^{*k} &= \varepsilon - E_i^k\end{aligned}\right\} \quad \text{for } i \in \mathcal{A}_0^k \cup \mathcal{A}_0^{*k} \tag{35a}$$

and

$$\left.\begin{aligned}\mu_i^k &= -\varepsilon - E_i^k \\ \mu_i^{*k} &= -\varepsilon + E_i^k\end{aligned}\right\} \quad \text{for } i \in \mathcal{A}_C^k \cup \mathcal{A}_C^{*k} \,. \tag{35b}$$

These multipliers are checked for positiveness with precision τ, and the variable with the most negative multiplier is transferred to $\mathcal{F}^k$ or $\mathcal{F}^{*k}$.

3.2 Regression with Variable Bias Term

If the bias term is treated as a variable, (26) can be minimized with respect to b (i.e., $\partial L_\mathrm{d}^*/\partial b = 0$) resulting in

$$\sum_{i=1}^N (\alpha_i - \alpha_i^*) = 0 \,. \tag{36}$$

Like in classification, this condition removes the last term from (29a) but must be treated as additional equality constraint:

$$\min_{\boldsymbol{\alpha},\,\boldsymbol{\alpha}^*} \quad J_\mathrm{d}(\boldsymbol{\alpha}, \boldsymbol{\alpha}^*) = \frac{1}{2}\sum_{i=1}^N\sum_{j=1}^N (\alpha_i - \alpha_i^*)(\alpha_j - \alpha_j^*)K_{ij} - \sum_{i=1}^N (\alpha_i - \alpha_i^*)y_i + \varepsilon\sum_{i=1}^N (\alpha_i + \alpha_i^*) \tag{37a}$$

$$\text{s.t.} \quad 0 \le \alpha_i^{(*)} \le C\,, \quad i = 1, \dots, N \tag{37b}$$

$$\sum_{i=1}^N (\alpha_i - \alpha_i^*) = 0 \tag{37c}$$

The Lagrangian of this QP problem is nearly identical to (30):

$$\begin{aligned}L_\mathrm{d}(\boldsymbol{\alpha}, \boldsymbol{\alpha}^*, \boldsymbol{\lambda}, \boldsymbol{\lambda}^*, \boldsymbol{\mu}, \boldsymbol{\mu}^*, \nu) &= \frac{1}{2}\sum_{i=1}^N\sum_{j=1}^N (\alpha_i - \alpha_i^*)(\alpha_j - \alpha_j^*)K_{ij} \\ &- \sum_{i=1}^N (\alpha_i - \alpha_i^*)y_i + \varepsilon\sum_{i=1}^N (\alpha_i + \alpha_i^*) - \nu\sum_{i=1}^N (\alpha_i - \alpha_i^*) \\ &- \sum_{i=1}^N \lambda_i\alpha_i - \sum_{i=1}^N \mu_i(C - \alpha_i) - \sum_{i=1}^N \lambda_i^*\alpha_i^* - \sum_{i=1}^N \mu_i^*(C - \alpha_i^*)\end{aligned} \tag{38}$$

Classification has already shown that the Lagrange multiplier ν of the equality constraint is basically the bias term ($\nu = -b$) that is treated as a variable. Compared to fixed b, (31) also comprises the equality constraint (37c), but the five cases (32) do not change. Consequently, the coefficients $a_i = \alpha_i - \alpha_i^*$ with $i \in \mathcal{F} \cup \mathcal{F}^*$ and the bias term b are computed by solving a block system having the same structure as (21):

$$\begin{pmatrix} \mathbf{H}^k & \mathbf{e} \\ \mathbf{e}^{\mathrm{T}} & 0 \end{pmatrix} \begin{pmatrix} \bar{\mathbf{a}}^k \\ b^k \end{pmatrix} = \begin{pmatrix} \mathbf{c}^k \\ d^k \end{pmatrix} \begin{matrix} \} \ p \text{ rows} \\ \} \ 1 \text{ row} \end{matrix} \tag{39}$$

with

$$d^k = - \sum_{j \in \mathcal{A}_C^k \cup \mathcal{A}_C^{*k}} a_j^k \quad \text{and} \quad \mathbf{e} = (1, \ldots, 1)^{\mathrm{T}} \tag{40}$$

i.e., the only difference is d^k which considers the indices in both $\mathcal{A}_C^k$ and $\mathcal{A}_C^{*k}$. This system can be solved by the algorithm derived in Sect. 2. The KKT conditions in step A2 remain exactly the same as (35).

4 Implementation Details

The active-set algorithm has been implemented as C MEX-file under MATLAB for classification and regression problems. It can handle both fixed and variable bias terms. Approximately the following memory is required:

- N floating point elements for the coefficient vector,
- N integer elements for the index vector,
- $N_{\mathrm{f}}(N_{\mathrm{f}} + 3)/2$ floating point elements for the triangular matrix and the right hand side of the linear system,

where N_{f} is the number of free variables in the optimum, i.e., those with $0 < \alpha_i^{(*)} < C$. As this number is unknown in the beginning, the algorithm starts with an initial amount of memory and increases it whenever variables are added. The index vector is needed to keep track of the stets $\mathcal{F}^{(*)}$, $\mathcal{A}_C^{(*)}$ and $\mathcal{A}_0^{(*)}$. It is also used as pivot vector for the Cholesky decomposition described in Sect. 4.1. Since most of the coefficients a_i will be zero in the optimum, the initial feasible solution is chosen as $a_i = 0$ for $i = 1, \ldots, N$. If shrinking and/or caching is activated, additional memory must be provided, see Sects. 4.4 and 4.5 for details.

Since all algorithms assume positive definite kernel functions, the kernel matrix has a Cholesky decomposition $\mathbf{H} = \mathbf{R}^{\mathrm{T}}\mathbf{R}$, where $\mathbf{R}$ is an upper triangular matrix. For a fixed bias term, the solution of the linear system in step A1 is then found by simple backsubstitution. For variable bias term, the block-algorithm described in Sect. 2.2 is used.

4.1 Cholesky Decomposition with Pivoting

Although the Cholesky decomposition is numerically stable, the active-set algorithm uses diagonal pivoting by default because $\mathbf{H}$ may be "nearly indefinite", i.e., it may become indefinite by round-off errors during the computation. This occurs e.g. for Gaussian kernels having large widths. There are two ways to cope with this problem: First, to use Cholesky decomposition with pivoting, and second, to slightly enlarge the diagonal elements to make $\mathbf{H}$ "more definite". The first case allows to extract the largest positive definite part of $\mathbf{H} = (h_{ij})$. All variables corresponding to the rest of the matrix are set to zero then.

Usually the Cholesky decomposition is computed element-wise using `axpy` operations defined in the BLAS [3]. However, the pivoting strategy needs the updated diagonal elements in each step, as they would be available if *outer product* updates were applied. Since these require many accesses to matrix elements, a mixed procedure is implemented that only updates the diagonal elements and makes use of `axpy` operations otherwise:

Compute for $i = 1, \ldots, p$*:*
  Find $k = \arg\max\{|\bar{h}_{ii}|, \ldots, |\bar{h}_{pp}|\}$.
  Swap rows and columns i *and* k *symmetrically.*
  Compute $r_{ii} = \sqrt{\bar{h}_{ii}}$.
  Compute for $j = i+1, \ldots, p$*:*

$$
\begin{aligned}
r_{ij} &= \left(h_{ij} - \sum_{k=1}^{i-1} r_{ki} r_{kj} \right) \Big/ r_{ii} \\
\bar{h}_{jj} &\leftarrow \bar{h}_{jj} - r_{ij}^2
\end{aligned}
$$

where p is the size of the system, $\bar{h}_{jj}$ are the updated diagonal elements and "$\leftarrow$" indicates the update process. The i-th step

$$
\begin{array}{c}
\begin{array}{cccccc} & & i & j & & \\ & & \downarrow & \longrightarrow & & \end{array} \\
\begin{pmatrix}
\bullet & \bullet & \bullet & \bullet & \bullet & \bullet \\
 & \bullet & \bullet & \bullet & \bullet & \bullet \\
 & & \circ & \circ & \circ & \circ \\
 & & & \times & * & * \\
 & & j \searrow & & \times & * \\
 & & & & & \times
\end{pmatrix} \begin{array}{c} \\ \\ \leftarrow i \\ \\ \\ \\ \end{array}
\end{array}
$$

of the algorithm computes the i-th row ($\circ$) of the matrix from the already finished elements ($\bullet$). The diagonal elements ($\times$) are updated whereas the rest ($*$) remains untouched. The result can be written as

$$
\mathbf{P}\mathbf{H}\mathbf{P}^{\mathrm{T}} = \mathbf{R}^{\mathrm{T}}\mathbf{R} \tag{41}
$$

with the permutation matrix $\mathbf{P}$. Of course the implementation uses the pivot vector described above instead of the complete matrix. Besides that, only the upper triangular part of $\mathbf{R}$ is stored, so that only memory for $p(p+1)/2$ elements is needed. This algorithm is almost as fast as the standard Cholesky decomposition.

4.2 Adding Variables

Since the active-set algorithm changes the active set by only one variable per step, it is reasonable to modify the existing Cholesky decomposition instead of computing it form scratch [2]. These techniques are faster but less accurate than the method described in Sect. 4.1, because they cannot be used with pivoting. The only way to cope with definiteness problems is to slightly enlarge the diagonal elements h_{jj}.

If a p-th variable is added to the linear system, a new column and a new row are appended to $\mathbf{H}$. As any element r_{ij} of the Cholesky decomposition is calculated solely from the diagonal element r_{ii} and the sub-columns i and j above the i-th row (see Sect. 4.1), only the last column needs to be computed:

Compute for i = 1,...,p:

$$r_{ip} = \left(h_{ip} - \sum_{k=1}^{i-1} r_{ki} r_{kp} \right) \Big/ r_{ii}$$

The columns $1, \ldots, p-1$ remain unchanged. This technique is only effective if the *last* column is appended. If an arbitrary column is inserted, elements of $\mathbf{R}$ need to be re-computed.

4.3 Removing Variables

Removing variables from an existing Cholesky decomposition is a more sophisticated task [2, 3]. For that, we introduce an unknown matrix $\mathbf{A} \in \mathbb{R}^{M \times p}$ with

$$\mathbf{H} = \mathbf{R}^{\mathrm{T}}\mathbf{R} = \mathbf{A}^{\mathrm{T}}\mathbf{A} \quad \text{and} \quad \mathbf{QA} = \begin{pmatrix} \mathbf{R} \\ \mathbf{0} \end{pmatrix}, \tag{42}$$

i.e., $\mathbf{R}$ also results from the QR decomposition of $\mathbf{A}$. Removing a variable from the Cholesky decomposition is equivalent to removing a column from $\mathbf{A}$:

$$\mathbf{Q}(\mathbf{a}_1 \ldots \mathbf{a}_{k-1}, \mathbf{a}_{k+1} \ldots \mathbf{a}_p) = \begin{pmatrix} \mathbf{r}_1 \ldots \mathbf{r}_{k-1}, \mathbf{r}_{k+1} \ldots \mathbf{r}_p \\ \mathbf{0} \end{pmatrix} \tag{43}$$

The non-zero part of the right hand side matrix is of size $p \times (p-1)$ now because the k-th column is missing. It is "nearly" an upper triangular matrix, only each of the columns $k+1, \ldots, p$ has one element below the diagonal:

$$
\begin{array}{c}
\begin{array}{cccccc}
 & & k{-}1 & k{+}1 & & \\
 & & \downarrow & \downarrow & &
\end{array} \\
\begin{pmatrix}
\bullet & \bullet & \bullet & \bullet & \bullet & \bullet \\
 & \bullet & \bullet & \bullet & \bullet & \bullet \\
 & & \bullet & \bullet & \bullet & \bullet \\
 & & & * & * & * \\
 & & & \times & * & * \\
 & & & & \times & * \\
 & & & & & \times
\end{pmatrix}
\begin{array}{l}
 \\ \\ \leftarrow k-1 \\ \\ \leftarrow k+1 \\ \\ \\
\end{array}
\end{array}
$$

The sub-diagonal elements are removed by *Givens rotations* $\mathbf{\Omega}_{k+1}, \ldots, \mathbf{\Omega}_p$:

$$\underbrace{\mathbf{\Omega}_p \cdots \mathbf{\Omega}_{k+1} \mathbf{Q}}_{\tilde{\mathbf{Q}}} \tilde{\mathbf{A}} = \begin{pmatrix} \tilde{\mathbf{R}} \\ \mathbf{0} \end{pmatrix} \tag{44}$$

$\tilde{\mathbf{R}}$ is the Cholesky factor of the reduced matrix $\tilde{\mathbf{H}}$, see [2, 3] for details.

However, it should be mentioned that modification techniques often do not lead to a strong acceleration. As long as N_f remains small, most of the computation time is spent to check the KKT conditions in $\mathcal{A}$ (during step A2 of the algorithm). For that, the algorithm uses Cholesky decomposition with pivoting when a variable is added to its linear system, and the above modification strategy when a variable is removed. Since only few reduction steps are performed repeatedly, round-off errors will not be propagated too much.

4.4 Shrinking the Problem Size

As pointed out above, checking the KKT conditions is the dominating factor of the computation time because the function values (9) or (28) need to be computed for all variables of the active set in each step. For that, the active-set algorithm uses two heuristics to accelerate the KKT check: shrinking the set of variables to be checked, and caching kernel function values (which will be described in the next section).

By default, step A2 of the algorithm checks *all* bounded variables. However, it can be observed that a variable fulfilling the KKT conditions for a number of iterations is likely to stay in the active set [1, 10]. The shrinking heuristic uses this observation to reduce number of KKT checks. It counts the number of consecutive successful KKT checks for each variable. If this number exceeds a given number s, then the variable is not checked again. Only if there are no variables left to be checked, a check of the complete active set is performed and the shrinking procedure starts again.

In experiments, small values of s (e.g., $s = 1, \ldots, 5$) have caused an acceleration up to a factor of 5. This shrinking heuristic requires an additional vector of N integer elements to count the KKT checks of each variable. If the correct active set is identified, shrinking does not change the solution. However, for low precisions τ it may happen that the algorithm chooses a different approximation of the active set, i.e., different support vectors.

4.5 Caching Kernel Values

Whereas the shrinking heuristic tries to *reduce* the number of function evaluations, the goal of a kernel cache is to *accelerate* the remaining ones. For that, as much kernel function values K_{ij} as possible are stored in a given chunk of memory to avoid re-calculation. Some algorithms also use a cache for the function values f_i (or prediction error values $E_i = f_i - y_i$, respectively), e.g. [8]. However, since the active-set algorithm changes the values of *all* free variables in each step, this type of cache would only be useful when the number of free variables remains small.

The kernel cache has a given maximum size and is organized as *row cache* [1, 10]. It stores a complete row of the kernel matrix for each support vector – as long as space is available. The row entries corresponding to the active set are exploited to compute (9) or (28) for the KKT check, whereas the remaining elements are used to rebuild the system matrix $\mathbf{H}$ when necessary. The following caching strategy has been implemented:

- If a variable becomes $a_i = 0$, then the according row is marked as free in the cache but not deleted.
- If a variable becomes $a_i \neq 0$, the algorithm first checks if the according row is already in the cache (possibly marked as free). Otherwise, the row is completely calculated and stored as long as space is available.
- When a row should be added, the algorithm first checks if the maximum number of rows is reached. Only if the cache is full, it starts to overwrite those rows that have been marked as free.

The kernel cache allows a trade-off between computation time and memory consumption. It requires $N \times m$ floating point elements for the kernel values (where m is the maximum number of rows that can be cached), and N integer elements for indexing purposes. It is most effective for kernel functions having high computational demands, e.g., Gaussians in high-dimensional input spaces. In these cases it usually speeds up the algorithm by a factor 5 or even more, see Sect. 5.2

4.6 Approximating the Solution

Active-set methods check the KKT conditions of the *complete* active set (apart from the shrinking heuristics) in each step. As pointed out above, this is a huge computational effort which is only reasonable for algorithms that make enough progress in each step. Typical working-set algorithms, on the other hand, avoid this complete check and follow the opposite strategy: They perform only small steps and therefore need to reduce the number of KKT evaluations to a minimum by additional heuristics.

The complete KKT check of active-set methods can be exploited to approximate the solution with a given number N_{SVmax} of support vectors. Remember that the N_{SV} support vectors are associated with

- N_{f} free variables $0 < \alpha_i^{(*)} < C$ (i.e., those with $i \in \mathcal{F}^{(*)}$).
- $N_{\mathrm{SV}} - N_{\mathrm{f}}$ upper bounded variables $\alpha_i^{(*)} = C$ (i.e., those with $i \in \mathcal{A}_C^{(*)}$).

The algorithm simply stops when at the end of step A3 a solution with more than N_{SVmax} support vectors is computed for the first time:

- If $N_{\mathrm{SV}}^k > N_{\mathrm{SVmax}}$ then stop with the previous solution.
- Otherwise accept the new solution and go to step A1.

The first case can only happen if in step A2 an $i \in \mathcal{A}_0^{(*)}$ was selected and in step A3 no variable is moved back to $\mathcal{A}_0^{(*)}$. All other cases do not increase the number of support vectors.

This heuristic approach does not always lead to a better approximation if more support vectors are allowed. However, experiments (like in Sect. 5.2) show that typically only a small fraction of support vectors significantly reduces the approximation error.

5 Results

This section shows experimental results for classification and regression. The proposed active-set method is compared with the well-established working-set method LIBSVM [1] for different problem settings. LIBSVM (Version 2.6) is chosen as a typical representative of working-set methods – other implementations like SMO [8] or ISDA [4] show similar characteristics. Both algorithms are available as MEX functions under MATLAB and were compiled with Microsoft Visual C/C++ 6.0. All experiments were done on a 800 MHz Pentium-III PC having 256 MB RAM.

Since the environmental conditions are identical for both algorithms, mainly the *computation time* is considered to measure the performance. By default, both use shrinking heuristics and have enough cache to store the complete kernel matrix if necessary. The influence of these acceleration techniques is examined in Sect. 5.2.

5.1 Classifying Demographic Data

The first example considers the "Adult" database from the *UCI machine learning repository* [6] that has been studied in several publications. The goal is to determine from 14 demographic features weather a person earns more than \$ 50,000 per year. All features have been normalized to $[-1, 1]$; nominal features were converted to numeric values before. In order to limit the computation time in the critical cases, a subset of 1000 samples has been selected as training data set. The SVMs use Gauss kernels with width $\sigma = 3$ and a precision of $\tau = 10^{-3}$ to check the KKT conditions.

Table 1 shows the results when the upper bound C is varied, e.g., to find the optimal C by cross-validation. Whereas the active-set method is nearly insensitive with respect to C, the computation time of LIBSVM differs by several magnitudes. Working-set methods typically perform better when the number N_{f} of *free* variables is small. The computation time of active-set methods mainly depends on the *complete* number N_{SV} of support vectors which roughly determines the number of iterations.

Table 1. Classification: Variation of C

	C	10^{-1}	10^{0}	10^{1}	10^{2}	10^{3}	10^{4}	10^{5}	10^{6}
Active Set	Time	8.7 s	7.4 s	4.3 s	5.4 s	8.4 s	12.3 s	11.1 s	10.8 s
	N_{SV}	494	480	422	389	379	364	357	337
	N_{f}	11	20	37	81	139	190	242	271
	Bias	0.9592	0.7829	0.0763	2.0514	3.5940	2.0549	−26.63	−95.56
Active Set ($b = 0$)	Time	7.9 s	6.1 s	3.7 s	5.7 s	8.6 s	11.5 s	11.9 s	9.6 s
	N_{SV}	510	481	422	391	378	364	360	339
	N_{f}	14	17	37	78	139	190	245	273
LIB SVM	Time	0.6 s	0.5 s	0.5 s	0.9 s	3.1 s	21.1 s	156.2 s	1198 s
	N_{SV}	496	481	422	390	379	366	356	334
	N_{f}	16	22	38	82	139	192	243	268
	Bias	0.9592	0.7826	0.0772	2.0554	3.5927	2.0960	−26.11	−107.56
Error	Train	24.4%	19.0%	16.6%	13.8%	11.4%	8.0%	5.1%	3.4%
	Test	24.6%	18.2%	16.9%	16.6%	17.9%	19.4%	21.5%	23.2%

Also a comparison between the standard SVM and the no-bias SVM (i.e., with bias term fixed at $b = 0$) can be found in Table 1. It shows that there is no need for a bias term when positive definite kernels are used. Although a missing bias usually leads to more support vectors, the results are very close to the standard SVM – even if the bias term takes large values. The errors on the training and testing data set are nearly identical for all three methods. Although the training error can be further reduced by in increasing C, the best generalization performance is achieved with $C = 10^2$ here. In that case LIBSVM is finds the solution very quickly as N_{f} is still small.

5.2 Estimating the Outlet Temperature of a Boiler

The following example applies the regression algorithm to a system identification problem. The goal is to estimate the outlet temperature T_{31} of a *high*

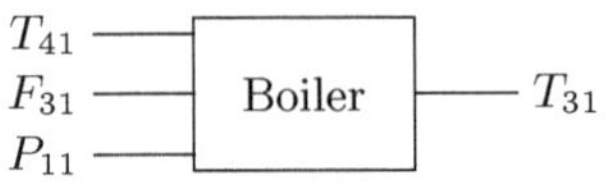

$$\begin{aligned} T_{31}(k) = f(&T_{41}(k), T_{41}(k-1), T_{41}(k-2), \\ &F_{31}(k), F_{31}(k-1), F_{31}(k-2), \\ &P_{11}(k), P_{11}(k-1), P_{11}(k-2), \\ &T_{31}(k-1), T_{31}(k-2)) \end{aligned} \tag{45}$$

Fig. 6. Block diagram and regression model of the boiler

efficiency (condensing) boiler from the system temperature T_{41}, the water flow F_{31} and the burner output P_{11} as inputs. Details about the data set under investigation can be found in [13] and [14]. Based on a theoretical analysis, second order dynamics are assumed for the output and all inputs, so the model has 11 regressors, see Fig. 6. For a sampling time of 30 s the training data set consists of 3344 samples, the validation data set of 2926 samples. Table 2 compares the active-set algorithm and LIBSVM when the upper bound C is varied. The SVM uses Gauss kernels having a width of $\sigma = 3$. The insensitivity zone is properly set to $\varepsilon = 0.01$, the precision used to check the KKT conditions is $\tau = 10^{-4}$. Both methods compute SVMs with variable bias term in order to make the results comparable. The RMSE is the *root-mean-square error* of the *predicted* output on the validation data set. The *simulation* error is not considered because models can be unstable for extreme settings of C.

Table 2. Regression: Variation of C for $\sigma = 3$ and $\tau = 10^{-4}$

	C	10^{-2}	10^{-1}	10^{0}	10^{1}	10^{2}	10^{3}	10^{4}	10^{5}
Active Set	Time	211.8 s	46.9 s	8.3 s	1.5 s	0.7 s	0.7 s	0.9 s	1.2 s
	RMSE	0.0330	0.0164	0.0097	0.0068	0.0062	0.0062	0.0064	0.0069
	N_{SV}	1938	954	427	143	91	87	92	116
	N_{f}	4	10	25	36	52	77	91	116
LIB SVM	Time	7.9 s	4.5 s	2.7 s	3.0 s	7.7 s	39.1 s	163.2 s	?
	RMSE	0.0330	0.0164	0.0097	0.0068	0.0062	0.0092	0.0064	?
	N_{SV}	1943	963	433	147	95	90	98	?
	N_{f}	10	23	35	45	56	80	97	?

Concerning computation time, Table 2 shows that LIBSVM can efficiently handle a large number N_{SV} of support vectors (with only few free ones) whereas the active-set method shows its strength if N_{SV} is small. For $C = 10^5$ LIBSVM converged extremely slow so that it was aborted after 12 hours. In this example, $C = 10^3$ is the optimal setting concerning support vectors and error. Also the active-set algorithm's memory consumption $\mathcal{O}(N_{\mathrm{f}}^2)$ (see Sect. 1) is not critical: When the number of support vectors increases, typically most of the Lagrange multipliers are bounded at C so that N_{f} remains small.

Table 3. Regression: Variation of σ for $C = 10^3$ and $\tau = 10^{-4}$

	σ	0.5	1	2	3	4	5	6	7
Active Set	Time	2.3 s	1.2 s	0.8 s	0.7 s	0.8 s	0.9 s	0.8 s	1.0 s
	RMSE	0.0278	0.0090	0.0064	0.0062	0.0061	0.0059	0.0059	0.0057
	N_{SV}	184	129	96	87	88	94	96	108
	N_{f}	184	129	94	77	60	49	40	39
LIB SVM	Time	4.1 s	13.7 s	25.1 s	38.2 s	31.8 s	22.4 s	15.4 s	13.1 s
	RMSE	0.0278	0.0091	0.0064	0.0062	0.0061	0.0070	0.0058	0.0057
	N_{SV}	196	134	96	90	92	102	99	110
	N_{f}	196	134	95	80	66	60	42	44
	cond(**H**)	$3{\cdot}10^5$	$8{\cdot}10^6$	$2{\cdot}10^8$	$3{\cdot}10^8$	$2{\cdot}10^8$	$1{\cdot}10^8$	$2{\cdot}10^8$	$2{\cdot}10^8$

A comparison with Table 1 confirms that the computation time for the active-set method mainly depends on the *number* N_{SV} of support vectors, whereas the *ratio* $N_{\mathrm{f}}/N_{\mathrm{SV}}$ has strong influence on working-set methods.

Table 3 examines a variation of the Gaussians' width σ for $C = 10^3$ and $\tau = 10^{-4}$. As expected, the computation time of the active-set algorithm is solely dependent on the number of support vectors. For large σ the computation times of LIBSVM decrease because the fraction of free variables gets smaller, whereas for small σ another effect can be observed: If the condition number the system matrix $\mathbf{H}$ in (33) or (39) decreases, the change in one variable has less effect on the other ones. For that, the computation time decreases although there are only free variables and their number even increases.

Table 4 compares the algorithms for different precisions τ in case of $\sigma = 5$ and $C = 10^2$. Both do not change the active set for precisions smaller then 10^{-5}. Whereas LIBSVM's computation time strongly increases, the active-set method does not need more time to meet a higher precision. Once the active set is found, active-set methods compute the solution with "full" precision, i.e., a smaller τ does not change the solution any more. For low precisions,

Table 4. Regression: Variation of τ for $\sigma = 5$ and $C = 10^2$

	τ	10^{-1}	10^{-2}	10^{-3}	10^{-4}	10^{-5}	10^{-6}	10^{-7}	10^{-8}
Active Set	Time	0.1 s	0.3 s	0.9 s	1.0 s	1.1 s	1.1 s	1.1 s	1.1 s
	RMSE	0.0248	0.0063	0.0059	0.0059	0.0059	0.0059	0.0059	0.0059
	N_{SV}	8	49	108	118	122	122	122	122
	N_{f}	7	17	23	33	39	39	39	39
LIB SVM	Time	0.2 s	2.9 s	4.5 s	4.9 s	7.7 s	9.9 s	18.9 s	25.9 s
	RMSE	0.0220	0.0060	0.0059	0.0058	0.0058	0.0058	0.0058	0.0058
	N_{SV}	30	90	119	121	123	123	123	123
	N_{f}	30	73	49	40	39	39	39	39

Table 5. Regression: Influence of shrinking and caching on the computation time for $\sigma = 5$, $C = 10^2$, $\tau = 10^{-4}$

Cached Rows	0	50	100	120	150	200
No shrinking	16.97 s	8.65 s	3.45 s	2.78 s	2.76 s	2.76 s
$s = 10$	5.91 s	3.30 s	1.63 s	1.37 s	1.35 s	1.35 s
$s = 3$	3.80 s	2.22 s	1.24 s	1.07 s	1.05 s	1.05 s
$s = 2$	3.45 s	2.13 s	1.18 s	1.05 s	1.02 s	1.02 s
$s = 1$	2.75 s	1.73 s	1.05 s	0.93 s	0.90 s	0.90 s

the active-set method produces more compact solutions, because it is able to stop earlier due to its complete KKT check in each iteration.

The influence of shrinking and caching is examined in Table 5 for $\sigma = 5$, $C = 10^2$ and $\tau = 10^{-4}$, which yields a SVM having $N_{\text{SV}} = 118$ support vectors. It confirms the estimates given in Sects. 4.4 and 4.5: Both shrinking and caching accelerate the algorithm by a factor of 6 in this example. Used in combination, they lead to a speed-up by nearly a factor of 20. If shrinking is activated, the cache has minor influence because less KKT checks have to be performed. Table 5 also shows that it is not necessary to spend cache for much more than N_{SV} rows, because this only saves the negligible time to search for free rows.

A final experiment demonstrates the approximation method described in Sect. 4.6. With the same settings as above ($\sigma = 5$, $C = 10^2$, $\tau = 10^{-4}$) the complete model contains 118 support vectors. However, Fig. 7 shows that the solution can be approximated with much less support vectors, e.g. 10–15 %.

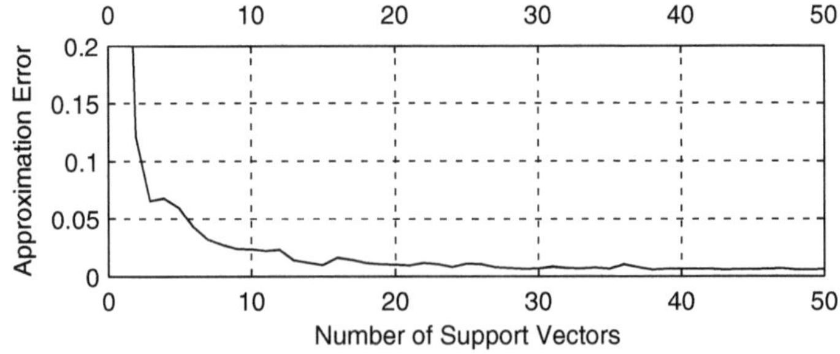

Fig. 7. Regression: Approximation of the solution

Whereas the objective function is still decreasing, more support vectors do not significantly reduce the approximation error.

6 Conclusions

An active-set algorithm has been proposed for SVM classification and regression tasks. The general strategy has been adapted to these problems for both fixed and variable bias terms. The result is a robust algorithm that requires approximately $\frac{1}{2}N_\mathrm{f}^2 + 2N$ elements of memory, where N_f is the number of free variables and N the number of data. Experimental results show that active-set methods are advantageous

- when the number of support vectors is small.
- when the fraction of bounded variables is small.
- when high precision is needed.
- when the problem is ill-conditioned.

Shrinking and caching heuristics can significantly accelerate the algorithm. Additionally, its KKT check can be exploited to approximate the solution with a reduced number of support vectors. Whereas the method is very robust to changes in the settings, it not should be overseen that working-set techniques like LIBSVM are still faster in certain cases and can handle larger data sets.

Currently, the algorithm changes the active set by only one variable per step, and (despite shrinking and caching) most of the computation time is spent to calculate the prediction errors E_i. Both problems can be improved by introducing *gradient projection* steps. If this technique is combined with iterative solvers, also a large number of free variables is possible. This may be a promising direction of future work on SVM optimization methods.

References

1. Chang CC, Lin CJ (2003) LIBSVM: A library for support vector machines. Technical report. National Taiwan University, Taipei, Taiwan
2. Gill PE et al. (1974) Methods for Modifying Matrix Computations. Mathematics of Computation 28(126):505–535
3. Golub GH, van Loan CF (1996) Matrix Computations. 3rd ed. The Johns Hopkins University Press, Baltimore, MD
4. Huang TM, Kecman V (2004) Bias Term b in SVMs again. In: Proceedings of the 12th European Symposium on Artificial Neural Networks (ESANN 2004), pp. 441–448, Bruges, Belgium
5. Mangasarian OL, Musicant DR (2001) Active set support vector machine classification. In: Leen TK, Tresp V, Dietterich TG (eds) Advances in Neural Information Processing Systems (NIPS 2000) Vol. 13, pp. 577–583. MIT Press, Cambridge, MA
6. Blake CL, Merz CJ (1998) UCI repository of machine learning databases. University of California, Irvine, http://www.ics.uci.edu/~mlearn/

7. Nocedal J, Wright SJ (1999) Numerical Optimization. Springer-Verlag, New York
8. Platt JC (1999) Fast training of support vector machines using sequential minimal optimization. In: Schölkopf B, Burges CJC, Smola AJ (eds) Advances in Kernel Methods – Support Vector Learning. MIT Press, Cambridge, MA
9. Poggio T et al. (2002) b. In: Winkler J, Niranjan M (eds) Uncertainty in Geometric Computations, pp. 131–141. Kluwer Academic Publishers, Boston
10. Schölkopf B, Smola AJ (2002) Lerning with Kernels. The MIT Press, Cambridge, MA
11. Vapnik VN (1995) The Nature of Statistical Learning Theory. Springer-Verlag, New York
12. Vishwanathan SVN, Smola AJ and Murty MN (2003) SimpleSVM. In: Proceedings of the 20th International Conference on Machine Learning (ICML 2003), pp. 760–767. Washington, DC
13. Vogt M, Kecman V (2004) An active-set algorithm for Support Vector Machines in nonlinear system identification. In: Proceedings of the 6th IFAC Symposium on Nonlinear Control Systems (NOLCOS 2004), pp. 495–500. Stuttgart, Germany
14. Vogt M, Spreitzer K, Kecman V (2003) Identification of a high efficiency boiler by Support Vector Machines without bias term. In: Proceedings of the 13th IFAC Symposium on System Identification (SYSID 2003), pp. 485–490. Rotterdam, The Netherlands

A The Karush-Kuhn-Tucker Conditions

A general constrained optimization problem is given by

$$\min_{\mathbf{a}} \quad J(\mathbf{a}) \tag{46a}$$
$$\text{s.t.} \quad \mathbf{F}(\mathbf{a}) \geq \mathbf{0} \tag{46b}$$
$$\mathbf{G}(\mathbf{a}) = \mathbf{0}\,. \tag{46c}$$

The Lagrangian of this problem is defined as

$$L(\mathbf{a}, \boldsymbol{\lambda}, \boldsymbol{\nu}) = J(\mathbf{a}) - \sum_i \lambda_i F_i(\mathbf{a}) - \sum_i \nu_i G_i(\mathbf{a})\,. \tag{47}$$

In the constrained optimum $(\mathbf{a}^*, \boldsymbol{\lambda}^*, \boldsymbol{\nu}^*)$ the following first-order necessary conditions [7] are satisfied for all i:

$$\nabla_{\mathbf{a}} L(\mathbf{a}^*, \boldsymbol{\lambda}^*, \boldsymbol{\nu}^*) = 0 \tag{48a}$$
$$F_i(\mathbf{a}^*) \geq 0 \tag{48b}$$
$$G_i(\mathbf{a}^*) = 0 \tag{48c}$$
$$\lambda_i^* \geq 0 \tag{48d}$$
$$\lambda_i^* F_i(\mathbf{a}^*) = 0 \tag{48e}$$
$$\nu_i^* G_i(\mathbf{a}^*) = 0 \tag{48f}$$

These are commonly referred to as *Karush-Kuhn-Tucker* conditions.

Theoretical and Practical Model Selection Methods for Support Vector Classifiers

D. Anguita[1], A. Boni[2], S. Ridella[1], F. Rivieccio[1], and D. Sterpi[1]

[1] DIBE – Dept. of Biophysical and Electronic Engineering, University of Genova, Via Opera Pia 11A, 16145, Genova, Italy.
{anguita,ridella,rivieccio,sterpi}@dibe.unige.it
[2] DIT – Dept. of Communication and Information Technology, University of Trento, Via Sommarive 14, 38050, Povo (TN), Italy.
andrea.boni@ing.unitn.it

Abstract. In this chapter, we revise several methods for SVM model selection, deriving from different approaches: some of them build on practical lines of reasoning but are not fully justified by a theoretical point of view; on the other hand, some methods rely on rigorous theoretical work but are of little help when applied to real-world problems, because the underlying hypotheses cannot be verified or the result of their application is uninformative. Our objective is to sketch some light on these issues by carefully analyze the most well-known methods and test some of them on standard benchmarks to evaluate their effectiveness.

Key words: model selection, generalization, cross validation, bootstrap, maximal discrepancy

1 Introduction

The selection of the appropriate Support Vector Machine (SVM) for solving a particular classification task is still an open problem. While the parameters of a SVM can be easily found by solving a quadratic programming problem, there are many proposals for identifying its hyperparameters (e.g. the kernel parameter or the regularization factor), but it is not clear yet which one is superior to the others.

A related problem is the evaluation of the generalization ability of the SVM. In fact, it is common use to select the optimal SVM (i.e. the optimal hyperparameters) by choosing the one with the lowest generalization error. However, there has been some criticism on this approach, because the true generalization error is obviously impossible to compute and it is necessary to resort to an upper bound of its value. Minimizing an upper bound of the error rate can be misleading and the actual value can be quite different from the true

D. Anguita et al.: *Theoretical and Practical Model Selection Methods for Support Vector Classifiers*, StudFuzz **177**, 159–179 (2005)
www.springerlink.com

one. On the other hand, an upper bound of the generalization error, if correctly derived, is of paramount importance for estimating the true applicability of the SVM to a particular classification task, especially on a real-world problem.

After introducing our notation in Sect. 2, we review, in the following section, some of the many methods available in the literature and describe precisely the underlying hypotheses or, in other words, when and how do the results hold, if applied for SVM model selection and error rate evaluation.

For this purpose, we put all the presented methods in the same framework, that is the probabilistic worst-case approach described by Vapnik [43], and present the error bounds, using a unique structure, as the sum of three terms: a training set dependent element (the *empirical error*), a complexity measure, which is often the quantity characterizing the method, and a penalization depending mainly from the training set cardinality. The three terms are not always present, but we believe that a common description of their structure is of practical help.

Some experimental trials and results are reported in Sect. 4, presenting the performance bounds related to various standard datasets.

The SVM algorithm implementation adopted for performing the experiments is called `cSVM` and has been developed during the last years by the authors. The code is written in `Fortran90` and is freely downloadable from the web pages of our laboratory (`http://www.smartlab.dibe.unige.it`).

2 The Normalized SVM

We recall here the main equations of the SVM for classification tasks: the purpose is to define our notation and not to describe the SVM itself. Therefore, we assume that the readers are familiar with the SVM. If this is not the case we refer them to the introductory chapter of this book or the vast amount of literature on this subject [10, 13, 14, 25, 41, 42, 43]. In the following text we will use the notation introduced by V. Vapnik [43].

The training set is composed of l patterns $\{\boldsymbol{x}_i, y_i\}_{i=1}^{l}$ with $\boldsymbol{x}_i \in \Re^n$ and $y_i \in \{-1, +1\}$. Usually the two classes are not perfectly balanced, so l^+ (l^-) will be the number of patterns for which $y_i = +1$ ($y_i = -1$) and $l^+ + l^- = l$. Obviously, if the two classes are perfectly balanced, then $l^+ = l^- = l/2$.

The SVM for classification is defined in the primal form as a perceptron, whose input space is nonlinearly mapped to a *feature* space, through a function $\boldsymbol{\Phi} : \Re^n \rightarrow \Re^N$, with $N \leq \infty$

$$\hat{y} = \operatorname{sign}\left(\boldsymbol{w} \cdot \boldsymbol{\Phi}(\boldsymbol{x}) + b\right). \tag{1}$$

The dual form of the above formulation allows to implicitly define the nonlinear mapping by means of positive definite kernel functions [22]:

Table 1. Some admissible kernels

Name	Kernel
Linear	$K(\boldsymbol{x}_1, \boldsymbol{x}_2) = \boldsymbol{x}_1 \cdot \boldsymbol{x}_2$
Normalized Gaussian	$K(\boldsymbol{x}_1, \boldsymbol{x}_2) = e^{-\frac{\gamma}{n}\|\boldsymbol{x}_1 - \boldsymbol{x}_2\|^2}$
Normalized Polynomial	$K(\boldsymbol{x}_1, \boldsymbol{x}_2) = \frac{(\boldsymbol{x}_1 \cdot \boldsymbol{x}_2 + n)^p}{\sqrt{(\boldsymbol{x}_1 \cdot \boldsymbol{x}_1 + n)^p}\sqrt{(\boldsymbol{x}_2 \cdot \boldsymbol{x}_2 + n)^p}}$

$$\hat{y} = \text{sign}\left(\sum_{i=1}^{l} y_i \alpha_i K(\boldsymbol{x}, \boldsymbol{x}_i) + b\right) \tag{2}$$

where appropriate kernels are summarized in Table 1.

Note that we make use of normalized kernels as they give better results [24]; a normalization respect to the dimensionality of the input space is also introduced. The variables γ and p are the kernel hyperparameters, which are chosen by the model selection procedure (see Sect. 3).

The primal formulation of the constrained quadratic programming problem, which must be solved for obtaining the SVM parameter, is:

$$\min_{\boldsymbol{w},\xi,b} E_{\mathcal{P}} = \frac{1}{2}\|\boldsymbol{w}\|^2 + C^+ \sum_{\substack{i=1 \\ y_i=+1}}^{l} \xi_i + C^- \sum_{\substack{i=1 \\ y_i=-1}}^{l} \xi_i \tag{3}$$

$$y_i\left(\boldsymbol{w} \cdot \boldsymbol{\Phi}(\boldsymbol{x}_i) + b\right) \geq 1 - \xi_i \quad \forall i = 1 \ldots l \tag{4}$$

$$\xi_i \geq 0 \quad \forall i = 1 \ldots l \tag{5}$$

where $C^+ = \frac{C\sqrt{\lambda}}{l}$ ($C^- = \frac{C}{l\sqrt{\lambda}}$) and C is another hyperparameter. This formulation is normalized respect to the number of patterns and allows the user to weight differently the positive and negative classes. In case of unbalanced classes, for example, a common heuristic is to weight them according to their cardinality [34]: $\lambda = \frac{C^+}{C^-} = \frac{l^-}{l^+}$.

The problem solved for obtaining $\boldsymbol{\alpha}$ is the usual dual formulation, which in our case is:

$$\min_{\boldsymbol{\alpha}} E_{\mathcal{D}} = \frac{1}{2}\boldsymbol{\alpha}^T \boldsymbol{Q} \boldsymbol{\alpha} - \sum_{i=1}^{l} \alpha_i \tag{6}$$

$$0 \leq \alpha_i \leq C^+ \quad \forall i = 1 \ldots l, y_i = +1 \tag{7}$$

$$0 \leq \alpha_i \leq C^- \quad \forall i = 1 \ldots l, y_i = -1 \tag{8}$$

$$\sum_{i=1}^{l} y_i \alpha_i = 0 \tag{9}$$

where $q_{ij} = K(\boldsymbol{x}_i, \boldsymbol{x}_j)$.

The algorithm used for solving the above problem is an improved SMO [27, 31, 35], translated in `Fortran90` from LIBSVM (the implementation of

SMO by C.-J. Lin [11]), which represents the current state of the art for SVM learning.

3 Generalization Error Estimate and Model Selection

The estimation of the generalization error is one of the most important issues in machine learning: roughly speaking, we are interested in estimating the probability that our learned machine will misclassify new patterns, assuming that the new data derives from the same (unknown) distribution underlying the original training set.

In the following text, we will use π to indicate the unknown generalization error and $\hat{\pi}$ will be its estimate. In particular, we are interested in a *worst-case* probabilistic setting: we want to find an upper bound of the true generalization error

$$\pi \leq \hat{\pi} \tag{10}$$

which holds with probability $1-\delta$, where δ is a user defined value (usually $\delta = 0.05$ or less, depending on the application). Note that this is slightly different from traditional statistical approaches, where the focus is on estimating an approximation of the error $\pi \approx \hat{\pi} \pm \Delta_\pi$, where Δ_π is a confidence interval.

The empirical error (i.e. the error performed on the training set) will be indicated by

$$\nu = \frac{1}{l} \sum_{i=1}^{l} I(y_i, \hat{y}_i) \tag{11}$$

where $I(y_i, \hat{y}_i) = 1$ if $y_i \neq \hat{y}_i$ and zero otherwise.

If possible, $\hat{\pi}$ will be indicated as the sum of three terms: (1) the *empirical error* ν, (2) a *correction term*, which takes in account the fact that ν is obviously an underestimation of the true error, and (3) a *confidence* term, which derives from the fact that the number of training patterns is finite.

The generalization estimate is closely related to the model selection, that is the procedure which allows to choose the optimal hyperparameters of the SVM (C, γ or p). The value $\hat{\pi}$ is computed for several hyperparameter values and the optimal SVM is selected as the one for which the minimum is attained.

Note that in many cases the bounds can be very loose and report a generalization error much greater than 1 (actually, any value greater than 0.5 is completely useless). However, as showed experimentally [1, 4, 18], the minimum of $\hat{\pi}$ often corresponds to the best hyperparameter choice, so its value can be of use for the model selection, if not for estimating the true generalization error.

In the following sections we will make use of some results from basic statistics and probability theory, which we recall here very briefly.

Let $X_i \in \{0, 1\}$ be a sample of a binary random variable X, $\pi = E[X]$ and $\nu = \frac{1}{l} \sum_{i=1}^{l} X_i$, then the Chernoff-Hoeffding bound [12, 26] states that

$$Pr\left\{\pi - \nu \geq \epsilon\right\} \leq e^{-2l\epsilon^2}, \tag{12}$$

that is, the sample mean converges in probability to the true one at an exponential rate.

It is also possible to have some information on the standard deviation of X: if the probability $Pr(X=1)=P$, then $\sigma = \sqrt{P(1-P)}$. As we are interested in computing σ, but we do not know P, it is possible to upper bound it

$$\sigma \leq \frac{1}{2} \tag{13}$$

(given that $0 \leq P \leq 1$) or estimate it from the samples:

$$\hat{\sigma} = \sqrt{\frac{1}{l-1}\sum_{i=1}^{l}\left(X_i - \frac{1}{l}\sum_{j=1}^{l}X_j\right)^2}. \tag{14}$$

Note that the upper bound (13) is always correct, even though it can be very loose, while the quality of the estimate (14) depends on the actual sample distribution and can be quite different from the true value for some extreme cases.

Traditionally, most generalization bounds are derived by applying the Chernoff-Hoeffding bound, but a better approach is to use an implicit form derived directly from the *cumulative binomial distribution* B_c of a binary random variable

$$B_c(e,l,\pi) = \sum_{i=0}^{e}\binom{l}{i}\pi^i(1-\pi)^{l-i} \tag{15}$$

which identifies the probability that l coin tosses, with a biased coin, will produce e or fewer heads. We can map the coin tosses to our problem, given that e can be considered the number of errors and, inverting B_c, we can bound the true error π, with a confidence δ [30], given the empirical error $\nu = \frac{e}{l}$:

$$\pi \leq B_c^{-1}(e,l,\delta) = \max_{\rho}\left\{\rho : B_c(e,l,\rho) \geq \delta\right\}. \tag{16}$$

The values computed through (16) are much more effective than the ones obtained through the Chernoff-Hoeffding bound. Unfortunately, (16) is in implicit form and does not allow to write explicitly an upper bound for π. For the sake of clarity, we will use (12) in the following text, but (16) will be used in the actual computations and experimental results.

Finally, we recall the DeMoivre-Laplace central limit theorem. Let $\overline{Y} = \frac{1}{l}(Y_1 + \ldots Y_l)$ be the average of l samples of a random variable deriving from a distribution with finite variance σ, then

$$\mathcal{P}\left(\frac{E\left[\overline{Y}\right] - \overline{Y}}{\sigma/\sqrt{l}}\right) \rightarrow \mathcal{N}(0,1) \quad \text{if} \quad n \rightarrow \infty \tag{17}$$

where $\mathcal{N}(0,1)$ is the zero mean and unit variance normal distribution. This result can be used for computing a confidence term for the generalization error:

$$Pr\left\{\pi - \nu \geq \epsilon\right\} = Pr\left\{\frac{\pi - \nu}{\sigma/\sqrt{l}} \geq \frac{\epsilon}{\sigma/\sqrt{l}}\right\} \approx Pr\left\{z \geq \frac{\epsilon}{\sigma/\sqrt{l}}\right\} \tag{18}$$

where z is normally distributed. Setting this probability equal to δ and solving for ϵ we obtain

$$\pi \leq \nu + \frac{\sigma}{\sqrt{l}} F^{-1}(1-\delta) \tag{19}$$

where $F^{-1}(\cdot)$ is the *inverse normal cumulative distribution function* [45]. Note that σ is unknown, therefore we must use a further approximation, such as (13) or (14), to replace its value.

The following sections describe several methods for performing the error estimation and the model selection, using the formulas mentioned above. However, it is important to note that the Chernoff-Hoeffding bound, expressed by (12), holds for any number of samples, while the approximation of (17) holds only asymptotically. As the cardinality of the training set is obviously finite, the results depending on the first method are exact, while in the second case they are only approximations. To avoid any confusion, we label the methods described in the following sections according to the approach used for their derivation: R indicates a rigorous result, that is a formula that holds even for a finite number of samples and for which all the underlying hypotheses are satisfied; H is used when a rigorous result is found but some heuristics is necessary to apply it in practice or not all the hypotheses are satisfied; finally, A indicates an estimation, which relies on asymptotic results and the assumption that the training samples are a good representation of the true data distribution (e.g. they allow for a reliable $\hat{\sigma}$ estimate). There are very well-known situations where these last assumptions do not hold [29], however, since the rigorous bounds can be very loose, the approximate methods are practically useful, if not theoretically fully justified.

Another subdivision of the methods presented here takes in account the use of an independent data set for assessing the generalization error. Despite the drawback of reducing the size of the training set for building an independent test set, in most cases this is the only way to avoid overly optimistic estimates. On the other hand, the most advanced methods try to estimate the generalization error directly from the empirical error: they represent the state-of-the-art of the research in machine learning, even though their practical effectiveness is yet to be verified.

Table 2 summarizes the generalization bounds considered in this work and detailed in the following sections. Each upper bound of the generalization error is composed by the sum of three terms named TRAIN, CORR and CONF: a void entry indicates that the corresponding term is not used in the computation. The last column indicates the underlying hypothesis in deriving the bounds.

Table 2. Generalization error upper bound estimates

Method	TRAIN	CORR	CONF	Bound
Training set	ν		$\frac{\hat{\sigma}}{\sqrt{l}}F^{-1}(1-\delta)$	A
Test set		ν_{test}	$\sqrt{\frac{-\ln\delta}{2m}}$	R
KCV		ν_{kcv}	$\sqrt{\frac{-k\ln\delta}{2l}}$	R
LOO		ν_{loo}	$\frac{\hat{\sigma}}{\sqrt{l}}F^{-1}(1-\delta)$	A
BOOT		ν_{boot}	$\frac{\hat{\sigma}_{\text{boot}}}{\sqrt{N_B}}F^{-1}(1-\delta)$	A
VC	ν	$\frac{\mathcal{E}}{2}\sqrt{1+\frac{4\nu}{\mathcal{E}}}$	$\frac{\mathcal{E}}{2}$	H
Margin		$\frac{2}{l}\left(h_{\text{eff}}\ln\frac{8el}{h_{\text{eff}}}\ln(32l)\right)$	$\frac{2}{l}\left(\ln\frac{l(16+\ln l)}{\delta}\right)$	R
MD	ν	$(1-2\overline{\nu})$	$3\sqrt{\frac{-\ln\delta}{2l}}$	H
Compression	ν	$\frac{d}{l-d}(\nu-\nu_d)$	$\sqrt{\frac{\ln\binom{l}{d}+\ln l-\ln\delta}{2(l-d)}}$	R

3.1 Training Set

In this case, the estimate of the generalization error is simply given by the error performed on the training set. This is an obvious underestimation of π, because there is a strong dependency between the errors, as all the data have been used for training the SVM. We can write

$$\hat{\pi}_{\text{train}} = \nu + \frac{\hat{\sigma}}{\sqrt{l}}F^{-1}(1-\delta)\,. \tag{20}$$

but it is recommended to avoid this method for estimating the generalization error or performing the model selection. If, for example, we choose a SVM with a gaussian kernel with a sufficiently large γ, then $\nu = 0$ even though $\pi \gg 0$.

3.2 Test Set

A test set of m patterns is not used for training purposes, but only to compute the error estimate. Using (12), it is possible to derive a theoretically sound upper bound for the generalization error:

$$\hat{\pi}_{\text{test}} = \nu_{\text{test}} + \sqrt{\frac{-\ln\delta}{2m}} \tag{21}$$

where ν_{test} is the error performed on the test set. The main problem of this approach is the waste of information due to the splitting of the data in two parts: the test set does not contribute to the learning process and the parameters of the SVM rely only on a subset of the available data. To circumvent this problem, it could be possible to retrain a new SVM on the entire dataset, without changing the hyperparameters found with the training-test data splitting. Unfortunately, there is no guarantee that this new SVM will perform as the original one.

Furthermore, different splittings can make the algorithm behave in different ways and severely affect the estimation. A better solution is to use a *resampling* technique as described in the following sections.

3.3 K-fold Cross Validation

The K-fold Cross Validation (KCV) technique is similar to the Test Set technique. The training set is divided in k parts consisting of k/l patterns each: $k-1$ ones are used for training, while the remaining one is used for testing. Then

$$\hat{\pi}_{kcv} = \nu_{\text{test}}^{(k)} + \sqrt{\frac{-k \ln \delta}{2l}} \tag{22}$$

where $\nu_{\text{test}}^{(k)} = \frac{k}{l} \sum_{i=1}^{l/k} I(y_i^{(k)}, \hat{y}_i^{(k)})$ and the superscript k indicates the part used as test set. However, differently from the Test Set technique, the procedure is usually iterated k times, using each one of the k parts as test set exactly once.

The idea behind this iteration is the improvement of the estimate of the empirical error, which becomes the average of the errors on each part of the training set

$$\nu_{kcv} = \frac{1}{k} \sum_{i=1}^{k} \nu_{\text{test}}^{(k)} \,. \tag{23}$$

Furthermore, this approach ensures that all the data is used for training, as well as for model selection purposes.

One could expect that the confidence term would improve in some way. Unfortunately, this is not the case [7, 28], because there is some statistical dependency between the training and test sets due to the K-fold procedure. However, it can be shown that the confidence term, given by (12), is still correct and (22) can be simply rewritten using ν_{kcv} instead of $\nu_{\text{test}}^{(k)}$. Note that we are not aware of any similar result, which holds also for (16), even though we will use it in practice.

While the previous result is rigorous, it suffers from a quite large confidence term, which can result in a pessimistic estimation of the generalization error. In general, the estimate can be improved using the asymptotic approximation instead of the Chernoff–Hoeffding bound; however, very recent results show that this kind of approach suffers from further problems and special care must be used in the splitting procedure [6].

As a final remark, note that common practice suggests $k = 5$ or $k = 10$: this is a good compromise between the improvement of the estimate and a large confidence term, which increases with k. Note also that the K-fold procedure could be iterated up to $\binom{l}{k}$ times without repeating the same training-test set splitting, but this approach is obviously infeasible.

There is still a last problem with KCV, which lies in the fact that this method finds k different SVMs (each one trained on a set of $(k-1)l/k$ samples) and it is not obvious how to combine them.

There are at least three possibilities: (1) retrain a SVM on the entire dataset using, eventually, the same hyperparameters found by KCV, (2) pick one trained SVM randomly, each time that a new sample arrives or (3) average in some way the k SVMs.

It is interesting to note that option (1), which could appear as the best solution and is often used by practitioners, is the less justified from a theoretical point of view. In this case, the generalization bound should take in account the behaviour of the algorithm when learning a different (larger) dataset, as for the Test Set method: this involves the computation of the VC dimension [5] and, therefore, severe practical difficulties. On the practical side, it is easy to verify that the hyperparameters of the trained SVMs must be adapted to the larger dataset in case of retraining and we are not aware of any reliable heuristic for this purpose.

Option (2) is obviously memory consuming, because k SVMs must be retained in the feed–forward phase, even though only one will be randomly selected for classifying a new data point. Note, however, that this is the most theoretically correct solution.

Method (3) can be implemented in different ways: the simplest one is to consider the output of the k SVMs and assign the class on which most of the SVM agree (or randomly if it happens to be a tie). Unfortunately, with this approach, all the trained SVMs must be memorized and applied to the new sample. We decided, instead, to build a new SVM by computing the average of the parameters of the k SVMs: as pointed out in the experimental section, this heuristic works well in practice and results in a large saving of both memory and computation time during the feed–forward phase.

3.4 Leave-One-Out

The Leave-one-out (LOO) procedure is analogous to the KCV, but with $k = l$. One of the patterns is selected as test set and the training is performed on the remaining $l-1$ ones; then, the procedure is iterated for all the patterns in the training set. The test set error is then defined as $\nu_{\text{loo}} = \frac{1}{l}\sum_{i=1}^{l} I(y_i^{\text{LOO}}, \hat{y}_i^{\text{LOO}})$ where LOO indicates the pattern deleted by the training set.

Unfortunately, the use of the Chernoff–Hoeffding bound is not theoretically justified in this case, because the test patterns are not independent. Therefore, a formula like (21), replacing m with l, would be wrong. At the same time, the correct use of the bound does not provide any useful information, because setting $k = l$ produce a fixed and overly pessimistic confidence term $(\sqrt{-\ln\delta})$.

Intuitively, however, the bound should be of some help, because the dependency between the test patterns is quite mild. The underlying essential concept for deriving a useful bound is the *stability* of the algorithm: if the algorithm does not depend heavily on the deletion of a particular pattern, then it is possible, at least in theory, to derive a bound similar to (21). This approach has been developed in [16, 17, 38] and applied to the SVM in [9]. Unfortunately, some hypothesis are not satisfied (e.g. the bound is valid only for $b = 0$ and for a particular cost function), but nevertheless, its formulation is interesting because resembles the Chernoff–Hoeffding bound. Using our notation and normalized kernels we can write:

$$\hat{\pi}_{\text{loo}}^{\mathcal{H}} = \nu_{\text{loo}} + 2C + (1 + 8lC)\sqrt{\frac{-\ln\delta}{2l}} \,. \tag{24}$$

It is clear, however, that the above bound is nontrivial only for very small values of C, which is a very odd choice for SVM training. Therefore, it is more effective to derive an asymptotic bound, which can be used in practice:

$$\hat{\pi}_{\text{loo}}^{\mathcal{A}} = \nu_{\text{loo}} + \frac{\hat{\sigma}}{\sqrt{l}} F^{-1}(1-\delta) \tag{25}$$

where $\hat{\sigma}$ is given by (14).

Note, however, that the same warnings of KCV apply to LOO: the variance estimate is not unbiased [6] and the LOO procedure finds l different trained SVMs. Fortunately, the last one is a minor concern because the training set of the LOO procedure differs from the original one only by one sample, therefore it reasonable to assume that the final SVM can be safely retrained on the entire dataset, using the same hyperparameters found with the LOO procedure.

As a final remark, note that we have neglected some LOO based methods, like the one proposed by Vapnik and Chapelle [44]. The main reason is that they provide an upper bound of the LOO error, while we are computing its exact value: the price to pay, in our case, is obviously a greater computational effort, but its value is more precise. In any case, both approaches suffer from the asymptotic nature of the LOO estimate.

3.5 Bootstrap

The Bootstrap technique [19] is similar in spirit to the KCV, but has a different training-test splitting technique. The training set is built by extracting l patterns *with replacement* from the original training set. Obviously, some of the patterns are picked up more than once, and some others are left out from the new training set: these ones can be used as an independent test set. The bootstrap theory shows that, on average, one third of the patterns ($l/e \approx 0.368l$) are left for the test set.

The new training set, as created with this procedure, is called a *bootstrap replicate* and up to $N_B = \binom{2l-1}{l}$ different replicates can be generated. In practice, $N_B = 1000$ or even less replicates suffice for performing a good error estimation [1].

The estimation of the generalization error is given by the average test error performed on each bootstrap replicate:

$$\nu_{\text{boot}} = \frac{1}{N_B} \sum_{i=1}^{N_B} \nu_{\text{test}}^{(i)} . \tag{26}$$

Unfortunately, there are no rigorous theoretical results for expressing the confidence term. The problem lies in the particular procedure used for building the training set, which makes impossible to use the Chernoff-Hoeffding bound. An approximate bound can be found by assuming that the distribution of the test errors is an approximation of the true error distribution. If we assume, by the law of large numbers, that this distribution is gaussian, then its standard error can be estimated by:

$$\hat{\sigma}_{\text{boot}} = \sqrt{\frac{1}{N_B - 1} \sum_{i=1}^{N_B} \left(\nu_{\text{test}}^{(i)} - \nu_{\text{test}}^{\text{boot}} \right)^2} . \tag{27}$$

and the bound is given by

$$\hat{\pi}_{\text{boot}} = \nu_{\text{boot}} + \frac{\hat{\sigma}_{\text{boot}}}{\sqrt{N_B}} F^{-1}(1-\delta) . \tag{28}$$

A possibly more accurate estimate can be obtained by avoiding the gaussian hypothesis and computing the δ-th percentile point of the test error distribution directly, or making use of nested bootstrap replicates [36], but we believe that the increased precision is not worth the much larger computational effort.

In the literature, the naive bootstrap described above is considered to be pessimistic biased, therefore some improvements have been suggested by taking in account not only the error performed on the test set, but also the one on the training set [20].

The first proposal is called the *bootstrap 632* and builds on the consideration that, on average, $0.632l$ patterns are used for training and $0.368l$ for testing. However, this estimate is overly optimistic because the training set error is often very low, due to overfitting.

A further proposal is the *bootstrap 632+*, which tries to balance the two terms according to an estimate of the amount of overfitting [33].

Our experience is that the two last techniques are not suited for the case of upper bounding the true generalization error of the SVM: even the 632+ version is too optimistic for our purposes. In fact, any estimation which lowers the test set error is quite dangerous, considering that the confidence term can be made negligible by simply increasing the number of bootstrap replicates.

As a last remark, note that, similarly to KCV, we are left with a number of different SVMs equal to the number of bootstrap replicates. However, the cardinality of each replicate is exactly the same as the original training set. Therefore, the final SVM can be safely computed by training it on the original training set, with the hyperparameters chosen by the model selection procedure.

3.6 VC-Bound

The SVM builds on the Vapnik–Chernovenkis (VC) theory [43], which provides distribution–independent bounds on the generalization ability of a learning machine. The bounds depend mainly on one quantity, the VC-dimension (h), which measures the complexity of the machine. The general bound can be written as

$$\hat{\pi}_{VC} = \nu + \frac{\mathcal{E}}{2}\sqrt{1 + \frac{4\nu}{\mathcal{E}}} + \frac{\mathcal{E}}{2} \tag{29}$$

where

$$\mathcal{E} = 4\frac{h\left(\ln\frac{2l}{h} + 1\right) - \ln\frac{\delta}{4}}{l} \ . \tag{30}$$

This form uses an improved version of the Chernoff-Hoeffding bound, which is tighter for $\nu \to 0$. The above formula could be used for our purposes by noting that the VC-dimension of a maximal margin perceptron, which corresponds to a linear SVM, is bounded by

$$h \leq R^2 \|\boldsymbol{w}\|^2 \ , \tag{31}$$

where R is the radius of the smallest hypersphere containing the data.

Unfortunately, the use of this bound for estimating the generalization ability of a nonlinear SVM is not theoretically justified. In fact, the nonlinear mapping is not taken in account by the computation of the VC-dimension: it can be shown, for example, that a SVM with a gaussian kernel has infinite VC-dimension for large values of γ [10]. Furthermore, the value of the margin in (31), should be derived in a data-independent way; in other words, a structure of nested sets of perceptrons with decreasing margin should be defined *before seeing the data*. Finally, (31) is derived in case of perfect training ($\nu = 0$), which is seldom the case in practice and surely an unnecessary constraint on the VC-bound.

Despite all these drawbacks, the bound can be of some practical use for model selection, if not for estimating the generalization error. In fact, the minimum of the estimated error as a function of the SVM hyperparameters often coincides with the optimal choice [1].

3.7 Margin Bound

The VC-theory has been extended to solve the drawbacks described in the previous section. In particular, the following bound depends on the margin of the separating hyperplane, a quantity which is computed *after* seeing the data [40]:

$$\hat{\pi}_m = \frac{2}{l}\left(h_{\text{eff}} \ln\left(\frac{8el}{h_{\text{eff}}}\right) \ln(32l)\right) + \frac{2}{l}\left(\ln\left(\frac{l(16+\ln l)}{\delta}\right)\right) \tag{32}$$

where

$$h_{\text{eff}} \leq 65\left(R\|\boldsymbol{w}\| + 3\sum_{i=1}^{l} \xi_i\right)^2 . \tag{33}$$

Note that the training error ν does not appear explicitly in the above formulas, but is implicitly contained in the computation of h_{eff}. In fact, $\nu \leq \frac{1}{l}\sum_{i=1}^{l} \xi_i$.

Unfortunately, the bound is too loose for being of any practical use, even though it gives some sort of justification on how the SVM works.

3.8 Maximal Discrepancy

The theory sketched in the two previous sections tries to derive generalization bounds by using the notion of *complexity* of the learning machine in a data-independent (h) or a data-dependent way (h_{eff}). In particular, for SVMs, the important element for computing its complexity is the *margin* $M = \frac{1}{\|\boldsymbol{w}\|}$.

The Maximal Discrepancy (MD) approach, instead, tries to measure the complexity of a learning machine using the training data itself and modifying it in a clever way. A new training set is built by flipping the targets of half of the training patterns, then the discrepancy in the machine behaviour, when learning the original and the modified data set, is selected as an indicator of the complexity of the machine itself when applied to solve the particular classification task.

Formally, this procedure gives rise to a generalization bound, which appears to be very promising [4]:

$$\hat{\pi}_{md} = \nu + (1 - 2\overline{\nu}) + 3\sqrt{\frac{-\ln \delta}{2l}} \tag{34}$$

where $\overline{\nu}$ is the error performed on the modified dataset (half of the target values have been flipped).

Note that, despite the theoretically soundness of this bound, its application to this case is not rigorously justified, because the SVM does not satisfy all the underlying hypotheses (see [4] for some insight on this issue), however it is one of the best methods, among the ones using only the information of the training set.

3.9 Compression Bound

A completely different approach to the error estimation problem makes use of the notion of *compressibility*. It can be applied to our case, because the SVM compresses the information carried by the training data by transferring it into a (usually smaller) number of patterns: the support vectors. It is a well-known fact that "compression" is related to "good generalization" [21] and recent results give some deeper insight on this topic [23]. We derive here a Compression Bound following [30], which suggests a very clever and simple way to deal with compression algorithms. Given d, the number of support vectors, we can consider the remaining $l-d$ samples as an independent "virtual" test set, because the SVM will find the same set of parameters even removing them from the training set, and, therefore, apply (16) on all the possible choices of d samples:

$$B_c^{-1}\left(e', l-d, \frac{\delta}{l\binom{l}{l-d}}\right) \tag{35}$$

where e' is the number of errors performed on the "virtual" test set.

By applying the Chernoff–Hoeffding formula we obtain an upper bound of the generalization error:

$$\hat{\pi}_{\text{comp}} = \nu + \frac{d}{l-d}\left(\nu - \nu_d\right) + \sqrt{\frac{\ln\binom{l}{d} + \ln l - \ln \delta}{2\left(l-d\right)}} \tag{36}$$

where ν_d is the error performed on the support vectors.

4 Practical Evaluation of Model Selection Methods

4.1 Experimental Setup

We tested the above described methods using 13 datasets, described in Table 3, which have been collected and prepared by G. Rätsch for the purpose of benchmarking machine learning algorithms [37].

The model space, which is searched for the optimal hyperparameters, is composed of 247 SVMs with Gaussian kernel, featuring a combination of 13 different error penalization values (C) and 19 different kernel widths (γ). More precisely, each of the considered models can be represented as a node in a 13×19 grid where the two hyperparameters take the following values:

$$\begin{aligned} C &= \{10^2, 10^{2.5}, 10^3, \ldots, 10^7, 10^{7.5}, 10^8\} \\ \gamma &= \{10^{-5}, 10^{-4.5}, 10^{-4}, \ldots, 10^3, 10^{3.5}, 10^4\} . \end{aligned}$$

Note that the sweep on the hyperparameters follows a logarithmic scale.

For each pair (C, γ), we estimated the generalization error with each one of the methods described in the previous section (except the VC Bound and the Margin Bound, which are known to be too pessimistic). If the method required an out-of-sample estimate (e.g. the Bootstrap or the Cross Validation), the samples were extracted from the training set.

Table 3. The datasets used for the experiments

Name	No. of Features	Training Samples	Test Samples
Banana	2	400	4900
Breast-Cancer	9	200	77
Diabetis	8	468	300
Flare-Solar	9	666	400
German	20	700	300
Heart	13	170	100
Image	18	1300	1010
Ringnorm	20	400	7000
Splice	60	1000	2175
Thyroid	5	140	75
Titanic	3	150	2051
Twonorm	20	400	7000
Waveform	21	400	4600

Each method identified an optimal SVM (the one with the lowest estimated error), which was subsequently used to classify the test set: the result of this classification was considered a good approximation of the *true* error, since none of the samples of the test set was used for training or model selection purposes.

Due to the large amount of computation needed to perform all the experiments, we used the ISAAC system in our laboratory, a cluster of P4–based machines [2], and carefully coded the implementation to make use of the vector instructions of the P4 CPU (see [3] for an example of such coding approach). Furthermore, we decided to use only one instance of the datasets, which were originally replicated by several random training-test splittings. Despite the use of this approach, the entire set of experiments took many weeks of cpu time to be completed.

We tested 7 different methods: the Bootstrap with 10 and 100 replicates (`BOOT10`, `BOOT100`), the Compression Bound (`COMP`), the K-fold Cross Validation (`KCV`) with $k = 9$ or $k = 10$, depending on the cardinality of the training set, the Leave-One-Out (`LOO`), the Maximal Discrepancy (`MD`), and the Test Set (`T30`), extracting 30% of the training data for performing the model selection.

For comparison purposes, we also selected the optimal SVM by learning the entire training set and identifying the hyperparameters with the lowest test set error: in this way we know a lower bound on the performance attainable by any model selection procedure.

Finally, we tested a fixed value of the hyperparameters, by setting $C = 1000$ and $\gamma = 1$ (`1000-1`), to check if the model selection procedure can be avoided, given the fact that all the data and the hyperparameters in the optimization problem are carefully normalized.

4.2 Discussion of the Experimental Results

The first question that we would like to answer is: which method selects the optimal hyperparameters of the SVM, that is, the model with the lowest error on the test set. The results of the experiments are summarized in Table 4.

Table 4. Model selection results. The values indicate the error performed on the test set (in percentage). The best figures are marked with "+", while the worst ones are marked with "–"

Dataset	BOOT10	BOOT100	COMP	KCV	LOO	MD	T30	1000/1	Test
Banana	10.53	10.53	16.74 –	10.65	10.53	12.31	12.84	10.51 +	10.00
Breast-	27.27 +	28.57	35.07 –	27.27 +	32.47	27.27 +	31.17	31.17	25.97
Diabetis	23.00 +	23.00 +	30.67 –	23.67	23.67	23.67	23.33	23.00 +	22.67
Flare-	33.00 +	33.00 +	33.75	33.75	33.75	34.00 –	33.25	33.00 +	32.75
German	22.67	22.67	30.00 –	23.67	22.33 +	30.00 –	23.00	23.30	21.33
Heart	18.00	18.00	31.00 –	15.00	15.00	13.00 +	17.00	21.00	13.00
Image	2.97 +	2.97 +	3.76	2.87 +	2.97 +	14.16 –	3.86	8.61	2.97
Ringn.	1.74 +	1.74 +	2.54	1.71 +	1.74 +	3.04 –	2.31	1.86	1.74
Splice	11.68	10.85 +	15.91	11.08	11.91	16.87 –	12.05	12.33	10.85
Thyroid	5.33	5.33	2.67 +	4.00	5.33	5.33	5.33	7.14 –	1.33
Titanic	22.72 –	22.38 +	22.72 –	22.72 –	22.72 –	22.38 +	22.72 –	22.72 –	22.38
Twon.	2.39 +	2.40	4.39 –	3.11	2.37 +	3.10	2.71	2.83	2.31
Wavef.	9.65	9.50 +	11.96 –	9.48 +	9.50 +	10.63	10.07	10.04	9.50

All the "classical" resampling methods (BOOT10, BOOT100, KCV, LOO) perform reasonably well, with a slight advantage of BOOT100 on the others. The T30 method, which is also a classical practitioner approach, clearly suffers from the dependency of the particular training-test data splitting: resampling methods, instead, appear more reliable in identifying the correct hyperparameters setting, because the training-test splitting is performed several times. The two methods based on the Statistical Learning Theory (COMP, MD) do not appear as effective as expected. In particular, method COMP is very poorly performing, while method MD shows a contradictory behaviour. It is interesting to note, however, that the MD method performs poorly only on the largest dataset (Image), while it selects a reasonably good model in all the other cases.

The unexpected result is that, setting the hyperparameters to a fixed value is not a bad idea. This approach performs reasonably well and does not require any computationally intensive search on the model space.

The KCV method is also worth of attention, because in three cases (Image, Ringnorm and Waveform) it produces a SVM that performs slightly better than the one obtained by optimizing the hyperparameters on the test set. This is possible because the SVM generated by the KCV is, in effect, an ensemble classifier, that is, the combination of 10 SVMs, each one trained on the 9/10th of the entire training set. The effect of combining the SVMs results in a boost of performance, as predicted also by theory [39].

In order to rank the methods described above we compute an average quality index $\mathcal{Q}_D$, which expresses the average deviation of each SVM from the optimal one. Given a model selection method, let $\mathcal{E}_S^i$ the error achieved by the selected SVM on the i-th training set ($i \in \mathcal{D}$) and $\mathcal{E}_T^i$ the error on the corresponding test set, then

$$\mathcal{Q}_D = \frac{1}{\mathrm{card}(\mathcal{D})} \sum_{i \in \mathcal{D}} \frac{\max\left(0, \mathcal{E}_S^i - \mathcal{E}_T^i\right)}{\mathcal{E}_T^i} \times 100\ (\%) \tag{37}$$

where, in our case, $\mathrm{card}(\mathcal{D}) = 13$.

The ranking of the methods, according to $\mathcal{Q}_D$, is given in Table 5. Note that, if we neglect the result on the Image dataset, the MD quality index improves to $\mathcal{Q}_D = 46.1\%$.

Table 5. Ranking of model selection methods, according to their ability to select the SVM with the lowest test error

Method	KCV	BOOT100	BOOT10	LOO	T30	COMP	1000/1	MD
$\mathcal{Q}_D$ (%)	21.8	28.2	28.6	28.6	37.7	50.5	59.5	71.5

The second issue that we want to address with the above experiments is the ability of each method to provide an effective estimate of the generalization error of the selected SVM.

Table 6 shows the estimates for each dataset, using the bounds summarized in Table 2.

These results clearly show why the estimation of the generalization error of a learning machine is still the holy grail of the research community. The methods relying on asymptotic assumptions (`BOOT10`, `BOOT100`, `LOO`) provide very good estimates, but in many cases they underestimate the true error because they do not take in account that the cardinality of the training set is finite. This behaviour is obviously unacceptable, in a worst-case setting, where we are interested in an upper bound of the error attainable by the classifier on future samples. On the other hand, the methods based on Statistical Learning Theory (`COMP`, `MD`) tend to overestimate the true error. In particular, `COMP` almost never provides a consistent value, giving an estimate greater than 50%, which represents a random classifier, most of the times. The `MD` method, instead, looks more promising because, despite its poor performance in absolute terms, it provides a consistent estimate most of the times. The `KCV` method lies in between the two approaches, while the training-test splitting methods (`T30`) shows to be unreliable, also in this case, because its performance depends heavily on the particular splitting.

A ranking of the methods can be computed, as in the previous case, by defining an average quality index $\mathcal{Q}_G$, which expresses the average deviation

Table 6. Generalization error estimates. The values indicate the estimate given by each method (in percentage), which must be compared with the true value of Table 4. The symbols "↓" and "↑" indicate an unconsistent value, that is an underestimation or a value greater than 50%, respectively. Among the consistent estimates, the best ones are marked with "+", while the worst ones are marked with "–"

Dataset	BOOT10	BOOT100	COMP	KCV	LOO	MD	T30
Banana	10.75	10.66 +	49.44 –	21.44	12.47	41.11	9.63 ↓
Breast-Cancer	28.30 +	25.44 ↓	66.78 ↑	45.56 –	27.37 ↓	60.96 ↑	30.38 ↓
Diabetis	24.34	22.55 ↓	69.28 ↑	34.69	23.81 +	42.83 –	24.82
Flare-Solar	34.25	33.07 +	85.02 ↑	42.48	33.66 ↓	48.16 –	39.40
German	25.58 +	25.80	72.89 ↑	34.17	26.36	43.88 –	29.65
Heart	18.61	18.46 +	66.12 ↑	39.56	19.85	45.22 –	14.51 ↓
Image	4.59	4.04	24.67 –	6.90	3.43 +	22.49	3.67 ↓
Ringnorm	1.11 ↓	1.71 ↓	35.69 –	11.31	1.82 +	24.11	3.89
Splice	14.36	13.18	63.15 ↑	18.72	13.27 +	25.10 –	15.90
Thyroid	5.46 +	4.60 ↓	29.11	29.67	6.16	47.46 –	14.24
Titanic	23.40 +	23.47	73.89 ↑	51.08 ↑	26.12	55.31 ↑	39.60 –
Twonorm	1.60 ↓	1.75 ↓	20.25	11.32	2.17 ↓	22.86 –	5.15 +
Waveform	11.08	11.24	50.38 ↑	21.44	10.80 +	37.86 –	15.70

of each method in predicting the generalization ability of the selected model. Let $\mathcal{E}_S^i$ the error achieved by the selected SVM on the i-th test set ($i \in \mathcal{D}$) and $\mathcal{E}_G^i$ the generalization error estimate, then

$$\mathcal{Q}_G = \frac{1}{\text{card}(\mathcal{D})} \sum_{i \in \mathcal{D}} \frac{|\mathcal{E}_S^i - \mathcal{E}_G^i|}{\mathcal{E}_S^i} \times 100\ (\%) \,. \tag{38}$$

The ranking of the methods, according to $\mathcal{Q}_G$, is given in Table 7.

Table 7. Ranking of model selection methods, according to their ability to provide a good generalization error estimate

Method	BOOT100	LOO	BOOT10	T30	KCV	MD	COMP
$\mathcal{Q}_G(\%)$	11.8	13.0	15.3	45.2	182.6	261.9	375.1

As a last remark, it is worthwhile mentioning the computational issues related to each method. The less demanding method, in terms of cpu time, is T30, which requires a single learning with only 7/10th of the training set samples. Then, COMP follows with a single learning of the entire training set. The method MD, instead, requires two learning phases of the entire training set and the learning of the random targets, for computing the Maximal Discrepancy, can be quite time consuming. The most demanding ones are obviously the resampling methods, which require 10 (BOOT10, KCV) or even 100 (BOOT100) different learning phases. Finally, the less efficient, in this respect, is the LOO method, whose computational requirements grow linearly with the cardinality of the training set and can be prohibitively expensive for large datasets, unless some method for reusing previous solutions, like the "alpha seeding" [15], is adopted.

5 Conclusion

We have reviewed and compared several methods for selecting the optimal hyperparameters of a SVM and estimating its generalization ability. Both classical and more modern ones (except the COMP method) can be used for model selection purposes, while the choice is much more difficult when dealing with the generalisation estimates.

Classical methods works quite well, but can be too optimistic due to the underlying asymptotic assumption, which they rely on. On the other hand, more modern methods, which have been developed for the non-asymptotic case, are too pessimistic and in many cases do not provide any useful result. It is interesting to note, however, that the MD method is the first one, after many years of research in Machine Learning, which is able to give consistent values. If, in the future, it will be possible to improve it by making it more reliable on the model selection procedure, through a resampling approach, and by lowering the pessimistic bias of the confidence term, it could become the method of choice for classification problems. Some preliminary results in this direction appear to be promising [8].

Until then, our suggestion is to use a classical resampling method, with relatively modest computational requirements, like BOOT10 or KCV, taking in account the caveats mentioned above.

References

1. Anguita, D., Boni, A., Ridella, S. (2000) Evaluating the generalization ability of Support Vector Machines through the Bootstrap. Neural Processing Letters, **11**, 51–58
2. Anguita, D., Bottini, N., Rivieccio, F., Scapolla, A.M. (2003) The ISAAC Server: a proposal for smart algorithms delivering. Proc. of EUNITE'03, 384–388
3. Anguita, D., Parodi, G., Zunino, R. (1994) An efficient implementation of BP on RISC-bases workstations. Neurocomputing, **6**, 57–65
4. Anguita, D., Ridella, S., Rivieccio, F., Zunino, R. (2003) Hyperparameter design criteria for support vector classifiers, Neurocomputing, **51**, 109–134
5. Anthony, M., Holden, S.B. (1998) Cross-validation for binary classification by real-valued functions: theoretical analysis. Proc. of the 11th Conf. on Computational Learning Theory, 218–229
6. Bengio, Y., Grandvalet, Y. (2004) No unbiased estimator of the variance of K-fold cross validation. In: Advances of Neural Processing Systems, **16**, The MIT Press

7. Blum, A., Kalai, A., Langford, J. (1999) Beating the hold-out: bounds for K-fold and progressive cross-validation. Proc. of the 12th Conf. on Computational Learning Theory, 203–208
8. Boucheron, S., Bousquet, O., Lugosi, G. Theory of classification: a survey of recent advances. Probability and Statistics, preprint
9. Bousquet, O., Elisseeff, A. (2002) Stability and generalization. Journal of Machine Learning Research, **2**, 499–526
10. Burges, C.J.C. (1998) A tutorial on Support Vector Machines for classification. Data Mining and Knowledge Discovery, **2**, 121–167
11. Chang, C.-C., Lin, C.-J. LIBSVM: a Library for Support Vector Machines. Dept. of Computer Science and Information Engineering, National Taiwan University, `http://csis.ntu.edu.tw/~cjlin`
12. Chernoff, H. (1952) A measure of asymptotic efficiency for tests of a hypothesis based on the sum of observations. Annals of Mathematical Statistics, **23**, 493–509
13. Cortes, C., Vapnik, V. (1995) Support Vector Networks. Machine Learning, **20**, 273–297
14. Cristianini, N., Shawe–Taylor, J. (2001) An introduction to Support Vector Machines. Cambridge University Press
15. De Coste, D., Wagstaff, K. (2000) Alpha seeding for support vector machines. Proc. of the 6th ACM SIGKDD Int. Conf. on Knowledge Discovery and Data Mining, 345–349
16. Devroye, L., Wagner, T. (1979) Distribution-free inequalities for the deleted and hold-out error estimates. IEEE Trans. on Information Theory, **25**, 202–207
17. Devroye, L., Wagner, T. (1979) Distribution-free performance bounds for potential function rules. IEEE Trans. on Information Theory, **25**, 601–604
18. Duan, K., Keerthi, S., Poo, A. (2001) Evaluation of simple performance measures for tuning SVM parameters. Tech. Rep. CD-01-11, University of Singapore
19. Efron, B., Tibshirani, R. (1993) An introduction to the bootstrap, Chapmann and Hall
20. Efron, B., Tibshirani, R. (1997) Improvements on cross-validation: the *632+* bootstrap method. J. Amer. Statist. Assoc., **92**, 548–560
21. Floyd, S., Warmuth, M. (1995) Sample compression, learnability, and the Vapnik-Chervonenkis dimension. Machine Learning, **21**, 269–304
22. Genton, M.G. (2001) Classes of kernels for machine learning: a statistics perspective. Journal of Machine Learning Research, **2**, 299–312
23. Graepel, T., Herbrich, R., Shawe–Taylor, J. (2000) Generalization error bounds for sparse linear classifiers. Proc. of the 13th Conf. on Computational Learning Theory, 298–303
24. Graf, A.B.A., Smola, A.J., Borer, S. (2003) Classification in a normalized feature space using support vector machine. IEEE Trans. on Neural Networks, **14**, 597–605
25. Herbrich, R. (2002) Learning Kernel Classifiers, The Mit Press
26. Hoeffding, W. (1963) Probability inequalities for sums of bounded random variables. American Statistical Association Journal, **58**, 13–30
27. Keerthi, S.S., Shevade, S.K., Bhattacharyya, C., Murthy, K.R.K. (2001) Improvements to Platt's SMO algorithm for SVM classifier design. Neural Computation, **13**, 637–649
28. Kalai, A. (2001) Probabilistic and on-line methods in machine learning. Tech. Rep. CMU-CS-01-132, Carnegie Mellon University

29. Kohavi, R. (1995) A study of cross-validation and boostrap for accuracy estimation and model selection. Proc. of the Int. Joint Conf. on Artificial Intelligence
30. Langford, J. (2002) Quantitatively tight sample bounds. PhD Thesis, Carnegie Mellon University
31. Lin, C.-J. (2002) Asymptotic convergence of an SMO algorithm without any assumption. IEEE Trans. on Neural Networks, **13**, 248–250
32. Luenberger, D.G. (1984) Linear and nonlinear programming. Addison–Wesley
33. Merler, S., Furlanello, C. (1997) Selection of tree-based classifiers with the bootstrap 632+ rule. Biometrical Journal, **39**, 1–14
34. Morik, K., Brockhausen, P., Joachims, T. (1999) Combining statistical learning with a knowledge-based approach: a case study in intensive care monitoring. Proc. of the 16th Int. Conf. on Machine Learning, 268–277
35. Platt, J. (1999) Fast training of support vector machines using sequential minimal optimization. In: Advances in Kernel Methods: Support Vector Learning, Schölkopf, B., Burges, C.J.C., Smola, A. (eds.), The MIT Press
36. Politis, D.N. (1998) Compute-intensive methods in statistical analysis. IEEE Signal Processing Magazine, **15**, 39–55
37. Rätsch, G., Onoda, T., Müller, K.-R. (2001) Soft margins for AdaBoost. Machine Learning, **42**, 287–320
38. Rogers, W., Wagner, T. (1978) A finite sample distribution-free performance bound for local discrimination rules. Annals of Statistics, **6**, 506–514
39. Schapire, R.E. (1990) The strength of weak learnability. Machine Learning, **5**, 197–227
40. Shawe–Taylor, J., Cristianini, N. (2000) Margin Distribution and Soft Margin. In: Advances in Large Margin Classifiers, Smola, A.J., Bartlett, P.L., Schölkopf, B., Schuurmans D. (eds.), The MIT Press
41. Smola, A.J., Bartlett, P.L., Schölkopf, B., Schuurmans D. (2000) Advances in large margin classifiers, The MIT Press
42. Schölkopf, B., Burges, C.J.C., Smola, A. (1999) Advances in kernel methods: Support Vector learning, The MIT Press
43. Vapnik, V. (1998) Statistical learning theory, Wiley, 1998
44. Vapnik, V., Chapelle, O. (2000) Bounds on error expectation for Support Vector Machines. Neural Computation, **12**, 2013–2036
45. Wichura, M.J. (1988) Algorithm AS241: the percentage points of the normal distribution. Applied Statistics, **37**, 477–484

Adaptive Discriminant and Quasiconformal Kernel Nearest Neighbor Classification

J. Peng[1], D.R. Heisterkamp[2], and H.K. Dai[2]

[1] Electrical Engineering and Computer Science Department, Tulane University, New Orleans, LA 70118
jp@eecs.tulane.edu
[2] Computer Science Department, Oklahoma State University, Stillwater, OK 74078
{doug,dai}@cs.okstate.edu

Abstract. Nearest neighbor classification assumes locally constant class conditional probabilities. This assumption becomes invalid in high dimensions due to the curse-of-dimensionality. Severe bias can be introduced under these conditions when using the nearest neighbor rule. We propose locally adaptive nearest neighbor classification methods to try to minimize bias. We use locally linear support vector machines as well as quasiconformal transformed kernels to estimate an effective metric for producing neighborhoods that are elongated along less discriminant feature dimensions and constricted along most discriminant ones. As a result, the class conditional probabilities can be expected to be approximately constant in the modified neighborhoods, whereby better classification performance can be achieved. The efficacy of our method is validated and compared against other competing techniques using a variety of data sets.

Key words: classification, nearest neighbors, quasiconformal mapping, kernel methods, SVMs

1 Introduction

In pattern classification, we are given l training samples $\{(\mathbf{x}_i, y_i)\}_{i=1}^{l}$, where the training samples consist of q feature measurements $\mathbf{x}_i = (x_{i1}, \ldots, x_{iq})^t \in \Re^q$ and the known class labels $y_i \in \{0, 1\}$. The goal is to induce a classifier $\hat{f} : \Re^q \rightarrow \{0, 1\}$ from the training samples that assigns the class label to a given query $\mathbf{x}'$.

A simple and attractive approach to this problem is nearest neighbor (NN) classification [10, 12, 13, 15, 17, 18, 25]. Such a method produces continuous and overlapping, rather than fixed, neighborhoods and uses a different neighborhood for each individual query so that all points in the neighborhood are

J. Peng, D.R. Heisterkamp, and H.K. Dai: *Adaptive Discriminant and Quasiconformal Kernel Nearest Neighbor Classification*, StudFuzz **177**, 181–203 (2005)
www.springerlink.com

close to the query. Furthermore, empirical evaluation to date shows that the NN rule is a rather robust method in a variety of applications. In addition, it has been shown [11] that the one NN rule has asymptotic error rate that is at most twice the Bayes error rate, independent of the distance metric used.

NN rules assume that locally the class (conditional) probabilities are approximately constant. However, this assumption is often invalid in practice due to the curse-of-dimensionality [4]. Severe bias[1] can be introduced in the NN rule in a high-dimensional space with finite samples. As such, the choice of a distance measure becomes crucial in determining the outcome of NN classification in high dimensional settings.

Figure 1 illustrates a case in point, where class boundaries are parallel to the coordinate axes. For query **a**, the vertical axis is more relevant, because a move along the that axis may change the class label, while for query **b**, the horizontal axis is more relevant. For query **c**, however, two axes are equally relevant. This implies that distance computation does not vary with equal strength or in the same proportion in all directions in the space emanating from the input query. Capturing such information, therefore, is of great importance to any classification procedure in a high dimensional space.

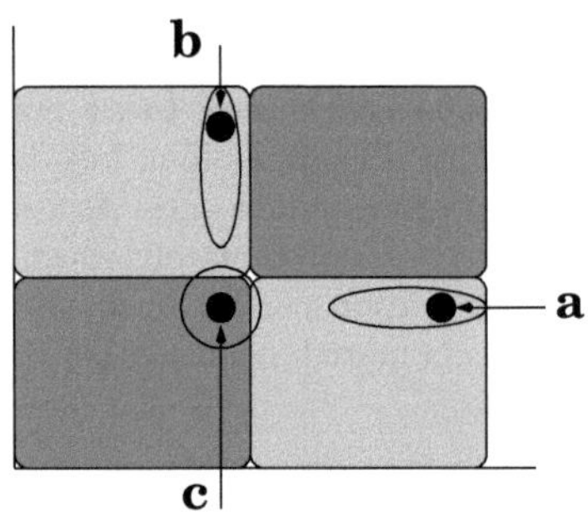

Fig. 1. Feature relevance varies with query locations

In this chapter we describe two approaches to adaptive NN classification to try to minimize bias in high dimensions. The first approach computes an effective local metric for computing neighborhoods by explicitly estimating feature relevance using locally linear support vector machines (SVMs). The second approach computes a local metric based on quasiconformal transformed kernels. In both approaches, the resulting neighborhoods are highly adaptive to query locations. Moreover, the neighborhoods are elongated along less relevant (discriminant) feature dimensions and constricted along most influential ones. As a result, the class conditional probabilities tend to be constant in the modified neighborhoods, whereby better classification performance can be obtained.

[1]Bias is defined as: $f - E\hat{f}$, where f represents the true target and E the expectation operator.

The rest of the chapter is organized as follows. Section 2 describes related work addressing issues of locally adaptive metric nearest neighbor classification. Section 3 presents adaptive nearest neighbor classifier design that uses locally linear SVMs to explicitly estimate feature relevance for nearest neighborhood computation. Section 4 introduces adaptive nearest neighbor classifier design that employs quasiconformal transformed kernels to compute more homogeneous nearest neighborhoods. It also shows that the discriminant adaptive nearest neighbor metric [13] is a special case of our adaptive quasiconformal kernel distance under appropriate conditions. Section 5 discusses how to determine procedural parameters used by our nearest neighbor classifiers. After that, Section 6 presents experimental results evaluating the efficacy of our adaptive nearest neighbor methods using a variety of real data. Finally, Sect. 7 concludes this chapter by pointing out possible extensions to the current work and future research directions.

2 Related Work

Friedman [12] describes an approach for learning local feature relevance that combines some of the best features of KNN learning and recursive partitioning. This approach recursively homes in on a query along the most (locally) relevant dimension, where local relevance is computed from a reduction in prediction error given the query's value along that dimension. This method performs well on a number of classification tasks. Let $\Pr(j|\mathbf{x}))$ and $\Pr(j|x_i = z_i)$ denote the probability of class j given a point $\mathbf{x}$ and the ith input variable of a point $\mathbf{x}$, respectively, and $\overline{\Pr}(j)$ and $\overline{Pr}(j|x_i = z_i)$ their corresponding expectation in the set $\{1, 2, \ldots, J\}$ of labels. The reduction in prediction error can be described by

$$I_i^2(\mathbf{z}) = \sum_{j=1}^{J} (\overline{\Pr}(j) - \overline{\Pr}(j|x_i = z_i)])^2 \, , \tag{1}$$

This measure reflects the influence of the ith input variable on the variation of $\Pr(j|\mathbf{x})$ at the particular point $x_i = z_i$. In this case, the most informative input variable is the one that gives the largest deviation from the average value of $\Pr(j|\mathbf{x})$. Notice that this is a greedy peeling strategy that at each step removes a subset of data points from further consideration, as in decision tree induction. As a result, changes in early splits, due to variability in parameter estimates, can have a significant impact on later splits, thereby producing high variance predictions.

In [13], Hastie and Tibshirani propose an adaptive nearest neighbor classification method based on linear discriminant analysis (LDA). The method computes a distance metric as a product of properly weighted within and between sum-of-squares matrices. They show that the resulting metric approximates the weighted *Chi-squared* distance between two points $\mathbf{x}$ and $\mathbf{x}'$ [13, 16, 22]

$$D(\mathbf{x}, \mathbf{x}') = \sum_{j=1}^{J} \frac{[\Pr(j|\mathbf{x}) - \Pr(j|\mathbf{x}')]^2}{\Pr(j|\mathbf{x}')} , \tag{2}$$

by a Taylor series expansion, given that class densities are Gaussian and have the same covariance matrix. While sound in theory, DANN may be limited in practice. The main concern is that in high dimensions we may never have sufficient data to fill in $q \times q$ matrices locally. We will show later that the metric proposed by Hastie and Tibshirani [13] is a special case of our more general quasiconformal kernel metric to be described in this chapter.

Amari and Wu [1] describe a method for improving SVM performance by increasing spatial resolution around the decision boundary surface based on the Riemannian geometry. The method first trains a SVM with an initial kernel that is then modified from the resulting set of support vectors and a quasiconformal mapping. A new SVM is built using the new kernel. Viewed under the same light, our goal is to expand the spatial resolution around samples whose class probabilities are different from the query and contract the spatial resolution around samples whose class probability distribution is similar to the query. The effect is to make the space around samples farther from or closer to the query, depending on their class (conditional) probability distributions.

Domeniconi et al. [10] describe an adaptive metric nearest neighbor method for improving the regular nearest neighbor procedure. The technique adaptively estimates local feature relevance at a given query by approximating the Chi-squared distance. The technique employs a "patient" averaging process to reduce variance. While the averaging process demonstrates robustness against noise variables, it is at the expense of increased computational complexity. Furthermore, the technique has several adjustable procedural parameters that must be determined at run time.

3 Adaptive Nearest Neighbor Classifiers Based on Locally Linear SVMs

3.1 Feature Relevance

Our technique is motivated as follows. In LDA (for $J = 2$), data are projected onto a single dimension where class label assignment is made for a given input query. From a set of training data $\{(\mathbf{x}_i, y_i)\}_{i=1}^{l}$, where $y_i \in \{0, 1\}$, this dimension is computed according to

$$\mathbf{w} = \mathbf{W}^{-1}(\bar{\mathbf{x}}_0 - \bar{\mathbf{x}}_1) , \tag{3}$$

where $\mathbf{W} = \sum_{j=0}^{1} \sum_{y_i=j} p_i(\mathbf{x}_i - \bar{\mathbf{x}}_j)(\mathbf{x}_i - \bar{\mathbf{x}}_j)^t$ denotes the within sum-of-squares matrix, $\bar{\mathbf{x}}_j$ the class means, and p_i the relative occurrence of $\mathbf{x}_i$ in class j. The vector $\mathbf{w} = (w_1, w_2, \ldots, w_q)^t$ represents the same direction as the discriminant in the Bayes classifier along which the data has the maximum separation when the two classes follow multivariate Gaussian distributions with the same covariance matrix. Furthermore, any direction, $\mathbf{\Theta}$, whose dot product with $\mathbf{w}$ is large, also carries discriminant information. The larger $|\mathbf{w} \cdot \mathbf{\Theta}|$ is, the more discriminant information that $\mathbf{\Theta}$ captures. State it differently, if we transform $\mathbf{\Theta}$ via

$$\tilde{\mathbf{\Theta}} = \mathbf{W}^{-1/2}\mathbf{\Theta} , \tag{4}$$

then in the transformed space, any direction $\tilde{\mathbf{\Theta}}$ close to $\mathbf{W}^{-1/2}(\bar{\mathbf{x}}_0 - \bar{\mathbf{x}}_1)$ carries discriminant information. More formally, let

$$J(\mathbf{w}) = \frac{\mathbf{w}^t\mathbf{B}\mathbf{w}}{\mathbf{w}^t\mathbf{W}\mathbf{w}} , \tag{5}$$

be the LDA criterion function maximized by $\mathbf{w}$ (3), where $\mathbf{B}$ is the between sum-of-squares matrix and computed according to

$$\mathbf{B} = (\bar{\mathbf{x}}_0 - \bar{\mathbf{x}}_1)(\bar{\mathbf{x}}_0 - \bar{\mathbf{x}}_1)^t . \tag{6}$$

If we let

$$\mathbf{B}^* = \mathbf{W}^{-1/2}\mathbf{B}\mathbf{W}^{-1/2} \tag{7}$$

be the between sum-of-squares matrix in the transformed space, then the criterion function (5) in the transformed space becomes

$$J^*(\tilde{\mathbf{\Theta}}) = \frac{\tilde{\mathbf{\Theta}}^t\mathbf{B}^*\tilde{\mathbf{\Theta}}}{\tilde{\mathbf{\Theta}}^t\tilde{\mathbf{\Theta}}} = \frac{(\tilde{\mathbf{w}}^t\tilde{\mathbf{\Theta}})^2}{\tilde{\mathbf{\Theta}}^t\tilde{\mathbf{\Theta}}} , \tag{8}$$

where $\tilde{\mathbf{w}} = \mathbf{W}^{-1/2}(\bar{\mathbf{x}}_0 - \bar{\mathbf{x}}_1)$ and $\tilde{\mathbf{\Theta}}$ by (4). Therefore, any direction $\tilde{\mathbf{\Theta}}$ that is close to $\mathbf{W}^{-1/2}(\bar{\mathbf{x}}_0 - \bar{\mathbf{x}}_1)$ in the transformed space computes higher values in J^*, thereby capturing discriminant information.

In particular, when $\mathbf{\Theta}$ is restricted to the feature axes, i.e.,

$$\mathbf{\Theta} \in \{\mathbf{e}_1, \ldots, \mathbf{e}_q\} ,$$

where $\mathbf{e}_i$ is a unit vector along the ith feature, the value of $|\mathbf{w} \cdot \mathbf{\Theta}|$, which is the magnitude of the projection of $\mathbf{w}$ along $\mathbf{\Theta}$, measures the degree of relevance of feature dimension $\mathbf{\Theta}$ in providing class discriminant information. When $\mathbf{\Theta} = \mathbf{e}_i$, we have $|\mathbf{w} \cdot \mathbf{\Theta}| = w_i$. It thus seems natural to associate

$$r_i = \frac{|w_i|}{\sum_{j=1}^{q} |w_j|} ,$$

with each dimension $\mathbf{\Theta}$ in a weighted nearest neighbor rule

$$D(\mathbf{x}, \mathbf{x}') = \sqrt{\sum_{i=1}^{q} r_i(x_i - x_i')^2} . \tag{9}$$

Now imagine for each input query we compute $\mathbf{w}$ locally, from which to induce a new neighborhood for the final classification of the query. In this case, large $|\mathbf{w} \cdot \mathbf{\Theta}|$ forces the shape of neighborhood to constrict along $\mathbf{\Theta}$, while small $|\mathbf{w} \cdot \mathbf{\Theta}|$ elongates the neighborhood along the $\mathbf{\Theta}$ direction. Figure 1 illustrates a case in point, where for query **a** the discriminant direction is parallel to the vertical axis, and as such, the shape of the neighborhood is squashed along that direction and elongated along the horizontal axis.

We use two-dimensional Gaussian data with two classes and substantial correlation, shown in Fig. 2, to illustrate neighborhood computation based on LDA. The number of data points for both classes is roughly the same (about 250). The (red) square, located at $(3.7, -2.9)$, represents the query. Figure 2(a) shows the 100 nearest neighbors (red squares) of the query found by the unweighted KNN method (simple Euclidean distance). The resulting shape of the neighborhood is circular, as expected. In contrast, Fig. 2(b) shows the 100 nearest neighbors of the query, computed by the technique described above. That is, the nearest neighbors shown in Fig. 2(a) are used to compute (3) and, hence, (9), with estimated new (normalized) weights: $r_1 = 0.3$ and $r_2 = 0.7$. As a result, the new (elliptical) neighborhood is elongated along the horizontal axis (the less important one) and constricted along the vertical axis (the more important one). The effect is that there is a sharp increase in the retrieved nearest neighbors that are in the same class as the query.

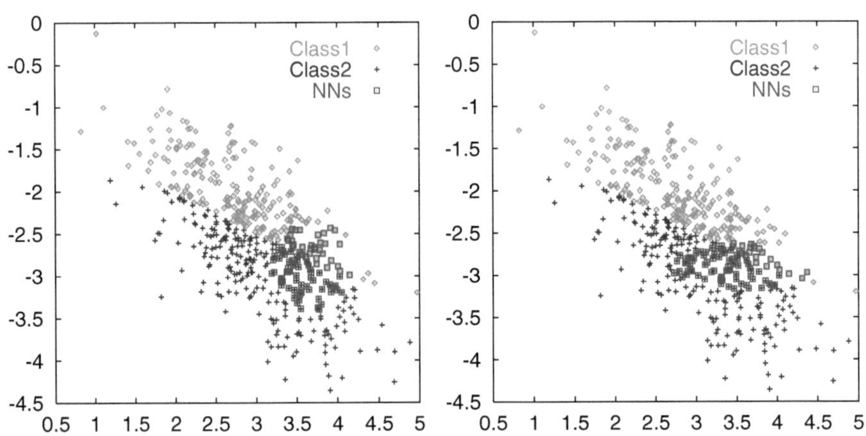

Fig. 2. Two-dimensional Gaussian data with two classes and substantial correlation. The *square* indicates the query. (**a**) Circular neighborhood (Euclidean distance). (**b**) Elliptical neighborhood, where features are weighted by LDA

This example demonstrates that even a simple problem in which a linear boundary roughly separates two classes can benefit from the feature relevance learning technique just described, especially when the query approaches the class boundary. It is important to note that for a given distance metric the shape of a neighborhood is fixed, independent of query locations. Furthermore, any distance calculation with equal contribution from each feature variable will always produce spherical neighborhoods. Only by capturing the relevant contribution of the feature variables can a desired neighborhood be realized that is highly customized to query locations.

While $\mathbf{w}$ points to a direction along which projected data can be well separated, the assumption on equal covariance structures for all classes is often invalid in practice. Computationally, if the dimension of the feature space is large, there will be insufficient data to locally estimate the $\Omega(q^2)$ elements of the within sum-of-squares matrices, thereby making them highly biased. Moreover, in high dimensions the within sum-of-squares matrices tend to be ill-conditioned. In such situations, one must decide at what threshold to zero small singular values in order to obtain better solutions? In addition, very often features tend to be independent locally. As we shall see later, weighting based on local LDA will be less effective when used without local clustering. This motivates us to consider the SVM based approach to feature relevance estimation.

3.2 Normal Vector Based on SVMs

The normal vector $\mathbf{w}$ based on SVMs can be explicitly written as

$$\mathbf{w} = \sum_{i \in SV} \alpha_i y_i \Phi(\mathbf{x}_i) \ , \tag{10}$$

where α_i's are Langrange coefficients. In a simple case in which Φ is the identity function on $\Re^q$, we have

$$\mathbf{w} = \sum_{i \in SV} \alpha_i y_i \mathbf{x}_i \ . \tag{11}$$

where SV represents the set of support vectors. Similar to LDA, the normal $\mathbf{w}$ points to the direction that is more discriminant and yields the maximum margin of separation between the data.

Let us revisit the normal computed by LDA (3). This normal is optimal (the same Bayes discriminant) under the assumption when the two classes follow multivariate Gaussian distributions with the same covariance matrix. Optimality breaks down, however, when the assumption is violated, which is often the case in practice. In contrast, SVMs compute the optimal (maximum margin) hyperplane (10) without such an assumption. This spells out the difference between the directions pointed to by the two normals, which has important implications on generalization performance.

Finally, we note that real data are often highly non-linear. In such situations, linear machines cannot be expected to work well. As such, $\mathbf{w}$ is unlikely to provide any useful discriminant information. On the other hand, piecewise local hyperplanes can approximate any decision boundaries, thereby enabling $\mathbf{w}$ to capture local discriminant information.

3.3 Discriminant Feature Relevance

Based on the above discussion, we now propose a measure of feature relevance for an input query $\mathbf{x}'$ as

$$\Upsilon_i(\mathbf{x}') = |w_i| \,, \tag{12}$$

where w_i denotes the ith component of $\mathbf{w}$ in (11) computed locally at $\mathbf{x}'$. One attractive property of (12) is that $\mathbf{w}$ enables Υ_i's to capture relevance information that may not otherwise be attainable should relevance estimates been conducted along each individual dimension one at a time, as in [9, 12].

The relative relevance, as a weighting scheme, can then be given by

$$r_i(\mathbf{x}') = (\Upsilon_i(\mathbf{x}'))^t \Big/ \sum_{j=1}^{q} (\Upsilon_j(\mathbf{x}'))^t \,. \tag{13}$$

where $t = 1, 2$, giving rise to linear and quadratic weightings, respectively. In this paper we employ the following exponential weighting scheme

$$r_i(\mathbf{x}') = \exp(C\Upsilon_i(\mathbf{x}')) \Big/ \sum_{j=1}^{q} \exp(C\Upsilon_j(\mathbf{x}')) \,, \tag{14}$$

where C is a parameter that can be chosen to maximize (minimize) the influence of Υ_i on r_i. When $C = 0$ we have $r_i = 1/q$, thereby ignoring any difference between the Υ_i's. On the other hand, when C is large a change in Υ_i will be exponentially reflected in r_i. The exponential weighting is more sensitive to changes in local feature relevance (12) and gives rise to better performance improvement. Moreover, exponential weighting is more stable because it prevents neighborhoods from extending infinitely in any direction, i.e., zero weight. This, however, can occur when either linear or quadratic weighting is used. Thus, (14) can be used as weights associated with features for weighted distance computation

$$D(\mathbf{x}, \mathbf{x}') = \sqrt{\sum_{i=1}^{q} r_i (x_i - x_i')^2} \,. \tag{15}$$

These weights enable the neighborhood to elongate along feature dimensions that run more or less parallel to the separating hyperplane, and, at the same time, to constrict along feature coordinates that have small angles with $\mathbf{w}$. This can be considered highly desirable in nearest neighbor search. Note that the technique is *query-based* because weightings depend on the query [3].

We desire that the parameter C in (14) increases with decreasing perpendicular distance between the input query and the decision boundary in an adaptive fashion. The advantage of doing so is that any difference among w_i's will be magnified exponentially in $\mathbf{r}$, thereby making the neighborhood highly elliptical as the input query approaches the decision boundary.

In general, however, the boundary is unknown. We can potentially solve the problem by computing the following: $|\mathbf{w} \cdot \mathbf{x} + b|$. After normalizing $\mathbf{w}$ to unit length, the above equation returns the perpendicular distance between $\mathbf{x}$ and the local separating hyperplance. We can set C to be inversely proportional to

$$\frac{|\mathbf{w} \cdot \mathbf{x} + b|}{||\mathbf{w}||} = |\mathbf{w} \cdot \mathbf{x} + b| \,. \tag{16}$$

In practice, we find it more effective to set C to a fixed constant. In the experiments reported here, C is determined through cross-validation.

Instead of axis parallel elongation and constriction, one attempts to use general Mahalanobis distance and have an ellipsoid whose main axis is parallel to the separating hyperplane, and whose width in other dimensions is determined by the distance of $\mathbf{x}$ from the hyperplane. The main concern with such an approach is that in high dimensions there may not be sufficient data to locally fill in $q \times q$ within sum-of-squares matrices. Moreover, very often features may be locally independent. Therefore, to effectively compute general Mahalanobis distance some sort of local clustering has to be done. In such situations, without local clustering, general Mahalanobis distance reduces to weighted Euclidean distance.

Let us examine the relevance measure (12) in the context of the Riemannian geometry proposed by Amari and Wu [1]. A large component of $\mathbf{w}$ along a direction $\boldsymbol{\Theta}$, i.e., a large value of $\boldsymbol{\Theta} \cdot \mathbf{w}$, implies that data points along that direction become far apart in terms of (15). Likewise, data points are moving closer to each other along directions that have a small value of dot product with $\mathbf{w}$. That is, (12) and (15) represent a local transformation that is judiciously chosen so as to increase spatial resolution along discriminant directions around the separating boundary. In contrast, the quasiconformal transformation introduced by Amari and Wu [1] does not attend to directions.

3.4 Neighborhood Morphing Algorithm

The neighborhood *mor*phing NN algorithm (MORF) is summarized in Fig. 3. At the beginning, a nearest neighborhood of K_N points around the query $\mathbf{x}'$ is computed using the simple Euclidean distance. From these K_N points a locally linear SVM is built (i.e., the mapping from the input space to the feature space is the identity mapping), whose $\mathbf{w}$ (normal to the separating hyperplance) is employed in (12) and (14) to obtain an exponential feature weighting scheme r_i. Finally, the resulting r_i (14) are used in (15) to compute K nearest neighbors at the query point $\mathbf{x}'$ to classify $\mathbf{x}'$. Note that if all K_N neighbors fall into one class, we can simply predict the query to have the same class label.

The bulk of the computational cost associated with the MORF algorithm is incurred by solving the quadratic programming problem to obtain local linear SVMs. This optimization problem can be bounded by $O(qK_N^2)$ [6]. Note that

Given a query point $\mathbf{x}'$, and procedural parameters K, K_N and C:

1. Initialize $\mathbf{r}$ in (15) to 1;
2. Compute the K_N nearest neighbors around $\mathbf{x}'$ using the weighted distance metric (15);
3. Build a local linear SVM from the K_N neighbors.
4. Update $\mathbf{r}$ according to (11) and (14).
5. At completion, use $\mathbf{r}$, hence (15), for K-nearest neighbor classification at the test point $\mathbf{x}'$.

Fig. 3. The MORF algorithm

throughout computation K_N remains fixed and is (usually) less than or equal to half the number of training points (l). For large but practical q, K_N can be viewed as a constant. While there is a cost associated with building local linear SVMs, the gain in performance over simple KNN outweighs this extra cost, as we shall see in the next section.

4 Adaptive Nearest Neighbor Classifiers Based on Quasiconformal Transformed Kernels

4.1 Kernel Distance

Another approach to locally adaptive NN classification is based on kernel distance. The kernel trick has been applied to numerous problems [8, 19, 20, 21]. Distance in the feature space may be calculated by means of the kernel [8, 24]. With $\mathbf{x}$ and $\mathbf{x}'$ in the input space then the (squared) feature space distance is

$$D(\mathbf{x}, \mathbf{x}') = ||\phi(\mathbf{x}) - \phi(\mathbf{x}')||^2 = K(\mathbf{x}, \mathbf{x}) - 2K(\mathbf{x}, \mathbf{x}') + K(\mathbf{x}', \mathbf{x}') \tag{17}$$

Since the dimensionality of the feature space may be very high, the meaning of the distance is not directly apparent. The image of the input space forms a submanifold in the feature space with the same dimensionality as the input space, thus what is available is a Riemannian metric [1, 8, 21]. The Riemannian metric tensor induced by a kernel is

$$g_{i,j}(\mathbf{z}) = \left(\frac{1}{2} \frac{\partial^2 K(\mathbf{x}, \mathbf{x})}{\partial x_i \partial x_j} - \frac{\partial^2 K(\mathbf{x}, \mathbf{z})}{\partial x_i \partial x_j} \right)_{\mathbf{x}=\mathbf{z}} . \tag{18}$$

4.2 Quasiconformal Kernel

It is a straight forward process to create a new kernel from existing kernels [8]. However, we desire to create a new kernel such that, for each input query

$\mathbf{x}'$, the class posterior probability in the neighborhood induced by the kernel metric tend to be homogeneous. We therefore look to conformal and quasiconformal mappings [2]. From the Riemannian metric (18), a conformal mapping is given

$$\tilde{g}_{i,j}(\mathbf{x}) = \Omega(\mathbf{x}) g_{i,j}(\mathbf{x}) , \tag{19}$$

where $\Omega(\mathbf{x})$ is a scalar function of $\mathbf{x}$. A conformal mapping keeps angles unchanged in the space.

An initial desire may be to use a conformal mapping. But in a higher-dimensional space, conformal mappings are limited to similarity transformations and spherical inversions [5], and hence it may be difficult to find a conformal mapping with the desired homogeneous property/properties. Quasiconformal mappings are more general than conformal mappings, containing conformal mappings as a special case. Hence, we seek quasiconformal mappings with the desired properties.

Previously, Amari and Wu [1] have modified a support vector machine with a quasiconformal mapping. Heisterkamp et al. [14] have used quasiconformal mappings to modify a kernel distance for content-based image retrieval. If $c(\mathbf{x})$ is a positive real valued function of $\mathbf{x}$, then a new kernel can be created by

$$\tilde{K}(\mathbf{x}, \mathbf{x}') = c(\mathbf{x}) c(\mathbf{x}') K(\mathbf{x}, \mathbf{x}') . \tag{20}$$

We call it a quasiconformal kernel. In fact, $\tilde{K}$ will be a valid kernel as long as $c(\mathbf{x})$ is real-valued [8].

The question becomes which $c(\mathbf{x})$ do we wish to use? We can change the Riemannian metric by the choice of $c(\mathbf{x})$. The metric $g_{i,j}$ associated with kernel K becomes the metric $\tilde{g}_{i,j}$ associated with kernel $\tilde{K}$ by the relationship [1]:

$$\tilde{g}_{i,j}(\mathbf{x}) = c_i(\mathbf{x}) c_j(\mathbf{x}) + c(\mathbf{x})^2 g_{i,j}(\mathbf{x}) , \tag{21}$$

where $c_i(\mathbf{x}) = \frac{\partial c(\mathbf{x})}{\partial x_i}$.

Amari and Wu [1] expanded the spatial resolution in the margin of a SVM by using the following

$$c(\mathbf{x}) = \sum_{i \in SV} \alpha_i e^{\frac{-||\mathbf{x} - \mathbf{x}_i||^2}{2\tau^2}} , \tag{22}$$

where $\mathbf{x}_i$ is the ith support vector, α_i is a positive number representing the contribution of $\mathbf{x}_i$, and τ is a free parameter. Since the support vectors are likely to be at the boundary of the margin, this creates an expansion of spatial resolution in the margin and a contraction elsewhere.

4.3 Adaptive Quasiconformal Kernel Nearest Neighbors

Our adaptive quasiconformal kernel nearest neighbor (AQK) algorithm is motivated as follows. From (17) and (20) the (squared) quasiconformal kernel distance can be written as

$$D(\mathbf{x},\mathbf{x}') = c(\mathbf{x})^2K(\mathbf{x},\mathbf{x}) - 2c(\mathbf{x})c(\mathbf{x}')K(\mathbf{x},\mathbf{x}') + c(\mathbf{x}')^2K(\mathbf{x}',\mathbf{x}') \,. \tag{23}$$

If the original kernel K is a radial kernel, then $K(\mathbf{x},\mathbf{x}) = 1$ and the distance becomes

$$\begin{aligned} D(\mathbf{x},\mathbf{x}') &= c(\mathbf{x})^2 - 2c(\mathbf{x})c(\mathbf{x}')K(\mathbf{x},\mathbf{x}') + c(\mathbf{x}')^2 \\ &= (c(\mathbf{x}) - c(\mathbf{x}'))^2 + 2c(\mathbf{x}')c(\mathbf{x})(1 - K(\mathbf{x},\mathbf{x}')) \,. \end{aligned} \tag{24}$$

Our goal is to produce neighborhoods where the class conditional probabilities tend to be homogeneous. That is, we want to expand the spatial resolution around samples whose class probabilities are different from the query and contract the spatial resolution around samples whose class probability distribution is similar to the query. The effect is to make the space around samples farther from or closer to the query, depending on their class (conditional) probability distributions. An appealing candidate for a sample $\mathbf{x}$ with a query $\mathbf{x}'$ is

$$c(\mathbf{x}) = \frac{\Pr(j_{\bar{m}}|\mathbf{x})}{\Pr(j_m|\mathbf{x}')} \,, \tag{25}$$

where $j_m = \arg\max_j \Pr(j|\mathbf{x}')$ and $j_{\bar{m}} = 1 - j_m$. It is based on the magnitude of the ratio of class conditional probabilities: the maximum probability class (j_m) of the query versus the complementary class ($j_{\bar{m}}$) of the sample.

Notice that:

1. The multiplier $c(\mathbf{x})$ for a sample $\mathbf{x}$ yields a contraction effect $c(\mathbf{x}) < 1$ when and only when $\Pr(j_m|\mathbf{x}) + \Pr(j_m|\mathbf{x}') > 1$, that is, both conditional (class) probabilities are consistent.
2. The multiplier $c(\mathbf{x}')$ for a query $\mathbf{x}'$ measures the degree of uncertainty in labeling $\mathbf{x}'$ with its maximum probability class.

Substituting $c(\mathbf{x})$ (25) into (24) and simplifying, we obtain

$$D(\mathbf{x},\mathbf{x}') = \left(\frac{\Pr(j_m|\mathbf{x}') - \Pr(j_m|\mathbf{x})}{\Pr(j_m|\mathbf{x}')}\right)^2 + 2c(x')c(x)(1 - K(x,x')) \tag{26}$$

To understand the distance (26) we look at the distance using a second-order Taylor series expansion of a general Gaussian kernel

$$K(\mathbf{x},\mathbf{x}') = e^{-\frac{1}{2}(\mathbf{x}-\mathbf{x}')^t\Sigma^{-1}(\mathbf{x}-\mathbf{x}')} \tag{27}$$

at the query point, $\mathbf{x}'$,

$$K(\mathbf{x},\mathbf{x}') \approx 1 - \frac{1}{2}(\mathbf{x}-\mathbf{x}')^t\Sigma^{-1}(\mathbf{x}-\mathbf{x}') \,. \tag{28}$$

Substituting the Taylors expansion into (26) yields

$$D(\mathbf{x},\mathbf{x}') \approx \frac{[\Pr(j_m|\mathbf{x}') - \Pr(j_m|\mathbf{x})]^2}{\Pr(j_m|\mathbf{x}')^2} + c(\mathbf{x}')c(\mathbf{x})(\mathbf{x}-\mathbf{x}')^t\Sigma^{-1}(\mathbf{x}-\mathbf{x}') \,. \tag{29}$$

The first term in the above equation is the *Chi-squared* (appropriately weighted) distance, while the second term is the weighted quadratic (Mahalanobis) distance. The distance (29) is more efficient computationally than (26) and achieves similar performance to that of (26), as we shall see later. When $c(\mathbf{x}) \approx c(\mathbf{x}')$, the quasiconformal kernel distance is reduced to the weighted Mahalanobis distance, with a weighting factor of $c(\mathbf{x}')c(\mathbf{x})$ depending on the degrees of class-consistency in $c(\mathbf{x})$ and of labeling-uncertainty in $c(\mathbf{x}')$.

There are two variants to the basic AQK algorithm (29) with which we have experimented. One is to set $\Sigma = \sigma^2 I$ in (27), where I is the identity matrix. In this case, the AQK distance (29) is reduced to

$$D(\mathbf{x}, \mathbf{x}') = \frac{[\Pr(j_m|\mathbf{x}') - \Pr(j_m|\mathbf{x})]^2}{\Pr(j_m|\mathbf{x}')^2} + \frac{c(\mathbf{x}')c(\mathbf{x})}{\sigma^2}\|\mathbf{x} - \mathbf{x}'\|^2 . \tag{30}$$

The second variant is driven by the fact that in practice it is more effective to assume Σ in (27) to be diagonal. This is particularly true when the dimension of the input space is large, since there will be insufficient data to locally estimate the $\Theta(q^2)$ elements of Σ. If we let $\Sigma = \Lambda$ we obtain

$$D(\mathbf{x}, \mathbf{x}') \approx \frac{[\Pr(j_m|\mathbf{x}') - \Pr(j_m|\mathbf{x})]^2}{\Pr(j_m|\mathbf{x}')^2} + c(\mathbf{x}')c(\mathbf{x})(\mathbf{x} - \mathbf{x}')^t \Lambda^{-1}(\mathbf{x} - \mathbf{x}') , \tag{31}$$

where the matrix Λ is the diagonal matrix with the diagonal entries of Σ.

4.4 Effect of Quasiconformal Mapping

Let us examine $c(\mathbf{x}')$ (25). Clearly, $0 \le c(\mathbf{x}') \le 1$. When $c(\mathbf{x}') \approx 0$ there is a high degree of certainty in labeling $\mathbf{x}'$, in which case $c(\mathbf{x}')$ is aggressive in modifying the Mahalanobis distance and applies a large contraction. On the other hand, when $c(\mathbf{x}') \approx 1$ there is a low degree of certainty. The *Chi-squared* term achieves little statistical information, in which case $c(\mathbf{x}')$ is cautious in modifying the Mahalanobis distance and applies little or no contraction.

Now consider the effect of $c(\mathbf{x})$ (25) on the distance. It is not difficult to show that

$$c(\mathbf{x}) = c(\mathbf{x}') \pm \sqrt{[\Pr(j_m|\mathbf{x}') - \Pr(j_m|\mathbf{x})]^2 / \Pr(j_m|\mathbf{x}')^2},$$

where $\pm$ represents the algebraic sign of $\Pr(j_m|\mathbf{x}') - \Pr(j_m|\mathbf{x})$. For a given $\mathbf{x}'$, $c(\mathbf{x}')$ is fixed. Thus the dilation/contraction of the Mahalanobis distance due to variations in $c(\mathbf{x})$ is proportional to the square root of the *Chi-squared* distance with the dilation/contraction determined by the direction of variation of $\Pr(j|\mathbf{x})$ from $\Pr(j|\mathbf{x}')$. That is, $c(\mathbf{x})$ attempts to compensate for the *Chi-squared* distance ignorance of the direction of variation of $\Pr(j|\mathbf{x})$ from $\Pr(j|\mathbf{x}')$ and is driving the neighborhood closer to homogeneous class conditional probabilities.

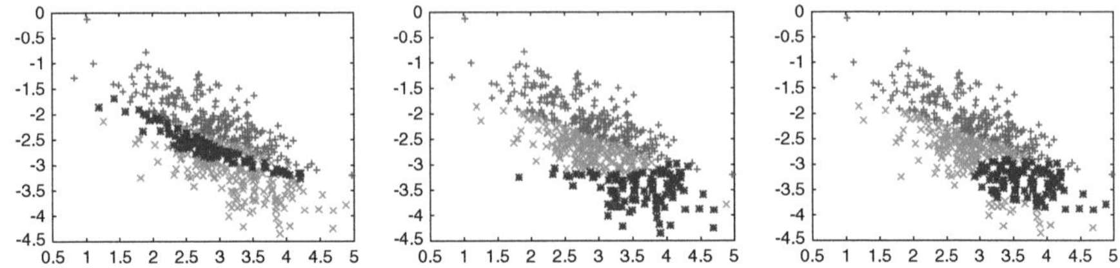

Fig. 4. *Left panel*: A nearest neighborhood of the query computed by the first term in (30). *Middle panel*: A neighborhood of the same query computed by the second term in (30). *Right panel*: A neighborhood computed by the two terms in (30)

One might argue that the metric (29) has the potential disadvantage of requiring the class probabilities (hence $c(\mathbf{x})$) as input. But these probabilities are the quantities that we are trying to estimate. If an initial estimate is required, then it seems logical to use an iterative process to improve the class probabilities, thereby increasing classification accuracy. Such a scheme, however, could potentially allow neighborhoods to extend infinitely in the complement of the subspace spanned by the sphered class centroids, which is dangerous. The left panel in Fig. 4 illustrates the case in point, where the neighborhood of the query is produced by the first term in (30). The neighborhood becomes highly non-local, as expected, because the distance is measured in the class probability space. The potential danger of such a neighborhood has also been predicted in [13]. Furthermore, even if such an iterative process is used to improve the class probabilities, as in the local Parzen Windows method to be described later in this chapter, their improvement in classification performance is not as pronounced as that produced by other competing methods, as we shall see later.

The middle panel in Fig. 4 shows a neighborhood of the same query computed by the second term in (30). While it is far less stretched along the subspace orthogonal to the space spanned by the sphered class centroids, it nevertheless is non-local. This is because this second term ignores the difference in the maximum likelihood probability (i.e., $\Pr(j_m|\mathbf{x})$) between a data point $\mathbf{x}$ and the query $\mathbf{x}'$. Instead, it favors data points whose $\Pr(j_{\bar{m}}|\mathbf{x})$ is small. Only by combining the two terms the desired neighborhood can be realized, as evidenced by the right panel in Fig. 4.

4.5 Estimation

Since both $\Pr(j_{\bar{m}}|\mathbf{x})$ and $\Pr(j_m|\mathbf{x}')$ in (25) are unknown, we must estimate them using training data in order for the distance (29) to be useful in practice. From a nearest neighborhood of K_N points around $\mathbf{x}'$ in the simple Euclidean distance, we take the maximum likelihood estimate for $\Pr(j)$. To estimate $p(\mathbf{x}|j)$, we use simple non-parametric density estimation: Parzen Windows estimate with Gaussian kernels [11, 23]. We place a Gaussian kernel over each point $\mathbf{x}_i$ in class j. The estimate $\hat{p}(\mathbf{x}|j)$ is then simply the average of the kernels. This type of technique is used in density-based nonparametric clustering analysis [7]. For simplicity, we use identical Gaussian kernels for all points with covariance $\Sigma = \tau^2 I$. More precisely,

$$\hat{p}(\mathbf{x}|j) = \frac{1}{|C_j|} \sum_{\mathbf{x}_i \in C_j} \frac{1}{\tau^q (2\pi)^{q/2}} e^{-\frac{1}{2\tau^2}(\mathbf{x}-\mathbf{x}_i)^t(\mathbf{x}-\mathbf{x}_i)} , \tag{32}$$

where C_j represents the set of training samples in class j. Together, $\hat{\Pr}(j)$ and $\hat{p}(\mathbf{x}|j)$ define $\hat{\Pr}(j|\mathbf{x})$ through the Bayes formula. Using the estimates in (32) and $\hat{\Pr}(j|\mathbf{x})$, we obtain an empirical estimate of (25) for each data point $\mathbf{x}$. To estimate the diagonal matrix Λ in (31), the strategy suggested in [13] can be followed.

4.6 Adaptive Quasiconformal Kernel Nearest Neighbor Algorithm

Figure 5 summaries our proposed adaptive quasiconformal kernel nearest neighbor classification algorithm. At the beginning, the simple Euclidean distance is used to compute an initial neighborhood of K_N points. The K_N points in the neighborhood are used to estimate $c(\mathbf{x})$ and Λ from which a new nearest neighborhood will be computed. This process (steps 2 and 3) is repeated until either a fixed number of iterations is reached or a stopping criterion is met. In our experiments, steps 2 and 3 are iterated only once. At completion, the AQK algorithm uses the resulting kernel distance to compute nearest neighbors at the query point $\mathbf{x}'$.

Assume that the procedural parameters for AQK has been determined. Then for a test point, AQK must first compute $l/5$ nearest neighbors that are used to estimate $c(\mathbf{x})$, which requires $O(l \log l)$ operations. To compute $c(\mathbf{x})$, it requires $O(l/5)$ calculations in the worst case. Therefore, the overall complexity of AQK is $O(l \log l + l)$.

Given a test point $\mathbf{x}'$.

1. Compute a nearest neighborhood of K_N points around $\mathbf{x}'$ in simple Euclidean distance.
2. Calculate $c(\mathbf{x})$ (32) and (25) and the weighted diagonal matrix Λ using the points in the neighborhood.
3. Compute a nearest neighborhood of K points around $\mathbf{x}'$ in the kernel distance (31).
4. Repeat steps 2 and 3 until a predefined stopping criterion is met.
5. At completion, use the kernel distance (31) for K-nearest neighbor classification at the test point $\mathbf{x}'$.

Fig. 5. The AQK algorithm

4.7 Relation to DANN Metric

Lemma 1. *If we assume $p(\mathbf{x}|j)$ in (25) to be Gaussian, then the DANN metric [13] is a limited approximation of the quasiconformal kernel distance (26).*

Proof. Let $p(\mathbf{x}|j)$ be Gaussian with mean μ_j and covariance Σ in (25) and (29), we obtain the first-order Taylor approximation to $\Pr(j_m|\mathbf{x})$ at $\mathbf{x}'$:

$$\Pr(j_m|\mathbf{x}) = \Pr(j_m|\mathbf{x}') - \Pr(j_m|\mathbf{x}')(\mu_{j_m} - \bar{\mu})^t\Sigma^{-1}(\mathbf{x} - \mathbf{x}'),$$

where $\bar{\mu} = \sum_j \Pr(j|\mathbf{x}')\mu_j$. Substituting this into the first term in (29) we obtain

$$\begin{aligned} D(\mathbf{x}, \mathbf{x}') = (\mathbf{x} - \mathbf{x}')^t[&\Sigma^{-1}(\mu_{j_m} - \bar{\mu})(\mu_{j_m} - \bar{\mu})^t\Sigma^{-1} \\ &+ \epsilon(\mathbf{x})\Sigma^{-1}](\mathbf{x} - \mathbf{x}')\ , \end{aligned} \tag{33}$$

where $\epsilon(\mathbf{x}) = c(\mathbf{x})c(\mathbf{x}')$. Estimating Σ by $\mathbf{W}$ (within sum-of-squares matrix)

$$\mathbf{W} = \sum_{j=1}^{J} \sum_{y_i=j} p_i(\mathbf{x}_i - \bar{\mathbf{x}}_j)(\mathbf{x}_i - \bar{\mathbf{x}}_j)^t\ , \tag{34}$$

where $\bar{\mathbf{x}}_i$ denotes the class means, and p_i the relative occurrence of $\bar{\mathbf{x}}_i$ in class j, and $(\mu_{j_m} - \bar{\mu})(\mu_{j_m} - \bar{\mu})^t$ by $\mathbf{B}$ (between sum-of-squares matrix) gives the following distance metric

$$\begin{aligned} D(\mathbf{x}, \mathbf{x}') &= (\mathbf{x} - \mathbf{x}')^t[\mathbf{W}^{-1}\mathbf{B}\mathbf{W}^{-1} + \epsilon(\mathbf{x})\mathbf{W}^{-1}](\mathbf{x} - \mathbf{x}') \\ &= (\mathbf{x} - \mathbf{x}')^t\mathbf{W}^{-1/2}[\mathbf{B}^* + \epsilon(\mathbf{x})\mathbf{I}]\mathbf{W}^{-1/2}(\mathbf{x} - \mathbf{x}')\ , \end{aligned} \tag{35}$$

where $\mathbf{B}^* = \mathbf{W}^{-1/2}\mathbf{B}\mathbf{W}^{-1/2}$ is the between sum-of-squares matrix in the sphered space. The term in (35)

$$\mathbf{W}^{-1/2}[\mathbf{B}^* + \epsilon(\mathbf{x})\mathbf{I}]\mathbf{W}^{-1/2}$$

is the DANN metric proposed by Hastie and Tibshirani (1996).

Note that the DANN metric is a local LDA metric (the first term in (35)). To prevent neighborhoods from being extended indefinitely in the null space of $\mathbf{B}$, Hastie and Tibshirani [13] argue for the addition of the second term in (35). Our derivation of the DANN metric here indicates that DANN implicitly computes a distance in the feature space via a quasiconformal mapping. As such, the second term in (35) represents the contribution from the original Gaussian kernel.

Note also that $\epsilon(\mathbf{x}) = \epsilon$ is a constant in the DANN metric [13]. Our derivation demonstrates how we can adaptively choose ϵ in desired ways. As a result, we expect AQK to outperform DANN in general, as we shall see in the next section.

5 Selection of Procedural Parameters

MORF has three adjustable procedural parameters:

- K_N: the number of nearest neighbors in $N_{K_N}(\mathbf{x}')$ for computing locally linear SVMs;
- K: the number of nearest neighbors to classify the input query $\mathbf{x}'$; and
- C: the positive factor for the exponential weighting scheme (14).

Similarly, three adjustable procedural parameters input to AQK are:

- K_N: the number of nearest neighbors in $N_{K_N}(\mathbf{x}')$ for estimation of the kernel distance;
- K: the number of nearest neighbors to classify the input query $\mathbf{x}'$; and
- τ: the *smoothing* parameter in (32) to calculate $c(\mathbf{x})$.

Notice that K is common to all NN rules. In the experiments, K is set to 5 for all nearest neighbor rules being compared. The parameter τ in AQK provides tradeoffs between bias and variance [23]. Smaller τ values give rise to sharper peaks for the density estimates, corresponding to low bias but high variance estimates of the density. We do not have a strong theory to guide its selection. In the experiments to be described later, the value of τ is determined experimentally through cross-validation. The value of C in MORF should increase as the input query moves close to the decision boundary, so that highly stretched neighborhoods will result.

The value of K_N should be large enough to support locally linear SVM computation, as in MORF, and ensure a good estimate of the kernel distance, as in AQK. Following the strategy suggested in [13], we use

$$K_N = \max\{l/5, 50\} .$$

The consistency requirement states that K_N should be a diminishing fraction of the number of training points. Also, the value of K_N should be larger for problems with high dimensionality.

Notice that the two variants of the AQK algorithm (31) have one additional procedural parameter, σ, that must be provided. The parameter σ is common to all SVMs using Gaussian kernels. Again its value is determined through cross-validation in our experiments.

6 Empirical Evaluation

In the following we compare several competing classification methods using a number of data sets:

1. MORF – The MORF algorithm based on locally linear SVMs described above.

2. AQK-k – The AQK algorithm using using the distance (26) with Gaussian kernels.
3. AQK-e – The AQK algorithm using distance (30).
4. AQK-Λ – The AQK algorithm using distance (31).
5. AQK-i – The AQK algorithm where the kernel in (24) is the identity, i.e., $K(\mathbf{x}, \mathbf{x}') = \mathbf{x} \cdot \mathbf{x}'$.
6. 5 NN – The simple five NN rule.
7. Machete [12] – A recursive "peeling" procedure, in which the input variable used for peeling at each step is the one that maximizes the estimated local relevance.
8. Scythe [12] – A generalization of the Machete algorithm, in which the input variables influence the peeling process in proportion to their estimated local relevance, rather than the winner-take-all strategy of Machete.
9. DANN – The discriminant adaptive NN rule [13].
10. Parzen – Local Parzen Windows method. A nearest neighborhood of K_N points around the query $\mathbf{x}'$ is used to estimate $\Pr(j|\mathbf{x})$ through (32) and the Bayes formula, from which the Bayes method is applied.

In all the experiments, the features are first normalized over the training data to have zero mean and unit variance, and the test data features are normalized using the corresponding training mean and variance. Procedural parameters for each method were determined empirically through cross-validation.

6.1 Data Sets

The data sets used were taken from the UCI Machine Learning Database Repository. They are: **Iris data** – $l = 100$ points, dimensionality $q = 4$ and the number of classes $J = 2$. **Sonar data** – $l = 208$ data points, $q = 60$ dimensions and $J = 2$ classes. **Ionosphere data** – $l = 351$ instances, $q = 34$ dimensions and $J = 2$ classes; **Liver data** – $l = 345$ instances, $q = 6$ dimensions and $J = 2$ classes; **Hepatitis data** – $l = 155$ instances, $q = 19$ dimensions and $J = 2$ classes; **Vote data** – $l = 232$ instances, $q = 16$ dimensions and $J = 2$ classes; **Pima data** – $l = 768$ samples, $q = 8$ dimensions and $J = 2$ classes; **OQ data** – $l = 1536$ instances of capital letters O and Q (randomly selected from 26 letter classes), $q = 16$ dimensions and $J = 2$ classes; and **Cancer data** – $l = 683$ instances, $q = 9$ dimensions and $J = 2$ classes.

6.2 Results

Table 1 shows the average error rates (μ) and corresponding standard deviations (σ) over 20 independent runs for the ten methods under consideration on the nine data sets. The average error rates for the Iris, Sonar, Vote, Ionosphere, Liver and Hepatitis data sets were based on 60% training and 40% testing, whereas the error rates for the remaining data sets were based on a random

Table 1. Average classification error rates for real data

	Iris		Sonar		Liver		Pima		Vote		OQ		Cancer		Ion		Hep	
	μ	σ	μ	σ	μ	σ	μ	σ	μ	σ	μ	σ	μ	σ	μ	σ	μ	σ
MORF	**4.8**	3.8	13.8	4.8	**28.2**	5.5	**24.9**	4.0	**2.0**	1.9	6.5	2.4	**3.2**	1.0	**7.0**	2.1	**13.1**	4.4
AQK-k	5.5	3.7	15.7	2.8	36.6	3.1	25.9	2.6	6.0	2.2	5.6	1.9	3.3	1.1	12.0	1.8	14.3	2.3
AQK-e	5.5	4.2	15.6	2.9	36.3	3.1	26.0	2.3	5.9	2.2	5.5	1.9	3.3	1.1	11.8	1.8	14.3	2.1
AQK-Λ	5.3	2.9	15.6	3.9	35.5	3.8	25.9	2.6	5.5	1.5	6.1	2.2	3.8	1.0	12.9	2.2	15.2	3.0
AQK-i	6.5	2.9	16.4	3.3	38.0	3.9	26.6	2.8	7.9	2.6	6.2	2.0	3.8	1.3	12.1	2.1	15.6	4.5
KNN	6.9	3.6	21.7	3.9	40.4	4.2	29.0	2.9	8.4	2.3	7.4	2.1	**3.2**	0.9	15.7	3.2	15.7	3.7
Machete	6.8	4.2	26.6	3.7	36.1	5.0	27.1	2.9	6.3	2.7	12.2	2.4	4.8	1.7	19.4	2.9	18.6	4.0
Scythe	6.0	3.5	22.4	4.5	37.0	4.3	27.6	2.5	5.9	2.8	9.3	2.5	3.6	0.9	12.8	2.6	17.9	4.3
DANN	6.9	3.8	**13.3**	3.7	36.2	4.1	27.6	2.9	5.4	1.9	**5.2**	2.0	3.5	1.3	12.6	2.3	15.1	4.2
Parzen	6.4	3.4	16.1	3.5	38.3	3.6	32.5	3.4	9.3	2.4	6.4	2.0	5.2	1.7	11.0	2.7	19.4	4.5

selection of 200 training data and 200 testing data (without replacement), since larger data sets are available in these cases.

Table 1 shows that MORF achieved the best performance in 7/9 of the real data sets, followed closely by AQK. For one of the remaining two data sets, MORF has the second best performance[2]. It should be clear that each method has its strengths and weaknesses. Therefore, it seems natural to ask the question of robustness. Following Friedman [12], we capture robustness by computing the ratio b_m of its error rate e_m and the smallest error rate over all methods being compared in a particular example:

$$b_m = e_m / \min_{1 \le k \le 9} e_k.$$

Thus, the best method m^* for that example has $b_{m^*} = 1$, and all other methods have larger values $b_m \geq 1$, for $m \neq m^*$. The larger the value of b_m, the worse the performance of the *mth* method is in relation to the best one for that example, among the methods being compared. The distribution of the b_m values for each method m over all the examples, therefore, seems to be a good indicator concerning its robustness. For example, if a particular method has an error rate close to the best in every problem, its b_m values should be densely distributed around the value 1. Any method whose b value distribution deviates from this ideal distribution reflect its lack of robustness.

As shown in Fig. 6, the spreads of the error distributions for MORF are narrow and close to 1. In particular, in 7/9 of the examples MORF's error rate was the best (median = 1.0). In 8/9 of them it was no worse than 3.8% higher

[2]The results of MORF are different from those reported in [18] because K_N in Fig. 3 is fixed here.

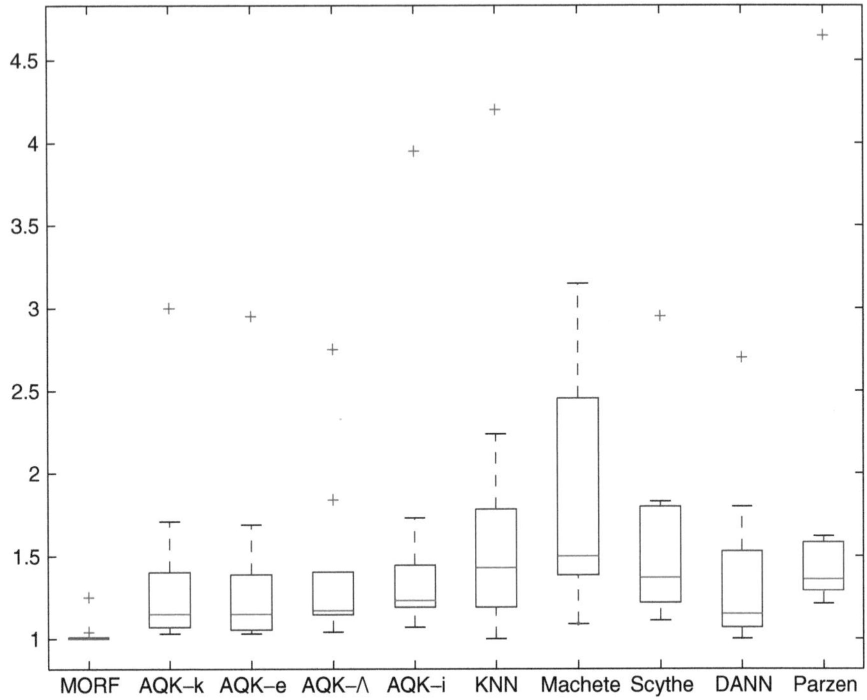

Fig. 6. Performance distributions

than the best error rate. In the worst case it was 25%. In contrast, Machete has the worst distribution, where the corresponding numbers are 235%, 277% and 315%. On the other hand, AQK showed performance characteristics similar to DANN on the data sets, as expected. Notice that there is a difference between the results reported here and that shown in [13]. The difference is due to the particular split of the data used in [13].

Figure 7 shows error rates relative to 5 NN across the nine real problems. On average, MORF is at least 30% better than 5 NN, and AQK is 20% better. AQK-k and AQK-e perform 3% worse than 5 NN in one example. Similarly, AQK-Λ and AQK-u are at most 18% worse. The results seem to demonstrate that both MORF and AQK obtained the most robust performance over these data sets. Similar characteristics were also observed for the MORF and AQK methods over simulated data sets we have experimented with.

It might be argued that the number of dimensions in the problems that we have experimented with is moderate. However, in the context of nearest neighbor classification, the number of dimensions by itself is not a critical factor. The critical factor is the local intrinsic dimensionality of the joint distribution of dimension values. This intrinsic dimensionality is often captured by the number of its singular values that are sufficiently large. When there are many features, it is highly likely that there exists a high degree of correlation

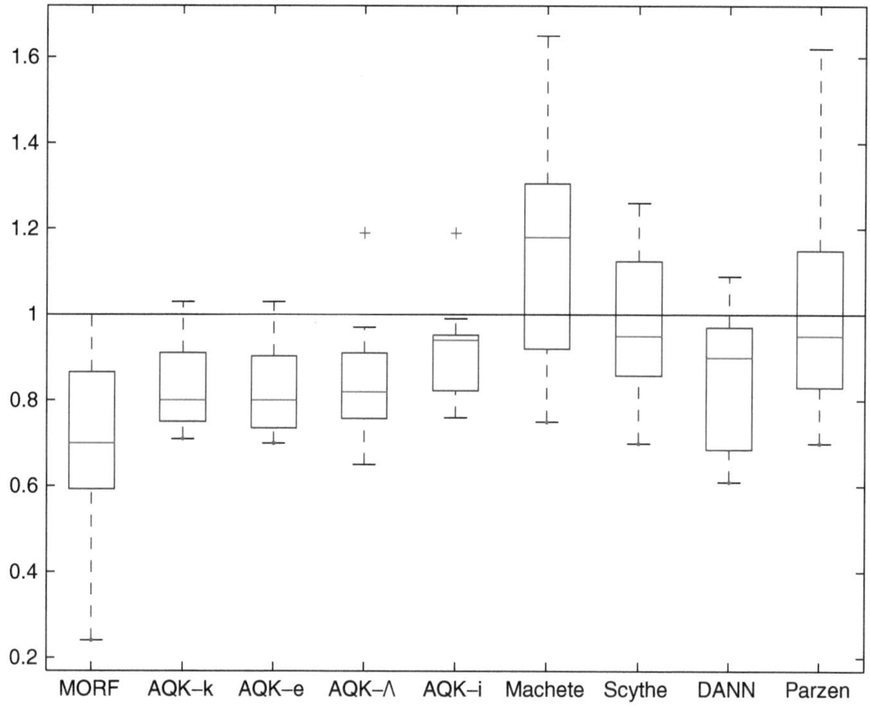

Fig. 7. Relative error rates of the methods across the nine real data sets. The error rate is divided by the error rate of 5 NN

among the features. Thus, the corresponding intrinsic dimensionality will be moderate. For example, in a typical vision problem such as face recognition, a subspace method such as PCA or KPCA is always applied first in order to capture such intrinsic dimensionality. In such cases, the technique presented here will likely be useful.

7 Summary and Conclusions

This chapter presents two locally adaptive nearest neighbor methods for effective pattern classification. MORF estimates a flexible metric for producing neighborhoods that are elongated along less relevant feature dimensions and constricted along most influential ones. On the other hand, AQK computes an adaptive distance based on quasiconformal transformed kernels. The effect of the distance is to move samples having similar class posterior probabilities to the query closer to it, while moving samples having different class posterior probabilities farther away from the query. As a result, the class conditional probabilities tend to be more homogeneous in the modified neighborhoods. The experimental results demonstrate that both the MORF and AQK algorithms can potentially improve the performance of KNN and recursive partitioning methods in some classification and data mining problems. The results are also in favor of MORF and AQK over similar competing methods such as Machete and DANN.

One limitation of both AQK and MORF is that they handle only two-class classification problems. One line of our future research is to extend both algorithms to multi-class problems. While SVM driven feature relevance learning is sound theoretically, the process is expensive computationally. We intend to explore margin-based feature relevance learning techniques to improve the efficiency of the MORF algorithm.

References

1. Amari, S., Wu., S. (1999) Improving support vector machine classifiers by modifying kernel functions. *Neural Networks*, 12(6):783–789.
2. Anderson, G.D., Vananamurthy, M.K., Vuorinen, M.K. (1997) *Conformal Invariants, Inequalities, and Quasiconformal Maps.* Canadian Mathematical Society Series of Monographs and Advanced Texts. John Wiley & Sons, Inc., New York.
3. Atkeson, C., Moore, A.W., Schaal, S. (1997) Locally weighted learning. *AI Review*, 11:11–73.
4. Bellman, R.E. (1961) *Adaptive Control Processes.* Princeton Univ. Press.
5. Blair, D.E. (2000) *Inversion Theory and Conformal Mapping.* American Mathematical Society.
6. Burges, C.J.C. (1998) A tutorial on support vector machines for pattern recognition. *Data Mining and Knowledge Discovery*, 2(2):121–167.
7. Comaniciu, D., Meer, P. (2002) Mean shift: A robust approach toward feature space analysis. *IEEE Trans. on Pattern Analysis and Machine Intelligence*, 24:603–619.
8. Cristianini, N., Shawe-Taylor, J. (2000) *An Introduction to Support Vector Machines and other kernel-based learning methods.* Cambridge University Press, Cambridge, UK.
9. Domeniconi, C., Peng, J., Gunopulos, D. (2001) An adaptive metric machine for pattern classification. In T. K. Leen, T. G. Dietterich, and V. Tresp, editors, *Advances in Neural Information Processing Systems*, volume 13, pages 458–464. The MIT Press.
10. Domeniconi, C., Peng, J., Gunopulos, D. (2002) Locally adaptive metric nearest neighbor classification. *IEEE Trans. on Pattern Analysis and Machine Intelligence*, 24(9):1281–1285.
11. Duda, R.O., Hart, P.E. (1973) *Pattern Classification and Scene Analysis.* John Wiley & Sons, Inc.
12. Friedman, J.H. (1994) *Flexible Metric Nearest Neighbor Classification.* Tech. Report, Dept. of Statistics, Stanford University.
13. Hastie, T., Tibshirani, R. (1996) Discriminant adaptive nearest neighbor classification. *IEEE Trans. on Pattern Analysis and Machine Intelligence*, 18(6):607–615.

14. Heisterkamp, D.R., Peng, J., Dai, H.K. (2001) An adaptive quasiconformal kernel metric for image retrieval. In *Proceedings of IEEE Computer Society Conference on Computer Vision and Pattern Recognition, Hawaii*, pp. 388–393.
15. Lowe, D.G. (1995) Similarity metric learning for a variable-kernel classifier. *Neural Computation*, 7(1):72–85.
16. Myles, J.P., Hand, D.J. (1990) The multi-class metric problem in nearestneighbor discrimination rules. *Pattern Recognition*, 723:1291–1297.
17. Peng, J., Heisterkamp, D.R., Dai, H.K. (2004) Adaptive quasiconformal kernel nearest neighbor classification. *IEEE Transactions on Pattern Analysis and Machine Intelligence*, 26(5):565–661.
18. Peng, J., Heisterkamp, D.R., Dai, H.K. (2004) Lda/svm driven nearest neighbor classification. *IEEE Transactions on Neural Networks*, 14(4):940–942.
19. Scholkopf, B. et al. (1999) Input space versus feature space in kernel-based methods. *IEEE Transactions on Neural Networks*, 10(5):1000–1017.
20. Schölkopf, B. (2001) The kernel trick for distances. In T. K. Leen, T. G. Dietterich, and V. Tresp, editors, *Advances in Neural Information Processing Systems*, volume 13, pp. 301–307. The MIT Press.
21. Schölkopf, B., Burges, C.J.C., Smola, A.J., editors. (1999) *Advances in kernel methods: support vector learning.* MIT Press, Cambridge, MA.
22. Short, R.D., Fukunaga, K. (1981) Optimal distance measure for nearest neighbor classification. *IEEE Transactions on Information Theory*, 27:622–627.
23. Tong, S., Koller, D. (2000) Restricted bayes optimal classifers. In *Proc. of AAAI.*
24. Vapnik, V. (1998) *Statistical learning theory.* Adaptive and learning systems for signal processing, communications, and control. Wiley, New York.
25. Wu, Y., Ianakiev, K.G., Govindaraju, V. (2001) Improvements in k-nearest neighbor classifications. In W. Kropatsch and N. Murshed, editors, *International Conference on Advances in pattern recognition*, volume 2, pages 222–229. Springer-Verlag.

Improving the Performance of the Support Vector Machine: Two Geometrical Scaling Methods

P. Williams, S. Wu, and J. Feng

Department of Informatics, University of Sussex, Falmer, Brighton BN1 9QH, UK

Abstract. In this chapter, we discuss two possible ways of improving the performance of the SVM, using geometric methods. The first adapts the kernel by magnifying the Riemannian metric in the neighborhood of the boundary, thereby increasing separation between the classes. The second method is concerned with optimal location of the separating boundary, given that the distributions of data on either side may have different scales.

Key words: kernel transformation, adaptive kernel, geometric scaling, generalization error, non-symmetric SVM

1 Introduction

The support vector machine (SVM) is a general method for pattern classification and regression proposed by Vapnik and co-authors [10]. It consists of two essential ideas, namely:

- to use a kernel function to map the original input data into a high-dimensional space so that two classes of data become linearly separable;
- to set the discrimination hyperplane in the middle of two classes.

Theoretical and experimental studies have proved that SVM methods can outperform conventional statistical approaches in term of minimizing the generalization error (see e.g. [3, 8]). In this chapter we review two geometrical scaling methods which attempt to improve the performance of the SVM further. These two methods concern two different ideas of scaling the SVM in order to reduce the generalization error.

The first approach concerns the scaling of the kernel function. From the geometrical point of view, the kernel mapping induces a Riemannian metric in the original input space [1, 2, 9]. Hence a good kernel should be one that can enlarge the separation between the two classes. To implement this idea, Amari

P. Williams, S. Wu, and J. Feng: *Improving the Performance of the Support Vector Machine: Two Geometrical Scaling Methods*, StudFuzz **177**, 205–218 (2005)
www.springerlink.com

and Wu [1, 9] propose a strategy which optimizes the kernel in a two-step procedure. In the first step of training, a primary kernel is used, whose training result provides information about where the separating boundary is roughly located. In the second step, the primary kernel is conformally scaled to magnify the Riemannian metric around the boundary, and hence the separation between the classes. In the original method proposed in [1, 9], the kernel is enlarged at the positions of the support vectors, which takes into account the fact that support vectors are in the vicinity of the boundary. This method, however, is susceptible to the distribution of data points. In the present study, we propose a different way for scaling kernel that directly acts on the distance to the boundary. Simulation shows that the new method works robustly.

The second approach to be reviewed concerns the optimal position for the discriminating hyperplane. The standard form of SVM chooses the separating boundary to be in the middle of two classes (more exactly, in the middle of the support vectors). By using extremal value theory in statistics, Feng and Williams [5] calculate the exact value of the generalization error in a one-dimensional separable case, and find that the optimal position is not necessarily to be at the mid-point, but instead it depends on the scales of the distances of the two classes of data with respect to the separating boundary. They further suggest how to use this knowledge to rescale SVM in order to achieve better generalization performance.

2 Scaling the Kernel Function

The SVM solution to a binary classification problem is given by a discriminant function of the form

$$f(\mathbf{x}) = \sum_{s \in SV} \alpha_s y_s K(\mathbf{x}_s, \mathbf{x}) + b \tag{1}$$

A new out-of-sample case is classified according to the sign of $f(\mathbf{x})$.[1]

The support vectors are, by definition, those $\mathbf{x}_i$ for which $\alpha_i > 0$. For separable problems each support vector $\mathbf{x}_s$ satisfies

$$f(\mathbf{x}_s) = y_s = \pm 1 \, .$$

In general, when the problem is not separable or is judged too costly to separate, a solution can always be found by bounding the multipliers α_i by the condition $\alpha_i \leq C$, for some (usually large) positive constant C. There are then two classes of support vector which satisfy the following distinguishing conditions:

$$\begin{aligned} \text{I}: \quad & y_s f(\mathbf{x}_s) = 1 \, , & 0 < \alpha_s < C \, ; \\ \text{II}: \quad & y_s f(\mathbf{x}_s) < 1 \, , & \alpha_s = C \, . \end{aligned}$$

[1] The significance of including or excluding a constant b term is discussed in [7].

Support vectors in the first class lie on the appropriate separating margin. Those in the second class lie on the wrong side (though they may be correctly classified in the sense that $\text{sign} f(\mathbf{x}_s) = y_s$). We shall call support vectors in the first class *true* support vectors and the others, by contrast, *bound*.

2.1 Kernel Geometry

It has been observed that the kernel $K(\mathbf{x}, \mathbf{x}')$ induces a Riemannian metric in the input space S [1, 9]. The metric tensor induced by K at $\mathbf{x} \in S$ is

$$g_{ij}(\mathbf{x}) = \frac{\partial}{\partial x_i} \frac{\partial}{\partial x'_j} K(\mathbf{x}, \mathbf{x}') \bigg|_{\mathbf{x}'=\mathbf{x}} . \tag{2}$$

This arises by considering K to correspond to the inner product

$$K(\mathbf{x}, \mathbf{x}') = \phi(\mathbf{x}) \cdot \phi(\mathbf{x}') \tag{3}$$

in some higher dimensional feature space H, where ϕ is a mapping of S into H (for further details see [4, p. 35]). The inner product metric in H then induces the Riemannian metric (2) in S via the mapping ϕ.

The volume element in S with respect to this metric is given by

$$dV = \sqrt{g(\mathbf{x})}\, dx_1 \cdots dx_n \tag{4}$$

where $g(\mathbf{x})$ is the determinant of the matrix whose (i, j)th element is $g_{ij}(\mathbf{x})$. The factor $\sqrt{g(\mathbf{x})}$, which we call the *magnification* factor, expresses how a local volume is expanded or contracted under the mapping ϕ. Amari and Wu [1, 9] suggest that it may be beneficial to increase the separation between sample points in S which are close to the separating boundary, by using a kernel $\tilde{K}$, whose corresponding mapping $\tilde{\phi}$ provides increased separation in H between such samples.

The problem is that the location of the boundary is initially unknown. Amari and Wu therefore suggest that the problem should first be solved in a standard way using some initial kernel K. It should then be solved a second time using a conformal transformation $\tilde{K}$ of the original kernel given by

$$\tilde{K}(\mathbf{x}, \mathbf{x}') = D(\mathbf{x}) K(\mathbf{x}, \mathbf{x}') D(\mathbf{x}') \tag{5}$$

for a suitably chosen positive function $D(\mathbf{x})$. It follows from (2) and (5) that the metric $\tilde{g}_{ij}(\mathbf{x})$ induced by $\tilde{K}$ is related to the original $g_{ij}(\mathbf{x})$ by

$$\begin{aligned} \tilde{g}_{ij}(\mathbf{x}) = {} & D(\mathbf{x})^2 g_{ij}(\mathbf{x}) + D_i(\mathbf{x}) K(\mathbf{x}, \mathbf{x}) D_j(\mathbf{x}) \\ & + D(\mathbf{x}) \{ K_i(\mathbf{x}, \mathbf{x}) D_j(\mathbf{x}) + K_j(\mathbf{x}, \mathbf{x}) D_i(\mathbf{x}) \} \end{aligned} \tag{6}$$

where $D_i(\mathbf{x}) = \partial D(\mathbf{x}) / \partial x_i$ and $K_i(\mathbf{x}, \mathbf{x}) = \partial K(\mathbf{x}, \mathbf{x}') / \partial x_i \,|_{\mathbf{x}'=\mathbf{x}}$. If $g_{ij}(\mathbf{x})$ is to be enlarged in the region of the class boundary, $D(\mathbf{x})$ needs to be largest in that vicinity, and its gradient needs to be small far away. Note that if D is chosen in this way, the resulting kernel $\tilde{K}$ becomes *data dependent.*

Amari and Wu consider the function

$$D(\mathbf{x}) = \sum_{i \in SV} e^{-\kappa_i \|\mathbf{x}-\mathbf{x}_i\|^2} \tag{7}$$

where κ_i are positive constants. The idea is that support vectors should normally be found close to the boundary, so that a magnification in the vicinity of support vectors should implement a magnification around the boundary. A possible difficulty is that, whilst this is correct for true support vectors, it need not be correct for bound ones.[2] Rather than attempt further refinement of the method embodied in (7), we shall describe here a more direct way of achieving the desired magnification.

2.2 New Approach

The idea here is to choose D so that it decays directly with distance, suitably measured, from the boundary determined by the first-pass solution using K. Specifically we consider

$$D(\mathbf{x}) = e^{-\kappa f(\mathbf{x})^2} \tag{8}$$

where f is given by (1) and κ is a positive constant. This takes its maximum value on the separating surface where $f(\mathbf{x}) = 0$, and decays to $e^{-\kappa}$ at the margins of the separating region where $f(\mathbf{x}) = \pm 1$, This is where the true support vectors lie. In the case where K is the simple inner product in S, the level sets of f and hence of D are just hyperplanes parallel to the separating hyperplane. In that case $|f(\mathbf{x})|$ measures perpendicular distance to the separating hyperplane, taking as unit the common distance of true support vectors from that hyperplane. In the general case the level sets are curved non-intersecting hypersurfaces.

3 Geometry and Magnification

To proceed further we need to consider specific forms for the kernel K.

3.1 RBF Kernels

Consider the Gaussian radial basis function kernel

$$K(\mathbf{x}, \mathbf{x}') = e^{-\|\mathbf{x}-\mathbf{x}'\|^2/2\sigma^2} . \tag{9}$$

---

[2]The method of choosing the κ_i in [9] attempts to meet this difficulty by making the decay rate roughly proportional to the local density of support vectors. Thus isolated support vectors are associated with a low decay rate, so that their influence is minimized.

This is of the general type where $K(\mathbf{x}, \mathbf{x}')$ depends on $\mathbf{x}$ and $\mathbf{x}'$ only through the norm their separation so that

$$K(\mathbf{x}, \mathbf{x}') = k\left(\|\mathbf{x} - \mathbf{x}'\|^2\right) . \tag{10}$$

Referring back to (2) it is straightforward to show that the induced metric is Euclidean with

$$g_{ij}(\mathbf{x}) = -2k'(0)\,\delta_{ij} . \tag{11}$$

In particular for the Gaussian kernel (9) where $k(\xi) = e^{-\xi/2\sigma^2}$ we have

$$g_{ij}(\mathbf{x}) = \frac{1}{\sigma^2}\,\delta_{ij} \tag{12}$$

so that $g(\mathbf{x}) = \det\{g_{ij}(\mathbf{x})\} = 1/\sigma^{2n}$ and hence the volume magnification is the constant

$$\sqrt{g(\mathbf{x})} = \frac{1}{\sigma^n} . \tag{13}$$

3.2 Inner Product Kernels

For another class of kernel, $K(\mathbf{x}, \mathbf{x}')$ depends on $\mathbf{x}$ and $\mathbf{x}'$ only through their inner product so that

$$K(\mathbf{x}, \mathbf{x}') = k(\mathbf{x} \cdot \mathbf{x}') . \tag{14}$$

A well known example is the inhomogeneous polynomial kernel

$$K(\mathbf{x}, \mathbf{x}') = (1 + \mathbf{x} \cdot \mathbf{x}')^d \tag{15}$$

for some positive integer d. For kernels of this type, it follows from (2) that the induced metric is

$$g_{ij}(\mathbf{x}) = k'(\|\mathbf{x}\|^2)\,\delta_{ij} + k''(\|\mathbf{x}\|^2)\,x_i x_j . \tag{16}$$

To evaluate the magnification factor, we need the following:

Lemma 1. *Suppose that* $\mathbf{a} = (a_1, \ldots, a_n)$ *is a vector and that the components* A_{ij} *of a matrix* $\mathbf{A}$ *are of the form* $A_{ij} = \alpha\delta_{ij} + \beta a_i a_j$. *Then* $\det \mathbf{A} = \alpha^{n-1}\left(\alpha + \beta\|\mathbf{a}\|^2\right)$.

It follows that, for kernels of the type (14), the magnification factor is

$$\sqrt{g(\mathbf{x})} = \sqrt{k'(\|\mathbf{x}\|^2)^n \left(1 + \frac{k''(\|\mathbf{x}\|^2)}{k'(\|\mathbf{x}\|^2)}\,\|\mathbf{x}\|^2\right)} \tag{17}$$

so that for the inhomogeneous polynomial kernel (15), where $k(\xi) = (1 + \xi)^d$,

$$\sqrt{g(\mathbf{x})} = \sqrt{d^n\,(1 + \|\mathbf{x}\|^2)^{n(d-1)-1}\,(1 + d\|\mathbf{x}\|^2)} . \tag{18}$$

For $d > 1$, the magnification factor (18) is a radial function, taking its minimum value at the origin and increasing, for $\|\mathbf{x}\| \gg 1$, as $\|\mathbf{x}\|^{n(d-1)}$. This suggests it might be most suitable, for binary classification, when one the classes forms a bounded cluster centered on the origin.

3.3 Conformal Kernel Transformations

To demonstrate the approach, we consider the case where the initial kernel K in (5) is the Gaussian RBF kernel (9). For illustration, consider the binary classification problem shown in Fig. 1, where 100 points have been selected at random in the square as a training set, and classified according to whether they fall above or below the curved boundary, which has been chosen as e^{-4x^2} up to a linear transform.

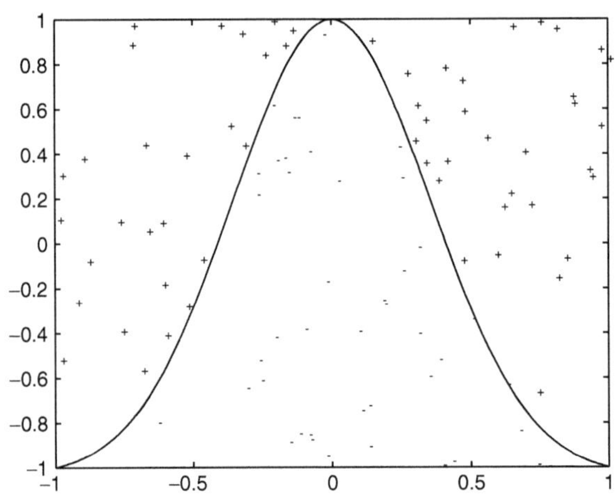

Fig. 1. A training set of 100 random points classified according to whether they lie above (+) or below (−) the Gaussian boundary shown

Our approach requires a first-pass solution using conventional methods. Using a Gaussian radial basis kernel with width 0.5 and soft-margin parameter $C = 10$, we obtain the solution shown in Fig. 2. This plots contours of the discriminant function f, which is of the form (1). For sufficiently large samples, the zero contour in Fig. 2 should coincide with the curve in Fig. 1.

To proceed with the second-pass we need to use the modified kernel given by (5) where K is given by (9) and D is given by (8). It is interesting first to calculate the general metric tensor $\tilde{g}_{ij}(\mathbf{x})$ when K is the Gaussian RBF kernel (9) and $\tilde{K}$ is derived from K by (5). Substituting in (6), and observing that in this case $K(\mathbf{x}, \mathbf{x}) = 1$ while $K_i(\mathbf{x}, \mathbf{x}) = K_j(\mathbf{x}, \mathbf{x}) = 0$, we obtain

$$\tilde{g}_{ij}(\mathbf{x}) = \frac{D(\mathbf{x})^2}{\sigma^2} \delta_{ij} + D_i(\mathbf{x}) D_j(\mathbf{x}) \, . \tag{19}$$

The $\tilde{g}_{ij}(\mathbf{x})$ in (19) are of the form considered in Lemma 1. Observing that $D_i(\mathbf{x})$ are the components of $\nabla D(\mathbf{x}) = D(\mathbf{x}) \nabla \log D(\mathbf{x})$, it follows that the ratio of the new to the old magnification factors is given by

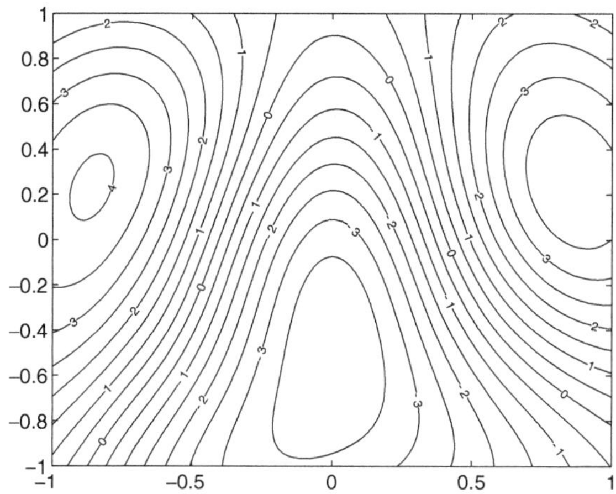

Fig. 2. First-pass SVM solution to the problem in Fig. 1 using a Gaussian kernel. The contours show the level sets of the discriminant function f defined by (1)

$$\sqrt{\frac{\tilde{g}(\mathbf{x})}{g(\mathbf{x})}} = D(\mathbf{x})^n \sqrt{1 + \sigma^2 \|\nabla \log D(\mathbf{x})\|^2} \,. \tag{20}$$

This is true for any positive scalar function $D(\mathbf{x})$. Let us now use the function given by (8) for which

$$\log D(\mathbf{x}) = -\kappa f(\mathbf{x})^2 \tag{21}$$

where f is the first-pass solution given by (1) and shown, for example, in Fig. 2. This gives

$$\sqrt{\frac{\tilde{g}(\mathbf{x})}{g(\mathbf{x})}} = \exp\left\{-n\kappa f(\mathbf{x})^2\right\} \sqrt{1 + 4\kappa^2\sigma^2 f(\mathbf{x})^2 \|\nabla f(\mathbf{x})\|^2} \,. \tag{22}$$

This means that

1. *the magnification is constant on the separating surface* $f(\mathbf{x}) = 0$*;*
2. *along contours of constant* $f(\mathbf{x}) \neq 0$*, the magnification is greatest where the contours are closest.*

The latter is because of the occurrence of $\|\nabla f(\mathbf{x})\|^2$ in (22). The gradient points uphill orthogonally to the local contour, hence in the direction of steepest ascent; the larger its magnitude, the steeper is the ascent, and hence the closer are the local contours. This character is illustrated in Fig. 3 which shows the magnification factor for the modified kernel based on the solution of Fig. 2. Notice that the magnification is low at distances remote from the boundary.

Solving the original problem again, but now using the modified kernel $\tilde{K}$, we obtain the solution shown in Fig. 4. Comparing this with the first-pass

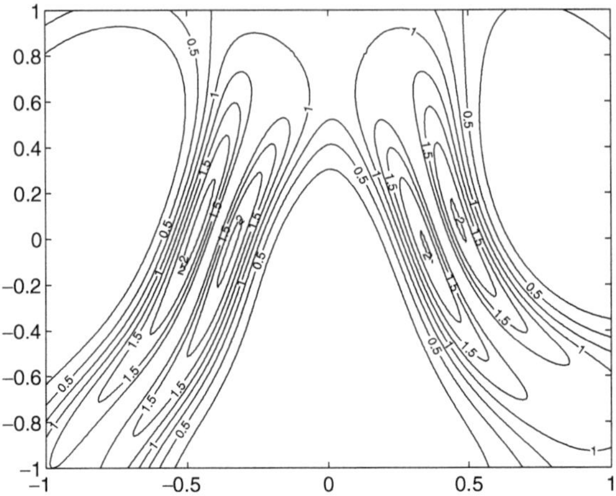

Fig. 3. Contours of the magnification factor (22) for the modified kernel using $D(\mathbf{x}) = \exp\{-\kappa f(\mathbf{x})^2\}$ with f defined by the solution of Fig. 2

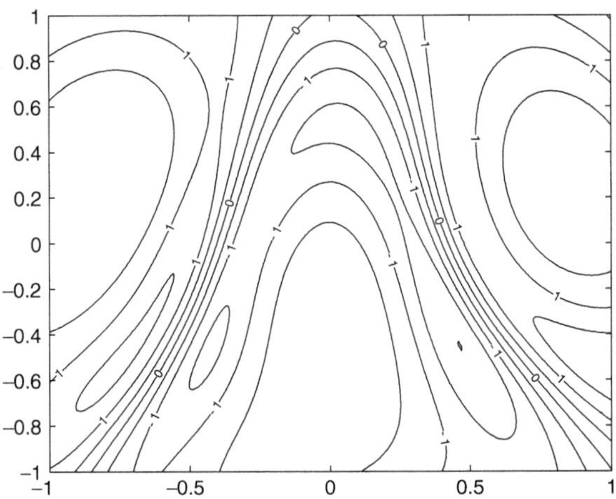

Fig. 4. Second-pass solution using the modified kernel

solution of Fig. 2, notice the steeper gradient in the vicinity of the boundary, and the relatively flat areas remote from the boundary.

In this instance the classification provided by the modified solution is little improvement on the original classification. This an accident of the choice of the training set shown in Fig. 1. We have repeated the experiment 10000 times, with a different choice of 100 training sites and 1000 test sites on each occasion, and have found an average of 14.5% improvement in classification

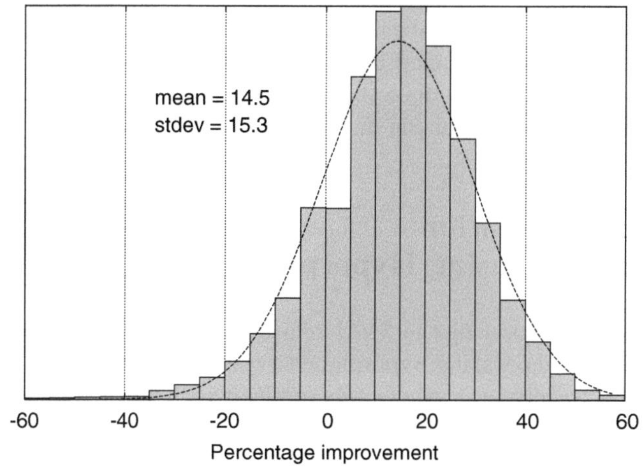

Fig. 5. Histogram of the percentage improvement in classification, over 10000 experiments, together with a normal curve with the same mean and standard deviation

performance.[3] A histogram of the percentage improvement, over the 10000 experiments, together with a normal curve with the same mean and standard deviation, is shown in Fig. 5.

3.4 Choice of κ

A presently unresolved issue is how best to make a systematic choice of κ. It is clear that κ is dimensionless, in the sense of being scale invariant. Suppose all input dimensions in the input space S are multiplied by a positive scalar a. To obtain the same results for the first-pass solution, a new $\sigma_a = a\sigma$ must be used in the Gaussian kernel (9). This leads to the first-pass solution f_a where $f_a(a\mathbf{x}) = f(\mathbf{x})$ with f being the initial solution using σ. It then follows from (5) and (8) that *provided κ is left unchanged* the rescaled second-pass solution automatically satisfies the corresponding covariance relation $\tilde{f}_a(a\mathbf{x}) = \tilde{f}(\mathbf{x})$ where $\tilde{f}$ was the original second-pass solution using σ.

It may appear that there is a relationship between κ and σ in the expression (22) for the magnification ratio. Using a corresponding notation, however, it is straightforward to show that the required covariance $\tilde{g}_a(a\mathbf{x})/g_a(a\mathbf{x}) = \tilde{g}(\mathbf{x})/g(\mathbf{x})$ also holds provided κ is left unchanged. The reason is that $\sigma\|\nabla f(\mathbf{x})\|$ is invariant under rescaling since a multiplies σ and divides $\nabla f(\mathbf{x})$.

Possibly κ should depend on the dimension n of the input space. This has not yet been investigated. In the trials reported above, it was found that a

[3]If there are 50 errors in 1000 for the original solution and 40 errors for the modified solution, we call this a 20% improvement. If there are 60 errors for the modified solution, we call it a -20% improvement.

suitable choice was $\kappa = 0.25$. We note that this is approximately the reciprocal of the maximum value obtained by f in the first pass solution.

In the following we introduce the second approach which concerns how to scale the optimal position of the discriminating hyperplane.

4 Scaling the Position of the Discriminating Hyperplane

The original motivation of the SVM relates to maximizing the margin (the distance from the separating hyperplane to the nearest example). The essence of the SVM is to rely on the set of examples which take extreme values, the so-called support vectors. But from the statistics of extreme values, we know that the disadvantage of such an approach is that information contained in most samples (not extreme) values is lost, so that such an approach would be expected to be less efficient than one which takes into account the lost information. These ideas were explored in [5]. We give here a summary of results.

To introduce the model, consider for simplicity a one-dimensional classification problem. Suppose that we have two populations, one of positive variables x and one of negative variables y, and that we observe t positive examples $x(1), \ldots, x(t) > 0$ and t negative examples $y(1), \ldots, y(t) < 0$. Since this case is separable, the SVM will use the threshold

$$z(t) = \frac{1}{2}\,\underline{x}(t) + \frac{1}{2}\,\overline{y}(t) \tag{23}$$

for classifying future cases, where

$$\underline{x}(t) = \min\{x(i) : i = 1, \ldots, t\}$$

is the minimum of the positive examples and

$$\overline{y}(t) = \max\{y(i) : i = 1, \ldots, t\}$$

is the maximum of the negative examples. A newly observed ξ will be classified as belonging to the x or y populations, depending on whether $\xi > z(t)$ or $\xi < z(t)$. This is pictured in Fig. 6.

4.1 Generalization Error

If a new ξ is observed, which may belong to either the x or y populations, an error occurs if ξ lies in the region between the dashed and solid lines shown in Fig. 6. The dashed line is fixed at the origin, but the solid line is located at the threshold $z(t)$ which, like ξ, is a random variable. A misclassification will occur if either $0 < \xi < z(t)$ or $z(t) < \xi < 0$.

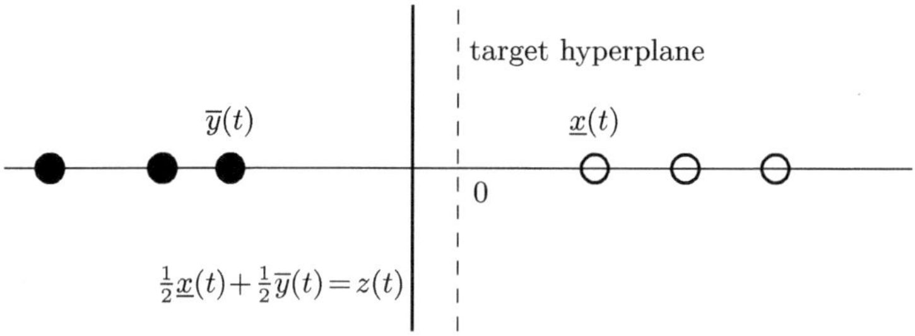

Fig. 6. Schematic representation of the one-dimensional support vector machine. The task is to separate the disks (*filled*) from the circles (*hollow*). The true separation is assumed to be given by the *dashed vertical line*. After learning t examples, the separating hyperplane for the support vector machine is at $z(t) = \frac{1}{2}\underline{x}(t) + \frac{1}{2}\overline{y}(t)$. The error region is then the region between the *dashed line* and the *solid line*

We define the *generalization error* $\varepsilon(t)$ to be the *probability* of misclassification. The generalization error $\varepsilon(t)$ is therefore a random variable whose distribution depends on the distributions of the x and y variables. In [5] it is shown that if there is an equal prior probability of ξ belonging to the x or y populations then, under a wide class of distributions for the x and y variables, the mean and variance of $\varepsilon(t)$, when defined in terms of the symmetric threshold (23), in the limit as $t \to \infty$ have the values

$$E\big(\varepsilon(t)\big) = 1/4t \tag{24}$$

$$\mathrm{var}\big(\varepsilon(t)\big) = 1/16t^2\ . \tag{25}$$

For example, suppose that the $x(t)$ are independent and uniformly distributed on the positive unit interval $[0, 1]$ and that the $y(t)$ are similarly distributed on the negative unit interval $[-1, 0]$. Then the exact value for the mean of the generalization error, for any $t > 0$, is in fact $t/(t + 1)(4t + 2) \approx 1/4t$. If the $x(t)$ have positive exponential distributions and the $y(t)$ have negative exponential distributions, the exact value for the mean is $1/(4t + 2) \approx 1/4t$. The generality of the limiting expressions (24) and (25) derives from results of extreme value theory [6, Chap. 1]. It is worth pointing out that results, such as (25), for the variance of the generalization error of the SVM have not previously been widely reported.

4.2 The Non-symmetric SVM

The threshold (23) follows the usual SVM practice of choosing the mid-point of the margin to separate the positive and negative examples. But if positive and negative examples are scaled differently in terms of their distance from the separating hyperplane, the mid-point may not be optimal. Let us therefore consider the general threshold

$$z(t) = \lambda\underline{x}(t) + \mu\overline{y}(t) \qquad (\lambda + \mu = 1)\ . \tag{26}$$

In separable cases (26) will correctly classify the observed examples for any $0 \leq \lambda \leq 1$. The symmetric SVM0 corresponds to $\lambda = 1/2$. The cases $\lambda = 0$ and $\lambda = 1$ were said in [5] to correspond to the "worst learning machine". We now calculate the distribution of the generalization error for the general threshold (26).

Note that the generalization error can be written as

$$\varepsilon_\lambda(t) = P\big(0 < \xi < z(t)\big)\, I(z(t) > 0) + P\big(z(t) < \xi < 0\big)\, I(z(t) < 0) \tag{27}$$

where $I(A)$ is the $\{0,1\}$-valued indicator function of the event A. To calculate the distribution of $\varepsilon_\lambda(t)$ we need to know the distributions of ξ and $z(t)$. To be specific, assume that each $x(i)$ has a positive exponential distribution with scale parameter a and each $y(i)$ has a negative exponential distribution with scale parameter b. It is then straightforward to show that $z(t)$ defined by (26) has an asymmetric Laplace distribution such that

$$P(z(t) > \zeta) = \left(\frac{\lambda a}{\lambda a + \mu b}\right) e^{-(t/\lambda a)\zeta} \qquad (\zeta > 0) \tag{28}$$

$$P(z(t) < \zeta) = \left(\frac{\mu b}{\lambda a + \mu b}\right) e^{(t/\mu b)\zeta} \qquad (\zeta < 0)\,. \tag{29}$$

Let us assume furthermore that a newly observed ξ has probability 1/2 of having the same distribution as either $x(i)$ or $y(i)$. In that case ξ also has an asymmetric Laplace distribution and (27) becomes

$$\varepsilon_\lambda(t) = \frac{1}{2}\left\{1 - e^{-z(t)/a}\right\} I(z(t) > 0) + \frac{1}{2}\left\{1 - e^{z(t)/b}\right\} I(z(t) < 0)\,. \tag{30}$$

Making use of (28) and (29) for the distribution of $z(t)$, it follows that for any $0 \leq p \leq 1$,

$$P\big(2\varepsilon_\lambda(t) > p\big) = \left(\frac{\lambda a}{\lambda a + \mu b}\right)(1-p)^{t/\lambda} + \left(\frac{\mu b}{\lambda a + \mu b}\right)(1-p)^{t/\mu} \tag{31}$$

which implies that $2\varepsilon_\lambda(t)$ has a mixture of Beta$(1, t/\lambda)$ and Beta$(1, t/\mu)$ distributions.[4] It follows that the mean of $2\varepsilon_\lambda(t)$ is

$$\left(\frac{\lambda a}{\lambda a + \mu b}\right)\left(\frac{\lambda}{t+\lambda}\right) + \left(\frac{\mu b}{\lambda a + \mu b}\right)\left(\frac{\mu}{t+\mu}\right) \tag{32}$$

so that for large t, since $\lambda, \mu \leq 1$, the expected generalization error has the limiting value

$$E\big(\varepsilon_\lambda(t)\big) = \frac{1}{2t}\left\{\frac{\lambda^2 a + \mu^2 b}{\lambda a + \mu b}\right\}\,. \tag{33}$$

[4]The error region always lies wholly to one side or other of the origin so that, under present assumptions, the probability that ξ lies in this region, and hence the value of the generalization error $\varepsilon_\lambda(t)$, is never more than 1/2.

A corresponding, though lengthier, expression can be found for the variance. Note that the limiting form (33) holds for a wide variety of distributions, for example if each $x(i)$ is uniformly distributed on $[0, a]$ and each $y(i)$ is uniformly distributed on $[-b, 0]$, compare [5].

Optimal Value of λ

What is the optimal value of λ if the aim is to minimize the expected generalization error given by (33)? The usual symmetric SVM chooses $\lambda = 1/2$. In that case we have

$$E\left(\varepsilon_{\frac{1}{2}}(t)\right) = \frac{1}{4t} \tag{34}$$

$$\operatorname{var}\left(\varepsilon_{\frac{1}{2}}(t)\right) = \frac{1}{16t^2} \tag{35}$$

as previously in (24) and (25). Interestingly, this shows that those results are independent of the scaling of the input distributions. However, if $a \neq b$, an improvement may be possible.

An alternative, which comes readily to mind, is to divide the margin in the inverse ratio of the two scales by using

$$\lambda^\dagger = \frac{b}{a+b}\,. \tag{36}$$

We then have

$$E(\varepsilon_{\lambda^\dagger}(t)) = \frac{1}{4t} \tag{37}$$

$$\operatorname{var}(\varepsilon_{\lambda^\dagger}(t)) = \frac{1}{16t^2}\left\{1 + \left(\frac{a-b}{a+b}\right)^2\right\}. \tag{38}$$

Notice that, for $\lambda = \lambda^\dagger$, the expected generalization error is unchanged, but the variance is increased, unless $a = b$.

It is easy to verify, however, that the minimum of (33) in fact occurs at

$$\lambda^* = \frac{\sqrt{b}}{\sqrt{a}+\sqrt{b}} \tag{39}$$

for which

$$E(\varepsilon_{\lambda^*}(t)) = \frac{1}{4t}\left\{1 - \left(\frac{\sqrt{a}-\sqrt{b}}{\sqrt{a}+\sqrt{b}}\right)^2\right\} \tag{40}$$

$$\operatorname{var}(\varepsilon_{\lambda^*}(t)) = \frac{1}{16t^2}\left\{1 - \left(\frac{\sqrt{a}-\sqrt{b}}{\sqrt{a}+\sqrt{b}}\right)^4\right\} \tag{41}$$

showing that both mean and variance are reduced for $\lambda = \lambda^*$ compared with $\lambda = 1/2$ or $\lambda = \lambda^\dagger$.

5 Conclusions

In this chapter we have introduced two methods for improving the performance of the SVM. One method is geometry-oriented, which concerns a data-dependent way to scale the kernel function so that the separation between two classes is enlarged. The other is statistics-motivated, which concerns how to optimize the position of the discriminating hyperplane based on the different scales of the two classes of data. Both methods have proved to be effective for reducing the generalization error of SVM. Combining the two methods together, we would expect a further reduction on the generalization error. This is currently under investigation.

Acknowledgement

Partially supported by grants from UK EPSRC (GR/R54569), (GR/S20574) and (GR/S30443).

References

1. Amari S, Si Wu (1999) Improving Support Vector Machine classifiers by modifying kernel functions. Neural Networks 12:783–789
2. Burges CJC (1999) Geometry and invariance in kernel based methods In: Burges C, Schölkopf B, Smola A (eds) Advances in Kernel Methods—Support Vector Learning, MIT Press, 89–116
3. Cristanini N, Shawe-Taylor J (2000) An Introduction to Support Vector Machines. Cambridge University Press, Cambridge, UK
4. Cucker F, Smale S (2001) On the mathematical foundations of learning. Bulletin of the AMS 39(1):1–49
5. Feng J, Williams P (2001) The generalization error of the symmetric and scaled Support Vector Machines. IEEE Transactions on Neural Networks 12(5):1255–1260
6. Leadbetter MR, Lindgren G, Rootzén H (1983) Extremes and Related Properties of Random Sequences and Processes. Springer-Verlag, New York
7. Poggio T, Mukherjee S, Rifkin R, Raklin A, Verri A (2002) B. In: Winkler J, Niranjan M (eds) Uncertainty in Geometric Computations. Kluwer Academic Publishers, 131–141
8. Schölkopf B, Smola A (2002) Learning with Kernels. MIT Press, UK
9. Si Wu, Amari S (2001) Conformal transformation of kernel functions: a data dependent way to improve Support Vector Machine classifiers. Neural Processing Letters 15:59–67
10. Vapnik V (1995) The Nature of Statistical Learning Theory. Springer, NY

An Accelerated Robust Support Vector Machine Algorithm

Q. Song, W.J. Hu and X.L. Yang

School of Electrical and Electronic Engineering, Nanyang Technological University Block S2, 50 Nanyang Avenue, Singapore 639798.
eqsong@ntu.edu.sg

Abstract. This chapter proposes an accelerated decomposition algorithm for robust support vector machine (SVM). Robust SVM aims at solving the overfitting problem when there is outlier in the training data set, which makes the decision surface less detored and results in sparse support vectors. Training of the robust SVM leads to a quadratic optimization problem with bound and linear constraint. Osuna provides a theorem which proves that the Standard SVM's Quadratic Programming (QP) problem can be broken down into a series of smaller QP sub-problems. This chapter derives the Kuhn-Tucker condition and decomposition algorithm for the robust SVM. Furthermore, a pre-selection technique is incorporated into the algorithm to speed up the calculation. The experiment using standard data sets shows that the accelerated decomposition algorithm makes the training process more efficient.

Key words: Support Vector Machines, Robust Support Vector Machine, Kernel Method, Decomposition Algorithm, Quadratic Programming

1 Introduction

Support Vector Machine (SVM) has been developed successfully to solve pattern recognition and nonlinear regression problems by Vapnik and other researchers [1, 2, 3] (here we call it standard SVM). It can be seen as an alternative training technique for Polynomial, Radial Basis Function and Multi-layer Perceptron classifiers. In practical situation the training data are often polluted by outlier[4], this makes the decision surface deviated from the optimal hyperplane severely, particularly, when the training data are misclassified as the wrong class accidently. Some techniques have been found to tackle the outlier problem, for example, the least square SVM [5] and adaptive margin SVM [6, 7, 8]. In [8] a robust SVM was proposed and the distance between each data point and the center of the respective class is used to calculate the adaptive margin which makes SVM less sensitive to the disturbance [6, 8].

Q. Song, W. Hu, and X. Yang: *An Accelerated Robust Support Vector Machine Algorithm*, StudFuzz **177**, 219–232 (2005)
www.springerlink.com

From the implementation point of view, training of the robust SVM is equivalent to solve a linearly constrained Quadratic Programming (QP) problem, in which the number of variables equal to the number of data points. The computing task becomes very challenge when the number of data is beyond a few thousands. Osuna [9] proposed a generalized decomposition strategy for the standard SVM, in which the original QP problem is replaced by a series of smaller sub-problems that is proved to be able to converge to a global optimum point. However, it's well known that the decomposition process heavily depends on the selection of a good working set of the data, which starts normally with a random subset in [9].

In this chapter we derive the Kuhn-Tucker [10] condition and decomposition algorithm for the robust SVM based on Osuna's original idea since the QP problem of the robust SVM is a convex optimization (it means that it's guaranteed to have a positive-semidefinite Hessian matrix and all constrains are linear). Furthermore, a pre-selection technique [11] is incorporated into the decomposition algorithm, which can find a better starting point of the training sets than the one selected randomly and reduce the number of iterations and training. Contrast to the adaptive margin algorithm in [6], we propose the accelerated method for the particular robust SVM algorithm in [8] because both of them depend on the distance between the training data and class centers in the feature space. It should also be noted that the pre-selection only considers the intersection area between two classes and the robust SVM treats every "remote" point as potential outlier . There is a potential risk for the pre-selection method that an outlier (wrong training data) may be easily captured in the early iteration due to its primary location to the intersection area. However, this will be discounted by the robust SVM since it is not sensitive to the outlier. The experiment result of Sect. 5.1 shows good result of the method in the presence of outliers, not only the number of support vectors but also the overall testing errors are reduced by the robust SVM with proper regularization parameter.

The rest of this chapter is organized as follows: Sect. II presents algorithm of the robust Support Vector Machine. In Sect. III, KT condition (Optimality Condition) and decomposition algorithm of the robust SVM are derived. The Pre-selection method is implemented and incorporated into the decomposition algorithm in Sect. IV. Finally experiments results are shown in Sect. V and the summary is given in Sect. VI.

2 Robust Support Vector Machine

Vapnik [1] shows support vector machine for pattern recognition problem which is represented by the prime optimization problem:

$$\begin{aligned} &\text{Minimize} \quad \Phi(\mathbf{w}) = \tfrac{1}{2}\mathbf{w}^T\mathbf{w} + C\sum_{i=1}^{l}\xi_i \\ &\text{Subject to} \quad y_i f(\mathbf{x}_i) \geq 1 - \xi_i \end{aligned} \tag{1}$$

where $f(\mathbf{x}_i) = \mathbf{w}\cdot\phi(\mathbf{x}_i)+b$, b is the bias, $\mathbf{w}$ is the weight of the kernel function, C is a constant for the slack variable $\{\xi_i\}_{i=1}^{l}$. One must admit some training errors to find the best tradeoff between training error and margin by choosing the appropriate value of C.

This leads to the following dual quadratic optimization problem (QP):

$$\begin{aligned} &\min_{\alpha} \Psi(\alpha) = \min_{\alpha} \frac{1}{2}\sum_{1}^{l}\sum_{1}^{l} y_i y_j K(\mathbf{x}_i, \mathbf{x}_j)\alpha_i\alpha_j - \sum_{1}^{l}\alpha_i \\ &\sum_{i=1}^{l} y_i\alpha_i = 0 \\ &0 \leq \alpha_i \leq C, \ \ \forall i \ , \end{aligned} \tag{2}$$

where $\alpha = (\alpha_1, \ldots, \alpha_l)^T$ is the Lagrangian for the optimization problem and y_i is the target of training sample $\mathbf{x}_i$. We call it standard Support Vector Machine (SVM).

To formulate prime problem of the robust SVM algorithm, the distance between each dada point and the center of the respective class is used to calculate the adaptive margin [8]. The classification accuracy is sacrificed to obtain smooth decision surface. A new slack variable $\lambda D^2(\mathbf{x}_i, \mathbf{x}^{*}_{y_i})$ for the robust SVM is introduced instead of $\{\xi_i\}_{i=1}^{\ell}$ in the standard SVM algorithm as follows:

$$\begin{aligned} &\text{Minimize} \quad \Phi(\mathbf{w}) = \frac{1}{2}\mathbf{w}^T\mathbf{w} \\ &\text{Subject to} \quad y_i f(\mathbf{x}_i) \geq 1 - \lambda D^2(\mathbf{x}_i, \mathbf{x}^*_{y_i}) \end{aligned} \tag{3}$$

where $\lambda \geq 0$ is a pre-selected parameter measuring the adaptation of the margin, and $D^2(\mathbf{x}_i, \mathbf{x}^*_{y_i})$ represents the normalized distance between each data point and the center of the respective class in the kernel space, which is calculated by

$$\begin{aligned} D^2(\mathbf{x}_i, \mathbf{x}^*_{y_i}) &= \left|\phi(\mathbf{x}_i) - \phi(\mathbf{x}^*_{y_i})\right|^2 / D^2_{\max} \\ &= [(\phi(\mathbf{x}_i)\cdot\phi(\mathbf{x}_i) - 2\phi(\mathbf{x}_i)\cdot\phi(\mathbf{x}^*_{y_i}) + \phi(\mathbf{x}^*_{y_i})\cdot\phi(\mathbf{x}^*_{y_i})]/D^2_{\max} \\ &= [k(\mathbf{x}_i, \mathbf{x}_i) - 2k(\mathbf{x}_i, \mathbf{x}^*_{y_i}) + k(\mathbf{x}^*_{y_i}, \mathbf{x}^*_{y_i})]/D^2_{\max} \end{aligned} \tag{4}$$

where $\{\phi(\mathbf{x}_i)\}_{i=1}^{h}$ $(h \leq l)$ denote a set of nonlinear transformations from the input space to the feature space, $k(\mathbf{x}_i, \mathbf{x}_j) = \phi(\mathbf{x}_i)\cdot\phi(\mathbf{x}_j)$ represents the inner-product kernel function, $D_{\max} = \max(D(\mathbf{x}_i, \mathbf{x}^*_{y_i}))$ is the maximum distance between the center and training data points of the respective class in the kernel space, $k(\mathbf{x}_i, \mathbf{x}^*_{y_i}) = \phi(\mathbf{x}_i)\cdot\phi(\mathbf{x}^*_{y_i})$ is the kernel function formed by the sample data and the center of the respective class in the feature space. For samples in class +1, $\phi(\mathbf{x}^*_{y_i}) = \frac{1}{n^+}\sum_{y_j=+1}\phi(\mathbf{x}_j)$, n^+ is the number of data points in class +1, in class −1, $\phi(\mathbf{x}^*_{y_i}) = \frac{1}{n^-}\sum_{y_j=-1}\phi(\mathbf{x}_j)$, n^- is the number of data points in class −1.

Accordingly, the dual formulation of optimization problem becomes

$$\begin{aligned} \min_{\alpha}\Psi(\alpha) = \min_{\alpha}\frac{1}{2}\sum_{1}^{\ell}\sum_{1}^{\ell} y_i y_j K(\mathbf{x}_i,\mathbf{x}_j)\alpha_i\alpha_j - \sum_{1}^{\ell}\alpha_i(1-\lambda D^2(\mathbf{x}_i,\mathbf{x}^*_{y_i})) \\ \sum_{i=1}^{l} y_i\alpha_i = 0 \\ \alpha_i \geq 0,\ \forall i \ , \end{aligned} \tag{5}$$

Comparing with the dual problem in the standard SVM, we may find that the only difference lies in the additional part in the $1-D^2(\mathbf{x}_i,\mathbf{x}^*_{y_i})$ in maximization functional $\Psi(\alpha)$. Here we conclude the effect of the parameter λ as follows:

1. If $\lambda = 0$ no adaptation of the margin is performed. The robust SVM becomes standard SVM with $C \to \infty$.
2. If $\lambda > 0$. The algorithm is robust against outliers. The support vector will be greatly influenced by the data that are the nearest point to the center of the respective class. The larger the parameter λ is, the nearer the support vectors will be towards the center of the respective class.

3 Decomposition Algorithm for Robust Support Vector Machine

This section presents the decomposition algorithm for the robust SVM. This strategy uses decomposition algorithm similar to that of standard SVM proposed by Osuna [9]. In each iteration the lagrange multiplier α_i are divided into two sets, that is, the set B of free variables (working set) and the set N of fixed variables (non-working set). To determine whether the algorithm has found the optimal solution α, consider that QP problem (5) is guaranteed to have a positive-semidefinite Hessian $Q_{ij} = y_i y_j k(\mathbf{x}_i,\mathbf{x}_j)$ and all constrains are linear (that means it's the convex optimization problem). The following Kuhn-Tucker conditions are necessary and sufficient for optimal solution of the QP problem:

$$\begin{aligned} \text{minimize}\quad & \Psi(\alpha) = -\alpha^T\gamma + \frac{1}{2}\alpha^T\mathbf{Q}\alpha \\ & \alpha^T\mathbf{y} = \mathbf{0} \\ & -\alpha \leq \mathbf{0}\ , \end{aligned} \tag{6}$$

where $\alpha = (\alpha_1, \ldots, \alpha_\ell)^T$,$Q_{ij} = y_i y_j k(\mathbf{x}_i, \mathbf{x}_j)$ and the adaptive margin multiplier is defined as $\gamma_i = 1 - \lambda D^2(\mathbf{x}_i, \mathbf{x}^\star)$. Denote the Lagrange multiplier for the equality constraint with β^{eq} and the Lagrange multipliers for the lower bounds with $\beta^{\mathbf{lo}}$, such that the KT conditions are defined as follows:

$$\begin{aligned}
\nabla\Psi(\alpha) + \beta^{eq}\mathbf{y} - \beta^{\mathbf{lo}} &= \mathbf{0} \\
\beta^{\mathbf{lo}} \cdot \alpha &= 0 \\
\beta^{eq} &\geq 0 \\
\beta^{\mathbf{lo}} &\geq \mathbf{0} \\
\alpha^T \mathbf{y} &= 0 \\
\alpha &\geq \mathbf{0}\,.
\end{aligned} \tag{7}$$

In order to derive further algebraic expressions from the optimality conditions (7), we consider the two possible values that each component of α should have as the following.

1. **Case:** $\alpha_i > 0$
   From the first two equation of (7), we have
   $$\nabla\Psi(\alpha) + \beta^{eq}\mathbf{y} = \mathbf{0} \tag{8}$$
   that means
   $$(\mathbf{Q}\alpha)_i + \beta^{eq} y_i = \gamma_i \tag{9}$$
   and then
   $$y_i \sum_j \alpha_j y_j k(\mathbf{x}_i, \mathbf{x}_j) + y_i \beta^{eq} = \gamma_i \tag{10}$$
   for $\alpha_i > 0$, the corresponding point $\mathbf{x}_i$ is called support vector which satisfies (see [1] Chap. 9, Sect. 9.5 for Kuhn-Tucker condition)
   $$y_i f(\mathbf{x}_i) = \gamma_i \tag{11}$$
   where $f(\mathbf{x}_i) = \sum_j \alpha_j y_j k(\mathbf{x}_i, \mathbf{x}_j) + b$ is the decision function. From (11) we obtain
   $$y_i \sum_j \alpha_j y_j k(\mathbf{x}_i, \mathbf{x}_j) + y_i b = \gamma_i\,. \tag{12}$$
   From (10) and (12) we get
   $$\beta^{eq} = b\,. \tag{13}$$
2. **Case:** $\alpha_i = 0$
   From the first two equation of (7), we have
   $$\nabla\Psi(\alpha) + \beta^{eq}\mathbf{y} - \beta^{\mathbf{lo}} = \mathbf{0} \tag{14}$$
   with $\beta^{\mathbf{lo}} \geq \mathbf{0}$, it follows
   $$\nabla\Psi(\alpha) + \beta^{eq}\mathbf{y} \geq \mathbf{0}\,. \tag{15}$$

From equation (13) we obtain

$$y_i f(\mathbf{x}_i) \geq \gamma_i \ . \tag{16}$$

A decomposition algorithm is proposed in this section which is similar to the approach by Osuna [9] and modified for the robust SVM algorithm. The main idea is iteration solution of the sub-problem and evaluation of the stop criteria for the algorithm. The optimality conditions derived above are essential for decomposition strategy to guarantee that at every iteration the objective function is improved. Then α is decomposed into two sub-vectors α_B and α_N, keeping fixed α_N and allowing changes only in α_B, thus defining the following sub-problem:

$$\begin{aligned} \text{Minimizing} \quad & W(\alpha_B) = -\alpha_B^T \gamma + \frac{1}{2}\left[\alpha_B^T \mathbf{Q}_{BB} \alpha_B + \alpha_B^T \mathbf{Q}_{BN} \alpha_N \right. \\ & \qquad \left. + \alpha_N^T \mathbf{Q}_{NB} \alpha_B + \ \alpha_N^T \mathbf{Q}_{NN} \alpha_N \right] - \alpha_N^T \gamma \\ \text{Subject to} \quad & \alpha_B^T \mathbf{y}_B + \alpha_N^T \mathbf{y}_N = 0 \\ & -\alpha_B \leq \mathbf{0} \ . \end{aligned} \tag{17}$$

where

$$\begin{pmatrix} \mathbf{Q}_{BB} & \mathbf{Q}_{BN} \\ \mathbf{Q}_{NB} & \mathbf{Q}_{NN} \end{pmatrix} \tag{18}$$

is a permutation of the matrix Q.

Using this decomposition we have the following propositions (proof referring to Sect. 3 in [9]): Moving a variable from B to N leaves the cost function unchanged, and the solution is feasible in the sub-problem. Moving variable that violate the optimality condition from N to B gives a strict improvement in the cost function when the sub-problem is re-optimized. Since the objective function is bounded, the algorithm must converge to the global optimal solution in a finite number of iterations.

4 Accelerated Method for the Decomposition Algorithm

The decomposition algorithm discussed above breaks large QP problem into a series of smaller QP sub-problems, as long as at least one sample that violates the KT conditions is added to the samples for the previous sub-problem, each step will reduce the overall objective function and maintain a feasible point that obeys all of the constraints.

It should be noted that the procedure begins with a random working set in the original approach [9]. Evidently, fast progress depends heavily on whether the algorithm can start with a good initial working set [12]. To do this, it is desirable to select a set of data, which are likely to be support vectors. From Fig. 1, it is clear that the support vectors are those points laying on the optimal hyperplanes and they are close to each other. Therefore, in [11],

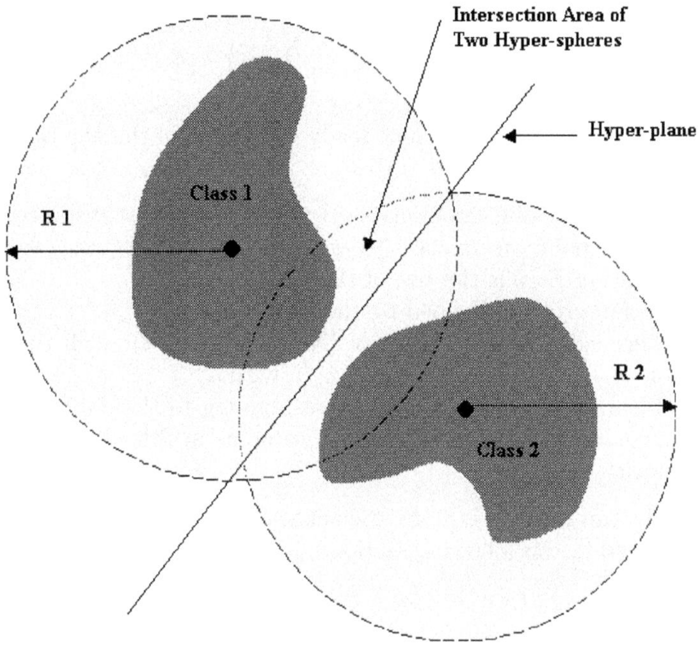

Fig. 1. Illustration of Pre-selection method. The data which are located in the intersection of the two hyper-spheres are to be selected as the working set and these data and most likely to be the support vectors

a pre-selection method is proposed and the distance between each data point and the center of the opposite class is evaluated as follows:

$$\begin{aligned} D_i &= \left\| \Phi(\mathbf{x}_i) - \frac{\Phi(\mathbf{x}_1) + \cdots + \Phi(\mathbf{x}_m)}{n^-} \right\|^2 \\ &= K(\mathbf{x}_i, \mathbf{x}_i) - \frac{2}{n^-} \sum_{j=1}^{n^-} K(\mathbf{x}_i, \mathbf{x}_j) \\ &\quad + \frac{\Phi(\mathbf{x}_1) + \cdots + \Phi(\mathbf{x}_m)}{n^-} \cdot \frac{\Phi(\mathbf{x}_1) + \cdots + \Phi(\mathbf{x}_m)}{n^-} , \end{aligned} \tag{19}$$

where $\mathbf{x}_i$ $i = 1, \ldots, n^+$ is the training data in class +1 and $\mathbf{x}_j$ $j = 1, \ldots, n^-$ is the training data in class −1 as defined in (3). For the training data in class +1, the third term of (19) is a constant and then can be dropped, such that

$$D_i = K(\mathbf{x}_i, \mathbf{x}_i) - \frac{2}{n^-} \sum_{j=1}^{n^-} K(\mathbf{x}_i, \mathbf{x}_j) . \tag{20}$$

The same simplification is added for the training data of class −1 as follows:

$$D_j = K(\mathbf{x}_j, \mathbf{x}_j) - \frac{2}{n^+} \sum_{i=1}^{n^+} K(\mathbf{x}_j, \mathbf{x}_i) \,. \tag{21}$$

Using the results above we are now ready to formulate the algorithm for the training data set

1. Choose $p/2$ data points from class $+1$ with the smallest distance value D_i and $p/2$ data points from class -1 with the smallest distance value D_j as the working set B (p is the size of the working set).
2. Solve the sub-problem defined by the variables in B.
3. While there exists some $\alpha_j, j \in N$ (Non-working set), such that the KT conditions are violated, replace $\alpha_i, i \in B$, with $\alpha_j, j \in N$ and solve the new sub-problem. Note that $\alpha_i, i \in B$ corresponding to the data points of the working set with the biggest distance value as in Step 1 will be replaced with priority.

Evidently, this method is more efficient and easy to be implemented, which is demonstrated in the following section.

5 Experiments

The following experiment evaluates the proposed accelerated decomposition method and the robust SVM. To show effectiveness of the robust SVM with the pre-selection method, we first study the bullet hole image recognition problem with outliers in the specific training data [8]. Secondly, we use public databases, such as the UCI mushroom data set[1] and the larger UCI "adult" data set to show some details of the accelerated training. The mushroom data set includes description of hypothetical samples corresponding 23 species of gilled mushrooms in the Agaricus and Lepiota Family. Each species is identified as definitely edible, definitely poisonous. The experiments are conducted on PIII 700 with 128Mb of RAM. The kernel function used here is RBF kernel $K(\mathbf{x}, y) = e^{-\|x-y\|^2/2\sigma^2}$ for all simulation results.

5.1 Advantages of the Robust SVM and Accelerated Decomposition Method

There are 300 training data and 200 testing data with 20 input features in the bullet hole image recognition system [8, 11]. Table 1 shows an study using different regularization parameters for both the robust SVM and the standard SVM. It shows that the overall testing error is smaller for the robust SVM compared with the standard SVM, while the number of support vectors is also

[1]Mushroom records drawn from The Audubon Society Field Guide to North American Mushrooms (1981). G. H. Lincoff (Press.), New York: Alfred A. Knopf. 27 April 1987.

Table 1. The influence of λ on the testing error and the number of support vectors (SVs) compared to the standard SVM

Algorithms	C	λ	No. of SVs	Test Error	Osuna	Accelerated
Standard SVM	0.1	–	95	2.6%	185.6s	104.5s
Standard SVM	1	–	60	2.4%	169.3s	92.1s
Standard SVM	10	–	35	2.1%	156.6s	87.7s
Standard SVM	20	–	25	2.2%	134.4s	69.8s
Standard SVM	50	–	18	3.2%	123.3s	59.4s
Standard SVM	100	–	18	3.8%	110.3s	54.2s
Standard SVM	500	–	19	5.8%	111.4s	60.3s
Standard SVM	1000	–	19	6.3%	113.4.2s	62.5s
Robust SVM	–	0.1	20	2.5%	124.5s	64s
Robust SVM	–	1	18	2.3%	119.4s	62.3s
Robust SVM	–	4	10	2.2%	103.7s	32.3s
Robust SVM	–	5	7	2.1%	90.9s	24.3s
Robust SVM	–	7	4	1.9%	89.2s	13.5s
Robust SVM	–	10	4	2.0%	91.2s	11.3s
Robust SVM	–	50	4	2.4%	93.3s	12.3s
Robust SVM	–	100	5	4.2%	97.5s	19.3s

reduced greatly to count the effect of outliers as λ increased. It also shows the total training time using Osuna's decomposition algorithm and the accelerated algorithm in the last two columns. The total training time includes training and pre-selection for the later. Since our main purpose here is to illustrate the advantage of the robust SVM in this section, therefore, both algorithms are trained to converge to a minimum point such that they produce almost the same testing error. Limited training and iteration times could produce different testing errors, which will be discussed in the next section.

Here we conclude the influence of λ:

- $\lambda = 0$, no adaption of the margin is performed. Robust support vector machine becomes standard support vector machine.
- $0 < \lambda \leq 1$, the robust SVM algorithm is not sensitive to the outlier located inside the region of separation but on the right side of the decision surface. The influence of the center of class is small, which means the number of support vectors and classification accuracy are almost the same as the standard SVM. Because the number of support vectors are not greatly reduced, we still choose almost the same size of working set.
- $\lambda > 1$, the robust SVM algorithm is not sensitive against the outlier falling on the wrong side of the decision surface. The support vectors will be greatly influenced by the data points that are the nearest points to the center of class. The larger the parameter λ is, the smaller the number of support vectors will be . The algorithm becomes more robust against the outliers and thus results in smoother decision surface. However, the

classification error may be increased correspondingly. Since the number of support vector is greatly reduced, we could accordingly choose smaller size of working set, in turn, the shorter training time.

It should be noted that one interesting point to combine the pre-selection method with the robust SVM is that both algorithms depend on the distance between the training data and class centers in the feature space, though the later treats every "remote" point as potential outlier and the former only concerns the intersection area between two classes. There is a potential risk for the pre-selection method that an outlier (wrong training data) may be easily captured in the early iteration due to its primary location to the intersection area. However, this will be discounted by the robust SVM since it is not sensitive to the outlier. This is why Table 1 shows good result of the method in the presence of outliers, not only the number of support vectors but also the overall testing errors are reduced by the robust SVM with proper regularization parameter.

Correspondingly, if we have some a priori knowledge of which training samples will turn out to be support vectors. We could start the training just on those data and get the accelerated algorithm by using the proposed pre-selection procedure as shown in the following.

5.2 Comparison of Training Time Using UCI Database

It should be pointed out that the robust SVM is useful against outliers, i.e. the wrong training data which is unknown and mixed with the correct training data [4]. Some public database, such as UCI may not contain outliers. Therefore, the robust SVM may produce larger testing error monotonically as the λ increasing in the absence of outliers and it does not make sense to compare different regularization parameters for the standard SVM and robust SVM. We shall rather concentrate on the training time and iterations with specific regularization parameters of the robust SVM or standard SVM without dealing with outliers.

Table 2 shows the simulation results of the Mushroom Predict problem with RBF kernel $\sigma = 2$ and $\lambda = 1.6$ (robust SVM)using the standard QP, general decomposition, and accelerated methods, respectively. We use the standard program in MATLAB toolbox as the QP solver for comparison purpose. Fig. 2 shows the relationship between training times and the number of data points by using the standard QP solver, general decomposition and accelerated method. The total time needed for accelerated method includes pre-selection time and training time. When the sample size becomes large, using of MATLAB's QP solver is not able to meet the computational requirement because of memory thrashing. However, one may notice that the decomposition method can greatly reduce the training time of the robust support vector machine.

With the positive parameter λ we get to know that the number of support vectors is much smaller than the number of training data. As if it is known

Table 2. Experiment result of Mushroom Prediction on different samples size using different numerical algorithms

Algorithms	Sample	Work Set	Training	Pre-selection	No. of SVs	Test Err	Iterations
QP	200	–	36.2s	–	37	9.75%	–
QP	400	–	274.5s	–	44	7.50%	–
QP	600	–	1128.3s	–	51	7.13%	–
QP	800	–	3120.0s	–	63	5.25%	–
QP	1000	–	–	–	–	–	–
Osuna	200	60	19.1s	–	32	9.75%	5
Osuna	400	70	58.2s	–	41	5.25%	6
Osuna	600	100	141.1s	–	48	3.38%	7
Osuna	800	150	325.5s	–	61	2.38%	8
Osuna	1000	200	546.7s	–	60	2.63%	7
Accelerated	200	60	12.9s	2.7s	34	8.88%	3
Accelerated	400	70	50.4s	10.9s	35	4.75%	3
Accelerated	600	100	109.5s	24.3s	50	3.25%	4
Accelerated	800	150	224.5s	43.2s	56	1.63%	4
Accelerated	1000	200	401.7s	67.3s	55	2.13%	4

a priori which of the training data will most likely be the support vectors, it will be sufficient to begin the training just on those samples and still get good result. In this connection the working set chosen by the pre-selection method presented in last section is a good choice. According Table 2 and Fig. 2, we show that the pre-selected working set can work as a better starting point for the general decomposition method rather than using a randomly selected working set. It reduces the number of steps needed to approximate an optimal solution and, in turn, the training time.

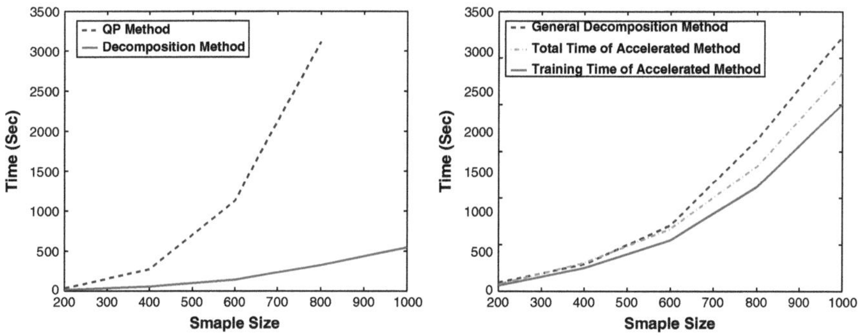

Fig. 2. Training times according Table (2). The *left* one is the comparison between Standard QP Method and General Decomposition Method. The *right* one is the comparison between General Decomposition Method and Accelerated Method

It should be addressed that different number of support vectors and accuracies are obtained in Table 2 since QP is a one-time solution and the decomposition methods depend on the number of iterations. Our observation shows that the QP algorithm with MATLAB toolbox achieves the lowest accuracy (even for a relatively small database) due to it's relatively poor numerical property, while the accelerated algorithm exhibits the best numerical property due to it's superior ability to capture most support vectors with few iterations. Because our target here is to compare training times of different algorithms, particularly, the decomposition algorithms for relatively large database. The iteration times are limited upon to certain points, which implies that further iterations could bring down the testing error gap between two decomposition algorithms.

To further confirm the accelerated algorithm for SVM. Another simulation was carried out by using the UCI "adult" data set [13]. The goal is to predict whether a household has an income greater than $50000. There are 6 continuous attributes and 8 nominal attributes. After discretization of the continuous attributes, there are 123 binary features, of which 14 features are true. The experiment results with different sample sizes by Pre-selection method and SVM light [14] has been shown in Table 3 with $\sigma = 10$ and $C = 100$ (both using the standard SVM to compare the total training time). The number of SVs and test accuracy has two values, the one on the left side is the value from Pre-selection method, and the other is from SVM light method. We also use the SVM light as a tool to solve the subproblem defined by the pre-selection method. We pursue an approximate process: let the size of working set be large enough to capture the SVs as many as possible and discard the rest of training samples. Therefore, it would fail because we can not capture all the SVs and this may result in a bad accuracy as shown in the last section . So we still check the KT condition on the rest samples, get the points which violate the KT condition and solve the subproblem again.

For the purpose to illustrate efficiency of the accelerated algorithm, we only solve the subproblem twice and the result shows that the accelerated algorithm achieves better accuracy and less total training time compared to

Table 3. Experiment results of "adult" data set (accelerated method and the SVM light method)

Size	Pre-selection	Training	Work Set	SVM Light Only	No. of SVs	Accuracy
1605	3.84s	5.03s	1000	7.82s	750/824	78.7/79.3%
2265	7.94s	11.34s	1400	18.41s	1027/1204	78.5/79.6%
3185	16.09s	24.85s	2000	40.59s	1427/1555	79.7/80.1%
4781	38.22s	45.82s	2900	87.66s	1914/2188	78.9/80.6%
6414	70.25s	140.34s	4000	234.36	2622/2893	79.1s/80.4%
11221	214.04s	1152.65s	7000	1152.65	4295/4827	78.3s/80.7%
16101	443.47s	5234.74s	10000	7205.13	6025/6796	78.4s/80.6%

the program using SVM light only (the second to fourth columns applied to the accelerated algorithm).

6 Conclusion

This chapter presents an accelerated decomposition algorithm for the robust support vector machine (SVM). The algorithm is based on a decomposition strategy with the extension of Kuhn-Tucker conditions for the robust SVM. Furthermore, a pre-selection method is developed for the decomposition algorithm which results in an efficient training process. The experiment using UCI mushroom data sets shows that the accelerated decomposition algorithm can make the training process more efficient with relatively bigger λ value and smaller number of support vectors.

References

1. Vapnik, V.N. (1998) Statistical Learning Theory., John Wiley&Sons, New York.
2. Burges, C.J.C. (1998) A Tutorial on Support Vector Machines for Pattern Recognition, Data Mining and Knowledge Discovery, 2(2), 955–974.
3. Cortes, C., Vapnik, V.N. (1995) Support Vector Networks, Machine Learning, 20, 273–297.
4. Chuang, C., Su, S., Jeng, J., Hsiao, C. (2002) Robust Support Vector Regression Networks for Function Approximation with Outliers, IEEE Tran. on Neural Netwroks, 13(6), 1322–1330.
5. Suykens, J.A.K., De Brabanter, J., Lukas, L., Vandewalle, J. (2002) Weighted Least Sqaures Support Vector Machines: Robustness and Sparse approaximation, Neurocomputing, 48, 85–105.
6. Herbrich, R., Weston, J. (1999) Adaptive Margin Support Vector Machines for Classification, The Ninth International Conference on Artificial Neural Network (ICANN 99), 2, 880–885.
7. Boser, B.E., Guyon, I.M., Vapnik, V.N. (1992) A Training Algorithm for Optimal Margin Classifier, In Proc. 5th ACM Workshop on Computational learning Theory, Pittsburgh, PA, 144–152.
8. Song, Q., Hu, W.J., Xie, W.F. (2002), Robust Support Vector Machine With Bulllet Hole Image Classification, IEEE Tran. on Systems, Man, and Cybernetics – Part C: Applications and Review, 32(4), 440–448.
9. Osuna, E., Freund, R., Girosi, F. (1997) An Improved Training Algorithm for Support Vector Machines, Proc. of IEEE NNSP's 97, Amelia Island.
10. Werner, J. (1984) Optimization-Theory and Applications, Vieweg.
11. Hu, W.J., Song, Q. (2001) A Pre-selection Method for Training of Support Vector Machines., Proc. of ANNIE 2001, St. Louis, Missouri, USA.
12. Hsu, C.W., Lin, C.J. (2002) A Simple Decomposition Method for Support Vector Machines, Machine Learning, 46(1–3), 291–314.

13. Platt, J.C. (1998) Sequential Minimum Optimization: A Fast algorithm for Training Support VEctor Machines, Tech Report, Microsoft Research.
14. Joachims, T. (1999) Making Large-scale SVM Learning Practical, In Adavances in Kernal Methods – Support Vector Learning, B. Scholkopf, C.J.C. Burges, and A.J. Smola, Eds., MIT Press, 169–184.

Fuzzy Support Vector Machines with Automatic Membership Setting

C.-fu Lin and S.-de Wang

Department of Electrical Engineering, National Taiwan University, Taipei, Taiwan

Abstract. Support vector machines like other classification approaches aim to learn the decision surface from the input points for classification problems or regression problems. In many applications, each input points may be associated with different weightings to reflect their relative strengths to conform to the decision surface. In our previous research, we applied a fuzzy membership to each input point and reformulate the support vector machines to be fuzzy support vector machines (FSVMs) such that different input points can make different contributions to the learning of the decision surface.

FSVMs provide a method for the classification problem with noises or outliers. However, there is no general rule to determine the membership of each data point. We can manually associate each data point with a fuzzy membership that can reflect their relative degrees as meaningful data. To enable automatic setting of memberships, we introduce two factors in training data points, the confident factor and the trashy factor, and automatically generate fuzzy memberships of training data points from a heuristic strategy by using these two factors and a mapping function. We investigate and compare two strategies in the experiments and the results show that the generalization error of FSVMs are comparable to other methods on benchmark datasets.

Key words: support vector machines, fuzzy membership, fuzzy SVM, noisy data traning

1 Introduction

Support vector machines (SVMs) make use of statistical learning techniques and have drawn much attention on this topic in recent years [1, 2, 3, 4]. This learning theory can be seen as an alternative training technique for polynomial, radial basis function and multi-layer perceptron classifiers [5]. SVMs are based on the idea of *structural risk minimization* (SRM) induction principle [6] that aims at minimizing a bound on the generalization error, rather than minimizing the mean square error. In many applications, SVMs have

C.-fu Lin and S.-de Wang: *Fuzzy Support Vector Machines with Automatic Membership Setting*, StudFuzz **177**, 233–254 (2005)
www.springerlink.com

been shown to provide better performance than traditional learning machines. SVMs can also been used as powerful tools for solving classification [7] and regression problems [8]. For the classification case, SVMs have been used for isolated handwritten digit recognition [1, 9], speaker identification [10, 11], face detection in images [11, 12], knowledge-based classifiers [13], and text categorization [14, 15]. For the regression estimation case, SVMs have been compared on benchmark time series prediction tests [16, 17], financial forecasting [18, 19], and the Boston housing problem [20].

When learning to solve the classification problem, SVMs find a separating hyperplane that maximizes the margin between two classes. Maximizing the margin is a quadratic programming (QP) problem and can be solved from its dual problem by introducing Lagrangian multipliers [21]. In most cases, searching suitable hyperplane in input space is a too restrictive application to be of practical use. The solution to this situation is mapping the input space into a higher dimension feature space and searching the optimal hyperplane in this feature space [22, 23]. Without any knowledge of the mapping, the SVMs find the optimal hyperplane by using the dot product functions in feature space that are called *kernels*. The solution of the optimal hyperplane can be written as a combination of a few input points that are called *support vectors*.

By introducing the ϵ-insensitive loss function and doing some small modifications in the formations of equations, the theory of SVMs can be easily applied in regression problems. The formulated equations of regression problems in SVMs are similar with the those of classification problems except the target variables.

Solving the quadratic programming problem with a dense, structured, positive semidefinite matrix is expensive in traditional quadratic programming algorithms [24]. Platt's Sequential Minimal Optimization (SMO) [25, 26] is a simple algorithm that quickly solves the SVMs QP problem without any extra matrix storage. LIBSVM [27], which is a simplification of both SMO and SVM^{light} [28], is provided as an integrated software for the implementation of support vector machines. These researches make the use of SVMs simple and easy.

More and more applications can be solved by using the SVMs techniques. However, in many applications, input points may not be appropriately assigned with the same importance in the training process. For the classification problem, some data points deserve to be treated more importantly so that SVMs can separate these points more correctly. For the regression problem, some data points corrupted by noises are less meaningful and the machine should better to discard them. Original SVMs do not consider the effect of noises or outliers and thus cannot treat differently to data points.

In our previous research, we applied a fuzzy membership to each input point of SVMs and reformulate SVMs into FSVMs such that different input points can make different contributions to the learning of decision surface. The proposed method enhances the SVMs in reducing the effect of outliers and noises in data points. FSVMs are suitable for applications in which data points have unmodeled characteristics.

For the classification problem, since the optimal hyperplane obtained by the SVM depends on only a small part of the data points, it may become sensitive to noises or outliers in the training set [29, 30]. FSVMs solve this kind of problems by introducing the fuzzy memberships of data points. We can treat the noises or outliers as less important and let these points have lower fuzzy membership. It is also based on the maximization of the margin like the classical SVMs, but uses fuzzy memberships to prevent noisy data points from making narrower margin. This equips FSVMs with the ability to train data with noises or outliers by setting lower fuzzy memberships to the data points that are considered as noises or outliers with higher probability.

We design a noise model that introduces two factors in training data points, the confident factor and the trashy factor, and automatically generates fuzzy memberships of training data points from a heuristic strategy by using these two factors and a mapping function. This model is used to estimate the probability that the data point is considered as noisy information and use this probability to tune the fuzzy membership in FSVMs. This simplifies the use of FSVMs in the training of data points with noises or outliers. The experiments show that the generalization error of FSVMs are comparable to other methods on benchmark datasets.

The rest of this chapter is organized as follows. A brief review of the theory of FSVMs will be given in Sect. 2. The training algorithm which reduces effects of noises or outliers in classification problems is illustrated in Sect. 3. Some concluding remarks are given in Sect. 4.

2 Fuzzy Support Vector Machines

In this section, we make a detail description about the idea and formulations of fuzzy support vector machines [31].

2.1 Fuzzy Property of Data Points

The theory of SVMs is a powerful tool for solving classification problems [7], but there are still some limitations of this theory. From the training set and formulations, each training point belongs to either one class or the other. For each class, we can easily check that all training points of this class are treated uniformly in the theory of SVMs.

In many real world applications, the effects of the training points are different. It is often that some training points are more important than others in the classification problem. We would require that the meaningful training points must be classified correctly and would not care about some training points like noises whether or not they are misclassified.

That is, each training point no more exactly belongs to one of the two classes. It may 90 percent belong to one class and 10 percent be meaningless, and it may 20 percent belong to one class and 80 percent be meaningless. In other words, there is a fuzzy membership $0 < s_i \leq 1$ associated with each training point $\mathbf{x}_i$. This fuzzy membership s_i can be regarded as the attitude of the corresponding training point toward one class in the classification problem and the value $(1 - s_i)$ can be regarded as the attitude of meaningless.

We found that this situation also occurred in the regression problems. The effects of the training points are the same in the standard regression algorithm of SVMs. The fuzzy membership s_i can be regarded as the importance of the corresponding training point in the regression problem. For example, in the time series prediction problem, we can associate the older training points with lower fuzzy memberships such that we can reduce the effect of the older training points in the optimization of regression function.

We extend the concept of SVMs with fuzzy membership and make it as fuzzy SVMs or FSVMs.

2.2 Reformulated SVMs for Classification Problems

Suppose we are given a set S of labeled training points with associated fuzzy membership

$$(y_1, \mathbf{x}_1, s_1), \ldots, (y_l, \mathbf{x}_l, s_l) \, . \tag{1}$$

Each training point $\mathbf{x}_i \in \mathcal{R}^N$ is given a label $y_i \in \{-1, 1\}$ and a fuzzy membership $\sigma \leq s_i \leq 1$ with $i = 1, \ldots, l$, and sufficient small $\sigma > 0$. Let $\mathbf{z} = \varphi(\mathbf{x})$ denote the corresponding feature space vector with a mapping φ from $\mathcal{R}^N$ to a feature space $\mathcal{Z}$.

Since the fuzzy membership s_i is the attitude of the corresponding point $\mathbf{x}_i$ toward one class and the parameter ξ_i is a measure of error in the SVMs, the term $s_i\xi_i$ is a measure of error with different weighting. The setting of fuzzy membership s_i is critical to the application of FSVMs. Although in the formulation of the problem we assume the fuzzy membership is given in advance, it is beneficial to have the parameters of membership being automatically setting up in the course of training. To this end, we design a noise model that introduces two factors in training data points, the confident factor and the trashy factor, and automatically generates fuzzy memberships of training data points from a heuristic strategy by using these two factors and a mapping function. This model is used to estimate the probability that the data point is considered as noisy data and can serve as an aide to tune the fuzzy membership in FSVMs. This simplifies the application of FSVMs in the training of noisy data points or data points polluted with outliers. The optimal hyperplane problem is regarded as the solution to

$$\text{minimize } \frac{1}{2}\mathbf{w}\cdot\mathbf{w} + C\sum_{i=1}^{l} s_i\xi_i \,, \tag{2}$$

$$\text{subject to } \begin{cases} y_i(\mathbf{w}\cdot\mathbf{z}_i + b) \geq 1 - \xi_i, & i = 1, \ldots, l \,, \\ \xi_i \geq 0, \quad i = 1, \ldots, l \,, \end{cases}$$

where C is a constant. It is noted that a smaller s_i reduces the effect of the parameter ξ_i in problem (2) such that the corresponding point $\mathbf{z}_i = \varphi(\mathbf{x}_i)$ is treated as less important.

To solve this optimization problem we construct the Lagrangian

$$\begin{aligned} L = \frac{1}{2}\mathbf{w}\cdot\mathbf{w} + C\sum_{i=1}^{l} s_i\xi_i - \sum_{i=1}^{l}\beta_i\xi_i \\ - \sum_{i=1}^{l}\alpha_i(y_i(\mathbf{w}\cdot\mathbf{z}_i + b) - 1 + \xi_i) \end{aligned} \tag{3}$$

and find the saddle point of L, where $\alpha_i \geq 0$ and $\beta_i \geq 0$. The parameters must satisfy the following conditions

$$\frac{\partial L}{\partial \mathbf{w}} = \mathbf{w} - \sum_{i=1}^{l}\alpha_i y_i \mathbf{z}_i = 0 \,, \tag{4}$$

$$\frac{\partial L}{\partial b} = -\sum_{i=1}^{l}\alpha_i y_i = 0 \,, \tag{5}$$

$$\frac{\partial L}{\partial \xi_i} = s_i C - \alpha_i - \beta_i = 0 \,. \tag{6}$$

Apply these conditions into the Lagrangian (3), the problem (2) can be transformed into

$$\text{maximize } \sum_{i=1}^{l}\alpha_i - \frac{1}{2}\sum_{i=1}^{l}\sum_{j=1}^{l}\alpha_i\alpha_j y_i y_j K(\mathbf{x}_i, \mathbf{x}_j) \,, \tag{7}$$

$$\text{subject to } \begin{cases} \sum_{i=1}^{l} \quad y_i\alpha_i = 0 \\ 0 \leq \alpha_i \leq s_i C, \quad i = 1, \ldots, l \,. \end{cases}$$

and the Karush-Kuhn-Tucker conditions are defined as

$$\bar{\alpha_i}(y_i(\bar{\mathbf{w}}\cdot\mathbf{z}_i + \bar{b}) - 1 + \bar{\xi_i}) = 0 \,, \quad i = 1, \ldots, l \,, \tag{8}$$

$$(s_i C - \bar{\alpha_i})\bar{\xi_i} = 0 \,, \quad i = 1, \ldots, l \,. \tag{9}$$

where $\bar{\alpha_i}$, $\bar{\mathbf{w}}$, and $\bar{b}$ denote a solution to the optimization problem (7).

The point $\mathbf{x}_i$ with the corresponding $\bar{\alpha_i} > 0$ is called a support vector. There are also two types of support vectors. The one with corresponding $0 < \bar{\alpha_i} < s_i C$ lies on the margin of the hyperplane. The one with corresponding $\bar{\alpha_i} = s_i C$ is misclassified. An important difference between SVMs and FSVMs is that the points with the same value of $\bar{\alpha_i}$ may indicate a different type of support vectors in FSVMs due to the factor s_i.

2.3 Reformulated SVMs for Regression Problems

Suppose we are given a set S of labeled training points with associated fuzzy membership

$$(y_1, \mathbf{x}_1, s_1), \ldots, (y_l, \mathbf{x}_l, s_l) . \tag{10}$$

Each training point $\mathbf{x}_i \in \mathcal{R}^N$ is given a label $y_i \in \mathcal{R}$ and a fuzzy membership $\sigma \le s_i \le 1$ with $i = 1, \ldots, l$, and sufficient small $\sigma > 0$. Let $\mathbf{z} = \varphi(\mathbf{x})$ denote the corresponding feature space vector with a mapping φ from $\mathcal{R}^N$ to a feature space $\mathcal{Z}$.

Since the fuzzy membership s_i is the importance of the corresponding point $\mathbf{x}_i$ and the parameter $\xi_i^{(*)}$ is a measure of error in the SVMs, the term $s_i\xi_i^{(*)}$ is a measure of error with different weighting. The regression problem is then regarded as the solution to

$$\text{minimize } \frac{1}{2}\mathbf{w}\cdot\mathbf{w} + C\sum_{i=1}^{l} s_i(\xi_i + \xi_i^*) \tag{11}$$

$$\text{subject to } \begin{cases} y_i - (\mathbf{w}\cdot\mathbf{z}_i + b) \le \epsilon + \xi_i , \\ (\mathbf{w}\cdot\mathbf{z}_i + b) - y_i \le \epsilon + \xi_i^* , \\ \xi_i, \xi_i^* \ge 0 . \end{cases}$$

where C is a constant. It is noted that a smaller s_i reduces the effect of the parameter ξ_i in problem (11) such that the corresponding point $\mathbf{z}_i = \varphi(\mathbf{x}_i)$ is treated as less important.

To solve this optimization problem we construct the Lagrangian

$$\begin{aligned} L = {} & \frac{1}{2}\mathbf{w}\cdot\mathbf{w} + C\sum_{i=1}^{l} s_i(\xi_i + \xi_i^*) \\ & - \sum_{i=1}^{l}(\eta_i\xi_i + \eta_i^*\xi_I^*) \\ & - \sum_{i=1}^{l}\alpha_i(\epsilon + \xi_i - y_i + \mathbf{w}\cdot\mathbf{z}_i + b) \\ & - \sum_{i=1}^{l}\alpha_i^*(\epsilon + \xi_i^* + y_i - \mathbf{w}\cdot\mathbf{z}_i - b) \end{aligned} \tag{12}$$

and find the saddle point of L, where $\alpha_i, \alpha_i^*, \eta_i, \eta_i^* \ge 0$. The parameters must satisfy the following conditions

$$\frac{\partial L}{\partial b} = \sum_{i=1}^{l}(\alpha_i^* - \alpha_i) = 0 \,, \tag{13}$$

$$\frac{\partial L}{\partial \mathbf{w}} = \mathbf{w} - \sum_{i=1}^{l}(\alpha_i - \alpha_i^*)\mathbf{x}_i = 0 \,. \tag{14}$$

$$\frac{\partial L}{\partial \xi_i^{(*)}} = s_i C - \alpha_i^{(*)} - \eta_i^{(*)} = 0 \tag{15}$$

Apply these conditions into the Lagrangian (12), the problem (11) can be transformed into

$$\text{maximize } -\frac{1}{2}\sum_{i,j=1}^{l}(\alpha_i - \alpha_i^*)(\alpha_j - \alpha_j^*)K(\mathbf{x}_i, \mathbf{x}_j) \tag{16}$$

$$-\epsilon\sum_{i=1}^{l}(\alpha_i + \alpha_i^*) + \sum_{i=1}^{l} y_i(\alpha_i - \alpha_i^*)$$

$$\text{subject to } \begin{cases} \sum_{i=1}^{l}(\alpha_i - \alpha_i^*) = 0 \,, \\ 0 \le \alpha_i^{(*)} \le s_i C, \quad i = 1, \ldots, l \,. \end{cases}$$

and the Karush-Kuhn-Tucker conditions are defined as

$$\bar{\alpha_i}(\epsilon + \bar{\xi_i} - y_i + \bar{\mathbf{w}} \cdot \mathbf{x}_i + \bar{b}) = 0 \,, \quad i = 1, \ldots, l \,, \tag{17}$$

$$\bar{\alpha_i^*}(\epsilon + \bar{\xi_i^*} + y_i - \bar{\mathbf{w}} \cdot \mathbf{x}_i - \bar{b}) = 0 \,, \quad i = 1, \ldots, l \,, \tag{18}$$

$$(s_i C - \bar{\alpha_i})\bar{\xi_i} = 0 \,, \quad i = 1, \ldots, l \,, \tag{19}$$

$$(s_i C - \bar{\alpha_i^*})\bar{\xi_i^*} = 0 \,, \quad i = 1, \ldots, l \,. \tag{20}$$

The point $\mathbf{x}_i$ with the corresponding $\bar{\alpha_i}^{(*)} > 0$ is called a support vector. There are also two types of support vectors. The one with corresponding $0 < \bar{\alpha_i}^{(*)} < s_i C$ lies on the ϵ-insensitive tube around the function f_R. The one with corresponding $\bar{\alpha_i}^{(*)} = s_i C$ is outside the tube. An important difference between SVMs and FSVMs is that the points with the same value of $\bar{\alpha_i}^{(*)}$ may indicate a different type of support vectors in FSVMs due to the factor s_i.

2.4 Dependence on the Fuzzy Membership

The only free parameter C in SVMs controls the trade-off between the maximization of margin and the amount of errors. In classification problems, a larger C makes the training of SVMs less misclassifications and narrower margin. The decrease of C makes SVMs ignore more training points and get wider margin. In regression problems, a larger C makes less amount of error in regression function and the decrease of C makes the regression flatter.

In FSVMs, we can set C to be a sufficient large value. It is the same as standard SVMs if we set all $s_i = 1$. With different value of s_i, we can control the trade-off of the respective training point $\mathbf{x}_i$ in the system. A smaller value of s_i makes the corresponding point $\mathbf{x}_i$ less important in the training.

There is only one free parameter in SVMs while the number of free parameters in FSVMs is equivalent to the number of training points.

2.5 Generating the Fuzzy Memberships

To choose the appropriate fuzzy memberships in a given problem is easy. First, the lower bound of fuzzy memberships must be defined, and second, we need to select the main property of data set and make connection between this property and fuzzy memberships.

Consider that we want to conduct the sequential learning problem. First, we choose $\sigma > 0$ as the lower bound of fuzzy memberships. Second, we identify that the time is the main property of this kind of problem and make fuzzy membership s_i be a function of time t_i

$$s_i = f(t_i) , \tag{21}$$

where $t_1 \leq \ldots \leq t_l$ is the time the point arrived in the system. We make the last point $\mathbf{x}_l$ be the most important and choose $s_l = f(t_l) = 1$, and make the first point $\mathbf{x}_1$ be the least important and choose $s_1 = f(t_1) = \sigma$. If we want to make fuzzy membership be a linear function of the time, we can select

$$s_i = f(t_i) = at_i + b . \tag{22}$$

By applying the boundary conditions, we can get

$$s_i = f(t_i) = \frac{1-\sigma}{t_l - t_1} t_i + \frac{t_l \sigma - t_1}{t_l - t_1} . \tag{23}$$

If we want to make fuzzy membership be a quadric function of the time, we can select

$$s_i = f(t_i) = a(t_i - b)^2 + c . \tag{24}$$

By applying the boundary conditions, we can get

$$s_i = f(t_i) = (1-\sigma)\left(\frac{t_i - t_1}{t_l - t_1}\right)^2 + \sigma . \tag{25}$$

2.6 Data with Time Property

Sequential learning and inference methods are important in many applications involving real-time signal processing [32]. For example, we would like to have a learning machine such that the points from recent past is given more weighting than the points far back in the past. For this purpose, we can select the fuzzy membership as a function of the time that the point generated and this kind of problem can be easily implemented by FSVMs.

2.6.1 A Simple Example

We consider a simple classification problem for example. Suppose we are given a sequence of training points

$$(y_1, \mathbf{x}_1, s_1, t_1), \ldots, (y_l, \mathbf{x}_l, s_l, t_l) , \tag{26}$$

where $t_1 \leq \ldots \leq t_l$ is the time the point arrived in the system. Let fuzzy membership s_i be a function of time t_i

$$s_i = f(t_i) \tag{27}$$

such that $s_1 = \sigma \leq \ldots \leq s_l = 1$.

The left part of Fig. 1 shows the result of SVMs and the right part of Fig. 1 shows the result of FSVMs by setting

$$s_i = f(t_i) = (1 - \sigma) \left(\frac{t_i - t_1}{t_l - t_1} \right)^2 + \sigma . \tag{28}$$

The underlined numbers are grouped as one class and the non-underlined numbers are grouped as the other class. The value of the number indicates the arrival sequence in the same interval. The smaller numbered data is the older one. We can easily check that the FSVMs classify the last ten points with high accuracy while the SVMs does not.

2.6.2 Financial Time Series Forecasting

The distribution of financial time series is changing over the time [33]. For this property, solving this kind of problem by FSVMs would be more feasible than by SVMs. Cao et al. [19] proposed an exponential function

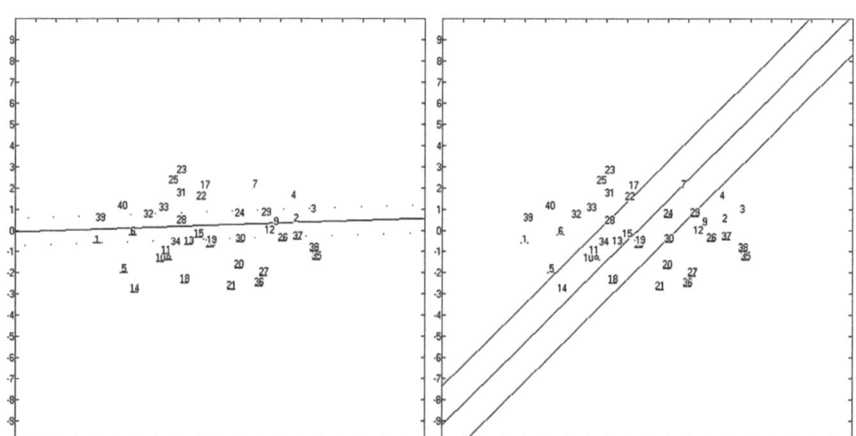

Fig. 1. The *left* part is the result of SVMs learning for data with time property and the *right* part is result of FSVMs learning for data with time property

$$s_i = \frac{1}{1 + \exp\left(a - 2a\left(\frac{t_i - t_1}{t_l - t_1}\right)\right)} , \tag{29}$$

which can be summarized as follows:

- When $a \to 0$, then $\lim_{a\to 0} s_i = 1/2$, the fuzzy membership s_i approaches 1/2. In this case, the same fuzzy memberships apply to all the training data points.
- When $a \to \infty$, then

$$\lim_{a\to\infty} s_i = \begin{cases} 0 \; t_i < (t_i + t_1)/2 , \\ 1 \; t_i > (t_i + t_1)/2 . \end{cases}$$

  In this case, the fuzzy memberships for the data points arrived in first half are reduced to zero, and the fuzzy memberships for the data points arrived in second half are equal to 1.
- When $a \in (0, \infty)$ and increases, the fuzzy memberships for the data points arrived in first half will become smaller, while the fuzzy memberships for the data points arrived in second half will become larger.

The simulation results in [19] demonstrated that FSVMs are effective in dealing with the structural change of financial time series.

2.7 Two Classes with Different Weighting

In some problems, we are more concerned about the one situation than the other. For example, in medical diagnosis problem we are more concerned about the accuracy of classifying a disease than that of no disease. The fault detection problem in materials also has such characteristic[34]. For example, given a point, if the machine says 1, it means that the point belongs to this class with very high accuracy, but if the machine says −1, it may belong to this class with lower accuracy or really belongs to another class. For this purpose, we can select the fuzzy membership as a function of respective class [5, 35].

Suppose we are given a sequence of training points

$$(y_1, \mathbf{x}_1, s_1), \ldots, (y_l, \mathbf{x}_l, s_l) . \tag{30}$$

Let fuzzy membership s_i be a function of class y_i

$$s_i = \begin{cases} s_+, & \text{if } y_i = 1 , \\ s_-, & \text{if } y_i = -1 . \end{cases} \tag{31}$$

The left part of Fig. 2 shows the result of SVMs and the right part of Fig. 2 shows the result of FSVMs by setting

$$s_i = \begin{cases} 1, & \text{if } y_i = 1 , \\ 0.1, & \text{if } y_i = -1 . \end{cases} \tag{32}$$

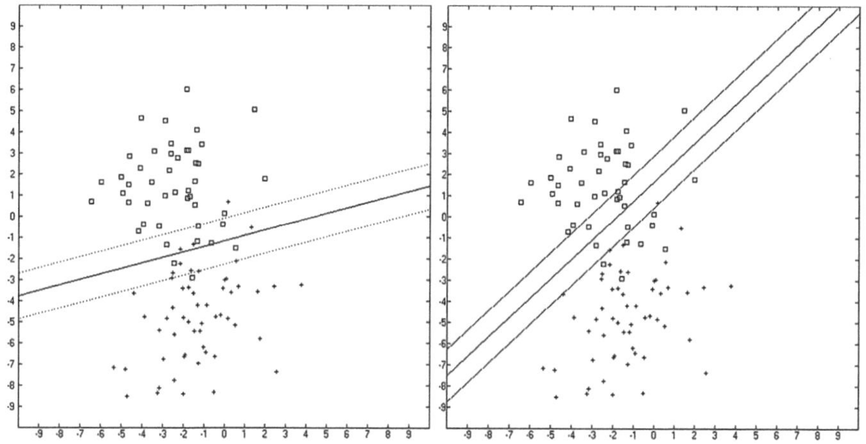

Fig. 2. The *left* part is the result of SVMs learning for data sets and the *right* part is the result of FSVMs learning for data sets with different weighting

The point $\mathbf{x}_i$ with $y_i = 1$ is indicated as cross, and the point $\mathbf{x}_i$ with $y_i = -1$ is indicated as square. In left part of Fig. 2, SVMs find the optimal hyperplane with errors appearing in each class. In right part of Fig. 2, we apply different fuzzy memberships to different classes, the FSVMs find the optimal hyperplane with errors appearing only in one class. We can easily check that FSVMs classify the class of cross with high accuracy and the class of square with low accuracy, while SVMs does not.

3 Reducing Effects of Noises in Classification Problems

Since the optimal hyperplane obtained by the SVM depends on only a small part of the data points, it may become sensitive to noises or outliers in the training set [29, 30]. To solve this problem, one approach is to do some preprocessing on training data to remove noises or outliers, and then use the remaining set to learn the decision function [36]. This method is hard to implement if we do not have enough knowledge about noises or outliers. In many real world applications, we are given a set of training data without knowledge about noises or outliers. There are some risks to remove the meaningful data points as noises or outliers.

There are many discussions in this topic and some of them show good performance. The theory of Leave-One-Out SVMs [37] (LOO-SVMs) is a modified version of SVMs. This approach differs from classical SVMs in that it is based on the maximization of the margin, but minimizes the expression given by the bound in an attempt to minimize the leave-one-out error. No free parameter makes this algorithm easy to use, but it lacks the flexibility of tuning the relative degree of outliers as meaningful data points. Its generalization, the theory of Adaptive Margin SVMs (AM-SVMs) [38], uses a parameter λ to adjust the margin for a given learning problem. It improves the flexibility of LOO-SVMs and shows better performance. The experiments in both of them show the robustness against outliers.

FSVMs solve this kind of problems by introducing the fuzzy memberships of data points. We can treat the noises or outliers as less importance and let these points have lower fuzzy membership. It is also based on the maximization of the margin like the classical SVMs, but uses fuzzy memberships to prevent noisy data points from making narrower margin. This equips FSVMs with the ability to train data spoiled with noises or outliers by setting lower fuzzy memberships to the data points that are considered as noises or outliers with higher probability.

We need to assume a noise model in the training data points, and then try and tune the fuzzy membership of each data point in the training. Without any knowledge of the distribution of data points, it is hard to associate the fuzzy membership to the data point.

In this section, we design a noise model that introduces two factors in training data points, the confident factor and the trashy factor, and automatically generates fuzzy memberships of training data points from a heuristic strategy by using these two factors and a mapping function [39]. This model is used to estimate the probability that the data point is considered as noisy information and use this probability to tune the fuzzy membership in FSVMs. This simplifies the use of FSVMs in the training of data points polluted with noises or outliers. The experiments show that the generalization error of FSVMs is comparable to other methods on benchmark datasets.

3.1 The Error Function

For efficient computation, the SVMs select the least absolute value to estimate the error function, that is $\sum_{i=1}^{l} \xi_i$, and use a regularization parameter C to balance between the minimization of the error function and the maximization of the margin of the optimal hyperplane. There are still some methods to estimate this error function. The LS-SVMs [40, 41] select the least square value and show the differences in the constraints and optimization processes. In the situations that the underlying error probability distribution can be estimated, we can use the maximum likelihood method to estimate the error function. Let ξ_i be i.i.d. with probability density function $p_e(\xi)$, $p_e(\xi) = 0$ if $\xi < 0$. The optimal hyperplane problem is then modified as the solution to the problem

$$\text{minimize} \quad \frac{1}{2}\mathbf{w} \cdot \mathbf{w} + C \sum_{i=1}^{l} \phi(\xi_i) \tag{33}$$

$$\text{subject to} \quad \begin{cases} y_i(\mathbf{w} \cdot \mathbf{z}_i + b) \geq 1 - \xi_i \,, & i = 1, \ldots, l \,, \\ \xi_i \geq 0 \,, \quad i = 1, \ldots, l \,, \end{cases}$$

where $\phi(\xi) = -\ln p_e(\xi)$. Clearly, $\phi(\xi) \propto \xi$ when $p_e(\xi) \propto e^{-\xi}$, that reduces the problem (33) to the original optimal hyperplane problem. Thus, with the different knowledge of the error probability model, a variety of error functions can be chosen and different optimal problems can be generated.

However, there are some critical issues in this kind of application. First, it is hard to implement the training program since solving the optimal hyperplane problem (33) is in general NP-complete [1]. Second, the error estimator ξ_i is related to the optimal hyperplane by

$$\xi_i = \begin{cases} 0 & \text{if } f_H(\mathbf{x}_i) \geq 1 \\ 1 - f_H(\mathbf{x}_i) & \text{otherwise} . \end{cases} \tag{34}$$

Therefore, one needs to use the correct error model in the optimization process, but one needs to know the underlying function to estimate the error model. In practice it is impossible to estimate the error distribution reliably without a good estimation of the underlying function. This is so called a "catch 22" situation [42]. Even though the probability density function $p_e(\xi)$ is unknown for almost all applications.

In contrast, in the cases where the noise distribution model of the data set is known, we assume $p_x(\mathbf{x})$ be the probability density function of the data point $\mathbf{x}$ that is not a noise. For the data point $\mathbf{x}_i$ with higher value of $p_x(\mathbf{x}_i)$, which means that this data point has higher probability to be a real data, it is expected that the data point would get a lower value of ξ_i in the training process since ξ_i stands for a penalty weight in error function. To achieve this purpose, we can modify the error function as

$$\sum_{i=1}^{l} p_x(\mathbf{x}_i)\xi_i \, . \tag{35}$$

Hence, the optimal hyperplane problem is then modified as the solution to the problem

$$\begin{aligned} &\text{minimize} \quad \frac{1}{2}\mathbf{w} \cdot \mathbf{w} + C \sum_{i=1}^{l} p_x(\mathbf{x}_i)\xi_i \\ &\text{subject to} \quad \begin{cases} y_i(\mathbf{w} \cdot \mathbf{z}_i + b) \geq 1 - \xi_i \, , & i = 1, \ldots, l \, , \\ \xi_i \geq 0 \, , \quad i = 1, \ldots, l \, . \end{cases} \end{aligned} \tag{36}$$

When the probability density function $p_x(\mathbf{x})$ in problem (36) can be viewed as some kind of fuzzy membership, we can simply replace $s_i = p_x(\mathbf{x}_i)$ in problem (2), such that we can solve problem (36) by using the algorithm of FSVMs.

3.2 The Noise Distribution Model

There exist many alternatives to setting up the fuzzy memberships in training data using FSVMs, depending on how much information contains in the data set. If the data points are already associated with the fuzzy memberships, we can just use this information in training FSVMs. If it is given a noise distribution model of the data set, we can set the fuzzy membership as the probability of the data point that is not spoiled by a noise, or as a function of it. In other words, let p_i be the probability of the data point $\mathbf{x}_i$ that is not spoiled by a noise. If there exists this kind of information in the training data, we can just assign the value $s_i = p_i$ or $s_i = f_p(p_i)$ as the fuzzy membership of each data point, and use these information to get the optimal hyperplane in the training of FSVMs. Since almost all applications lack this information, we need some other methods to predict this probability.

Suppose we are given a heuristic function $h(\mathbf{x})$ that is highly relevant to the probability density function $p_x(\mathbf{x})$. For this assumption, we can build a relationship between the probability density function $p_x(\mathbf{x})$ and the heuristic function $h(\mathbf{x})$, that is defined as

$$p_x(\mathbf{x}) = \begin{cases} 1 & \text{if } h(\mathbf{x}_i) > h_C \\ \sigma & \text{if } h(\mathbf{x}_i) < h_T \\ \sigma + (1-\sigma)\left(\frac{h(\mathbf{x})-h_T}{h_C-h_T}\right)^d & \text{otherwise} \end{cases} \tag{37}$$

where h_C is the confident factor and h_T is the trashy factor. These two factors control the mapping region between $p_x(\mathbf{x})$ and $h(\mathbf{x})$, and d is the parameter that controls the degree of mapping function as shown in Fig. 3.

The training points are divided into three regions by the confident factor h_C and trashy factor h_T. If the data point, whose heuristic value $h(\mathbf{x})$ is bigger than the confident factor h_C, lies in the region of $h(\mathbf{x}) > h_C$, it can be viewed as valid examples with high confidence and the fuzzy membership is equal

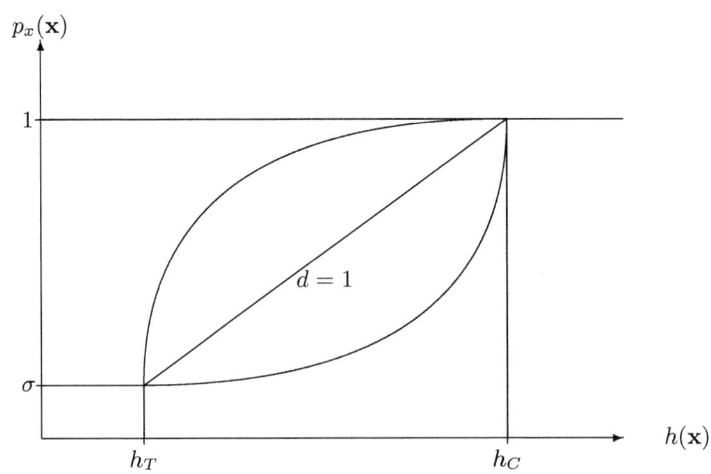

Fig. 3. The mapping between the probability density function $p_x(\mathbf{x})$ and the heuristic function $h(\mathbf{x})$

to 1. In contract, if the data point, whose heuristic value $h(\mathbf{x})$ is lower than the trashy factor h_T, lies in the region of $h(\mathbf{x}) < h_T$, it can be highly thought as noisy data and the fuzzy membership is assigned to lowest value σ. The data points in the rest region are considered as noisy ones with different probabilities and can make different distributions in the training process. There is no enough knowledge to choose proper function of this mapping. For simplicity, the polynomial function is selected in this mapping and the parameter d is used to control the degree of mapping.

3.3 The Heuristic Function

As for steps, discriminating between noisy data and noiseless ones, we propose two strategies: one is based on kernel-target alignment and the other is using k-NN.

3.3.1 Strategy of Using Kernel-Target Alignment

The idea of kernel-target alignment is introduced in [43]. Let $f_K(\mathbf{x}_i, y_i) = \sum_{j=1}^{l} y_i y_j K(\mathbf{x}_i, \mathbf{x}_j)$. The kernel-target alignment is defined as

$$A_{KT} = \frac{\sum_{i=1}^{l} f_K(\mathbf{x}_i, y_i)}{l\sqrt{\sum_{i,j=1}^{l} K^2(\mathbf{x}_i, \mathbf{x}_j)}} \tag{38}$$

This definition provides a method for selecting kernel parameters and the experimental results show that adapting the kernel to improve alignment on the training data enhances the alignment on the test data, thus improved classification accuracy.

In order to discover some relation between the noise distribution and the data point, we simply focus on the value $f_K(\mathbf{x}_i, y_i)$. Suppose $K(\mathbf{x}_i, \mathbf{x}_j)$ is a kind of distance measure between data points $\mathbf{x}_i$ and $\mathbf{x}_j$ in feature space $\mathcal{F}$. For example, by using the RBF kernel $K(\mathbf{x}_i, \mathbf{x}_j) = e^{-\gamma\|\mathbf{x}_i - \mathbf{x}_j\|^2}$, the data points live on the surface of a hypersphere in feature space $\mathcal{F}$ as shown in Fig. 4.Then $K(\mathbf{x}_i, \mathbf{x}_j) = \varphi(\mathbf{x}_i) \cdot \varphi(\mathbf{x}_j)$ is the cosine of the angle between $\varphi(\mathbf{x}_i)$ and $\varphi(\mathbf{x}_j)$. For the outlier $\varphi(\mathbf{x}_1)$ and the representative $\varphi(\mathbf{x}_2)$, we have

$$\begin{aligned} f_K(\mathbf{x}_1, y_1) &= \sum_{y_i = y_1} K(\mathbf{x}_1, \mathbf{x}_i) - \sum_{y_i \neq y_1} K(\mathbf{x}_1, \mathbf{x}_i) \\ f_K(\mathbf{x}_2, y_2) &= \sum_{y_i = y_2} K(\mathbf{x}_2, \mathbf{x}_i) - \sum_{y_i \neq y_2} K(\mathbf{x}_2, \mathbf{x}_i) \,. \end{aligned} \tag{39}$$

We can easily check the following

$$\begin{aligned} \sum_{y_i = y_1} K(\mathbf{x}_1, \mathbf{x}_i) &< \sum_{y_i = y_2} K(\mathbf{x}_2, \mathbf{x}_i) \\ \sum_{y_i \neq y_1} K(\mathbf{x}_1, \mathbf{x}_i) &> \sum_{y_i \neq y_2} K(\mathbf{x}_2, \mathbf{x}_i) \,, \end{aligned} \tag{40}$$

such that the value $f_K(\mathbf{x}_1, y_1)$ is lower than $f_K(\mathbf{x}_2, y_2)$.

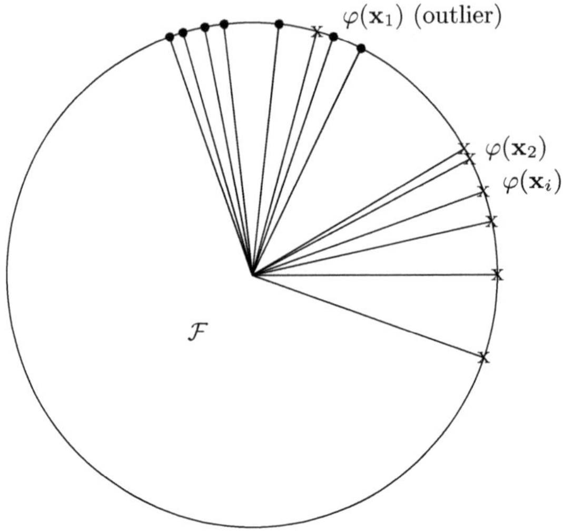

Fig. 4. The value $f_K(\mathbf{x}_1, y_1)$ is lower than $f_K(\mathbf{x}_2, y_2)$ in the RBF kernel

We observe this situation and assume that the data point $\mathbf{x}_i$ with lower value of $f_K(\mathbf{x}_i, y_i)$ can be considered as outlier and should make less contribution of the classification accuracy. Hence, we can use the function $f_K(\mathbf{x}, y)$ as a heuristic function $h(\mathbf{x})$.

This heuristic function assumes that a data point will be considered as a noisy one with high probability if this data point is more closer to the other class than its class. For a more theoretic discussion, let $D_\pm(\mathbf{x})$ be the mean distance between the data point $\mathbf{x}$ and data points $\mathbf{x}_i$ with $y_i = \pm 1$, which is defined as

$$D_\pm(\mathbf{x}) = \frac{1}{l_\pm} \sum_{y_i=\pm 1} \|\mathbf{x} - \mathbf{x}_i\|^2 , \tag{41}$$

where $l_\pm$ is the number of data points with $y_i = \pm 1$, respectively. Then the value $y_k(D_+(\mathbf{x}_k) - D_-(\mathbf{x}_k))$ can be considered as an indication of a noise. For the same case in feature space, $D_\pm(\mathbf{x})$ is reformulated as

$$\begin{aligned} D_\pm(\mathbf{x}) &= \frac{1}{l_\pm} \sum_{y_i=\pm 1} \|\varphi(\mathbf{x}) - \varphi(\mathbf{x}_i)\|^2 \\ &= \frac{1}{l_\pm} \sum_{y_i=\pm 1} (K(\mathbf{x}, \mathbf{x}) - 2K(\mathbf{x}, \mathbf{x}_i) + K(\mathbf{x}_i, \mathbf{x}_i)) \\ &= K(\mathbf{x}, \mathbf{x}) + \frac{1}{l_\pm} \sum_{y_i=\pm 1} (K(\mathbf{x}_i, \mathbf{x}_i) - 2K(\mathbf{x}, \mathbf{x}_i)) . \end{aligned} \tag{42}$$

Assume that $l_+ \simeq l_-$, we can replace $l_\pm$ by $l/2$, and the value of $K(\mathbf{x}, \mathbf{x})$ is 1 for the RBF kernel. Then

$$
\begin{aligned}
y_k(D_+(\mathbf{x}_k) - D_-(\mathbf{x}_k)) &= y_k\left(\frac{2}{l}\sum_{y_i=1}(1-2K(\mathbf{x}_k,\mathbf{x}_i)) \right. \\
&\qquad \left. -\frac{2}{l}\sum_{y_i=-1}(1-2K(\mathbf{x}_k,\mathbf{x}_i))\right) \\
&= \frac{4y_k}{l}\sum_i -y_iK(\mathbf{x}_k,\mathbf{x}_i) \\
&= -\frac{4}{l}f_K(\mathbf{x}_k,y_k)\ ,
\end{aligned} \tag{43}
$$

which is reduced to the heuristic function $f_K(\mathbf{x}_k, y_k)$.

3.3.2 Strategy of Using k-NN

For each data point $\mathbf{x}_i$, we can find a set S_i^k that consists of k nearest neighbors of $\mathbf{x}_i$. Let n_i be the number of data points in the set S_i^k that the class label is the same as the class label of data point $\mathbf{x}_i$. It is reasonable to assume that the data point with lower value of n_i is more probable as noisy data. It is trivial to select the heuristic function $h(\mathbf{x}_i) = n_i$. But for the data points that are near the margin of two classes, the value n_i of these points may be lower. It will get poor performance if we set these data points with lower fuzzy memberships. In order to avoid this situation, the confident factor h_C, which controls the threshold of which data point needs to reduce its fuzzy membership, will be carefully chosen.

3.4 The Overall Procedure

There is no explicit way to solve the problem of choosing parameters for SVMs. The use of a gradient descent algorithm over the set of parameters by minimizing some estimates of the generalization error of SVMs is discussed in [44]. On the other hand, the exhaustive search or the grid search is the popular method to choose the parameters, but it becomes intractable in this application as number of parameters is growing.

In order to select parameters in this kind of problem, we divide the training procedure into two main parts and propose the following procedures.

1. Use the original algorithm of SVMs to get the optimal kernel parameters and the regularization parameter C.
2. Fix the kernel parameters and the regularization parameter C, that are obtained in the previous procedure, and find the other parameters in FSVMs.
   (a) Define the heuristic function $h(\mathbf{x})$
   (b) Use the exhaustive search or the grid search to choose the confident factor h_C, the trashy factor h_T, the mapping degree d, and the fuzzy membership lower bound σ.

3.5 Experiments

In these simulations, we use the RBF kernel as

$$K(\mathbf{x}_i, \mathbf{x}_j) = e^{-\gamma\|\mathbf{x}_i - \mathbf{x}_j\|^2} . \tag{44}$$

We conducted computer simulations of SVMs and FSVMs using the same data sets as in [45]. Each data set is split into 100 sample sets of training and test sets. For each sample set, the test set is independent of training set. For each data set, we train and test the first 5 sample sets iteratively to find the parameters of the best average test error. Then we use these parameters to train and test the whole sample sets iteratively and get the average test error. Since there are more parameters than the original algorithm of SVMs, we use two procedures to find the parameters as described in the previous section. In the first procedure, we search the kernel parameters and C using the original algorithm of SVMs. In the second procedure, we fix the kernel parameters and C that are found in the first stage, and search the parameters of the fuzzy membership mapping function.

To find the parameters of strategy using kernel-target alignment, we first fix $h_C = \max_i f_K(\mathbf{x}_i, y_i)$ and $h_T = \min_i f_K(\mathbf{x}_i, y_i)$, and perform a two-dimensional search of parameters σ and d. The value of σ is chosen from 0.1 to 0.9 step by 0.1. For some case, we also compare the result of $\sigma = 0.01$. The value of d is chosen from 2^{-8} to 2^8 multiply by 2. Then, we fix σ and d, and perform a two-dimensional search of parameters h_C and h_T. The value of h_C is chosen such that 0%, 10%, 20%, 30%, 40%, and 50% of data points have the value of fuzzy membership as 1. The value of h_T is chosen such that 0%, 10%, 20%, 30%, 40%, and 50% of data points have the value of fuzzy membership as σ.

To find the parameters of strategy using k-NN, we just perform a two-dimensional search of parameters σ and k. We fix the value $h_C = k/2$, $h_T = 0$, and $d = 1$, since we don't find much gain or loss when we choose other values of these two parameters such that we skip searching for saving time. The value of σ is chosen from 0.1 to 0.9 stepped by 0.1. For some case, we also compare the result of $\sigma = 0.01$. The value of k is chosen from 2^1 to 2^8 multiplied by 2.

Table 1 shows the results of our simulations. For comparison with SVMs, FSVMs with kernel-target alignment perform better in 9 data sets, and FSVMs with k-NN perform better in 5 data sets. By checking the average training error of SVMs in each data set, we find that FSVMs perform well in the data set when the average training error is high. These results show that our algorithm can improve the performance of SVMs when the data set contains noisy data.

Table 1. The test error of SVMs, FSVMs using strategy of kernel-target alignment (KT), and FSVMs using strategy of k-NN (k-NN), and the average training error of SVMs (TR) on 13 datasets

	SVMs	KT	k-NN	TR
Banana	11.5 ± 0.7	**10.4 ± 0.5**	11.4 ± 0.6	6.7
B. Cancer	26.0 ± 4.7	25.3 ± 4.4	**25.2 ± 4.1**	18.3
Diabetes	23.5 ± 1.7	**23.3 ± 1.7**	23.5 ± 1.7	19.4
F. Solar	**32.4 ± 1.8**	32.4 ± 1.8	32.4 ± 1.8	32.6
German	23.6 ± 2.1	**23.3 ± 2.3**	23.6 ± 2.1	16.2
Heart	16.0 ± 3.3	**15.2 ± 3.1**	15.5 ± 3.4	12.8
Image	3.0 ± 0.6	**2.9 ± 0.7**	–	1.3
Ringnorm	1.7 ± 0.1	–	–	0.0
Splice	**10.9 ± 0.7**	–	–	0.0
Thyroid	4.8 ± 2.2	**4.7 ± 2.3**	–	0.4
Titanic	22.4 ± 1.0	**22.3 ± 0.9**	**22.3 ± 1.1**	19.6
Twonorm	3.0 ± 0.2	**2.4 ± 0.1**	2.9 ± 0.2	0.4
Waveform	**9.9 ± 0.4**	9.9 ± 0.4	–	3.5

4 Conclusions

In this chapter, we reviewed the concept of fuzzy support vector machines and proposed training procedures for FSVMs. By associating the data points with fuzzy memberships, FSVMs train data points with different memberships in learning the decision function. However, the extra freedom in selecting the membership poses an issue to learning. Thus, systematic methods are required for the applicability of the FSVMs. The proposed training procedures along with two strategies for setting fuzzy membership can effectively solve the membership selection problem. This makes FSVMs more feasible in the application of reducing the effects of noises or outliers. The experiments show that the performance is better in the applications with the noisy data.

It is still an issue that FSVMs should select a proper fuzzy model for a given specific problem. Some problems may involve different domains that are outside the discipline of learning techniques. For example, the problem of economical trend prediction may work better with both the domain knowledge of economics and the learning technique of computer scientists. The illustrated examples in this chapter show only the basic application of FSVMs. More versatile applications are expected in the near future.

References

1. Cortes, C. and Vapnik, V. (1995) Support vector networks. Machine Learning, **20**, 273–297

2. Vapnik, V. (1995) The Nature of Statistical Learning Theory. New York: Springer
3. Vapnik, V. (1998) Statistical Learning Theory. New York: Wiley
4. Scholkopf, B., Mika, S., Burges, C., Knirsch, P., Muller, K.-R., Ratsch, G. and A. Smola, (1999) Input space vs. feature space in kernel-based methods. IEEE Transactions on Neural Networks, **10**, 5, 1000–1017
5. Osuna, E., Freund, R., and Girosi, F. (1996) Support vector machines: Training and applications. Tech. Rep. AIM-1602, MIT A.I. Lab.
6. Vapnik, V. (1982) Estimation of Dependences Based on Empirical Data. Springer-Verlag
7. Burges, C. J. C. (1998) A tutorial on support vector machines for pattern recognition. Data Mining and Knowledge Discovery, **2**, 2, 121–167
8. Smola, A. and B. Scholkopf (1998) A tutorial on support vector regression. Tech. Rep. NC2-TR-1998-030, Neural and Computational Learning II
9. Burges, C. J. C., Scholkopf, B. (1997) Improving the accuracy and speed of support vector learning machines. in Advances in Neural Information Processing Systems 9 (M. Mozer, M. Jordan, and T. Petsche, eds.), 375–381, Cambridge, MA: MIT Press
10. Schmidt, M. (1996) Identifying speaker with support vector networks. in Interface '96 Proceedings, (Sydney)
11. Ben-Yacoub, S., Abdeljaoued, Y., Mayoraz E. (1999) Fusion of face and speech data for person identity verification. IEEE Transactions on Neural Networks, **10**, 5, 1065–1074
12. Osuna, E., Freund, R., Girosi, F. (1997) An improved training algorithm for support vector machines. in 1997 IEEE Workshop on Neural Networks for Signal Processing, 276–285
13. Fung, G., Mangasarian, O. L., Shavlik, J. (2002) Knowledge-based support vector machine classifiers. in Advances in Neural Information Processing
14. Joachims, T. (1998) Text categorization with support vector machines: learning with many relevant features. in Proceedings of ECML-98, 10th European Conference on Machine Learning (C. Nedellec and C. Rouveirol, eds.), (Chemnitz, DE), 137–142, Springer Verlag, Heidelberg, DE
15. Crammer, K., Singer, Y. (2000) On the learnability and design of output codes for multiclass problems. in Computational Learning Theory, 35–46
16. Muller, K.-R., Smola, A., Ratsch, G., Scholkopf, B., Kohlmorgen, J., Vapnik, V. (1997) Predicting time series with support vector machines. in Articial Neural Networks – ICANN97 (W. Gerstner, A. Germond, M. Hasler, and J.-D. Nicoud, eds.), 999–1004
17. Mukherjee, S., Osuna, E., Girosi, F. (1997) Nonlinear prediction of chaotic time series using support vector machines. in 1997 IEEE Workshop on Neural Networks for Signal Processing, 511–519
18. Tay, F. E. H., Cao, L. (2001) Application of support vector machines in financial time series forecasting. Omega, **29**, 309–317
19. Cao, L., J. Chua, K. S., Guan, L. K. (2003) c-ascending support vector machines for financial time series forecasting. in 2003 International Conference on Computational Intelligence for Financial Engineering (CIFEr2003), (Hong Kong), 317–323
20. Drucker, H., Burges, C. J. C., Kaufman, L., Smola, A., Vapnik, V. (1997) Support vector regression machines. in Advances in Neural Information Processing Systems, **9**, p. 155, The MIT Press

21. Fletcher, R. (1987) Practical methods of optimization. Chichester and New York: John Wiley and Sons
22. Aizerman, M., Braverman, E., Rozonoer, L. (1964) Theoretical foundations of the potential function method in pattern recognition learning. Automations and Remote Control, **25** 821–837
23. Nilsson, N. J. (1965) Learning machines: Foundations of trainable pattern classifying systems. McGraw-Hill
24. Kaufman, L. (1998) Solving the quadratic programming problem arising in support vector classification. in Advances in Kernel Methods: Support Vector Learning (B. SchLolkopf, C. Burges, and A. Smola, eds.), 147–168, Cambridge, MA: MIT Press
25. Platt, J. (1998) Sequential minimal optimization: A fast algorithm for training support vector machines. Tech. Rep. 98-14, Microsoft Research, Washington
26. Platt, J. (1998) Fast training of support vector machines using sequential minimal optimization. in Advances in Kernel Methods: Support Vector Learning (B. SchLolkopf, C. Burges, and A. Smola, eds.), 185–208, Cambridge, MA: MIT Press
27. Chang, C.-C., Lin, C.-J. (2001) Libsvm: a library for support vector machines. 2001. Software avaiable at http://www.csie.ntu.edu.tw/ cjlin/libsvm/
28. Platt, J. (1998) Making large-scale svm learning practical. in Advances in Kernel Methods: Support Vector Learning (B. SchLolkopf, C. Burges, and A. Smola, eds.), 169–184, Cambridge, MA: MIT Press
29. Boser, B. E., Guyon, I., Vapnik, V. (1992) A training algorithm for optimal margin classifiers. in Computational Learing Theory, 144–152
30. Zhang, X. (1999) Using class-center vectors to build support vector machines. In 1999 IEEE Workshop on Neural Networks for Signal Processing, 3–11
31. Lin, C.-F., Wang, S.-D. (2002) Fuzzy support vector machines. IEEE Transactions on Neural Networks, **13**, 464–471
32. Freitas, N. D., Milo, M., Clarkson, P., Niranjan, M., Gee, A. (1999) Sequential support vector machines. In 1999 IEEE Workshop on Neural Networks for Signal Processing, 31–40
33. Yaser, S. A., Atiya, A. F. (1996) Introduction to financial forecasting. Applied Intelligence, **6**, 205–213
34. Lee, K. K., Gunn, S. R., Harris, C. J., Reed, P. A. S. (2001) Classification of unbalanced data with transparent kernels. in International Joint Conference on Neural Networks (IJCNN 01), **4**, 2445–2450
35. Quang, A. T., Zhang, Q.-L., Li, X. (2002) Evolving support vector machine parameters. in 2002 International Conference on Machine Learning and Cybernetics, **1**, 548–551
36. Cao, L. J., Lee, H. P., Chong, W. K. (2003) Modified support vector novelty detector using training data with outliers. Pattern Recognition Letters, **24**, 2479–2487
37. Weston, J. (1999) Leave-one-out support vector machines. in Proceedings of the Sixteenth International Joint Conference on Artificial Intelligence, IJCAI 99 (T. Dean, ed.), 727–733, Morgan Kaufmann
38. Weston, J., Herbrich, R. (2000) Adaptive margin support vector machines. in Advances in Large Margin Classifiers, 281–295, Cambridge, MA: MIT Press
39. Lin, C.-F., Wang, S.-D. (2003) Training algorithms for fuzzy support vector machines with noisy data. in 2003 IEEE Workshop on Neural Networks for Signal Processing

40. Suykens, J. A. K., Vandewalle, J. (1999) Least squares support vector machine classifiers. Neural Processing Letters, **9**, 293–300
41. Chua, K. S. (2003) Efficient computations for large least square support vector machine classifiers. Pattern Recognition Letters, **24**, 75–80
42. Chen, D. S., Jain, R. C. (1994) A robust back propagation learning algorithm for function approximation. IEEE Transactions on Neural Networks, **5**, 467–479
43. Cristianini, N., Shawe-Taylor, Elisseeff, J., A., Kandola, J. (2002) On kernel-target alignment. in Advances in Neural Information Processing Systems 14, 367–373, MIT Press
44. Chapelle, O., Vapnik, V., Bousquet, O., Mukherjee, S. (2002) Choosing multiple parameters for support vector machines. Machine Learning, **46**, no. 1–3, 131–159
45. Ratsch, G., Onoda, T., Muller, K.-R. (2001) Soft margins for AdaBoost. Machine Learning, **42**, 287–320

Iterative Single Data Algorithm for Training Kernel Machines from Huge Data Sets: Theory and Performance

V. Kecman[1], T.-M. Huang[1], and M. Vogt[2]

[1] School of Engineering, The University of Auckland, Auckland, New Zealand
v.kecman@auckland.ac.nz,
[2] Institute of Automatic Control, TU Darmstadt, Darmstadt, Germany
mvogt@iat.tu-darmstadt.de

Abstract. The chapter introduces the latest developments and results of Iterative Single Data Algorithm (ISDA) for solving large-scale support vector machines (SVMs) problems. First, the equality of a Kernel AdaTron (KA) method (originating from a gradient ascent learning approach) and the Sequential Minimal Optimization (SMO) learning algorithm (based on an analytic quadratic programming step for a model without bias term b) in designing SVMs with positive definite kernels is shown for both the nonlinear classification and the nonlinear regression tasks. The chapter also introduces the classic Gauss-Seidel procedure and its derivative known as the successive over-relaxation algorithm as viable (and usually faster) training algorithms. The convergence theorem for these related iterative algorithms is proven. The second part of the chapter presents the effects and the methods of incorporating explicit bias term b into the ISDA. The algorithms shown here implement the single training data based iteration routine (a.k.a. per-pattern learning). This makes the proposed ISDAs remarkably quick. The final solution in a dual domain is not an approximate one, but it is the optimal set of dual variables which would have been obtained by using any of existing and proven QP problem solvers if they only could deal with huge data sets.

Key words: machine learning, huge data set, support vector machines, kernel machines, iterative single data algorithm

1 Introduction

One of the mainstream research fields in learning from empirical data by support vector machines (SVMs), and solving both the classification and the regression problems, is an implementation of the incremental learning schemes when the training data set is huge. The challenge of applying SVMs on huge data sets comes from the fact that the amount of computer memory required

V. Kecman, T.-M. Huang, and M. Vogt: *Iterative Single Data Algorithm for Training Kernel Machines from Huge Data Sets: Theory and Performance*, StudFuzz **177**, 255–274 (2005)
www.springerlink.com

for a standard quadratic programming (QP) solver grows exponentially as the size of the problem increased. Among several candidates that avoid the use of standard QP solvers, the two learning approaches which recently have drawn the attention are the Iterative Single Data Algorithms (ISDAs), and the sequential minimal optimization (SMO) [9, 12, 17, 23].

The ISDAs work on one data point at a time (per-pattern based learning) towards the optimal solution. The Kernel AdaTron (KA) is the earliest ISDA for SVMs, which uses kernel functions to map data into SVMs' high dimensional feature space [7] and performs AdaTron learning [1] in the feature space. The Platt's SMO algorithm is an extreme case of the decomposition methods developed in [10, 15], which works on a working set of two data points at a time. Because of the fact that the solution for working set of two can be found analytically, SMO algorithm does not invoke standard QP solvers. Due to its analytical foundation the SMO approach is particularly popular and at the moment the widest used, analyzed and still heavily developing algorithm. At the same time, the KA although providing similar results in solving classification problems (in terms of both the accuracy and the training computation time required) did not attract that many devotees. There are two basic reasons for that. First, until recently [22], the KA seemed to be restricted to the classification problems only and second, it "lacked" the fleur of the strong theory (despite its beautiful "simplicity" and strong convergence proofs). The KA is based on a gradient ascent technique and this fact might have also distracted some researchers being aware of problems with gradient ascent approaches faced with possibly ill-conditioned kernel matrix. In the next section, for a missing bias term b, we derive and show the equality of two seemingly different ISDAs, which are a KA method and a without-bias version of SMO learning algorithm [23] in designing the SVMs having positive definite kernels. The equality is valid for both the nonlinear classification and the nonlinear regression tasks, and it sheds a new light on these seemingly different learning approaches. We also introduce other learning techniques related to the two mentioned approaches, such as the classic Gauss-Seidel coordinate ascent procedure and its derivative known as the successive over-relaxation algorithm as a viable and usually faster training algorithms for performing nonlinear classification and regression tasks. In the third section, we derive and show how explicit bias term b can be incorporated into the ISDAs derived in the second section of this chapter. Finally, the comparison in performance between different ISDAs derived in this chapter and the popular SVM software LIBSVM [2] is presented. The goal of this chapter is to show how the latest developments in ISDA can lead to the remarkable tool for solving large-scale SVMs as well as to present the effect of an explicit bias term b within the ISDA.

In order to have a good understanding on these algorithms, it is necessary to review the optimization problem induced from SVMs. The problem to solve in SVMs classification is [3, 4, 20, 21]

$$\min \frac{1}{2}\mathbf{w}^T\mathbf{w}, \quad i = 1, \ldots, l\,, \tag{1}$$

$$\text{s.t.}\; y_i\left[\mathbf{w}^T\Phi(\mathbf{x}_i) + b\right] \geq 1 \quad i = 1, \ldots, l\,, \tag{2}$$

which can be transformed into its dual form by minimizing the primal Lagrangian

$$L_p(\mathbf{w}, b, \alpha) = \frac{1}{2}\mathbf{w}^T\mathbf{w} - \sum_{i=1}^{l} \alpha_i \left\{y_i\left[\mathbf{w}^T\Phi(\mathbf{x}_i) + b\right] - 1\right\}\,, \tag{3}$$

in respect to $\mathbf{w}$ and b by using $\partial L_p/\partial\mathbf{w} = 0$ and $\partial L_p/\partial b = 0$, i.e., by exploiting

$$\frac{\partial L_p}{\partial \mathbf{w}} = 0 \quad \mathbf{w} = \sum_{i=1}^{l} \alpha_i y_i \Phi(\mathbf{x}_i)\,, \tag{4}$$

$$\frac{\partial L_p}{\partial b} = 0 \quad \sum_{i=1}^{l} \alpha_i y_i = 0\,. \tag{5}$$

The standard change to a dual problem is to substitute $\mathbf{w}$ from (4) into the primal Lagrangian (3) and this leads to a dual Lagrangian problem below,

$$L_d(\alpha) = \sum_{i=1}^{l} \alpha_i - \frac{1}{2}\sum_{i,j=1}^{l} y_i y_j \alpha_i \alpha_j K(\mathbf{x}_i, \mathbf{x}_j) - \sum_{i=1}^{l} \alpha_i y_i b\,, \tag{6}$$

subject to the box constraints (7) where the scalar $K(\mathbf{x}_i, \mathbf{x}_j) = \Phi(\mathbf{x}_i)^T\Phi(\mathbf{x}_j)$. In the standard SVMs formulation, (5) is used to eliminate the last term of (6) that should be solved subject to the following constraints

$$\alpha_i \geq 0, \quad i = 1, \ldots, l \quad \text{and} \tag{7}$$

$$\sum_{i=1}^{l} \alpha_i y_i = 0\,. \tag{8}$$

As a result the dual function to be maximized is (9) with box constraints (7) and equality constraint (8).

$$L_d(\alpha) = \sum_{i=1}^{l} \alpha_i - \frac{1}{2}\sum_{i,j=1}^{l} y_i y_j \alpha_i \alpha_j K(\mathbf{x}_i, \mathbf{x}_j)\,. \tag{9}$$

An important point to remember is that without the bias term b in the SVMs model, the equality constraint (8) does not exist. This association between bias b and (8) is explored extensively to develop ISDA schemes in the rest of the chapter. Because of the noise, or due to the generic class' features, there will be an overlapping of training data points. Nothing, but constraints, in solving (9) changes and, for the overlapping classes, they are

$$C \geq \alpha_i \geq 0, \quad i = 1, \ldots, l \quad \text{and} \tag{10}$$

$$\sum_{i=1}^{l} \alpha_i y_i = 0 \,, \tag{11}$$

where $0 < C < \infty$, is a penalty parameter trading off the size of a margin with a number of misclassifications. This formulation is often referred to as the soft margin classifier.

In the case of the *nonlinear regression* the learning problem is the maximization of a dual Lagrangian below

$$\begin{aligned} L_d(\alpha,\, \alpha^*) = & -\varepsilon \sum_{i=1}^{l} (\alpha_i^* + \alpha_i) + \sum_{i=1}^{l} (\alpha_i - \alpha_i^*) y_i \\ & - \frac{1}{2} \sum_{i,j=1}^{l} (\alpha_i - \alpha_i^*)(\alpha_j - \alpha_j^*) K(\mathbf{x}_i, \mathbf{x}_j) \end{aligned} \tag{12}$$

$$\text{s.t.} \sum_{i=1}^{l} \alpha_i = \sum_{i=1}^{l} \alpha_i^*, \tag{13}$$

$$0 \leq \alpha_i^* \leq C, \; 0 \leq \alpha_i \leq C, \quad i = 1, \ldots, l \,. \tag{14}$$

Again, the equality constraint (13) is the result of including bias term in the SVMs model.

2 Iterative Single Data Algorithm for Positive Definite Kernels without Bias Term b

In terms of representational capability, when applying Gaussian kernels, SVMs are similar to radial basis function networks. At the end of the learning, they produce a decision function of the following form

$$f(\mathbf{x}) = \sum_{j=1}^{l} v_j K(\mathbf{x}, \mathbf{x}_j) + b \,. \tag{15}$$

However, it is well known that *positive definite kernels* (such as the most popular and the most widely used RBF Gaussian kernels as well as the complete polynomial ones) do not require bias term b [6, 12]. This means that the SVM learning problems should maximize (9) with box constraints (10) in classification and maximize (12) with box constraints (14) in regression. In this section, the KA and the SMO algorithms will be presented for such a fixed (i.e., no-) bias design problem and compared for the classification and regression cases. The equality of two learning schemes and resulting models will be established. Originally, in [18], the SMO *classification* algorithm was developed for solving (9) including the equality constraint (8) related to the bias b. In these early publications (on the classification tasks only) the case when bias b is fixed variable was also mentioned but the detailed analysis of a fixed bias update was not accomplished. The algorithms here extend and develop a new method to regression problems too.

2.1 Iterative Single Data Algorithm without Bias Term b in Classification

2.1.1 Kernel AdaTron in Classification

The classic AdaTron algorithm as given in [1] is developed for a linear classifier. As mentioned previously, the KA is a variant of the classic AdaTron algorithm in the feature space of SVMs. The KA algorithm solves the maximization of the dual Lagrangian (9) by implementing the gradient ascent algorithm. The update $\Delta\alpha_i$ of the dual variables α_i is given as

$$\Delta\alpha_i = \eta\frac{\partial L_d}{\partial\alpha_i} = \eta\left(1 - y_i\sum_{j=1}^{l}\alpha_j y_j K(\mathbf{x}_i, \mathbf{x}_j)\right) = \eta\left(1 - y_i f_i\right) , \tag{16a}$$

where f_i is the value of the decision function f at the point $\mathbf{x}_i$, i.e., $f_i = \sum_{j=1}^{l}\alpha_j y_j K(\mathbf{x}_i, \mathbf{x}_j)$, and y_i denotes the value of the desired target (or the class' label) which is either $+1$ or -1. The update of the dual variables α_i is given as

$$\alpha_i \leftarrow \min(\max(0, \alpha_i + \Delta\alpha_i), C) \quad (i = 1, \ldots, l) . \tag{16b}$$

In other words, the dual variables α_i are clipped to zero if $(\alpha_i + \Delta\alpha_i) < 0$. In the case of the soft nonlinear classifier $(C < \infty)\alpha_i$ are clipped between zero and C, $(0 \leq \alpha_i \leq C)$. The algorithm converges from any initial setting for the Lagrange multipliers α_i.

2.1.2 SMO without Bias Term in Classification

Recently [23] derived the update rule for multipliers α_i that includes a detailed analysis of the Karush-Kuhn-Tucker (KKT) conditions for checking the optimality of the solution. (As referred above, a fixed bias update was mentioned only in Platt's papers). The following update rule for α_i for a no-bias SMO algorithm was proposed

$$\Delta\alpha_i = -\frac{y_i E_i}{K(\mathbf{x}_i, \mathbf{x}_i)} = -\frac{y_i f_i - 1}{K(\mathbf{x}_i, \mathbf{x}_i)} = \frac{1 - y_i f_i}{K(\mathbf{x}_i, \mathbf{x}_i)} , \tag{17}$$

where $E_i = f_i - y_i$ denotes the difference between the value of the decision function f at the point $\mathbf{x}_i$ and the desired target (label) y_i. Note the equality of (16a) and (17) when the learning rate in (16a) is chosen to be $\eta_i = 1/K(\mathbf{x}_i, \mathbf{x}_i)$. The important part of the SMO algorithm is to check the KKT conditions with precision τ (e.g., $\tau = 10^{-3}$) in each step. An update is performed only if

$$\begin{aligned} &\alpha_i < C \ \wedge \ y_i E_i < -\tau \ , \quad \text{or} \\ &\alpha_i > 0 \ \wedge \ y_i E_i > \tau \ . \end{aligned} \tag{17a}$$

After an update, the same clipping operation as in (16b) is performed

$$\alpha_i \leftarrow \min(\max(0, \alpha_i + \Delta\alpha_i), C) \quad (i = 1, \ldots, l) \ . \tag{17b}$$

It is the nonlinear clipping operation in (16b) and in (17b) that strictly equals the KA and the SMO without-bias-term algorithm in solving nonlinear classification problems. This fact sheds new light on both algorithms. This equality is not that obvious in the case of a "classic" SMO algorithm with bias term due to the heuristics involved in the selection of active points which should ensure the largest increase of the dual Lagrangian L_d during the iterative optimization steps.

2.2 Iterative Single Data Algorithm without Bias Term b in Regression

Similarly to the case of classification, for the models without bias term b, there is a strict equality between the KA and the SMO algorithm when positive definite kernels are used for nonlinear regression.

2.2.1 Kernel AdaTron in Regression

The first extension of the Kernel AdaTron algorithm for regression is presented in [22] as the following gradient ascent update rules for α_i and α_i^*

$$\begin{aligned} \Delta\alpha_i &= \eta_i \frac{\partial L_d}{\partial \alpha_i} = \eta_i \left(y_i - \varepsilon - \sum_{j=1}^{l} (\alpha_j - \alpha_j^*) K(\mathbf{x}_j, \mathbf{x}_i) \right) = \eta_i \left(y_i - \varepsilon - f_i \right) \\ &= -\eta_i (E_i + \varepsilon), \end{aligned} \tag{18a}$$

$$\begin{aligned} \Delta\alpha_i^* &= \eta_i \frac{\partial L_d}{\partial \alpha_i^*} = \eta_i \left(-y_i - \varepsilon + \sum_{j=1}^{l} (\alpha_j - \alpha_j^*) K(\mathbf{x}_j, \mathbf{x}_i) \right) = \eta_i \left(-y_i - \varepsilon + f_i \right) \\ &= \eta_i \left(E_i - \varepsilon \right) \ , \end{aligned} \tag{18b}$$

where y_i is the measured value for the input $\mathbf{x}_i$, ε is the prescribed insensitivity zone, and $E_i = f_i - y_i$ stands for the difference between the regression function f at the point $\mathbf{x}_i$ and the desired target value y_i at this point. The calculation of the gradient above does not take into account the geometric reality that no training data can be on both sides of the tube. In other words, it does not use the fact that either α_i or α_i^* or both will be nonzero. i.e., that $\alpha_i \alpha_i^* = 0$ must be fulfilled in each iteration step. Below we derive the gradients of the dual Lagrangian L_d accounting for geometry. This new formulation of the KA algorithm strictly equals the SMO method and it is given as

$$
\begin{aligned}
\frac{\partial L_d}{\partial \alpha_i} &= -K(\mathbf{x}_i, \mathbf{x}_i)\alpha_i - \sum_{j=1, j\neq i}^{l} (\alpha_j - \alpha_j^*)K(\mathbf{x}_j, \mathbf{x}_i) + y_i \\
&\quad - \varepsilon + K(\mathbf{x}_i, \mathbf{x}_i)\alpha_i^* - K(\mathbf{x}_i, \mathbf{x}_i)\alpha_i^* \\
&= -K(\mathbf{x}_i, \mathbf{x}_i)\alpha_i^* - (\alpha_i - \alpha_i^*)K(\mathbf{x}_i, \mathbf{x}_i) \\
&\quad - \sum_{j=1, j\neq i}^{l} (\alpha_j - \alpha_j^*)K(\mathbf{x}_j, \mathbf{x}_i) + y_i - \varepsilon \qquad \text{(19a)} \\
&= -K(\mathbf{x}_i, \mathbf{x}_i)\alpha_i^* + y_i - \varepsilon - f_i \\
&= -\left(K(\mathbf{x}_i, \mathbf{x}_i)\alpha_i^* + E_i + \varepsilon\right) .
\end{aligned}
$$

For the α_i^* multipliers, the value of the gradient is

$$
\frac{\partial L_d}{\partial \alpha_i} = -K(\mathbf{x}_i, \mathbf{x}_i)\alpha_i + E_i - \varepsilon . \qquad \text{(19b)}
$$

The update value for α_i is now

$$
\Delta\alpha_i = \eta_i \frac{\partial L_d}{\partial \alpha_i} = -\eta_i \left(K(x_i, x_i)\alpha_i^* + E_i + \varepsilon\right), \qquad \text{(20a)}
$$

$$
\alpha_i \leftarrow \alpha_i + \Delta\alpha_i = \alpha_i + \eta_i \frac{\partial L_d}{\partial \alpha_i} = \alpha_i - \eta_i \left(K(\mathbf{x}_i, \mathbf{x}_i)\alpha_i^* + E_i + \varepsilon\right) \qquad \text{(20b)}
$$

For the learning rate $\eta_i = 1/K(\mathbf{x}_i, \mathbf{x}_i)$ the gradient ascent learning KA is defined as,

$$
\alpha_i \leftarrow \alpha_i - \alpha_i^* - \frac{E_i + \varepsilon}{K(\mathbf{x}_i, \mathbf{x}_i)} . \qquad \text{(21a)}
$$

Similarly, the update rule for α_i^* is

$$
\alpha_i^* \leftarrow \alpha_i^* - \alpha_i + \frac{E_i - \varepsilon}{K(\mathbf{x}_i, \mathbf{x}_i)} . \qquad \text{(21b)}
$$

Same as in the classification, α_i and α_i^* are clipped between zero and C,

$$
\alpha_i \leftarrow \min(\max(0, \alpha_i + \Delta\alpha_i), C), \quad (i = 1, \ldots, l), \qquad \text{(22a)}
$$

$$
\alpha_i^* \leftarrow \min(\max(0, \alpha_i^* + \Delta\alpha_i^*), C), \quad (i = 1, \ldots, l) . \qquad \text{(22b)}
$$

2.2.2 SMO without Bias Term b in Regression

The first algorithm for the SMO without-bias-term in regression (together with a detailed analysis of the KKT conditions for checking the optimality of the solution) is derived in [23]. The following learning rules for the Lagrange multipliers α_i and α_i^* updates were proposed

$$
\alpha_i \leftarrow \alpha_i - \alpha_i^* - \frac{E_i + \varepsilon}{K(\mathbf{x}_i, \mathbf{x}_i)} , \qquad \text{(23a)}
$$

$$
\alpha_i^* \leftarrow \alpha_i^* - \alpha_i + \frac{E_i - \varepsilon}{K(\mathbf{x}_i, \mathbf{x}_i)} . \qquad \text{(23b)}
$$

The equality of (21a, b) and (23a, b) is obvious when the learning rate, as presented above in (21a, b), is chosen to be $\eta_i = 1/K(\mathbf{x}_i, \mathbf{x}_i)$. Note that in both the classification and the regression, the optimal learning rate is not necessarily equal for all training data pairs. For a Gaussian kernel, $\eta = 1$ is same for all data points, and for a complete nth order polynomial each data point has different learning rate $\eta_i = 1/K(\mathbf{x}_i, \mathbf{x}_i)$. Similar to classification, a joint update of α_i and α_i^* is performed only if the KKT conditions are violated by at least τ , i.e. if

$$\begin{aligned} &\alpha_i < C \ \wedge \ \varepsilon + E_i < -\tau \ , \quad \text{or} \\ &\alpha_i > 0 \ \wedge \ \varepsilon + E_i > \tau \ , \quad \text{or} \\ &\alpha_i^* < C \ \wedge \ \varepsilon - E_i < -\tau \ , \quad \text{or} \\ &\alpha_i^* > 0 \ \wedge \ \varepsilon - E_i > \tau \ . \end{aligned} \tag{24}$$

After the changes, the same clipping operations as defined in (22) are performed

$$\alpha_i \leftarrow \min(\max(0, \alpha_i + \Delta\alpha_i), C) \quad (i = 1, \ldots, l) \ , \tag{25a}$$

$$\alpha_i^* \leftarrow \min(\max(0, \alpha_i^* + \Delta\alpha_i^*), C) \quad (i = 1, \ldots, l) \ . \tag{25b}$$

The KA learning as formulated in this section and the SMO algorithm without bias term for solving regression tasks are strictly equal in terms of both the number of iterations required and the final values of the Lagrange multipliers. The equality is strict despite the fact that the implementation is slightly different. In every iteration step, namely, the KA algorithm updates both weights α_i and α_i^* without any checking whether the KKT conditions are fulfilled or not, while the SMO performs an update according to (24).

2.3 The Coordinate Ascent Based Learning for Nonlinear Classification and Regression Tasks

When positive definite kernels are used, the learning problem for both tasks is same. In a vector-matrix notation, in a dual space, the learning is represented as:

$$\max \quad L_d(\alpha) = -0.5\alpha^T \mathbf{K} \alpha + \mathbf{f}^T \alpha \ , \tag{26}$$

$$\text{s.t.} \quad 0 \leq \alpha_i \leq C, \quad (i = 1, \ldots, n) \ , \tag{27}$$

where, in the classification $n = l$ and the matrix $\mathbf{K}$ is an (l, l) symmetric positive definite matrix, while in regression $n = 2l$ and $\mathbf{K}$ is a $(2l, 2l)$ symmetric semipositive definite one. Note that the constraints (27) define a convex subspace over which the convex dual Lagrangian should be maximized. It is very well known that the vector $\boldsymbol{\alpha}$ may be looked at as the *iterative* solution of a system of linear equations

$$\mathbf{K}\alpha = \mathbf{f} \ , \tag{28}$$

subject to the same constraints given by (27), namely $0 \leq \alpha_i \leq C$, $(i = 1, \ldots, n)$.

Thus, it may seem natural to solve (28), subject to (27), by applying some of the well known and established techniques for solving a general linear system of equations. The size of training data set and the constraints (27) eliminate direct techniques. Hence, one has to resort to the *iterative approaches* in solving the problems above. There are three possible iterative avenues that can be followed. They are; the use of the Non-Negative Least Squares (NNLS) technique [13], application of the Non-Negative Conjugate Gradient (NNCG) method [8] and the implementation of the Gauss-Seidel i.e., the related Successive Over-Relaxation technique. The first two methods solve for the non-negative constraints only, but the bound $\alpha_i <= C$, for solving "soft" tasks can be readily incorporated into both the NNLS and NNCG. In the case of nonlinear regression, one can apply NNLS and NNCG by taking $C = \infty$ and compensating (i.e. smoothing or "softening" the solution) by increasing the sensitivity zone ε.

Here we show how to extend the application of Gauss-Seidel and Successive Over-Relaxation to both the nonlinear classification and to the nonlinear regression tasks. The Gauss-Seidel method solves (28) by using the ith equation to update the ith unknown doing it iteratively, i.e., starting in the kth step with the first equation to compute the α_1^{k+1}, then the second equation is used to calculate the α_2^{k+1} by using new α_1^{k+1} and $\alpha_i^k (i > 2)$ and so on. The iterative learning takes the following form,

$$\begin{aligned} \alpha_i^{k+1} &= \left(f_i - \sum_{j=1}^{i-1} K_{ij}\alpha_j^{k+1} - \sum_{j=i+1}^{n} K_{ij}\alpha_j^k \right) \Bigg/ K_{ii} \\ &= \alpha_i^k - \frac{1}{K_{ii}} \left(\sum_{j=1}^{i-1} K_{ij}\alpha_j^{k+1} + \sum_{j=i}^{n} K_{ij}\alpha_j^k - f_i \right) \\ &= \alpha_i^k + \frac{1}{K_{ii}} \left. \frac{\partial L_d}{\partial \alpha_i} \right|_{k+1}, \end{aligned} \tag{29}$$

where we use the fact that the term within a second bracket (called the residual r_i in mathematics' references) is the ith element of the gradient of a dual Lagrangian L_d given in (26) at the $k+1$th iteration step. The (29) above shows that Gauss-Seidel method is a *coordinate* gradient ascent procedure as well as the KA and the SMO are. *The KA and SMO for positive definite kernels equal the Gauss-Seidel!* Note that the optimal learning rate used in both the KA algorithm and in the SMO without-bias-term approach is exactly equal to the coefficient $1/K_{ii}$ in a Gauss-Seidel method. Based on this equality, the convergence theorem for the KA, SMO and Gauss-Seidel (i.e., successive over-relaxation) in solving (26) subject to constraints (27) can be stated and proved as follows:

Theorem: *For SVMs with positive definite kernels, the iterative learning algorithms KA i.e., SMO i.e., Gauss-Seidel i.e., successive over-relaxation, in solving nonlinear classification and regression tasks (26) subject to constraints (27), converge starting from any initial choice of* $\boldsymbol{\alpha}_0$.

Proof: The proof is based on the very well known theorem of convergence of the Gauss-Seidel method for symmetric positive definite matrices in solving (28) without constraints [16]. First note that for positive definite kernels, the matrix $\mathbf{K}$ created by terms $y_i y_j K(\mathbf{x}_i, \mathbf{x}_j)$ in the second sum in (9), and involved in solving classification problem, is also positive definite. In regression tasks $\mathbf{K}$ is a symmetric positive semidefinite (meaning still convex) matrix, which after a mild regularization given as ($\mathbf{K} \leftarrow \boldsymbol{K} + \lambda\mathbf{I}$, $\lambda \sim$ 1e $-$ 12) becomes positive definite one. (Note that the proof in the case of regression does not need regularization at all, but there is no space here to go into these details). Hence, the learning without constraints (27) converges, starting from any initial point $\boldsymbol{\alpha}_0$, and each point in an n-dimensional search space for multipliers α_i is a viable starting point ensuring a convergence of the algorithm to the maximum of a dual Lagrangian L_d. This, naturally, includes all the (starting) points within, or on a boundary of, any convex subspace of a search space ensuring the convergence of the algorithm to the maximum of a dual Lagrangian L_d over the given subspace. The constraints imposed by (27) preventing variables α_i to be negative or bigger than C, and implemented by the clipping operators above, define such a convex subspace. Thus, each "clipped" multiplier value α_i defines a new starting point of the algorithm guaranteeing the convergence to the maximum of L_d over the subspace defined by (27). For a convex constraining subspace such a constrained maximum is unique.

Due to the lack of the space we do not go into the discussion on the convergence rate here and we leave it to some other occasion. It should be only mentioned that both KA and SMO (i.e. Gauss-Seidel and successive over-relaxation) for positive definite kernels have been successfully applied for many problems (see references given here, as well as many other, benchmarking the mentioned methods on various data sets). Finally, let us just mention that the standard extension of the Gauss-Seidel method is the method of successive over-relaxation that can reduce the number of iterations required by proper choice of a relaxation parameter ω significantly. The successive over-relaxation method uses the following updating rule

$$\alpha_i^{k+1} = \alpha_i^k - \omega \frac{1}{K_{ii}} \left(\sum_{j=1}^{i-1} K_{ij} \alpha_j^{k+1} + \sum_{j=i}^{n} K_{ij} \alpha_j^k - f_i \right) = \alpha_i^k + \omega \frac{1}{K_{ii}} \left. \frac{\partial L_d}{\partial \alpha_i} \right|_{k+1}, \tag{30}$$

and similarly to the KA, SMO, and Gauss-Seidel its convergence is guaranteed.

2.4 Discussions

Both the KA and the SMO algorithms were recently developed and introduced as alternatives to solve quadratic programming problem while training support vector machines on huge data sets. It was shown that when using positive definite kernels the two algorithms are identical in their analytic form and numerical implementation. In addition, for positive definite kernels both algorithms are strictly identical with a classic iterative Gauss-Seidel (optimal coordinate ascent) learning and its extension successive over-relaxation. Until now, these facts were blurred mainly due to different pace in posing the learning problems and due to the "heavy" heuristics involved in the SMO implementation that shadowed an insight into the possible identity of the methods. It is shown that in the so-called no-bias SVMs, both the KA and the SMO procedure are the coordinate ascent based methods and can be classified as ISDA. Hence, they are the inheritors of all good and bad "genes" of a gradient approach and both algorithms have same performance.

In the next section, the ISDAs with explicit bias term b will be presented. The motivations for incorporating bias term into the ISDAs are to improve the versatility and the performance of the algorithms. The ISDAs without bias term developed in this section can only deal with positive definite kernel, which may be a limitation in applications where positive semi-definite kernel such as a linear kernel is more desirable. As it will be discussed shortly, ISDAs with explicit bias term b also seems to be faster in terms of training time.

3 Iterative Single Data Algorithms with an Explicit Bias Term b

Before presenting iterative algorithms with bias term b, we discuss some recent presentations of the bias b utilization. As mentioned previously, for positive definite kernels there is no need for bias b. However, one can use it and this means implementing a different kernel. In [19] it was also shown that when using positive definite kernels, one can choose between two types of solutions for both classification and regression. The first one uses the model without bias term (i.e.,$f(\mathbf{x}) = \sum_{j=1}^{l} v_j K(\mathbf{x}, \mathbf{x}_j)$), while the second SVM uses an explicit bias term b. For the second one $f(\mathbf{x}) = \sum_{i=1}^{l} v_i K(\mathbf{x}, \mathbf{x}_i) + b$, and it was shown that $f(\mathbf{x})$ is a function resulting from a minimization of the functional shown below

$$I[f] = \sum_{j=1}^{l} V(y_j, f(\mathbf{x}_j)) + \lambda \|f\|_{K*}^2 , \tag{31}$$

where $K^* = K - a$ (for an appropriate constant a) and K is an original kernel function (more details can be found in [19]). This means that by adding a constant term to a positive definite kernel function K, one obtains the solution to the functional $I[f]$ where K^* is a conditionally positive definite kernel.

Interestingly, similar type of model was also presented in [14]. However, their formulation is done for the classification problems only. They reformulated the optimization problem by adding the $b^2/2$ term to the cost function $\|\mathbf{w}\|^2/2$. This is equivalent to an addition of 1 to each element of the original kernel matrix $\mathbf{K}$. As a result, they changed the original classification dual problem to the optimization of the following one

$$L_d(\alpha) = \sum_{i=1}^{l} \alpha_i - \frac{1}{2} \sum_{i,j=1}^{l} y_i y_j \alpha_i \alpha_j (K(\mathbf{x}_i, \mathbf{x}_j) + 1) \ . \tag{32}$$

3.1 Iterative Single Data Algorithms for SVMs Classification with Bias Term b

In the previous section, for the SVMs' models when positive definite kernels are used without a bias term b, the learning algorithms for classification and regression (in a dual domain) were solved with box constraints only, originating from minimization of a primal Lagrangian in respect to the weights w_i. However, there remains an open question – how to apply the proposed ISDA scheme for the SVMs that do use explicit bias term b. Such general nonlinear SVMs in classification and regression tasks are given below,

$$f(\mathbf{x}_i) = \sum_{j=1}^{l} y_j \alpha_j \mathbf{\Phi}(\mathbf{x}_i)^T \mathbf{\Phi}(\mathbf{x}_j) + b = \sum_{j=1}^{l} v_j K(\mathbf{x}_i, \mathbf{x}_j) + b \ , \tag{33a}$$

$$f(\mathbf{x}_i) = \sum_{j=1}^{l} (\alpha_j^* - \alpha_j) \mathbf{\Phi}(\mathbf{x}_i)^T \mathbf{\Phi}(\mathbf{x}_j) + b = \sum_{j=1}^{l} v_j K(\mathbf{x}_i, \mathbf{x}_j) + b \ , \tag{33b}$$

where $\mathbf{\Phi}(\mathbf{x}_i)$ is the l-dimensional vector that maps n-dimensional input vector $\mathbf{x}_i$ into the feature space. Note that $\mathbf{\Phi}(\mathbf{x}_i)$ could be infinite dimensional and we do not have necessarily to know either the $\mathbf{\Phi}(\mathbf{x}_i)$ or the weight vector $\mathbf{w}$. (Note also that for a classification model in (33a), we usually take the sign of $f(\mathbf{x})$ but this is of lesser importance now). For the SVMs' models (33), there are also *the equality constraints* originating from minimizing the primal objective function in respect to the bias b as given in (8) for classification and (13) for regression. The motivation for developing the ISDAs for the SVMs with an explicit bias term b originates from the fact that the use of an explicit bias b *seems to lead to the SVMs with less support vectors.* This fact can often be very useful for both the data (information) compression and the speed of learning. Below, we present an iterative learning algorithm for the classification SVMs (33a) with an explicit bias b, subjected to the equality constraints (8). (*The same procedure is developed for the regression SVMs but due to the space constraints we do not go into these details here. However we give some relevant hints for the regression SVMs with bias b shortly*).

There are *three major avenues* (procedures, algorithms) possible in solving the dual problem (6), (7) and (8).

The first one is the standard SVMs algorithm which imposes the equality constraints (8) during the optimization and in this way ensures that the solution never leaves a feasible region. In this case the last term in (6) vanishes. After the dual problem is solved, the bias term is calculated by using *unbounded* Lagrange multipliers α_i [11, 20] as follows

$$b = \frac{1}{\#\ unboundSVecs} \left(\sum_{i=1}^{\#\ UnboundSVecs} \left(y_i - \sum_{j=1}^{l} \alpha_j y_j K(\mathbf{x}_i, \mathbf{x}_j) \right) \right) . \tag{34}$$

Note that in a standard SMO iterative scheme the minimal number of training data points enforcing (8) and ensuring staying in a feasible region is two.

Below, we show *two more possible ways* how the ISDA works for the SVMs containing an explicit bias term b too. In *the first method*, the cost function (1) is augmented with the term $0.5kb^2$ (where $k \geq 0$) and this step changes the primal Lagrangian (3) into the following one

$$L_p(\mathbf{w}, b, \alpha) = \frac{1}{2}\mathbf{w}^T\mathbf{w} - \sum_{i=1}^{l} \alpha_i \left\{ y_i \left[\mathbf{w}^T \Phi(\mathbf{x}_i) + b \right] - 1 \right\} + \frac{k}{2} b^2 . \tag{35}$$

Equation (5) also changes as given below

$$\frac{\partial L_p}{\partial b} = 0 \quad b = \frac{1}{k} \sum_{i=1}^{l} \alpha_i y_i . \tag{36}$$

After forming (35) as well as using (36) and (4), one obtains the dual problem without an explicit bias b,

$$\begin{aligned} L_d(\alpha) &= \sum_{i=1}^{l} \alpha_i - \frac{1}{2} \sum_{i,j=1}^{l} y_i y_j \alpha_i \alpha_j K(\mathbf{x}_i, \mathbf{x}_j) \\ &\quad - \frac{1}{k} \sum_{i,j=1}^{l} \alpha_i y_i \alpha_j y_j + \frac{1}{2k} \sum_{i,j=1}^{l} \alpha_i y_i \alpha_j y_j \\ &= \sum_{i=1}^{l} \alpha_i - \frac{1}{2} \sum_{i,j=1}^{l} y_i y_j \alpha_i \alpha_j \left(K(\mathbf{x}_i, \mathbf{x}_j) + \frac{1}{k} \right) . \end{aligned} \tag{37}$$

Actually, the optimization of a dual Lagrangian is reformulated for the SVMs with a bias b by applying "tiny" changes of $1/k$ only to the original matrix $\mathbf{K}$ as illustrated in (37). Hence, for the nonlinear classification problems ISDA stands for an *iterative* solving of the following linear system

$$\mathbf{K}_k \alpha = \mathbf{1}_l , \tag{38a}$$

$$\text{s.t. } 0 \leq \alpha_i \leq C, \quad i = 1, \ldots, l , \tag{38b}$$

where $K_k(\mathbf{x}_i, \mathbf{x}_j) = y_i y_j (K(\mathbf{x}_i, \ \mathbf{x}_j) + 1/k)$, $\mathbf{1}_l$ is an l-dimensional vector containing ones and C is a penalty factor equal to infinity for a hard margin classifier. Note that during the updates of α_i, the bias term b must not be used because it is implicitly incorporated within the $\mathbf{K}_k$ matrix. Only after the solution vector $\boldsymbol{\alpha}$ in (38) is found, the bias b should be calculated either by using *unbounded* Lagrange multipliers α_i as given in (34), or by implementing the equality constraints from $\partial L_p / \partial b = 0$ and given in (36) as

$$b = \frac{1}{k} \sum_{i=1}^{\#\, SVecs} \alpha_j y_j \, . \tag{39}$$

Note, however, that all the Lagrange multipliers, meaning both bounded (clipped to C) and unbounded (smaller than C) must be used in (39). Both equations, (34) and (39), result in the same value for the bias b. Thus, using the SVMs with an explicit bias term b means that, in the ISDA proposed above, the original kernel is changed, i.e., another kernel function is used. This means that the alpha values will be different for each k chosen, and so will be the value for b. The final SVM as given in (33a) is produced by original kernels. Namely, $f(\mathbf{x})$ is obtained by adding the sum of the weighted original kernel values and corresponding bias b. The approach of adding a small change to the kernel function can also be associated with a classic penalty function method in optimization as follows below.

To illustrate the idea of the penalty function, let us consider the problem of maximizing a function $f(x)$ subject to an equality constraint $g(x) = 0$. To solve this problem using classical penalty function method, the following quadratic penalty function is formulated,

$$\max \ P(x, \rho) = f(x) - \frac{1}{2} \rho \left\| g(x) \right\|_2^2 \, , \tag{40}$$

where ρ is the penalty parameter and $\|g(x)\|_2^2$ is the square of the L_2 norm of the function $g(x)$. As the penalty parameter ρ increases towards infinity, the size of the $g(x)$ is pushed towards zero, hence the equality constraint $g(x) = 0$ is fulfilled. Now, let us consider the standard SVMs' dual problem, which is maximizing (9) subject to box constraints (10) and the equality constraint (11). By applying the classical penalty method (40) to the equality constraint (11), we can form the following quadratic penalty function.

$$\begin{aligned} P(x, \rho) &= L_d(\alpha) - \frac{1}{2} \rho \left\| \sum_{i=1}^{l} \alpha_i y_i \right\|_2^2 \\ &= \sum_{i=1}^{l} \alpha_i - \frac{1}{2} \sum_{i,j=1}^{l} y_i y_j \alpha_i \alpha_j K(\mathbf{x}_i, \mathbf{x}_j) - \frac{1}{2} \rho \sum_{i,j=1}^{l} y_i y_j \alpha_i \alpha_j \\ &= \sum_{i=1}^{l} \alpha_i - \frac{1}{2} \sum_{i,j=1}^{l} y_i y_j \alpha_i \alpha_j (K(\mathbf{x}_i, \mathbf{x}_j) + \rho) \, . \end{aligned} \tag{41}$$

The expression above is exactly equal to (37) when ρ equals $1/k$. Thus, the parameter $1/k$ in (37) for the first method of adding bias into the ISDAs can be regarded as a penalty parameter of enforcing equality constraint (11) in the original SVMs dual problem. Also, for a large value of $1/k$, the solution will have a small L_2 norm of (11). In other words, as k approaches zero a bias b converges to the solution of the standard QP method that enforces the equality constraints. However, we do not use the ISDA with small parameter k values here, because the condition number of the matrix $\mathbf{K}_k$ increases as $1/k$ rises. Furthermore, the strict fulfilment of (11) may not be needed in obtaining a good SVM. In the next section, it will be shown that in classifying the MNIST data with Gaussian kernels, the value $k = 10$ proved to be a very good one justifying all the reasons for its introduction (fast learning, small number of support vectors and good generalization).

The second method in implementing the ISDA for SVMs with the bias term b is to work with original cost function (1) and keep imposing the equality constraints during the iterations as suggested in [22]. The learning starts with $b = 0$ and after each *epoch* the bias b is updated by applying a secant method as follows

$$b^k = b^{k-1} - \omega^{k-1} \frac{b^{k-1} - b^{k-2}}{\omega^{k-1} - \omega^{k-2}}\,, \tag{42}$$

where $\omega = \sum_{i=1}^{l} \alpha_i y_i$ represents the value of equality constraint after each epoch. In the case of the regression SVMs, (42) is used by implementing the corresponding regression's equality constraints, namely $\omega = \sum_{i=1}^{l} (\alpha_i - \alpha_i^*)$. This is different from [22] where an iterative update after each data pair is proposed. In our SVMs regression experiments such an updating led to an unstable learning. Also, in an addition to changing expression for ω, both the $\mathbf{K}$ matrix, which is now $(2l, 2l)$ matrix, and the right hand side of (38a) which becomes $(2l, 1)$ vector, should be changed too and formed as given in [12].

3.2 Performance of the Iterative Single Data Algorithm and Comparisons

To measure the relative performance of different ISDAs, we ran all the algorithms with RBF Gaussian kernels on a MNIST dataset with 576-dimensional inputs [5], and compared the performance of our ISDAs with LIBSVM V2.4 [2] which is one of the fastest and the most popular SVM solvers at the moment based on the SMO type of an algorithm. The MNIST dataset consists of 60,000 training and 10,000 test data pairs. To make sure that the comparison is based purely on the nature of the algorithm rather than on the differences in implementation, our encoding of the algorithms are the same as LIBSVM's ones in terms of caching strategy (LRU–Least Recent Used), data structure, heuristics for shrinking and stopping criterions. The only significant difference is that instead of two heuristic rules for selecting and updating two data points at each iteration step aiming at the maximal improvement of the dual objective function, our ISDA selects the worse KKT violator only and updates its α_i at each step.

Also, in order to speed up the LIBSVM's training process, we modified the original LIBSVM routine to perform faster by reducing the numbers of complete KKT checking without any deterioration of accuracy. All the routines were written and compiled in Visual C++ 6.0, and all simulations were run on a 2.4 GHz P4 processor PC with 1.5 Gigabyte of memory under the operating system Windows XP Professional. The shape parameter σ^2 of an RBF Gaussian kernel and the penalty factor C are set to be 0.3 and 10 [5]. The stopping criterion τ and the size of the cache used are 0.01 and 250 Megabytes. The simulation results of different ISDAs against both LIBSVM are presented in Tables 1 and 2, and in a Fig. 1.

The first and the second column of the tables show the performance of the original and modified LIBSVM respectively. The last three columns show the results for single data point learning algorithms with various values of constant $1/k$ added to the kernel matrix in (12a). For $k = \infty$, ISDA is equivalent to the SVMs without bias term, and for $k = 1$, it is the same as the classification formulation proposed in [14].

Table 1 illustrates the running time for each algorithm. The ISDA with $k = 10$ was the quickest and required the shortest average time (T_{10}) to complete the training. The average time needed for the original LIBSVM is almost $2T_{10}$ and the average time for a modified version of LIBSVM is 10.3% bigger than T_{10}. This is contributed mostly to the simplicity of the ISDA. One may think that the improvement achieved is minor, but it is important to consider the fact that approximately more than 50% of the CPU time is spent on the final checking of the KKT conditions in all simulations. During

Table 1. Simulation time for different algorithms

	LIBSVM Original	LIBSVM Modified	Iterative $k = 1$	Single Data $k = 10$	Algorithm (ISDA) $k = \infty$
Class	Time (sec)	Time (sec)	Time (sec)	Time (sec)	Time (sec)
0	1606	885	800	794	1004
1	740	465	490	491	855
2	2377	1311	1398	1181	1296
3	2321	1307	1318	1160	1513
4	1997	1125	1206	1028	1235
5	2311	1289	1295	1143	1328
6	1474	818	808	754	1045
7	2027	1156	2137	1026	1250
8	2591	1499	1631	1321	1764
9	2255	1266	1410	1185	1651
Time, hr	5.5	3.1	3.5	**2.8**	3.6
Time Increase	+95.3%	+10.3%	+23.9%	**0**	+28.3%

Table 2. Number of support vectors for each algorithm

	LIBSVM Original	LIBSVM Modified	Iterative $k = 1$	Single Data $k = 10$	Algorithm (ISDA) $k = \infty$
Class	#SV (BSV)	#SV (BSV)	#SV (BSV)	#SV (BSV)	#SV (BSV)
0	2172 (0)	2172 (0)	2162 (0)	2132 (0)	2682 (0)
1	1440 (4)	1440 (4)	1429 (4)	1453 (4)	2373 (4)
2	3055 (0)	3055 (0)	3047 (0)	3017 (0)	3327 (0)
3	2902 (0)	2902 (0)	2888 (0)	2897 (0)	3723 (0)
4	2641 (0)	2641 (0)	2623 (0)	2601 (0)	3096 (0)
5	2900 (0)	2900 (0)	2884 (0)	2856 (0)	3275 (0)
6	2055 (0)	2055 (0)	2042 (0)	2037 (0)	2761 (0)
7	2651 (4)	2651 (4)	3315 (4)	2609 (4)	3139 (4)
8	3222 (0)	3222 (0)	3267 (0)	3226 (0)	4224 (0)
9	2702 (2)	2702 (2)	2733 (2)	2756 (2)	3914 (2)
Av. # SVs	2574	2574	2639	**2558**	3151

BSV = Bounded Support Vectors

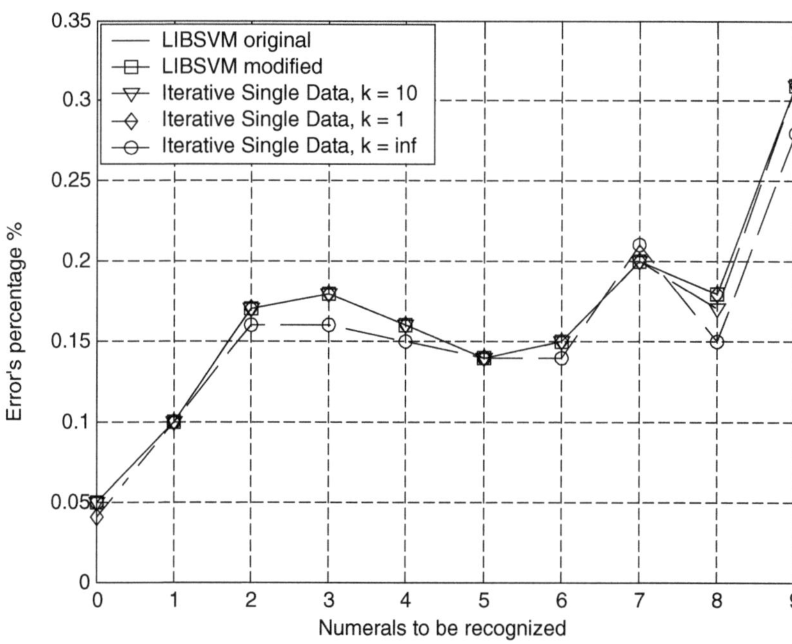

Fig. 1. The percentage of an error on the test data

the checking, the algorithm must calculate the output of the model at each datum in order to evaluate the KKT violations. This process is unavoidable if one wants to ensure the solution's global convergence, i.e. that *all the data* do satisfy the KKT conditions with precision τ indeed. Therefore, the reduction of time spent on iterations is approximately double the figures shown. Note that the ISDA slows down for $k < 10$ here. This is a consequence of the fact that with a decrease in k there is an increase of the condition number of a matrix $\mathbf{K}_k$, which leads to more iterations in solving (38). At the same time, implementing the no-bias SVMs, i.e., working with $k = \infty$, also slows the learning down due to an increase in the number of support vectors needed when working without bias b.

Table 2 presents the numbers of support vectors selected. For the ISDA, the numbers reduce significantly when the explicit bias term b is included. One can compare the numbers of SVs for the case without the bias b ($k = \infty$) and the ones when an explicit bias b is used (cases with $k = 1$ and $k = 10$). Because identifying less support vectors speeds the overall training definitely up, the SVMs implementations with an explicit bias b are faster than the version without bias.

In terms of a generalization, or a performance on a test data set, all algorithms had very similar results and this demonstrates that the ISDAs produce models that are as good as the standard QP, i.e., SMO based, algorithms (see Fig. 1).

The percentages of the errors on the test data are shown in Fig. 1. Notice the extremely low error percentages on the test data sets for all numerals.

3.3 Discussions

In final part of this chapter, we demonstrate the use, the calculation and the effect of incorporating an explicit bias term b in the SVMs trained with the ISDA. The simulation results show that models generated by ISDAs (either with or without the bias term b) are as good as the standard SMO based algorithms in terms of a generalization performance. Moreover, ISDAs with an appropriate k value are faster than the standard SMO algorithms on large scale classification problems ($k = 10$ worked particularly well in all our simulations using Gaussian RBF kernels). This is due to both the simplicity of ISDAs and the decrease in the number of SVs chosen after an inclusion of an explicit bias b in the model. The simplicity of ISDAs is the consequence of the fact that the equality constraints (8) do not need to be fulfilled during the training stage. In this way, the *second choice heuristics is avoided* during the iterations. Thus, the ISDA is an extremely good tool for solving large scale SVMs problems containing huge training data sets because it is faster than, and it delivers "same" generalization results as, the other standard QP (SMO) based algorithms. The fact that an introduction of an explicit bias b means solving the problem with different kernel suggests that it may be hard to tell in advance for what kind of previously unknown multivariable decision (regression) function the models with bias b may perform better, or may be more suitable, than the ones without it. As it is often the case, the real experimental results, their comparisons and the new theoretical developments should probably be able to tell one day. As for the single data based learning approach presented here, the future work will focus on the development of even faster training algorithms.

References

1. Anlauf, J. K., Biehl, M., The AdaTron – an adaptive perceptron algorithm. *Europhysics Letters*, 10(7), pp. 687–692, 1989
2. Chang, C., Lin, C., LIBSVM : A library for support vector machines, (available at: http://www.csie.ntu.edu.tw/~cjlin/libsvm/), 2003
3. Cherkassky, V., Mulier, F., *Learning From Data: Concepts, Theory and Methods*, John Wiley & Sons, New York, NY, 1998
4. Cristianini, N., Shawe-Taylor, J., *An introduction to Support Vector Machines and other kernel-based learning methods*, Cambridge University Press, Cambridge, UK, 2000
5. Dong, X., Krzyzak, A., Suen, C. Y., A fast SVM training algorithm, International Journal of Pattern Recognition and Artificial Intelligence, Vol. 17, No. 3, pp. 367–384, 2003
6. Evgeniou, T., Pontil, M., Poggio, T., Regularization networks and support vector machines, Advances in Computational Mathematics, 13, pp. 1–50, 2000
7. Frieß, T.-T., Cristianini, N., Campbell, I. C. G., The Kernel-Adatron: a Fast and Simple Learning Procedure for Support Vector Machines. In Shavlik, J., editor, *Proceedings of the 15th International Conference on Machine Learning*, Morgan Kaufmann, pp. 188–196, San Francisco, CA, 1998
8. Hestenes, M. Conjugate Direction Method In Optimization: Application of Mathematics, Vol. 12, Springer-Verlag New York, Heidelberg, 1980
9. Huang, T.-M., Kecman, V., Bias Term b in SVMs Again, Proc. of ESANN 2004, 12th European Symposium on Artificial Neural Networks, Bruges, Belgium, (downloadable from http://www.support-vector.ws), 2004
10. Joachims, T. (1999). Making Large-scale SVM learning practical. *Advances in Kernel Methods- Support Vector Learning.* B. Schlkopf, Smola, A. J., and Burges, C. J. C. Cambridge, M.A., MIT Press: 169–184
11. Kecman, V., *Learning and Soft Computing, Support Vector Machines, Neural Networks, and Fuzzy Logic Models,* The MIT Press, Cambridge, MA, (See http://www.support-vector.ws), 2001
12. Kecman, V., Vogt, M., Huang, T.-M., On the Equality of Kernel AdaTron and Sequential Minimal Optimization in Classification and Regression Tasks and Alike Algorithms for Kernel Machines, *Proc. of the 11th European Symposium on Artificial Neural Networks*, ESANN 2003, pp. 215–222, Bruges, Belgium, (downloadable from http://www.support-vector.ws), 2003
13. Lawson, C. I., Hanson, R. J., *Solving Least Squares Problems*, Prentice-Hall, Englewood Cliffs, N.J., 1974
14. Mangasarian, O. L., Musicant, D. R., Successive Overrelaxation for Support Vector Machines, IEEE Trans. Neural Networks, 11(4), 1003–1008, 1999
15. Osuna E, Freund R, Girosi F, An Improved Training Algorithm for Support Vector Machines. In *Neural Networks for Signal Processing VII, Proceedings of the 1997 Signal Processing Society Workshop*, pp. 276–285, 1997
16. Ostrowski, A. M., Solutions of Equations and Systems of Equations, 2nd ed., Academic Press, New York, 1966

17. Platt, J. Sequential Minimal Optimization: A Fast Algorithm for Training Support Vector Machines, Microsoft Research Technical Report MSR-TR-98-14, 1998
18. Platt, J. C., Fast Training of Support Vector Machines using Sequential Minimal Optimization. *Chap. 12 in Advances in Kernel Methods – Support Vector Learning*, edited by B. Schölkopf, C. Burges, A. Smola, The MIT Press, Cambridge, MA, 1999
19. Poggio, T., Mukherjee, S., Rifkin, R., Rakhlin, A., Verri, A., *b*, CBCL Paper #198/AI Memo# 2001-011, Massachusetts Institute of Technology, Cambridge, MA, 2001, also it is a Chapter 11 in "Uncertainty in Geometric Computations", pp. 131–141, Eds., J. Winkler and M. Niranjan, Kluwer Academic Publishers, Boston, MA, 2002
20. Schölkopf, B., Smola, A., *Learning with Kernels – Support Vector Machines, Optimization, and Beyond*, The MIT Press, Cambridge, MA, 2002
21. Vapnik, V. N., *The Nature of Statistical Learning Theory*, Springer Verlag Inc, New York, NY, 1995
22. Veropoulos, K., *Machine Learning Approaches to Medical Decision Making*, PhD Thesis, The University of Bristol, Bristol, UK, 2001
23. Vogt, M., SMO Algorithms for Support Vector Machines without Bias, Institute Report, Institute of Automatic Control, TU Darmstadt, Darmstadt, Germany, (Available at http://www.iat.tu-darmstadt.de/~vogt), 2002
24. Vapnik, V. N., 1995. The Nature of Statistical Learning Theory, Springer Verlag Inc, New York, NY
25. Vapnik, V., S. Golowich, A. Smola. 1997. Support vector method for function approximation, regression estimation, and signal processing, In Advances in Neural Information Processing Systems 9, MIT Press, Cambridge, MA
26. Vapnik, V. N., 1998. Statistical Learning Theory, J. Wiley & Sons, Inc., New York, NY

Kernel Discriminant Learning with Application to Face Recognition

J. Lu[1], K.N. Plataniotis[2], and A.N. Venetsanopoulos[3]

Bell Canada Multimedia Laboratory
The Edward S. Rogers Sr. Department of Electrical and Computer Engineering
University of Toronto, Toronto, M5S 3G4, Ontario, Canada
{juwei[1],kostas[2],anv[3]}@dsp.toronto.edu

Abstract. When applied to high-dimensional pattern classification tasks such as face recognition, traditional kernel discriminant analysis methods often suffer from two problems: (1) small training sample size compared to the dimensionality of the sample (or mapped kernel feature) space, and (2) high computational complexity. In this chapter, we introduce a new kernel discriminant learning method, which attempts to deal with the two problems by using regularization and subspace decomposition techniques. The proposed method is tested by extensive experiments performed on real face databases. The obtained results indicate that the method outperforms, in terms of classification accuracy, existing kernel methods, such as kernel Principal Component Analysis and kernel Linear Discriminant Analysis, at a significantly reduced computational cost.

Key words: Statistical Discriminant Analysis, Kernel Machines, Small Sample Size, Nonlinear Feature Extraction, Face Recognition

1 Introduction

Statistical learning theory tells us essentially that the difficulty of an estimation problem increases drastically with the dimensionality J of the sample space, since in principle, as a function of J, one needs exponentially many patterns to sample the space properly [18, 32]. Unfortunately, in many practical tasks such as face recognition, the number of available training samples per subject is usually much smaller than the dimensionality of the sample space. For instance, a canonical example used for face recognition is a 112×92 image, which exists in a 10304-dimensional real space. Nevertheless, the number of examples per class available for learning is not more than ten in most cases. This results in the so-called *small sample size* (SSS) problem, which is known to have significant influences on the performance of a statistical pattern recognition system (see e.g. [3, 5, 9, 12, 13, 16, 21, 33, 34]).

J. Lu, K.N. Plataniotis, and A.N. Venetsanopoulos: *Kernel Discriminant Learning with Application to Face Recognition*, StudFuzz **177**, 275–296 (2005)
www.springerlink.com

When it comes to statistical discriminant learning tasks such as Linear Discriminant Analysis (LDA), the SSS problem often gives rise to high variance in the estimation for the between- and within-class scatter matrices, which are either poorly- or ill-posed. To address the problem, one popular approach is to introduce an intermediate Principal Component Analysis (PCA) step to remove the null spaces of the two scatter matrices. LDA is then performed in the lower dimensional PCA subspace, as it was done for example in [3, 29]. However, it has been shown that the discarded null spaces may contain significant discriminatory information [10]. To prevent this from happening, solutions without a separate PCA step, called *direct* LDA (D-LDA) approaches have been presented recently in [5, 12, 34]. The underlying principle behind the approaches is that the information residing in (or close to) the null space of the within-class scatter matrix is more significant for discriminant tasks than the information out of (or far away from) the null space. Generally, the null space of a matrix is determined by its zero eigenvalues. However, due to insufficient training samples, it is very difficult to identify the true null eigenvalues. As a result, high variance is often introduced in the estimation for the zero (or very small) eigenvalues of the within-class scatter matrix. Note that the eigenvectors corresponding to these eigenvalues are considered the most significant feature bases in the D-LDA approaches [5, 12, 34].

In this chapter, we study statistical discriminant learning algorithms in some high-dimensional feature space, mapped from the input sample space by the so-called "kernel machine" technique [18, 22, 25, 32]. In the feature space, it is hoped that the distribution of the mapped data is simplified, so that traditional linear methods can perform well. A problem with the idea is that the dimensionality of the feature space may be extremely higher than that of the sample space, resulting in the introduction of the SSS problem, or the worse if it has existed. In addition, kernel-based algorithms are generally much more computationally expensive compared to their linear counterparts. To address these problems, we introduce a regularized discriminant analysis method in the kernel feature space. This method deals with the SSS problem under the D-LDA framework of [12, 34]. Nevertheless, it is based on a modified Fisher's discriminant criterion specifically designed to avoid the unstable problem with the approach of [34]. Also, a side-effect of the design is that the computational complexity is significantly reduced compared to other two popular kernel methods, kernel PCA (KPCA) [26] and kernel LDA (GDA) [2]. The effectiveness of the presented method is demonstrated in the face recognition application.

2 Kernel-based Statistical Pattern Analysis

In the statistical pattern recognition tasks, the problem of feature extraction can be stated as follows: Assume that we have a training set, $\mathcal{Z} = \{\mathcal{Z}_i\}_{i=1}^{C}$, containing C classes with each class $\mathcal{Z}_i = \{\mathbf{z}_{ij}\}_{j=1}^{C_i}$ consisting of a number of examples $\mathbf{z}_{ij} \in \mathbb{R}^J$, where $\mathbb{R}^J$ denotes the J-dimensional real space. Taking as input such a set $\mathcal{Z}$, the objective of learning is to find, based on optimization of certain separability criteria, a transformation φ which produces a feature representation $\mathbf{y}_{ij} = \varphi(\mathbf{z}_{ij})$, $\mathbf{y}_{ij} \in \mathbb{R}^M$, intrinsic to the objects of these examples with enhanced discriminatory power.

2.1 Input Sample Space *vs* Kernel Feature Space

The kernel machines provide an elegant way of designing nonlinear algorithms by reducing them to linear ones in some high-dimensional feature space $\mathbb{F}$ nonlinearly related to the input sample space $\mathbb{R}^J$:

$$\phi : \mathbf{z} \in \mathbb{R}^J \rightarrow \phi(\mathbf{z}) \in \mathbb{F} \tag{1}$$

The idea can be illustrated by a toy example depicted in Fig. 1, where two-dimensional input samples, say $\mathbf{z} = [z_1, z_2]$, are mapped to a three-dimensional feature space through a nonlinear transform: $\phi : \mathbf{z} = [z_1, z_2] \rightarrow \phi(\mathbf{z}) = [x_1, x_2, x_3] := \left[z_1^2, \sqrt{2}z_1 z_2, z_2^2\right]$ [27]. It can be seen from Fig. 1 that in the sample space, a nonlinear ellipsoidal decision boundary is needed to separate classes A and B, in contrast with this, the two classes become linearly separable in the higher-dimensional feature space.

The feature space $\mathbb{F}$ could be regarded as a "linearization space" [1]. However, to reach this goal, its dimensionality could be arbitrarily large, possibly infinite. Fortunately, the exact $\phi(\mathbf{z})$ is not needed and the feature space can become implicit by using kernel machines. The trick behind the methods is to replace dot products in $\mathbb{F}$ with a kernel function in the input space $\mathbb{R}^J$ so that the nonlinear mapping is performed implicitly in $\mathbb{R}^J$. Let us come back to the

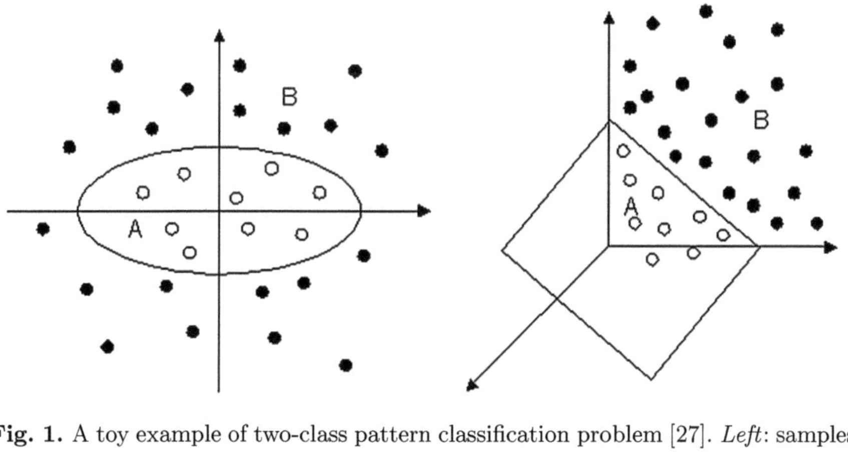

Fig. 1. A toy example of two-class pattern classification problem [27]. *Left*: samples lie in the 2-D input space, where it needs a nonlinear ellipsoidal decision boundary to separate classes A and B. *Right*: Samples are mapped to a 3-D feature space, where a linear hyperplane can separate the two classes

toy example of Fig. 1, where the feature space is spanned by the second-order monomials of the input sample. Let $\mathbf{z}_i \in \mathbb{R}^2$ and $\mathbf{z}_j \in \mathbb{R}^2$ be two examples in the input space, and the dot product of their feature vectors $\phi(\mathbf{z}_i) \in \mathbb{F}$ and $\phi(\mathbf{z}_j) \in \mathbb{F}$ can be computed by the following kernel function, $k(\mathbf{z}_i, \mathbf{z}_j)$, defined in $\mathbb{R}^2$,

$$\begin{aligned}\phi(\mathbf{z}_i) \cdot \phi(\mathbf{z}_j) &= \left[z_{i1}^2, \sqrt{2} z_{i1} z_{i2}, z_{i2}^2\right] \left[z_{j1}^2, \sqrt{2} z_{j1} z_{j2}, z_{j2}^2\right]^T \\ &= \left(\left[z_{i1}, z_{i2}\right] \left[z_{j1}, z_{j2}\right]^T\right)^2 = (\mathbf{z}_i \cdot \mathbf{z}_j)^2 =: k(\mathbf{z}_i, \mathbf{z}_j) \qquad (2)\end{aligned}$$

From this example, it can be seen that the central issue to generalize a linear learning algorithm to its kernel version is to reformulate all the computations of the algorithm in the feature space in the form of dot product. Based on the properties of the kernel functions used, the kernel generation gives rise to neural-network structures, splines, Gaussian, Polynomial or Fourier expansions, etc. Any function satisfying Mercer's condition [17] can be used as a kernel. Table 1 lists some of the most widely used kernel functions, and more sophisticated kernels can be found in [24, 27, 28, 36].

Table 1. Some of the most widely used kernel functions, where $\mathbf{z}_1, \mathbf{z}_2 \in \mathbb{R}^J$

Gaussian RBF	$k(\mathbf{z}_1, \mathbf{z}_2) = \exp\left(\frac{-\|\mathbf{z}_1 - \mathbf{z}_2\|^2}{\sigma^2}\right)$, $\sigma \in \mathbb{R}$
Polynomial	$k(\mathbf{z}_1, \mathbf{z}_2) = (a(\mathbf{z}_1 \cdot \mathbf{z}_2) + b)^d$, $a \in \mathbb{R}$, $b \in \mathbb{R}$, $d \in \mathbb{N}$
Sigmoidal	$k(\mathbf{z}_1, \mathbf{z}_2) = \tanh(a(\mathbf{z}_1 \cdot \mathbf{z}_2) + b)$, $a \in \mathbb{R}$, $b \in \mathbb{R}$
Inverse multiquadric	$1/\sqrt{\|\mathbf{z}_1 - \mathbf{z}_2\|^2 + \sigma^2}$, $\sigma \in \mathbb{R}$

2.2 Kernel Principal Component Analysis (KPCA)

To find principal components of a non convex distribution, the classic PCA has been generalized to the *kernel* PCA (KPCA) [26]. Given the nonlinear mapping of (1), the covariance matrix of the training sample $\mathcal{Z}$ in the feature space $\mathbb{F}$ can be expressed as

$$\tilde{\mathbf{S}}_{\text{cov}} = \frac{1}{N} \sum_{i=1}^{C} \sum_{j=1}^{C_i} (\phi(\mathbf{z}_{ij}) - \bar{\phi})(\phi(\mathbf{z}_{ij}) - \bar{\phi})^T \qquad (3)$$

where $N = \sum_{i=1}^{C} C_i$, and $\bar{\phi} = \frac{1}{N} \sum_{i=1}^{C} \sum_{j=1}^{C_i} \phi(\mathbf{z}_{ij})$ is the average of the ensemble in $\mathbb{F}$. The KPCA is actually a classic PCA performed in the feature space $\mathbb{F}$. Let $\tilde{\mathbf{g}}_m \in \mathbb{F}$ $(m = 1, 2, \ldots, M)$ be the first M most significant eigenvectors of $\tilde{\mathbf{S}}_{\text{cov}}$, and they form a low-dimensional subspace, called "KPCA subspace" in $\mathbb{F}$. All these $\{\tilde{\mathbf{g}}_m\}_{m=1}^{M}$ lie in the span of $\{\phi(\mathbf{z}_{ij})\}_{\mathbf{z}_{ij} \in \mathcal{Z}}$, and have $\tilde{\mathbf{g}}_m = \sum_{i=1}^{C} \sum_{j=1}^{C_i} a_{ij} \phi(\mathbf{z}_{ij})$, where a_{ij} are the linear combination coefficients.

For any input pattern $\mathbf{z}$, its nonlinear principal components can be obtained by the dot product, $\mathbf{y}_m = \tilde{\mathbf{g}}_m \cdot (\phi(\mathbf{z}) - \bar{\phi})$, computed indirectly through a kernel function $k()$.

2.3 Generalized Discriminant Analysis (GDA)

As such, *Generalized Discriminant Analysis* (GDA, also known as kernel LDA) [2] is a process to extract a nonlinear discriminant feature representation by performing a classic LDA in the high-dimensional feature space $\mathbb{F}$. Let $\tilde{\mathbf{S}}_b$ and $\tilde{\mathbf{S}}_w$ be the between- and within-class scatter matrices in the feature space $\mathbb{F}$ respectively, and they have following expressions:

$$\tilde{\mathbf{S}}_b = \frac{1}{N}\sum_{i=1}^{C} C_i(\bar{\phi}_i - \bar{\phi})(\bar{\phi}_i - \bar{\phi})^T \tag{4}$$

$$\tilde{\mathbf{S}}_w = \frac{1}{N}\sum_{i=1}^{C}\sum_{j=1}^{C_i} (\phi(\mathbf{z}_{ij}) - \bar{\phi}_i)(\phi(\mathbf{z}_{ij}) - \bar{\phi}_i)^T \tag{5}$$

where $\bar{\phi}_i = \frac{1}{C_i}\sum_{j=1}^{C_i}\phi(\mathbf{z}_{ij})$ is the mean of class $\mathcal{Z}_i$. In the same way as LDA, GDA determines a set of optimal nonlinear discriminant basis vectors by maximizing the standard Fisher's criterion:

$$\tilde{\Psi} = \arg\max_{\tilde{\Psi}} \frac{|\tilde{\Psi}^T\tilde{\mathbf{S}}_b\tilde{\Psi}|}{|\tilde{\Psi}^T\tilde{\mathbf{S}}_w\tilde{\Psi}|}, \quad \tilde{\Psi} = [\tilde{\psi}_1, \ldots, \tilde{\psi}_M], \quad \tilde{\psi}_m \in \mathbb{F} \tag{6}$$

Similar to KPCA, the GDA-based feature representation of an input pattern $\mathbf{z}$ can be obtained by a linear projection in $\mathbb{F}$, $\mathbf{y}_m = \tilde{\psi}_m \cdot \mathbf{z}$.

From the above presentation, it can be seen that KPCA and GDA are based on the exactly same optimization criteria to their linear counterparts, PCA and LDA. Especially, KPCA and GDA reduce to PCA and LDA, respectively, when $\phi(\mathbf{z}) = \mathbf{z}$. As we know, LDA optimizes the low-dimensional representation of the objects with focus on the most discriminant feature extraction while PCA achieves simply object reconstruction in a least-square sense. The difference may lead to significantly different orientations of feature bases as shown in Fig. 2: **Left**, where it is not difficult to see that the representation obtained by PCA is entirely unsuitable for the task of separating the two classes. As a result, it is generally believed that when it comes to solving problems of pattern classification, the LDA-based feature representation is usually superior to the PCA-based one [3, 5, 34].

3 Discriminant Learning in Small-Sample-Size Scenarios

For simplicity, we start the discussion with the linear case of discriminant learning, i.e. LDA, which optimizes the criterion of (6) in the sample space $\mathbb{R}^J$. This is equivalent to setting $\phi(\mathbf{z}) = \mathbf{z}$ during the GDA process.

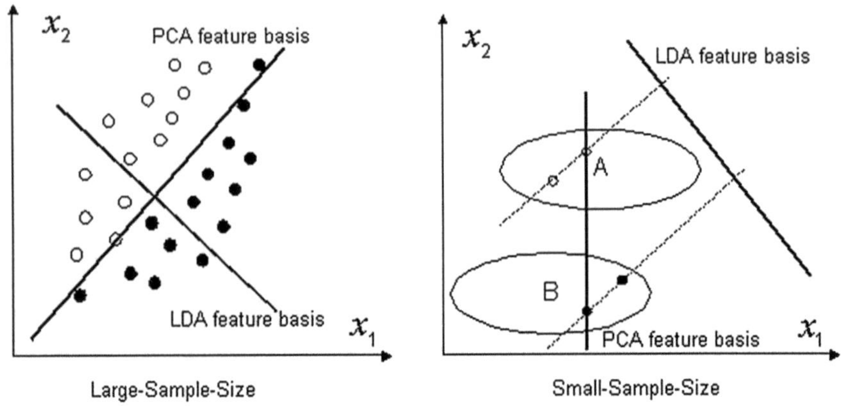

Fig. 2. PCA vs LDA in different learning scenarios. *Left*: given a large size sample of two classes, LDA finds a much better feature basis than PCA for the classification task. *Right*: given a small size sample of two classes, LDA gets over-fitting, and is outperformed by PCA [15]

3.1 The Small-Sample-Size (SSS) Problem

As mentioned in Sect. 1, the so-called Small-Sample-Size (SSS) problem is often introduced when LDA is carried out in some high-dimensional space. Compared to the PCA solution, the LDA solution is much more susceptible to the SSS problem given the same training set, since the latter requires many more training samples than the former due to the increased number of parameters needed to be estimated [33]. Especially when the number of available training samples is less than the dimensionality of the space, the two scatter matrix estimates, $\tilde{\mathbf{S}}_b$ and $\tilde{\mathbf{S}}_w$, are highly ill-posed and singular. As a result, the general belief that LDA is superior to PCA in the context of pattern classification may not be correct in the SSS scenarios [15]. The phenomenon of LDA over-fitting the training data in the SSS settings can be illustrated by a simple example shown in Fig. 2: **Right**, where PCA yields a superior feature basis for the purpose of pattern classification [15].

3.2 Where are the Optimal Discriminant Features?

When $\tilde{\mathbf{S}}_w$ is non-singular, the basis vectors $\tilde{\Psi}$ sought in (6) correspond to the first M most significant eigenvectors of $(\tilde{\mathbf{S}}_w^{-1}\tilde{\mathbf{S}}_b)$, where the "significant" means that the eigenvalues corresponding to these eigenvectors are the first M largest ones. However, due to the SSS problem, often an extremely singular $\tilde{\mathbf{S}}_w$ is generated when $N \ll J$. Let us assume that $\mathcal{A}$ and $\mathcal{B}$ represent the null spaces of $\tilde{\mathbf{S}}_b$ and $\tilde{\mathbf{S}}_w$ respectively, while $\mathcal{A}' = \mathbb{R}^J - \mathcal{A}$ and $\mathcal{B}' = \mathbb{R}^J - \mathcal{B}$ denote the orthogonal complements of $\mathcal{A}$ and $\mathcal{B}$. Traditional approaches attempt to solve the problem by utilizing an intermediate PCA step to remove $\mathcal{A}$ and

$\mathcal{B}$. LDA is then performed in the lower dimensional PCA subspace, as it was done for example in [3, 29]. Nevertheless, it should be noted at this point that the maximum of the ratio in (6) can be reached only when $|\tilde{\Psi}^T\tilde{\mathbf{S}}_w\tilde{\Psi}| = 0$ and $|\tilde{\Psi}^T\tilde{\mathbf{S}}_b\tilde{\Psi}| \neq 0$. This means that the discarded null space $\mathcal{B}$ may contain the most significant discriminatory information. On the other hand, there is no significant information, in terms of the maximization in (6), to be lost if $\mathcal{A}$ is discarded. It is not difficult to see at this point that when $\tilde{\Psi} \in \mathcal{A}$, the ratio $\frac{|\tilde{\Psi}^T\tilde{\mathbf{S}}_b\tilde{\Psi}|}{|\tilde{\Psi}^T\tilde{\mathbf{S}}_w\tilde{\Psi}|}$ drops to its minimum value, 0. Therefore, many researchers consider the intersection space $(\mathcal{A}' \cap \mathcal{B})$ to be spanned by the optimal discriminant feature bases [5, 10].

Based on the above ideas, Yu and Yang proposed the so-called direct LDA (YD-LDA) approach in order to prevent the removal of useful discriminant information contained in the null space $\mathcal{B}$ [34]. However, it has been recently found that the YD-LDA performance may deteriorate rapidly due to two problems that may be encountered when the SSS problem becomes severe [13]. One problem is that the zero eigenvalues of the within-class scatter matrix are used as possible divisors, so that the YD-LDA process can not be carried out. The other is that the worse of the SSS situations may significantly increase the variance in the estimation for the small eigenvalues of the within-class scatter matrix, while the importance of the eigenvectors corresponding to these small eigenvalues is dramatically exaggerated.

The discussions given in these two sections are based on LDA carried out in the sample space $\mathbb{R}^J$. When LDA comes to the feature space $\mathbb{F}$, it is not difficult to see that the SSS problem becomes worse essentially due to the much higher dimensionality. However, GDA, following traditional approach, attempts to solve the problem simply by removing the two null spaces, $\mathcal{A}$ and $\mathcal{B}$. As a result, it can be known from the above analysis that some significant discriminant information may be lost inevitably due to such a process.

4 Regularized Kernel Discriminant Learning (R-KDA)

To address the problems with the GDA and YD-LDA methods in the SSS scenarios, a *regularized kernel discriminant analysis* method, named R-KDA, is developed here.

4.1 A Regularized Fisher's Criterion

To this end, we first introduce a regularized Fisher's criterion [14]. The criterion, which is utilized in this work instead of the conventional one (6), can be expressed as follows:

$$\tilde{\Psi} = \arg\max_{\tilde{\Psi}} \frac{|\tilde{\Psi}^T\tilde{\mathbf{S}}_b\tilde{\Psi}|}{|\eta(\tilde{\Psi}^T\tilde{\mathbf{S}}_b\tilde{\Psi}) + (\tilde{\Psi}^T\tilde{\mathbf{S}}_w\tilde{\Psi})|} \tag{7}$$

where $0 \leq \eta \leq 1$ is a regularization parameter. Although (7) looks different from (6), it can be shown that the modified criterion is exactly equivalent to the conventional one by the following theorem.

Theorem 1. *Let $\mathbb{R}^J$ denote the J-dimensional real space, and suppose that $\forall \psi \in \mathbb{R}^J$, $u(\psi) \geq 0$, $v(\psi) \geq 0$, $u(\psi) + v(\psi) > 0$ and $0 \leq \eta \leq 1$. Let $q_1(\psi) = \frac{u(\psi)}{v(\psi)}$ and $q_2(\psi) = \frac{u(\psi)}{\eta \cdot u(\psi)+v(\psi)}$. Then, $q_1(\psi)$ has the maximum (including positive infinity) at point $\psi^* \in \mathbb{R}^J$ iff $q_2(\psi)$ has the maximum at point ψ^*.*

Proof. Since $u(\psi) \geq 0$, $v(\psi) \geq 0$ and $0 \leq \eta \leq 1$, we have $0 \leq q_1(\psi) \leq +\infty$ and $0 \leq q_2(\psi) \leq \frac{1}{\eta}$.

1. If $\eta = 0$, then $q_1(\psi) = q_2(\psi)$.
2. If $0 < \eta \leq 1$ and $v(\psi) = 0$, then $q_1(\psi) = +\infty$ and $q_2(\psi) = \frac{1}{\eta}$.
3. If $0 < \eta \leq 1$ and $v(\psi) > 0$, then

$$q_2(\psi) = \frac{\frac{u(\psi)}{v(\psi)}}{1 + \eta\frac{u(\psi)}{v(\psi)}} = \frac{q_1(\psi)}{1 + \eta q_1(\psi)} = \frac{1}{\eta}\left(1 - \frac{1}{1 + \eta q_1(\psi)}\right) \tag{8}$$

It can be seen from (8) that $q_2(\psi)$ increases *iff* $q_1(\psi)$ increases.

Combining the above three cases, the theorem is proven.

The regularized Fisher's criterion is a function of the parameter η, which controls the strength of regularization. Within the variation range of η, two extremes should be noted. In one extreme where $\eta = 0$, the modified Fisher's criterion is reduced to the conventional one with no regularization. In contrast with this, strong regularization is introduced in another extreme where $\eta = 1$. In this case, (7) becomes $\tilde{\Psi} = \arg\max_{\tilde{\Psi}} \frac{|\tilde{\Psi}^T \tilde{\mathbf{S}}_b \tilde{\Psi}|}{|(\tilde{\Psi}^T \tilde{\mathbf{S}}_b \tilde{\Psi}) + (\tilde{\Psi}^T \tilde{\mathbf{S}}_w \tilde{\Psi})|}$, which as a variant of the original Fisher's criterion has been also widely used for example in [5, 10, 11, 12]. Among these examples, the method of [12] is a D-LDA variant with $\eta = 1$ (hereafter JD-LDA). The advantages of introducing the regularization will be seen during the development of the R-KDA method proposed in the following sections.

4.2 Eigen-analysis of $\tilde{\mathbf{S}}_b$ in the Feature Space $\mathbb{F}$

Following the D-LDA framework of [11, 12], we start by solving the eigenvalue problem of $\tilde{\mathbf{S}}_b$, which can be rewritten here as follows,

$$\tilde{\mathbf{S}}_b = \sum_{i=1}^{C} \left(\sqrt{\frac{C_i}{N}}\left(\bar{\phi}_i - \bar{\phi}\right)\right)\left(\sqrt{\frac{C_i}{N}}\left(\bar{\phi}_i - \bar{\phi}\right)\right)^T = \sum_{i=1}^{C} \tilde{\bar{\phi}}_i \tilde{\bar{\phi}}_i^T = \tilde{\Phi}_b \tilde{\Phi}_b^T \tag{9}$$

where $\tilde{\bar{\phi}}_i = \sqrt{\frac{C_i}{N}}(\bar{\phi}_i - \bar{\phi})$ and $\tilde{\Phi}_b = [\tilde{\bar{\phi}}_1, \dots, \tilde{\bar{\phi}}_c]$. Since the dimensionality of the feature space $\mathbb{F}$, denoted as J', could be arbitrarily large or possibly infinite, it is intractable to directly compute the eigenvectors of the $(J' \times J')$

matrix $\tilde{\mathbf{S}}_b$. Fortunately, the first m ($\leq C-1$) most significant eigenvectors of $\tilde{\mathbf{S}}_b$, corresponding to non-zero eigenvalues, can be indirectly derived from the eigenvectors of the matrix $\tilde{\Phi}_b^T\tilde{\Phi}_b$ (with size $C\times C$) [11].

To this end, we assume that there exists a kernel function $k(\mathbf{z}_i,\mathbf{z}_j)=\phi(\mathbf{z}_i)\cdot\phi(\mathbf{z}_j)$ for any $\phi(\mathbf{z}_i),\phi(\mathbf{z}_j)\in\mathbb{F}$, and then define an $N\times N$ dot product matrix $\mathbf{K}$,

$$\mathbf{K}=(K_{lh})_{\substack{l=1,\ldots,C\\h=1,\ldots,C}} \quad\text{with } K_{lh}=(k_{ij})_{\substack{i=1,\ldots,C_l\\j=1,\ldots,C_h}} \tag{10}$$

where $k_{ij}=k(\mathbf{z}_{li},\mathbf{z}_{hj})=\phi_{li}\cdot\phi_{hj}$, $\phi_{li}=\phi(\mathbf{z}_{li})$ and $\phi_{hj}=\phi(\mathbf{z}_{hj})$. The matrix $\mathbf{K}$ allows us to express $\tilde{\Phi}_b^T\tilde{\Phi}_b$ as follows [11]:

$$\begin{aligned}\tilde{\Phi}_b^T\tilde{\Phi}_b=\frac{1}{N}\mathbf{B}\cdot\Big(&\mathbf{A}_{NC}^T\cdot\mathbf{K}\cdot\mathbf{A}_{NC}-\frac{1}{N}(\mathbf{A}_{NC}^T\cdot\mathbf{K}\cdot\mathbf{1}_{NC})\\&-\frac{1}{N}(\mathbf{1}_{NC}^T\cdot\mathbf{K}\cdot\mathbf{A}_{NC})+\frac{1}{N^2}(\mathbf{1}_{NC}^T\cdot\mathbf{K}\cdot\mathbf{1}_{NC})\Big)\cdot\mathbf{B}\end{aligned} \tag{11}$$

where $\mathbf{B}=\mathbf{diag}\left[\sqrt{C_1},\ldots,\sqrt{C_c}\right]$, $\mathbf{1}_{NC}$ is an $N\times C$ matrix with terms all equal to one, $\mathbf{A}_{NC}=\mathbf{diag}\left[\mathbf{a}_{c_1},\ldots,\mathbf{a}_{c_c}\right]$ is an $N\times C$ block diagonal matrix, and $\mathbf{a}_{c_i}$ is a $C_i\times 1$ vector with all terms equal to: $\frac{1}{C_i}$.

Let $\tilde{\lambda}_i$ and $\tilde{\mathbf{e}}_i$ ($i=1,\ldots,C$) be the i-th eigenvalue and its corresponding eigenvector of $\tilde{\Phi}_b^T\tilde{\Phi}_b$, sorted in **decreasing** order of the eigenvalues. Since $(\tilde{\Phi}_b\tilde{\Phi}_b^T)(\tilde{\Phi}_b\tilde{\mathbf{e}}_i)=\tilde{\lambda}_i(\tilde{\Phi}_b\tilde{\mathbf{e}}_i)$, $\tilde{\mathbf{v}}_i=\tilde{\Phi}_b\tilde{\mathbf{e}}_i$ is an eigenvector of $\tilde{\mathbf{S}}_b$. In order to remove the null space of $\tilde{\mathbf{S}}_b$, we only use its first m ($\leq C-1$) eigenvectors: $\tilde{\mathbf{V}}=[\tilde{\mathbf{v}}_1,\ldots,\tilde{\mathbf{v}}_m]=\tilde{\Phi}_b\tilde{\mathbf{E}}_m$ with $\tilde{\mathbf{E}}_m=[\tilde{\mathbf{e}}_1,\ldots,\tilde{\mathbf{e}}_m]$, whose corresponding eigenvalues are greater than 0. It is not difficult to see that $\tilde{\mathbf{V}}^T\tilde{\mathbf{S}}_b\tilde{\mathbf{V}}=\tilde{\Lambda}_b$, with $\tilde{\Lambda}_b=\mathbf{diag}[\tilde{\lambda}_1^2,\ldots,\tilde{\lambda}_m^2]$, an $(m\times m)$ diagonal matrix.

4.3 Eigen-analysis of $\tilde{\mathbf{S}}_w$ in the Feature Space $\mathbb{F}$

Let $\tilde{\mathbf{U}}=\tilde{\mathbf{V}}\tilde{\Lambda}_b^{-1/2}$, each column vector of which lies in the feature space $\mathbb{F}$. Projecting both $\tilde{\mathbf{S}}_b$ and $\tilde{\mathbf{S}}_w$ into the subspace spanned by $\tilde{\mathbf{U}}$, it can be easily seen that $\tilde{\mathbf{U}}^T\tilde{\mathbf{S}}_b\tilde{\mathbf{U}}=\mathbf{I}$, an $(m\times m)$ identity matrix, while $\tilde{\mathbf{U}}^T\tilde{\mathbf{S}}_w\tilde{\mathbf{U}}$ can be expanded as:

$$\tilde{\mathbf{U}}^T\tilde{\mathbf{S}}_w\tilde{\mathbf{U}}=(\tilde{\mathbf{E}}_m\tilde{\Lambda}_b^{-1/2})^T\left(\tilde{\Phi}_b^T\tilde{\mathbf{S}}_w\tilde{\Phi}_b\right)\left(\tilde{\mathbf{E}}_m\tilde{\Lambda}_b^{-1/2}\right) \tag{12}$$

Using the kernel matrix $\mathbf{K}$, a closed form expression of $\tilde{\Phi}_b^T\tilde{\mathbf{S}}_w\tilde{\Phi}_b$ can be obtained as follows [11],

$$\begin{aligned}\tilde{\Phi}_b^T\tilde{\mathbf{S}}_w\tilde{\Phi}_b=\frac{1}{N^2}\mathbf{B}\cdot(&\mathbf{A}_{NC}^T\cdot\hat{\mathbf{K}}\cdot\mathbf{A}_{NC}-\frac{1}{N}\left(\mathbf{A}_{NC}^T\cdot\hat{\mathbf{K}}\cdot\mathbf{1}_{NC}\right)\\&-\frac{1}{N}(\mathbf{1}_{NC}^T\cdot\hat{\mathbf{K}}\cdot\mathbf{A}_{NC})+\frac{1}{N^2}\left(\mathbf{1}_{NC}^T\cdot\hat{\mathbf{K}}\cdot\mathbf{1}_{NC})\right)\cdot\mathbf{B}\end{aligned} \tag{13}$$

where $\hat{\mathbf{K}}=\mathbf{K}\cdot(\mathbf{I}-\mathbf{W})\cdot\mathbf{K}$, $\mathbf{W}=\mathbf{diag}[\mathbf{w}_1,\ldots,\mathbf{w}_c]$ is an $N\times N$ block diagonal matrix, and $\mathbf{w}_i$ is a $C_i\times C_i$ matrix with terms all equal to: $\frac{1}{C_i}$.

We proceed by diagonalizing $\tilde{\mathbf{U}}^T\tilde{\mathbf{S}}_w\tilde{\mathbf{U}}$, a tractable matrix with size $m \times m$. Let $\tilde{\mathbf{p}}_i$ be the i-th eigenvector of $\tilde{\mathbf{U}}^T\tilde{\mathbf{S}}_w\tilde{\mathbf{U}}$, where $i = 1, \ldots, m$, sorted in **increasing** order of its corresponding eigenvalue $\tilde{\lambda}'_i$. In the set of ordered eigenvectors, those corresponding to the smallest eigenvalues minimize the denominator of (7), and should be considered the most discriminative features. Let $\tilde{\mathbf{P}}_M = [\tilde{\mathbf{p}}_1, \ldots, \tilde{\mathbf{p}}_M]$ and $\tilde{\Lambda}_w = \mathbf{diag}[\tilde{\lambda}'_1, \ldots, \tilde{\lambda}'_M]$ be the selected $M(\leq m)$ eigenvectors and their corresponding eigenvalues, respectively. Then, the sought solution can be derived through $\tilde{\Gamma} = \tilde{\mathbf{U}}\tilde{\mathbf{P}}_M(\eta\mathbf{I} + \tilde{\Lambda}_w)^{-1/2}$, which is a set of optimal nonlinear discriminant feature bases.

4.4 Dimensionality Reduction and Feature Extraction

For any input pattern $\mathbf{z}$, its projection into the subspace spanned by the set of feature bases, $\tilde{\Gamma}$, derived in Sect. 4.3, can be computed by

$$\mathbf{y} = \tilde{\Gamma}^T\phi(\mathbf{z}) = \left(\tilde{\mathbf{E}}_m \cdot \tilde{\Lambda}_b^{-1/2} \cdot \tilde{\mathbf{P}}_M \cdot (\eta\mathbf{I} + \tilde{\Lambda}_w)^{-1/2}\right)^T \left(\tilde{\Phi}_b^T\phi(\mathbf{z})\right) \tag{14}$$

where $\tilde{\Phi}_b^T\phi(\mathbf{z}) = [\tilde{\bar{\phi}}_1 \cdots \tilde{\bar{\phi}}_c]^T\phi(\mathbf{z})$. We introduce an $(N \times 1)$ kernel vector,

$$\nu(\phi(\mathbf{z})) = \left[\phi_{11}^T\phi(\mathbf{z})\ \phi_{12}^T\phi(\mathbf{z}) \cdots \phi_{c(c_c-1)}^T\phi(\mathbf{z})\ \phi_{cc_c}^T\phi(\mathbf{z})\right]^T, \tag{15}$$

which is obtained by dot products of $\phi(\mathbf{z})$ and each mapped training sample $\phi(\mathbf{z}_{ij})$ in $\mathbb{F}$. Reformulating (14) by using the kernel vector, we obtain

$$\mathbf{y} = \Theta \cdot \nu(\phi(\mathbf{z})) \tag{16}$$

where

$$\Theta = \frac{1}{\sqrt{N}}\left(\tilde{\mathbf{E}}_m \cdot \tilde{\Lambda}_b^{-1/2} \cdot \tilde{\mathbf{P}}_M \cdot (\eta\mathbf{I} + \tilde{\Lambda}_w)^{-1/2}\right)^T \cdot \mathbf{B} \cdot \left(\mathbf{A}_{NC}^T - \frac{1}{N}\mathbf{1}_{NC}^T\right) \tag{17}$$

is an $(M \times N)$ matrix that can be computed off-line. Thus, through (16), a low-dimensional nonlinear representation $(\mathbf{y})$ of $\mathbf{z}$ with enhanced discriminant power has been introduced. The detailed steps to implement the R-KDA method are summarized in Fig. 3.

5 Comments

In this section ,we discuss the main properties and advantages of the proposed R-KDA method.

Firstly, R-KDA effectively deals with the SSS problem in the high-dimensional feature space by employing the regularized Fisher's criterion and the D-LDA subspace technique. It can be seen that R-KDA reduces to kernel YD-LDA and kernel JD-LDA (also called KDDA [11]) when $\eta = 0$ and $\eta = 1$,

Input: A training set $\mathcal{Z}$ with C classes: $\mathcal{Z} = \{\mathcal{Z}_i\}_{i=1}^{C}$, each class containing $\mathcal{Z}_i = \{\mathbf{z}_{ij}\}_{j=1}^{C_i}$ examples, and the regularization parameter η.
Output: The matrix Θ; For an input example $\mathbf{z}$, its R-KDA based feature representation $\mathbf{y}$.
Algorithm:
Step 1. Compute the kernel matrix $\mathbf{K}$ using (10).
Step 2. Compute $\tilde{\Phi}_b^T \tilde{\Phi}_b$ using (11), and find $\tilde{\mathbf{E}}_m$ and $\tilde{\Lambda}_b$ from $\tilde{\Phi}_b^T \tilde{\Phi}_b$ in the way shown in Sect. (4.2).
Step 3. Compute $\tilde{\mathbf{U}}^T \tilde{\mathbf{S}}_w \tilde{\mathbf{U}}$ using (12) and (13), and find $\tilde{\mathbf{P}}_M$ and $\tilde{\Lambda}_w$ from $\tilde{\mathbf{U}}^T \tilde{\mathbf{S}}_w \tilde{\mathbf{U}}$ in the way depicted in Sect. (4.3);
Step 4. Compute Θ using (17).
Step 5. Compute the kernel vector of the input $\mathbf{z}$, $\nu(\phi(\mathbf{z}))$, using (15).
Step 6. The optimal nonlinear discriminant feature representation of $\mathbf{z}$ can be obtained by $\mathbf{y} = \Theta \cdot \nu(\phi(\mathbf{z}))$.

Fig. 3. R-KDA pseudo-code implementation (Matlab code is available by contacting the authors)

respectively. Varying the values of η within $[0, 1]$ leads to a set of intermediate kernel D-LDA variants between kernel YD-LDA and KDDA. Since the subspace spanned by $\tilde{\Psi}$ may contain the intersection space $(\mathbf{A}' \cap \mathbf{B})$, it is possible that there exist zero or very small eigenvalues in $\tilde{\Lambda}_w$, which have been shown to be high variance for estimation in the SSS environments [7]. As a result, any bias arising from the eigenvectors corresponding to these eigenvalues is dramatically exaggerated due to the normalization process $(\tilde{\mathbf{P}}_M \tilde{\Lambda}_w^{-1/2})$. Against the effect, the introduction of the regularization helps to decrease the importance of these highly unstable eigenvectors, thereby reducing the overall variance. Also, there may exist the zero eigenvalues in $\tilde{\Lambda}_w$, which are used as divisors in YD-LDA due to $\eta = 0$. However, it is not difficult to see that the problem can be avoided in the R-KDA solution, $\tilde{\Psi} = \tilde{\mathbf{U}}\tilde{\mathbf{P}}_M(\eta\mathbf{I} + \tilde{\Lambda}_w)^{-1/2}$, simply by setting the parameter $\eta > 0$. In this way, R-KDA can exactly extract the optimal discriminant features from both inside and outside of $\tilde{\mathbf{S}}_w$'s null space, while avoiding the risk of experiencing high variance in estimating the scatter matrices at the same time. This point makes R-KDA significantly different from existing nonlinear discriminant analysis methods such as GDA in the SSS situations.

In GDA, to remove the null space of $\tilde{\mathbf{S}}_w$, it is required to compute the pseudo inverse of the kernel matrix $\mathbf{K}$, which could be extremely ill-conditioned when certain kernels or kernel parameters are used. Pseudo inversion is based on inversion of the nonzero eigenvalues. Due to round-off errors, it is not easy to identify the true null eigenvalues. As a result, numerical stability problems often occur [22]. However, it can be seen from the derivation of R-KDA that such problems are avoided in R-KDA. The improvement can be observed also in experimental results reported in Figs. 8–9: **Left**.

In GDA, both the two eigen-decompositions of $\tilde{\mathbf{S}}_b$ and $\tilde{\mathbf{S}}_w$ have to be implemented in the feature space $\mathbb{F}$. In contrast with this, it can be seen from Sect. 4.3 that the eigen-decomposition of $\tilde{\mathbf{S}}_w$ is replaced by that of $\tilde{\mathbf{U}}^T \tilde{\mathbf{S}}_w \tilde{\mathbf{U}}$, which is an $(m \times m)$ matrix with $m \leq C - 1$. Also, it should be noted at this point that it generally requires much more computational costs to implement an eigen-decomposition for $\tilde{\mathbf{S}}_w$ than $\tilde{\mathbf{S}}_b$, due to $C \ll N$ in most cases. Therefore, based on the two factors, it is not difficult to see that the computational complexity of R-KDA is significantly reduced compared to GDA. This point is demonstrated by the face recognition experiment reported in Sect. 6.2, where R-KDA is approximately 20 times faster than GDA.

6 Experimental Results

Two sets of experiments are included here to illustrate the effectiveness of the R-KDA method in different learning scenarios. The first experiment is conducted on Fisher's iris data [6] to assess the performance of R-KDA in traditional large-sample-size situations. Then, R-KDA is applied to face recognition tasks in the second experiment, where various SSS settings are introduced. In addition to R-KDA, other two kernel-based feature extraction methods, KPCA and GDA, are implemented to provide a comparison of performance, in terms of classification error and computational cost.

6.1 Fisher's Iris Data

The iris flower data set originally comes from Fisher's work [6]. The set consists of $N = 150$ iris specimens of $C = 3$ species (classes). Each specimen is represented by a four-dimensional vector, describing four parameters, sepal length/width and petal length/width. Among the three classes, one is linearly separable from the other two, while the latter are not linearly separable from each other. Due to $J(= 4) \ll N$, there is no SSS problem introduced in this case, and thus we set $\eta = 0.001$ for R-KDA.

Firstly, it is of interest to observe how R-KDA linearizes and simplifies the complicated data distribution as GDA did in [2]. To this end, four types of feature bases are generalized from the iris set by utilizing the LDA, KPCA, R-KDA and GDA algorithms, respectively. These feature bases form four subspaces, accordingly. Then, all the examples are projected to the four subspaces. For each example, its projections in the first two most significant feature bases of each subspace are visualized in Fig. 4. As analyzed in Sect. 2.3, the PCA-based features are optimized with focus on object reconstruction. Not surprisingly, it can be seen from Fig. 4 that the subjects are not separable in the KPCA subspace, even with the introduction of nonlinear kernel. Unlike the PCA approaches, LDA optimizes the feature representation based on separability criteria. However, subject to the limitation of linearity, the two non-separable classes remain non-separable in the LDA subspace. In contrast

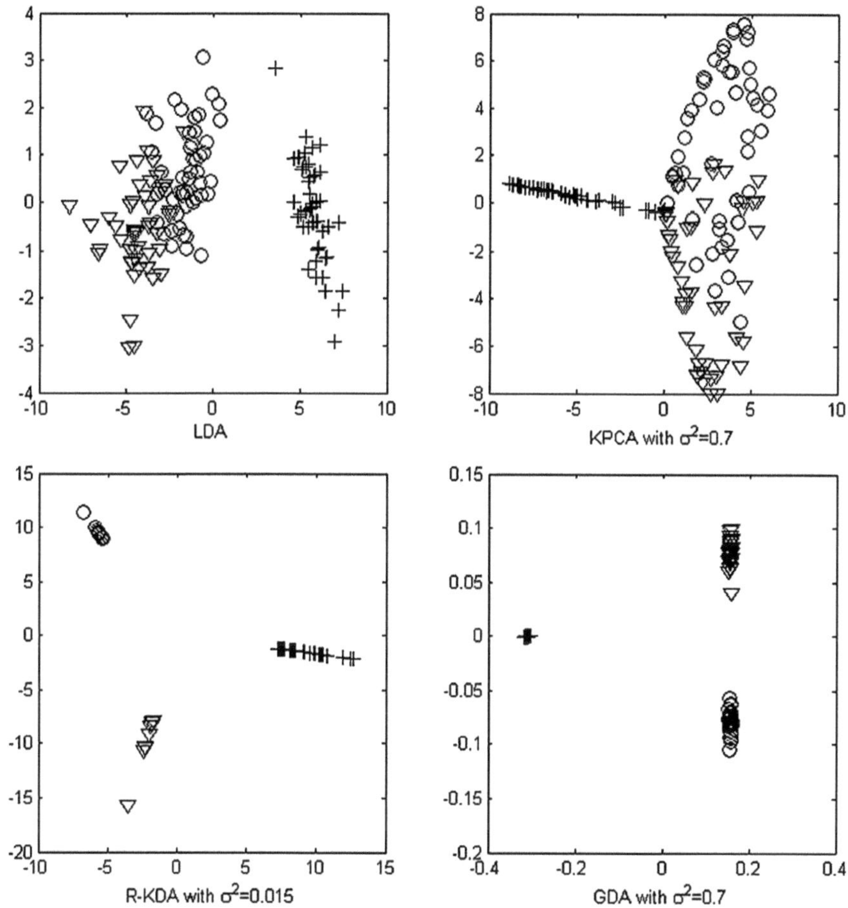

Fig. 4. Iris data are project to four feature spaces obtained by LDA, KPCA, R-KDA and GDA respectively. LDA is derived from R-KDA by using a polynomial kernel with degree one, while all other three kernel methods use a RBF kernel

to this, we can see the linearization property in the R-KDA and GDA subspaces, where all of classes are well linearly separable when a RBF kernel with appropriate parameters is used.

Also, we examine the classification error rate (CER) of the three kernel feature extraction algorithms compared here with the so-called "leave one out" test method. Following the recommendation in [2], a RBF kernel with $\sigma^2 = 0.7$ is used for all these algorithms in this experiment. The CERs obtained by GDA and R-KDA are only 7.33% and 6% respectively, while the CER of KPCA with the same feature number ($M = 2$) to the formers goes up to 20%. The two experiments conducted on the iris data indicate that the performance of

R-KDA is comparable to that of GDA in the large-sample-size learning scenarios, although the former is designed specifically to address the SSS problem.

6.2 Face Recognition

Face Recognition Evaluation Design

Face recognition is one of current most challenging applications in the pattern recognition literature [4, 23, 30, 31, 35]. In this work, the algorithms are evaluated with two widely used face databases, UMIST [8] and FERET [19]. The UMIST repository is a multi-view database, consisting of 575 images of 20 people, each covering a wide range of poses from profile to frontal views [8]. The FERET database has been considered current most comprehensive and representative face database [19, 20]. For the convenience of preprocessing, we only choose a medium-size subset of the database. The subset consists of 1147 images of 120 people, each one having at least 6 samples so that we can generalize a set of SSS learning tasks. These images cover a wide range of variations in illumination and facial expression/details with pose angles less than 30 degrees. Figs. 5–6 depict some examples from the two databases. For computational convenience, each image is represented as a column vector of length $J = 10304$ for UMIST and $J = 17154$ for FERET.

The SSS problem is defined in terms of the number of available training samples per subject, L. Thus the value of L has a significant influence on the required strength of regularization. To study the sensitivity of the performance, in terms of *correct recognition rate* (CRR), to L, five tests were performed with various L values ranging from $L = 2$ to $L = 6$. For a particular L, any database evaluated here is randomly partitioned into two subsets: a training set and a test set. The training set is composed of $(L \times C)$ samples: L images per person were randomly chosen. The remaining $(N - L \times C)$ images were used to form the test set. There is no overlapping between the two subsets. To enhance the accuracy of the assessment, five runs of such a partition were executed, and all the results reported below have been averaged over the five runs.

Fig. 5. Some samples of four people come from the UMIST database

Fig. 6. Some samples of eight people come from the normalized FERET database

CRRs with Varying Regularization Parameter

In this experiment, we examine the performance of R-KDA with varying regularization parameter values in different SSS scenarios, $L = 2 \sim 4$. For simplicity, R-KDA is only tested with a linear polynomial kernel in the FERET subset. Figure 7 depicts the obtained CRRs as a function of (M, η), where M is the number of feature vectors used.

The parameter η controls the strength of regularization, which balances the tradeoff between variance and bias in the estimation for the zero or small eigenvalues of the within-class scatter matrix. Varying the η values within $[0, 1]$ leads to a set of intermediate kernel D-LDA variants between kernel YD-LDA and KDDA. In theory, kernel YD-LDA with no bias introduced should be the best performer among these variants if sufficient training samples are available. It can be observed at this point from Fig. 7 that the CRR peaks gradually moved from the right side toward the left side ($\eta = 0$) that is the case of kernel YD-LDA as L increases. Small values of η have been good enough for the regularization requirement in many cases ($L \geq 4$) as shown in Fig. 7. However, it also can be seen that kernel YD-LDA performed poorly when $L = 2, 3$. This should be attributed to the high variance in the estimate of $\tilde{\mathbf{S}}_w$ due to insufficient training samples. In these cases, even $\tilde{\mathbf{U}}^T \tilde{\mathbf{S}}_w \tilde{\mathbf{U}}$ is singular or close to singular, and the resulting effect is to dramatically exaggerate the importance associated with the eigenvectors corresponding to the smallest eigenvalues. Against the effect, the introduction of regularization helps to decrease the larger eigenvalues and increase the smaller ones, thereby counteracting for some extent the bias. This is also why KDDA outperforms kernel YD-LDA when L is small.

Performance Comparison with KPCA and GDA

This experiment compares the performance of the R-KDA algorithms, in terms of the CRR and the computational cost, to the KPCA and GDA algorithms. For simplicity, only the RBF kernel is tested in this work, and the classification is performed with the nearest neighbor rule.

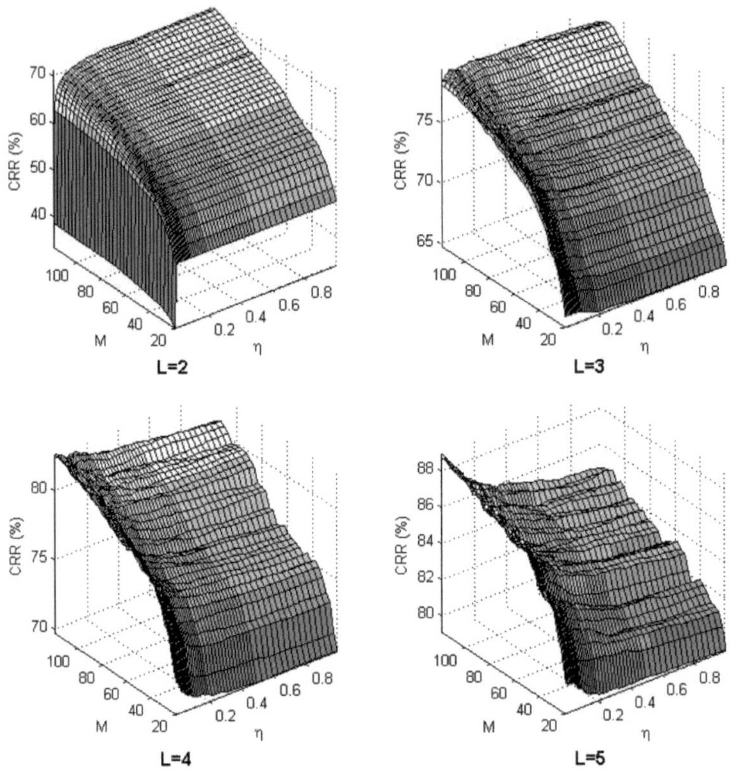

Fig. 7. CRRs(M, η) obtained by R-KDA with a linear polynomial kernel

Tables 2–3 depict a quantitative comparison of the best CRRs with corresponding parameter values (σ^{2*}, M^*), found by the three methods in the UMIST and FERET databases, each one having introduced five SSS cases from $L = 2$ to $L = 6$. In addition to σ^2 and M, R-KDA's performance is affected by the regularization parameter, η. Considering the high computational cost of searching the best η^*, we simply set $\eta = 1.0$ for the $L = 2$ cases and

Table 2. Comparison of the best found CRRs (%) with corresponding parameter values in the UMIST database

Methods	KPCA			GDA			R-KDA			
	CRR	σ^{2*}	M^*	CRR	σ^*	M^*	CRR	σ^*	M^*	η
$L = 2$	57.91	2.11×10^7	34	62.92	1.34×10^8	19	66.73	1.5×10^8	14	1.0
$L = 3$	69.67	5.33×10^7	58	76.00	3.72×10^7	18	80.97	1.5×10^8	14	0.001
$L = 4$	78.02	6.94×10^7	78	84.20	5.33×10^7	19	89.17	1.5×10^8	11	0.001
$L = 5$	84.67	2.11×10^7	95	90.32	5.33×10^7	19	93.01	1.34×10^8	13	0.001
$L = 6$	87.91	6.94×10^7	119	92.97	6.94×10^7	19	95.30	1.5×10^8	14	0.001

Table 3. Comparison of the best found CRRs (%) with corresponding parameter values in the FERET database

Methods	KPCA			GDA			R-KDA			
	CRR	σ^{2*}	M^*	CRR	σ^*	M^*	CRR	σ^*	M^*	η
$L = 2$	60.93	2.34×10^5	238	71.18	2.68×10^4	118	73.38	3.0×10^5	102	1.0
$L = 3$	67.32	7.44×10^3	358	80.58	2.68×10^4	118	85.51	3.0×10^5	106	0.001
$L = 4$	71.39	2.34×10^5	468	85.07	2.68×10^4	118	88.34	3.0×10^5	108	0.001
$L = 5$	75.32	2.03×10^4	590	88.48	2.68×10^4	118	91.96	2.34×10^5	104	0.001
$L = 6$	77.85	2.03×10^4	716	90.21	2.03×10^4	118	92.74	3.0×10^5	110	0.001

$\eta = 0.001$ for other cases based on the observation and analysis of the results in Sect. 6.2. Also, the CRRs as a function of σ^2 and M respectively in several representative UMIST cases are shown in Figs. 8–9. From these results, it can be seen that R-KDA is the top performer in all the experimental cases. On

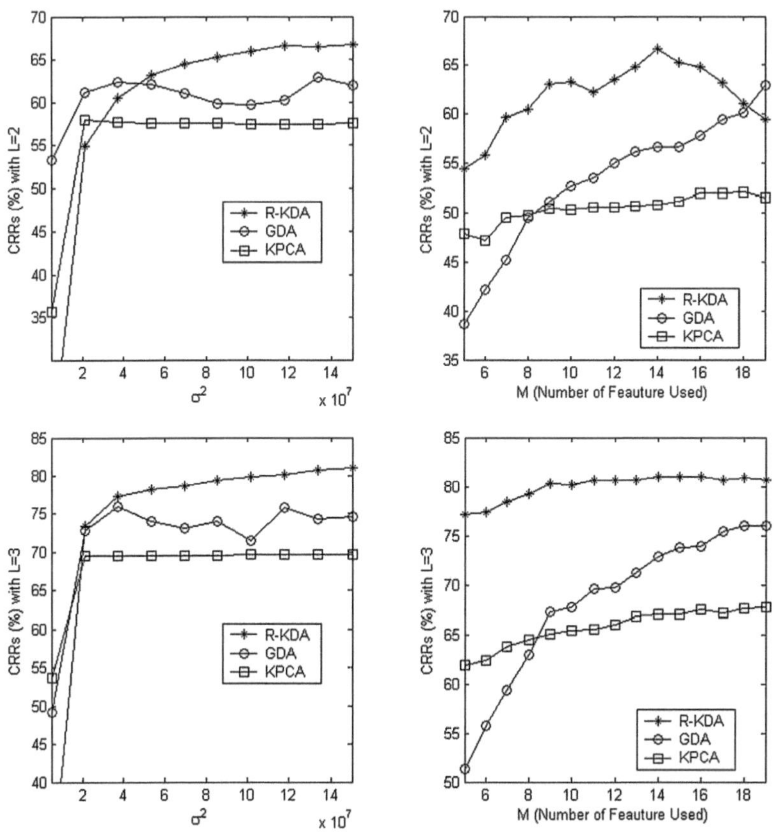

Fig. 8. A comparison of CRRs based on the RBF kernel function in the UMIST cases of $L = 2 \sim 3$. *Left*: CRRs as a function of σ^2 with the best found M^*. *Right*: CRRs as a function of M with the best found σ^{2*}

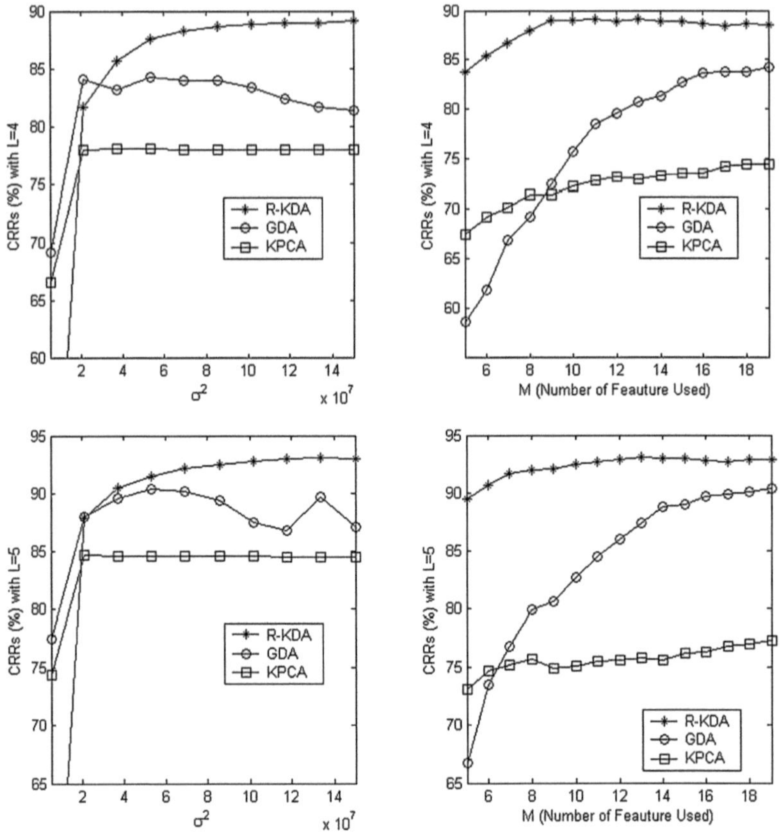

Fig. 9. A comparison of CRRs based on the RBF kernel function in the UMIST cases of $L = 4 \sim 5$. *Left*: CRRs as a function of σ^2 with the best found M^*. *Right*: CRRs as a function of M with the best found σ^{2*}

average, R-KDA leads KPCA and GDA up to 9.4% and 3.8% in the UMIST database, and 15.8% and 3.3% in the FERET database. It should be also noted that Figs. 8–9: **Left** reveal the numerical stability problems existing in practical implementations of GDA. Comparing GDA to R-KDA, we can see that the later is more stable and predictable, resulting in a cost-effective determination of parameter values during the training phase.

In addition to the CRR, it is of interest to compare the performance with respect to the computational complexity. For each of the methods evaluated here, the simulation process consists of (1) a training stage that includes all operations performed in the training set; (2) a test stage for the CRR determination. The computational times consumed by these methods with the parameter configuration depicted in Tables 2–3 are reported in Table 4. T_{trn} and T_{tst} are the amounts of time spent on training and testing respectively. The simulation studies reported in this work were implemented on a personal

Table 4. A comparison of computational times, $T_{trn} + T_{tst}$ (Seconds)

DBS	Methods	$L = 2$	$L = 3$	$L = 4$	$L = 5$	$L = 6$
UMIST	KPCA	0.8 + 11.3	2.1 + 12.0	4.5 + 16.9	7.3 + 25.2	8.5 + 19.5
	GDA	6.2 + 44.9	14.2 + 65.1	25.7 + 83.9	40.7 + 101.1	55.9 + 109.0
	R-KDA	0.3 + 2.2	0.7 + 3.2	1.2 + 4.0	2.0 + 4.7	2.8 + 5.4
FERET	KPCA	76 + 203	134 + 205	320 + 323	375 + 254	526 + 245
	GDA	392 + 750	905 + 1014	1641 + 1156	2662 + 1198	3861 + 1121
	R-KDA	19 + 38	42 + 50	76 + 58	117 + 57	170 + 57

Table 5. A comparison of the computational time of KPCA or GDA over that of R-KDA, $\xi_{trn} + \xi_{tst}$

DBS	Methods	$L = 2$	$L = 3$	$L = 4$	$L = 5$	$L = 6$	Aver.
UMIST	KPCA	2.6+5.1	2.9+3.8	3.6+4.2	3.8+5.4	3.0+3.6	3.2+4.4
	GDA	19.4+20.2	20.0+20.4	20.6+21.1	20.8+21.5	20.1+20.0	20.2+20.6
FERET	KPCA	4.0+5.3	3.2+4.1	4.2+5.5	3.2+4.4	3.1+4.3	3.5+4.7
	GDA	20.8+19.6	21.5+20.2	21.6+19.8	22.7+20.9	22.7+19.6	21.9+20.0

computer system equipped with a 2.0 GHz Intel Pentium 4 processor and 1.0 GB RAM. All programs are written in Matlab v6.5 and executed in MS Windows 2000. For the convenience of comparison, we introduce a quantitative statistic in Table 5 regarding the computational time of KPCA or GDA over that of R-KDA, $\xi_{trn}(\cdot) = T_{trn}(\cdot)/T_{trn}(\text{R-KDA})$ and $\xi_{tst}(\cdot) = T_{tst}(\cdot)/T_{tst}(\text{R-KDA})$. As analyzed in Sect. 5, the computational cost of R-KDA should be less than that of GDA. It can be observed clearly at this point from Table 5 that R-KDA is approximately 20 times faster than GDA in both the training and test phases. Moreover, R-KDA is more than 3 times in training and 4 times in testing faster than KPCA. The higher computational complexity of KPCA is due to the significantly larger feature number used, M^* as shown in Tables 2–3. The advantage of R-KDA in computation is particularly important for the practical face recognition tasks, where algorithms are often required to deal with huge scale databases.

7 Conclusion

Due to the extremely high dimensionality of the kernel feature spaces, the SSS problem is often encountered when traditional kernel discriminant analysis methods are applied to many practical tasks such as face recognition. To address the problem, a regularized kernel discriminant analysis method is introduced in this chapter. The proposed method is based a novel regularized Fisher's discriminant criterion, which is particularly robust against the SSS problem compared to the original one used in traditional linear/kernel discriminant analysis methods. It has been also shown that a series of traditional LDA variants and their kernel versions including the recently introduced YD-LDA,

JD-LDA and KDDA can be derived from the proposed framework by adjusting the regularization and kernel parameters. Experimental results obtained in the face recognition tasks indicate that the CRR performance of the proposed R-KDA algorithm is overall superior to those obtained by the KPCA or GDA approaches in various SSS situations. Also, the R-KDA method has significantly less computational complexity than the GDA method. This point has been demonstrated in the face recognition experiments, where R-KDA is approximately 20 times faster than GDA in both the training and test phases.

In conclusion, the R-KDA algorithm provides a general pattern recognition framework for nonlinear feature extraction from high-dimensional input patterns in the SSS situations. We expect that in addition to face recognition, R-KDA will provide excellent performance in applications where classification tasks are routinely performed, such as content-based image indexing and retrieval, video and audio classification.

Acknowledgements

Portions of the research in this dissertation use the FERET database of facial images collected under the FERET program [19]. We would like to thank the FERET Technical Agent, the U.S. National Institute of Standards and Technology (NIST) for providing the FERET database. Also, We would like to thank Dr. Daniel Graham and Dr. Nigel Allinson for providing the UMIST face database [8].

References

1. Aizerman, M. A., Braverman, E. M., Rozonoér, L. I. (1964) Theoretical foundations of the potential function method in pattern recognition learning. *Automation and Remote Control*, 25:821–837.
2. Baudat, G., Anouar, F. (2000) Generalized discriminant analysis using a kernel approach. *Neural Computation*, 12:2385–2404.
3. Belhumeur, P. N., Hespanha, J. P., Kriegman, D. J. (1997) Eigenfaces vs. Fisherfaces: recognition using class specific linear projection. *IEEE Transactions on Pattern Analysis and Machine Intelligence*, 19(7):711–720.
4. Chellappa, R., Wilson, C., Sirohey, S. (1995) Human and machine recognition of faces: A survey. *The Proceedings of the IEEE*, 83:705–740.
5. Chen, L.-F., Liao, H.-Y. M., Ko, M.-T., Lin, J.-C., Yu., G.-J. (2000) A new LDA-based face recognition system which can solve the small sample size problem. *Pattern Recognition*, 33:1713–1726.
6. Fisher, R. (1936) The use of multiple measures in taxonomic problems. *Ann. Eugenics*, 7:179–188.
7. Friedman, J. H. (1989) Regularized discriminant analysis. *Journal of the American Statistical Association*, 84:165–175.

8. Graham, D. B., Allinson, N. M. (1998) Characterizing virtual eigensignatures for general purpose face recognition. In Wechsler, H., Phillips, P. J., Bruce, V., Soulie, F. F., Huang, T. S., editors, *Face Recognition: From Theory to Applications, NATO ASI Series F, Computer and Systems Sciences*, 163:446–456.
9. Kanal, L., Chandrasekaran, B. (1971) On dimensionality and sample size in statistical pattern classification. *Pattern Recognition*, 3:238–255.
10. Liu, K., Cheng, Y., Yang, J., Liu, X. (1992) An efficient algorithm for foley-sammon optimal set of discriminant vectors by algebraic method. *Int. J. Pattern Recog. Artif. Intell.*, 6:817–829.
11. Lu, J., Plataniotis, K., Venetsanopoulos, A. (2003) Face recognition using kernel direct discriminant analysis algorithms. *IEEE Transactions on Neural Networks*, 14(1):117–126.
12. Lu, J., Plataniotis, K., Venetsanopoulos, A. (2003) Face recognition using LDA based algorithms. *IEEE Transactions on Neural Networks*, 14(1):195–200.
13. Lu, J., Plataniotis, K., Venetsanopoulos, A. (2003) Regularized discriminant analysis for the small sample size problem in face recognition. *Pattern Recognition Letter*, 24(16):3079–3087, December.
14. Lu, J., Plataniotis, K., Venetsanopoulos, A. (2005) Regularization studies of linear discriminant analysis in small sample size scenarios with application to face recognition. *Pattern Recognition Letter*, 26(2):181–191 January 2005.
15. Martínez, A. M., Kak, A. C. (2001) PCA versus LDA. *IEEE Transactions on Pattern Analysis and Machine Intelligence*, 23(2):228–233.
16. McLachlan, G. (1992) *Discriminant Analysis and Statistical Pattern Recognition.* Wiley, New York.
17. Mercer, J. (1909) Functions of positive and negative type and their connection with the theory of integral equations. *Philos. Trans. Roy. Soc. London*, A 209:415–446.
18. Müller, K. R., Mika, S., Rätsch, G., Tsuda, K., Schölkopf, B. (2001) An introduction to kernel-based learning algorithms. *IEEE Transactions on Neural Networks*, 12(2):181–201, March.
19. Phillips, P. J., Moon, H., Rizvi, S. A., Rauss, P. J. (2000) The FERET evaluation methodology for face-recognition algorithms. *IEEE Transactions on Pattern Analysis and Machine Intelligence*, 22(10):1090–1104.
20. Phillips, P. J., Wechsler, H., Huang, J., Rauss, P. (1998) The FERET database and evaluation procedure for face recognition algorithms. *Image and Vision Computing J*, 16(5):295–306.
21. Raudys, S. J., Jain, A. K. (1991) Small sample size effects in statistical pattern recognition: Recommendations for practitioners. *IEEE Transactions on Pattern Analysis and Machine Intelligence*, 13(3):252–264.
22. Ruiz, A., Teruel, P. L. de. (2001) Nonlinear kernel-based statistical pattern analysis. *IEEE Transactions on Neural Networks*, 12(1):16–32, January.
23. Samal, A., Iyengar, P. A. (1992) Automatic recognition and analysis of human faces and facial expressions: A survey. *Pattern Recognition*, 25:65–77.
24. Schölkopf, B. (1997) *Support Vector Learning.* Oldenbourg-Verlag, Munich, Germany.
25. Schölkopf, B., Burges, C., Smola, A. J. (1999) *Advances in Kernel Methods – Support Vector Learning.* MIT Press, Cambridge, MA.
26. Schölkopf, B., Smola, A., Müller, K. R. (1999) Nonlinear component analysis as a kernel eigenvalue problem. *Neural Computation*, 10:1299–1319.

27. Schölkopf, B., Smola, A. J. (2001) *Learning with Kernels.* MA: MIT Press, Cambridge.
28. Smola, A. J., Schölkopf, B., Müller, K. R. (1998) The connection between regularization operators and support vector kernels. *Neural Networks*, 11:637–649.
29. Swets, D. L., Weng, J. (1996) Using discriminant eigenfeatures for image retrieval. *IEEE Transactions on Pattern Analysis and Machine Intelligence*, 18:831–836.
30. Turk, M. (2001) A random walk through eigenspace. *IEICE Trans. Inf. & Syst.*, E84-D(12):1586–1695, December.
31. Valentin, D., Alice, H. A., Toole, J. O., Cottrell, G. W. (1994) Connectionist models of face processing: A survey. *Pattern Recognition*, 27(9):1209–1230.
32. Vapnik, V. N. (1995) *The Nature of Statistical Learning Theory.* Springer-Verlag, New York.
33. Wald, P., Kronmal, R. (1977) Discriminant functions when covariance are unequal and sample sizes are moderate. *Biometrics*, 33:479–484.
34. Yu, H., Yang, J. (2001) A direct LDA algorithm for high-dimensional data – with application to face recognition. *Pattern Recognition*, 34:2067–2070.
35. Zhao, W., Chellappa, R., Phillips, P., Rosenfeld, A. (2003) Face recognition: A literature survey. *ACM Computing Surveys*, 35(4):399–458, December.
36. Zien, A., Rätsch, G., Mika, S., Schölkopf, B., Lengauer, T., Müller, K.-R. (2000) Engineering support vector machine kernels that recognize translation initiation sites in dna. *Bioinformatics*, 16:799–807.

Fast Color Texture-Based Object Detection in Images: Application to License Plate Localization

K.I. Kim[1], K. Jung[2], and H.J. Kim[3]

[1] A. I. Lab, Korea Advanced Institute of Science and Technology, Taejon, Korea
[2] School of Media, Soongsil University, Seoul, Korea
[3] Computer Engineering Dept., Kyungpook National University, Taegoo, Korea

Abstract. The current chapter presents a color texture-based method for object detection in images. A support vector machine (SVM) is used to classify each pixel in the image into object of interest and background based on localized color texture patterns. The main problem in this approach is high run-time complexity of SVMs. To alleviate this problem, two methods are proposed. Firstly, an artificial neural network (ANN) is adopted to make the problem linearly separable. Training an ANN on a given problem to achieve low training error and taking up to the last hidden layer replaces the kernel map in nonlinear SVMs, which is a major computational burden in SVMs. As such, the resulting color texture analyzer is embedded in the continuously adaptive mean shift algorithm (CAMShift), which then automatically identifies regions of interest in a coarse-to-fine manner. Consequently, the combination of CAMShift and SVMs produces robust and efficient object detection, as time-consuming color texture analyses of less relevant pixels are restricted, leaving only a small part of the input image to be analyzed. To demonstrate the validity of the proposed technique, a vehicle license plate (LP) localization system is developed and experiments conducted with a variety of images.

Key words: object detection, license plate recognition, color texture classification, support vector machines, neural networks

1 Introduction

The detection of objects in a complex background is a well studied, but unresolved problem. A number of different techniques for detecting objects in a complex background have already been proposed [14, 19, 22, 29], however, the problem has not yet been completely solved. Typically, the challenge of object detection is an equivalent problem of segmenting a given image into object regions or background regions, based on discriminating between associated

K.I. Kim, K. Jung, and H.J. Kim: *Fast Color Texture-based Object Detection in Images: Application to License Plate Localization*, StudFuzz **177**, 297–320 (2005)
www.springerlink.com

image features. Therefore, the success of a detection method is inextricably tied to the types of features used and the reliability with which these features are extracted and classified [14]. The features commonly used for this kind of task include color, texture, shape, and various combinations thereof, where color texture is one of the best candidates, as outdoor scenes are rich in color and texture.

Accordingly, the current chapter presents a generic framework for object detection based on color texture. The framework is demonstrated on, and in part motivated by, the task of vehicle license plate (LP) detection. As such, a Korean vehicle LP localization system is developed that can identify LPs with an arbitrary size and perspective under moderate amounts of change in the illumination. Since the underlying technique is fairly general, it can also be used for detecting objects in other problem domains, where the object of interest may not be perfectly planar or rigid. Consequently, as regards the problem of object detection, LPs can henceforth be regarded as a general class of objects of interest. The following outlines the challenges involved in LP detection and gives a brief overview of previous work related to the objectives of the present study.

1.1 License Plate Detection Problem

LP detection is interesting because it is usually the first step of an automatic LP recognition system with possible applications in traffic surveillance systems, automated parking lots, etc.

Korean LPs are a rectangular plate with two rows of white characters embossed on a green background. Although a human observer can effectively locate plates in any background, automatic LP detection is a challenging problem, as LPs can have a significantly variable appearance in an image:

1. Variations in shape due to distortion of the plate and differences in the characters and digits embossed on the plate (Fig. 1a)
2. Variations in luminance due to similar yet definitely different surface reflectance properties, changes in the illumination conditions, haze, dust on the plate, blurring during image acquisition, etc. (Fig. 1a).
3. Size variations and perspective distortions caused by a change in the sensor (camera) placement relative to the vehicle (Fig. 1b).

1.2 Previous Work

While comprehensive surveys of image segmentation and object detection in various application domains can be found in [9, 14, 19, 22, 29], the current section focuses on an overview of LP detection methods. A number of approaches have already been proposed for developing LP detection systems. For example, color (gray level)-based methods utilize the fact that LPs (when disregarding the embossed characters) in images often exhibit a unique and

(a)

(b)

Fig. 1. Example LP images with different zoom, perspective, and various other imaging conditions

homogenous color (gray level). In this case, an input image is segmented according to the color (gray level) homogeneity, then the color or shape of each segment is analyzed. Kim et al. [15] adopt genetic algorithms (GAs) for color segmentation and search for green rectangular regions as LPs. To make the system insensitive to noise and variations in the illumination conditions, the consistency of labeling between neighboring pixels is emphasized during the color segmentation. Lee et al. [18] utilize artificial neural networks (ANNs) to estimate the surface color of LPs from given samples of LP images. All the pixels in the image are filtered by an ANN and the *greenness* of each pixel calculated. An LP region is then identified by verifying a green rectangular region based on structural features. Crucial to the success of color (or gray level)-based methods is the color (gray level) segmentation stage. However, currently available solutions do not provide a high degree of accuracy in outdoor scenes, as color values are often affected by illumination.

In contrast, edge-based methods use the contrast between the characters on the LP and their background (plate region) as related to gray levels. As such, LPs are found by searching for regions with such a high contrast. Draghici [8] searches for regions with a high edge magnitude and then verifies them by examining the presence of rectangular boundaries. In [10], Gao and Zhou compute the gradient magnitude and local variance in an image. Then, those regions with a high edge magnitude and high edge variance are identified as LP regions. Although efficient and effective in simple images, edge-based methods cannot be applied to a complex image, where a background region can also show a high edge magnitude or variance.

Based on the assumption that an LP region consists of dark characters on a light background, Cui and Huang [7] apply spatial thresholding to an input image based on a Markov random field (MRF), then detect characters (LPs) according to the spatial edge variances. Similarly, Naito et al. [20] also apply adaptive thresholding to an image, segment the resulting binary image using priori knowledge of the character sizes in LPs, and then detect a string of characters based on the geometrical property (character arrangement) of LPs. While the reported results with clean images are promising, the performance of these methods might be degraded when the LPs are partially shaded or stained, as shown in Fig. 1a, because binarization may not correctly separate the characters from the background.

Another type of approach stems from the well-known method of (color) texture analysis. Park et al. [23] adopt an ANN to analyze the color textural properties of horizontal and vertical cross-sections of LPs in an image and perform a projection profile analysis on the classification result to generate LP bounding boxes. Barroso et al. [1] utilize the textural properties of LPs occurring at a horizontal cross-section (referred to as *signature* in [1]). In [3], Brugge et al. utilize discrete time cellular ANNs (DT-CNNs) for analyzing the textural properties of LPs. They also attempt to combine a texture-based method with an edge-based method. Texture-based methods are known to perform well even with noisy or degraded LPs and be relatively insensitive to variations in the illumination conditions, however, they are also time-consuming, as texture classification is inherently computationally dense.

1.3 Overview of Present Work

The current work proposes a color texture-based method for LP detection in images. The following areas are crucial to the success of color texture-based LP detection: construction of (1) a classifier that can discriminate between the color textures associated with different classes (LP and non-LP) and (2) an object segmentation module that can operate on the classification results obtained from (1). In the proposed method, a support vector machine (SVM) is used as the color texture classifier. SVMs are a natural choice due to their robustness, even with a lack of training examples. The previous successful application of SVMs in texture classification [16] and other related problems [17, 21] also provided further motivation to use SVMs as the classifier for identifying object regions. The main problem of SVM, however, is its high runtime complexity which is mainly caused by expanding the solution in terms of *kernel map*. Here we propose to use an ANN as a replacement of kernel map. Training an ANN on a given problem to achieve low training error and taking up to the last hidden layer makes the given problem linearly separable and enables application of linear SVMs. To avoid the resulting overfitting of the ANN *feature extractor*, optimal brain surgeon (OBS) algorithm is adopted as an implicit regularization.

After classification, an LP score image is generated where each pixel represents the probability of the corresponding pixel in the input image being part of an LP. The segmentation module then identifies the bounding boxes of LPs by applying the continuously adaptive mean shift algorithm (CAMShift), the effectiveness of which has already been demonstrated in face detection [2]. This combination of CAMShift and SVMs produces robust and efficient LP detection through restricting the color texture classification of less relevant pixels. Accordingly, only a small part of the input image is actually texture-analyzed. In addition to producing a better detection performance than existing techniques, the processing time for the proposed method is also much shorter (average 1.1 sec. for 320 × 240-sized image).

2 Color Texture-based Object Detection

The proposed method considers LP detection as a color texture classification, where problem-specific knowledge is available prior to classification. Since this knowledge (the number and type of color textures) is often available in the form of example patterns, the classification can be supervised. An SVM as a trainable classifier is adopted for this task. Specifically, the system uses a small window to scan an input image, then classifies the pixel located at the center of the window into *plate* or *non-plate* (*background*) by analyzing its color and texture properties using an SVM. To facilitate the detection of LPs on different scales, a pyramid of images is generated from the original image by gradually changing the resolution at each level. The classification results are hypothesized at each level and then fused to the original scale. To reduce the processing time, CAMShift is adopted as a mechanism for automatically selecting the region of interest (ROI). Then, the LP bounding boxes are located by analyzing only these ROIs. Fig. 2 summarizes the LP detection process.

The rest of this section is organized as follows: Sect. 2.1 presents a brief overview of SVMs and describes the basic idea of using SVMs for color texture classification. Next, Sect. 2.2 discusses the use of ANN for high-speed SVM

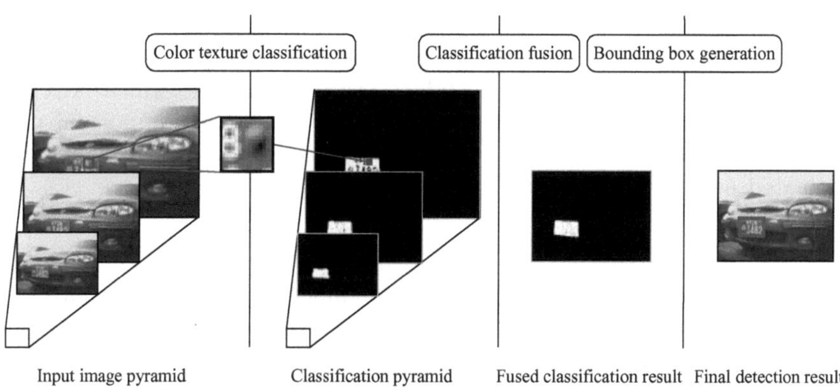

Fig. 2. Top-level process for LP detection system

classification. Sect. 2.3 discusses the fusion of the classification results on different scales. Finally, Sect. 2.4 outlines the object segmentation process based on CAMShift.

2.1 Support Vector Machines for Color Texture Classification

For the pattern classification problem, from a given set of labeled training examples $(\mathbf{x}_i, y_i) \in \mathbf{R}^N \times \{\pm 1\}, i = 1, \ldots, l$, an SVM constructs a linear classifier by determining the separating hyperplane that has maximum distance to the closest points of the training set (called *margin*). The appeal of SVMs lies in their strong connection to the underlying statistical learning theory. According to the structural risk minimization principle [28], a function that can classify training data accurately and which belongs to a set of functions with the lowest capacity (particularly in the VC-dimension) will generalize best, regardless of the dimensionality of the input space. In the case of separating hyperplanes, the VC-dimension h is upper bounded by a term depending on the margin Δ and the radius of the smallest sphere R including all the data points as follows [28]

$$h \leq \min\left(\left[\frac{R^2}{\Delta^2}\right], N\right) + 1^1 \tag{1}$$

Accordingly, SVMs approximately implement SRM by maximizing Δ for a fixed R (since the example set is fixed). The solution of an SVM is obtained by solving a QP problem which shows that the classifier is represented equivalently as either

$$f(\mathbf{x}) = \operatorname{sgn}(\mathbf{x} \cdot \mathbf{w} + b) \tag{2}$$

or

$$f(\mathbf{x}) = \operatorname{sgn}\left(\sum_{i=1}^{l*} y_i \alpha_i \mathbf{x}_i^* \cdot \mathbf{x} + b\right) , \tag{3}$$

where $\mathbf{x}_i^*$'s are a subset of training points lying on the margin (called *support vectors* (SVs)).

The basic idea of nonlinear SVMs is to project the data into a high-dimensional *Reproducing Kernel Hilbert Space* (RKHS) F which is related to the input space by a nonlinear map $\Phi : \mathbf{R}^N \rightarrow F$ [26]. An important property of an RKHS is that the inner product of two points mapped by Φ can be evaluated using *kernel functions*

$$k(\mathbf{x}, \mathbf{y}) = \Phi(\mathbf{x}) \cdot \Phi(\mathbf{y}) , \tag{4}$$

which allows us to compute the value of the inner product without having to carry out the map Φ explicitly. By replacing each occurrence of inner product

[1] This bound (called *radius margin bound*) holds in only linearly separable case. For the generalization of this bound for linearly non-separable case, readers are referred to Sect. 4.3.3 of [6].

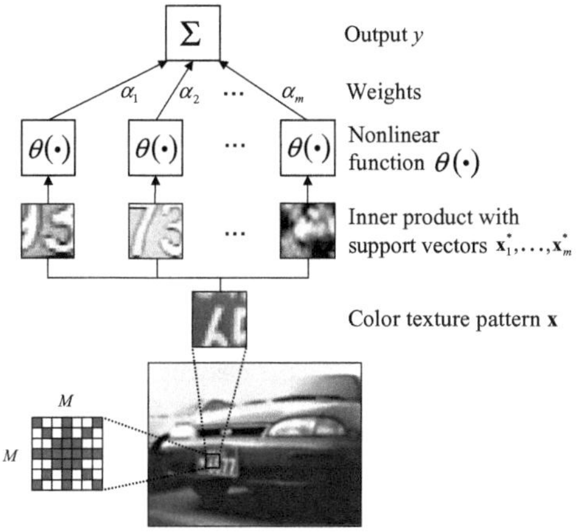

Fig. 3. Architecture of color texture classifier

in (3) with kernel function (4), the SVM can be compactly represented even in very high-dimensional (possibly infinite) spaces:

$$f(\mathbf{x}) = \operatorname{sgn}\left(\sum_{i=1}^{l*} y_i \alpha_i k\left(\mathbf{x}_i^*, \mathbf{x}\right) + b\right) . \tag{5}$$

Figure 3 shows the architecture of an SVM with a polynomial kernel ($k(\mathbf{x}, \mathbf{y}) = (\mathbf{x} \cdot \mathbf{y} + 1)^p$) as the base color texture classifier[2]. It shows a two-layer network architecture: the input layer is made up of source nodes that connect the SVM to its environment. Its activation $\mathbf{x}$ comes from an $M \times M$ window in the input image (color coordinate values in a color space e.g. RGB, HSL, etc.). However, instead of using all the pixels in the window, a configuration for autoregressive features (shaded pixels in Fig. 3) is used [16]. This reduces the size of the feature vector (from $3 \times M^2$ to $3 \times (4M - 3)$) and results in an improved generalization performance and classification speed. The hidden layer calculates the inner products between the input and SVs and applies a nonlinear function ($\theta(x) = (x + 1)^p$). The sign of the output y, obtained by weighting the activation of the hidden layer, then represents the class of the central pixel in the input window. For training, +1 was assigned as the plate class and −1 as the non-plate class. Consequently, if the SVM output for an input pattern is positive, it is classified as a plate.

[2] The architecture of nonlinear SVM can be viewed from two different perspectives: one is the linear classifier lying in F and is parameterized by ($\mathbf{w}$,b) (cf. (2)). Another is the linear classifier lying in the space spanned by the *empirical kernel map* with respect to the training data ($k(\mathbf{x}_1, \cdot), \ldots, k(\mathbf{x}_l, \cdot)$) and is parametereized by ($\alpha_1, \ldots, \alpha_{l*}$,$b$) (cf. (3)). The second viewpoint characterizes the SVM as a two-layer ANN.

To gain an insight into the performance of SVMs, an SVM with degree 3 polynomial kernel was trained on approximately 20,000 plate and non-plate patterns. The training set was initialized with 2,000 patterns sampled from a database of 100 vehicle images and is augmented by performing bootstrapping [27] on the same database. The trained SVM was composed of around 1,700 SVs and showed 2.7% training error rate. For testing, another set of 10,000 plate and non-plate patterns were collected from 350 vehicle images which are distinct from the images used in training. The SVM achieved 4.4% error rate with processing time of 1.7 sec. on 10,000 patterns (which corresponds to approximately 24 sec. for processing a 320×240-sized image). In terms of classification accuracy, the SVM was superior to several other existing methods (cf. Sect. 4) and is suitable to LP detection application. However, the processing time of the SVM is far below the acceptable level. The main computational burden lies in evaluating the kernel map in (5) which is in proportional to the dimensionality of the input space and the number of SVs. In the case of polynomial kernel, this implies 1,700 inner product operations in processing a single pattern.

There are already several methods to reduce the runtime complexity of SVM classifiers, including reduced set method [26], cascade architecture of SVMs [13], etc. The preliminary experiment with reduced set method has shown that at least 400 SVs were required to retain a moderate reduction of generalization performance[3], which is still not enough to realize a practical LP detection system. The application of cascade method is beyond the scope of this work. However, it should be noted that cascade architecture is independent of a specific classification method and can be directly applied to the method proposed in this work.

2.2 Combining Artificial Neural Networks with Support Vector Machines

Recall that the basic idea of the nonlinear SVM is to cast the problem into a space where the problem is linearly separable and then use linear SVMs in that space. This idea is validated by Cover's theorem on the separability of patterns [5, 12]:

A complex pattern-classification problem cast in a high-dimensional space nonlinearly is more likely to be linearly separable than in a low-dimensional space

The main advantage of using high-dimensional spaces to make the problem linearly separable is that it enables analysis to stay within the class of linear

[3] The reduced set method tries to find an approximate solution which is expanded in a small set of vectors (called *reduced set*). Finding the reduced set is a nonlinear optimization problem which in this work, is solved using a gradient-based method. For details, readers are referred to [26]. The less number of SVs than 400 produced significant increase in the error rate.

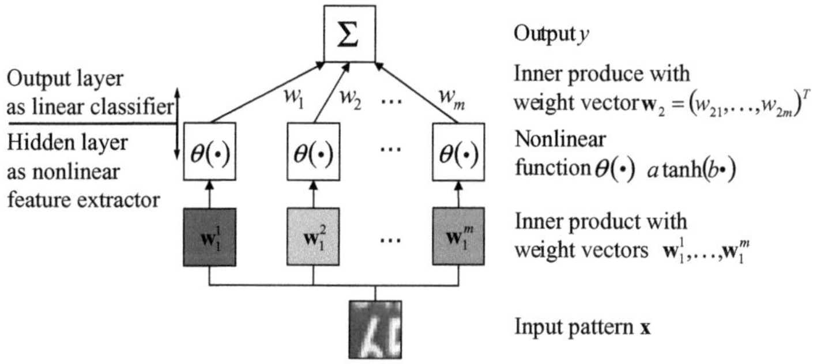

Fig. 4. Example of two-layer ANN (single hidden layer)

functions and accordingly leads to the convex optimization problem. On the other hand, as a disadvantage, it requires the solution to be represented in kernel expansion since one can hardly manipulate the solution directly in such spaces.

The basic intuition in using the ANN is to cast the problem into a moderately low-dimensional space where the problem is still (almost) linearly separable. In other words, we perform Φ explicitly and construct the linear SVM in terms of the direct solution (2) rather than the kernel representation (3) in the range of Φ. This reduces the computational cost when the dimensionality of the corresponding feature space is lower than the number of SVs. As demonstrated in many practical applications [12], ANNs has an ability to find a local minimum in the empirical error surface. If it once achieves an acceptably low error rate, it is guaranteed that the problem in the space constructed by up to the last hidden layer is (almost) linearly separable since the output layer is simply a linear classifier (Fig. 4). The final classifier is then obtained by replacing the output layer of the ANN with a linear SVM.

However, this strategy alone does not work as exemplified in Fig. 5: for a two dimensional toy problem, an ANN with two hidden layers of size 30 and 2 is trained: (a) plots the training examples in the input space while (b) plots the activation of the last hidden layer corresponding to training examples in (a). The ANN feature extractor (up to the last hidden layer) did make the problem linearly separable. At the same time, it clustered training examples almost perfectly by mapping them (close) to two cluster centers according to class labels. Certainly, the trained ANN failed to generalize. This is an example of overfitting occurred during the feature extraction. In this case, even the SVM with capacity control capability does not have any chance to generalize better than the ANN since all the information contained in the training set is already lost during the feature extraction.

From the model selection viewpoint, choosing the feature map Φ is an optimization problem where one controls the kernel parameter (and equivalently the Φ) to minimize the cross validation error or a generalization error bound. Within this framework, one might try to train simultaneously both the feature

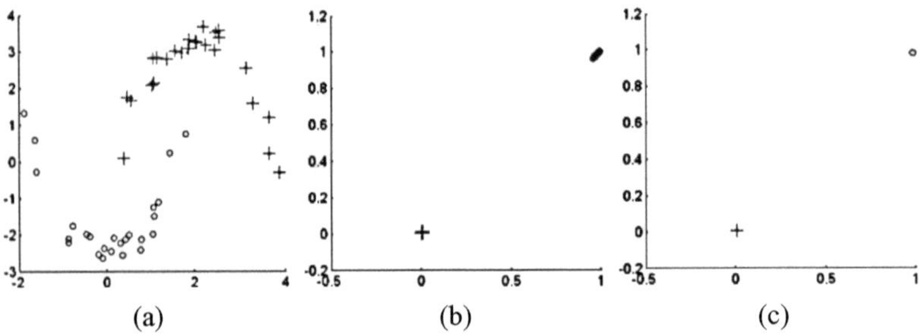

Fig. 5. Example of overfitting occurred during feature extraction: (**a**) input data, (**b**) activation of the last hidden layer corresponding to (**a**), and (**c**) extreme case of overfitting

extractor Φ and the classifier based on the unified criterion of the generalization error bound. However, we argue that this method is not applicable to the ANN as Φ. Here we give two examples of error bounds which are commonly used in model selection.

In terms of the radius margin bound (1), the model selection problem is to minimize the capacity h by controlling both the margin Δ and the radius of the sphere R. Since the problem of estimating Δ from a fixed set of training examples is convex, both the Δ and R are solely determined by choosing Φ.

Let us define a map Φ which maps all the training examples into two points corresponding to their class labels, respectively (Fig. 5(c)). Certainly, there are infinitely many extensions of Φ for unseen data points, most of which may generalize poorly. However, for the radius margin bound, all these extensions of Φ are equally optimal (cf. (1)).

The *span bound* [4] provides an efficient estimation of the *leave-one-out* (LOO) error based on the geometrical analysis of the *span of SVs*. It is characterized by the cost of removing an SV and approximating the original solution based on the linear combination of remaining SVs. Again, the Φ in Fig. 5(c) is optimal in the sense that all training examples became duplicates of one of two SVs in F and accordingly the removal of one SV (one duplicate) does not affect the solution[4].

Certainly, the example map in Fig. 5(c) is unrealistic and is even not optimal for all existing bounds. However it clarifies the inherent limitation in using error bounds for choosing Φ, especially when it is chosen from a large class of functions (containing very complex functions). It should be noted that this is essentially the same problem to that of choosing the classifier in a fixed

[4] It should be noted that the map in Fig. 5(c) is optimal even in terms of the true LOO error.

feature space F, whose solution suggests controlling the complexity[5] of Φ to avoid overfitting[6].

There are already several methods developed for reducing the complexity of ANNs including early stopping of training, weight decaying, etc. We rely on the network pruning approach where one starts with a large network with high complexity and prunes the network based on the criteria of minimizing the *damage* of network. This type of algorithms does not have direct relation to any existing complexity control methods (e.g., regularization). However, they are simple and have advantage of providing the decomposition of choosing Φ into two sub-problems: 1) obtaining a linearly separable map and 2) reducing the complexity of the map while retaining the linear separability.

For the first objective, an ANN with one hidden layer of size 47 is trained based on the *back-propagation* algorithm. The use of only one hidden layer as the feature extractor is supported by the architecture of the kernelized SVM where only one hidden layer (depending on the type of kernel function) is often enough to guarantee the linear separability (cf. Fig. 3). The number of hidden nodes was chosen based on trial and error: a fairly large network of 100 hidden nodes was initially constructed and is reduced by removing nodes by ones while retaining the training error similar to that of the SVM with polynomial kernel.

The trained ANN is then pruned based on the OBS algorithm which evaluates the damage of the network based on the quadratic approximation of increase of training error and prunes it to obtain the minimal damage. Here we give a brief review of OBS. For more detail, readers are referred to [11, 12].

The basic idea of OBS is to use the second order approximation of error surfaces. Suppose for a given configuration of weights $\mathbf{w}$ (cf. Fig. 4), the cost function E in terms of empirical error is represented based on *Taylor series* about $\mathbf{w}$:

$$E(\mathbf{w}+\Delta\mathbf{w}) = E(\mathbf{w}) + \mathbf{g}^T(\mathbf{w})\,\Delta\mathbf{w} + \frac{1}{2}\Delta\mathbf{w}^T\mathbf{H}(\mathbf{w})\,\Delta\mathbf{w} + O(\|\Delta\mathbf{w}\|^3)\,,$$

where $\Delta\mathbf{w}$ is a perturbation applied to the operating point $\mathbf{w}$ and $\mathbf{g}(\mathbf{w})$ and $\mathbf{H}(\mathbf{w})$ are the gradient vector and the Hessian matrix evaluated at the point $\mathbf{w}$, respectively. Assuming that the $\mathbf{w}$ is located at the local optima and ignoring the third and all higher order terms. We get the approximation of increase in the error E based on $\Delta\mathbf{w}$ as follows

[5] While there are various existing notion of complexity of function class including VC-dimension, smoothness, the number of parameters, etc., no specific definition of complexity is imposed on the class of Φ, since we will not directly minimize any of them. Accoridngly, the "complexity" here should be understood in abstract sense and not be confused with VC-dimension.

[6] The reported excellent performances of model selection methods based on error bounds [4, 28] can be explained by the fact that the class of functions where the kernel is chosen was small (e.g., polynomial kernels or Gaussian kernels with one or few parameters).

$$\Delta E = E(\mathbf{w} + \Delta\mathbf{w}) - E(\mathbf{w}) \approx \frac{1}{2}\Delta\mathbf{w}^T \mathbf{H}(\mathbf{w})\Delta\mathbf{w} . \quad (6)$$

The goal of OBS is to find the index i of a particular weight w_i which manimize ΔE when w_i is set to zero. The elimination of w_i is equivalent to the condition

$$\Delta w_i + w_i = 0 . \quad (7)$$

To solve this constrained optimization problem, we construct the *Lagrangian*

$$S_i = \frac{1}{2}\Delta\mathbf{w}^T \mathbf{H}(\mathbf{w})\Delta\mathbf{w} - \lambda(\Delta w_i + w_i) ,$$

where λ is the *Lagrange multiplier*. Taking the derivative of S_i with respect to $\Delta\mathbf{w}$, applying the constraint of (7), and using matrix inversion, the optimum change in the weight vector $\mathbf{w}$ and resulting change in error (the optimal value of S_i) are obtained as

$$\Delta\mathbf{w} = -\frac{w_i}{[\mathbf{H}^{-1}]_{i,i}}\mathbf{H}^{-1}\mathbf{1}_i \quad (8)$$

and

$$S_i^* = \frac{w_i^2}{2[\mathbf{H}^{-1}]_{i,i}} , \quad (9)$$

respectively, where $[\mathbf{H}^{-1}]_{i,i}$ is the (i,i)-th element of the inverse of the Hessian matrix $\mathbf{H}^{-1}$ and $\mathbf{1}_i$ is the unit vector whose elements are all zero except for the i-th element.

The OBS procedure for constructing the feature extractor Φ is summarized in Fig. 6.

Although the OBS procedure does not have any direct correspondence to the regularization framework, the ANN obtained from the OBS will henceforth be referred to as the *regularized* ANN.

2.3 Fusion of Classification on Different Scales

As outlined in Sect. 2.2, no color texture-specific feature extractor is utilized. Instead, the classifier receives the color values of small input windows as a pattern for classification. While already proven as effective for general texture classification [16], this approach has an important shortcoming as regards

1. Compute $\mathbf{H}^{-1}$.
2. Find the index i that gives the smallest *saliency* value S_i^*. If the S_i^* is larger than a given threshold, then go to step 4.
3. Use the i from step 2 to update all weights according to (8). Go to step 2.
4. Retrain the network based on standard back propagation algorithm.

Fig. 6. OBS procedure for pruning ANN

object detection: it lacks of any explicit mechanism to adapt to variations in the texture scale. This can be dealt with by constructing a sufficiently large set of training patterns sampled on different scales. However, it is often more efficient to deal with the problem by representing the object of interest as size-invariant.

Accordingly, a pyramidal approach is adopted for this purpose, where the classifier is trained on a single scale. Thereafter, in the classification stage, a pyramid of images is generated by gradually changing the resolution of the input image, and the classification is performed on each pyramid level. A mechanism is then required to fuse the resulting pyramid of classification results (classification hypotheses at each level) into a single scale classification. This is actually a classifier combination problem and many methods have already been proposed for this purpose, including voting [25], ANN-based arbitration [25], boosting [12], etc. The current study uses a method similar to ANN-based arbitration, which has already been shown to be effective in face detection [25].

The basic idea is to train a new classifier as an arbitrator to collect the outputs of the classifiers on each level. For each location of interest (i, j) in the original image scale, the arbitrator examines the corresponding locations on each level of the output pyramid, together with their spatial neighborhood (i', j') ($|i' - i| \leq N$ and $|j' - j| \leq N$, where N defines the neighborhood relations and is empirically determined to be 2). A linear SVM is used as the arbitrator and receives the normalized output of the color texture classifier in an $N \times N$ window around the location of interest on each scale. The normalization is then simply performed by applying a sigmoid function to the classifier. Since this method is similar to that of ANN-based arbitration, readers are referred to [25] for more detail. Figure 7 shows examples of color texture classification performed by a trained classifier.

2.4 Continuously Adaptive Mean Shift Algorithm for Object Segmentation

In most object detection applications, the detection process needs to be fast and efficient so that objects can be detected in real-time, while consuming as few system resources as possible. However, many (color) texture-based object detection methods suffer from the considerable computation involved. The majority of this computation lies in texture classification, therefore, if the number of calls for texture classification can be reduced, this will save computation time. The proposed speed-up approach achieves this based on the following two observations:

1. LPs (and some other classes of object) form smooth boundaries and
2. the LP size usually does not dominate the image.

The first observation indicates that LP regions in an image usually include an aggregate of pixels that exhibit LP-specific characteristics (color texture).

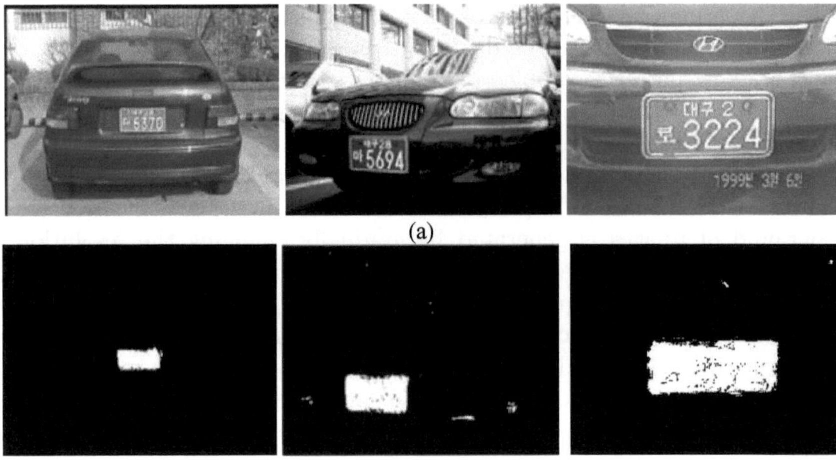

Fig. 7. Examples of color texture classification: (**a**) input images and (**b**) classification results where LP pixels are marked as *white*

As such, this facilitates the use of the *coarse-to-fine* approach for local feature-based object detection: first, the ROI related to a possible object region is selected based on a coarse level of classification (sub-sampled classification of image pixels). Then, only the pixels in the ROI are classified on a finer level. This can significantly reduce the processing time when the object size does not dominate the image size (as supported by the second observation). It should be noted that the prerequisite for this approach is that the object of interest must be characterized with local features (e.g. color, texture, color texture, etc). Accordingly, features representing the holistic characteristics of an object, such as the contour or geometric moments can not be directly applied.

The implementation of this approach is borrowed from well-developed face detection methodologies. CAMShift was originally developed by Bradski [2] to detect and track faces in a video stream. As a modification of the mean shift algorithm that climbs the gradient of a probability distribution to find the dominant mode, CAMShift locates faces by seeking the modes of the flesh probability distribution[7]. The distribution is defined as a two-dimensional image $\{y_{i,j}\}_{i,j=1,\dots,IW,IH}$ (IW: image width, IH: image height) whose entry $y_{i,j}$ represents the probability of a pixel $x_{i,j}$ in the original image $\{x_{i,j}\}_{i,j=1,\dots,IW,IH}$ being part of a face, and is obtained by matching $x_{i,j}$ with a facial color model. Then, from the initial search window, CAMShift iteratively changes the location and size of the window to fit its contents (or flesh probability distribution within the window) during the search process.

[7] Actually, it is not a probability distribution, because its entries do not total 1. However, this is not generally a problem with the objective of peak (mode) detection.

More specifically, the size of the window varies with respect to the sum of the flesh probabilities within the window, and the center of the window is moved to the mean of this local probability distribution. For the purpose of locating the facial bounding box, the shape of the search window is set as a rectangle. After finishing the iteration, the finalized search window itself then represents the bounding box of the face in the image. For a more detailed description of CAMShift, readers are referred to [2].

The proposed method simply replaces the flesh probability $y_{i,j}$ with a *LP probability* $z_{i,j}$ obtained by performing a color texture analysis on the input $x_{i,j}$, and operating CAMShift on $\{z_{i,j}\}_{i,j=1,\ldots,IW,IH}$. For this purpose, the output of the classifier (scale arbitrator: cf. Sect. 2.2) is converted into a probability by applying a sigmoid activation function based on Platt's method [24].

As a gradient ascent algorithm, CAMShift can get stuck on local optima. Therefore, to resolve this, CAMShift is used in parallel with different initial window positions, thereby also facilitating the detection of multiple objects in an image.

One important advantage of using CAMShift in color texture-based object detection is that it does not necessarily require all the pixels in the input image to be classified. Since CAMShift utilizes a local gradient, only the probability distribution (or classification result) within the window is required for iteration. Furthermore, since the window size varies in proportion to the probabilities within the window, the search windows initially located outside the LP region diminish, while the windows located within the LP region grow. This is actually a mechanism for the automatic selection of the ROI.

The parameters controlled in CAMShift at iteration t are the position $x(t)$, $y(t)$, width $w(t)$, height $h(t)$, and orientation $\theta(t)$ of the search window. x and y can be simply computed using moments:

$$x = M_{10}/M_{00} \quad \text{and} \quad y = M_{01}/M_{00} \ , \tag{10}$$

where M_{ab} is the $(a+b)$-th moment as defined by

$$M_{ab}(W) = \sum_{i,j \in W} i^a j^b z_{i,j} \ .$$

w and h are estimated by considering the two eigenvectors and their corresponding eigenvalues of the correlation matrix $\mathbf{R}$ of the probability distribution within the window[8]. These variables can be calculated using up to the second order moments as follows [2]:

[8] Since the input space is 2D, $\mathbf{R}$ is 2×2 along with the existence of two (normal) eigenvectors: the first gives the direction of the maximal scatter, while the second gives the related perpendicular direction (assuming that the eigenvectors are sorted in a descending order of their eigenvalue size). The corresponding eigenvalues then indicate the degrees of scatter along the direction of the corresponding eigenvectors

$$
\begin{aligned}
w &= 2\alpha_w \sqrt{\left((a+c) + \sqrt{b^2 + (a-c)^2}\right)/2} \\
h &= 2\alpha_h \sqrt{\left((a+c) - \sqrt{b^2 + (a-c)^2}\right)/2}\,,
\end{aligned} \tag{11}
$$

where the intermediate variables a, b, and c are

$a = M_{20}/M_{00} - x^2$, $b = 2\,(M_{11}/M_{00} - xy)$, and $c = M_{02}/M_{00} - y^2$ and α_w and α_h are constants determined as 1.5. These (rather large) values of α_w and α_h enable the window to grow as long as the major content of the window is LP pixels and thereby explore a potentially large object area during iteration. When the iteration terminates, the final window size is re-estimated using the new coefficient values of $\alpha_w = \alpha_h = 1.2$.

Similarly, the orientation θ can also be estimated by considering the first eigenvector and corresponding eigenvalue of the probability distribution and then calculated using up to the second order moments as follows:

$$
\theta = \arctan\left(\frac{b}{a-c}\right)/2\,. \tag{12}
$$

The terminal condition for iteration is that for each parameter, the difference between the two parameters of $x(t+1) - x(t), y(t+1) - y(t), w(t+1) - w(t), h(t+1) - h(t)$, and $\theta(t+1) - \theta(t)$ in two consecutive iterations $(t+1)$ and (t) is less than the predefined thresholds T_x, T_y, T_w, T_h, and T_θ respectively.

During the CAMShift iteration, search windows can overlap each other. In this case, they are examined as to whether they are originally a single object or multiple objects. This is performed by checking the degree of overlap between the two windows, which is measured using the size of the overlap divided by the size of each window.

Supposing that D_α and D_β are the areas covered by two windows α and β, then the degree of overlap between α and β is defined as

$$
\Lambda(\alpha, \beta) = \max\left(\text{size}\,(D_\alpha \cap D_\beta)/\text{size}\,(D_\alpha), \text{size}\,(D_\alpha \cap D_\beta)/\text{size}\,(D_\beta)\right)\,,
$$

---

[12]. Accordingly, the estimated orientation and width (height for similar manner) should simply be regarded as the principal axis and its variance of the object pixels. While generally this may not be serious problem, it should be noted that in the case of LPs, the estimated parameters (especially the orientation) may not exactly correspond to the actual width, height, and orientation of the LPs in the image and the accuracy may be proportional to the elongatedness of the LPs in the image. To obtain exact parameters, domain specific post-processing methods should be utilized.

1. Set up initial locations and sizes of search windows Ws in image.
   For each W, repeat steps 2 to 4 until terminal condition is satisfied.
2. Generate LP probability distribution within W using SVM.
3. Estimate parameters (location, size, orientation) of W using (10) to (12) with $\alpha_w = \alpha_h = 1.5$.
4. Modify W according to estimated parameters.
5. Re-estimate sizes of Ws using (11) with $\alpha_w = \alpha_h = 1.2$.
6. Output bounding boxes of Ws.

Fig. 8. CAMShift for object detection

where size (λ) counts the number of pixels within λ. Then, α and β are determined to be

a single object if $T_0 \leq \Lambda(\alpha, \beta)$
multiple objects otherwise,
where T_0 is the threshold set at 0.5.

As such, in the CAMShift iteration, every pair of overlapping windows is checked and those pairs identified as a single object are merged to form a single large encompassing window. After finishing the CAMShift iteration, any small windows are eliminated, as they are usually false detections.

Figure 8 summarizes the operation of CAMShift for LP detection. It should be noted that, in the case of overlapping windows, the classification results are cached so that the classification of a particular pixel is only performed once for an entire image. Figure 9 shows an example of LP detection using CAMShift. A considerable number of pixels (91.3% of all the pixels in the image) are excluded from the color texture analysis in the given image.

3 Experimental Results

The proposed method was tested using an LP image database of 450 images, of which 200 images were of stationary vehicles taken in parking lots, while the remaining 250 images were of moving vehicles on a road. The images included LPs with varying appearances in terms of size, orientation, perspective, illumination conditions, etc. The resolution of the images ranged from 240×320 to 1024×1024 and the sizes of the LPs in these images ranged from about 79×38 to 390×185. All the images were represented using a 24-bit RGB color system [15]. For training the ANN+SVM classifier, 20,000 training examples which were used in training the base SVM classifier (Sect. 2.1) were used: the ANN was firstly trained on random selection of 10,000 patterns. Then, the linear SVM was trained on the output of the ANN feature extractor on whole 20,000 patterns (including 10,000 patterns used to train the ANN). The size (number of weights) of the ANN feature extractor was initially 5,781 (123×47) which were reduced to 843 after OBS procedure where the stopping

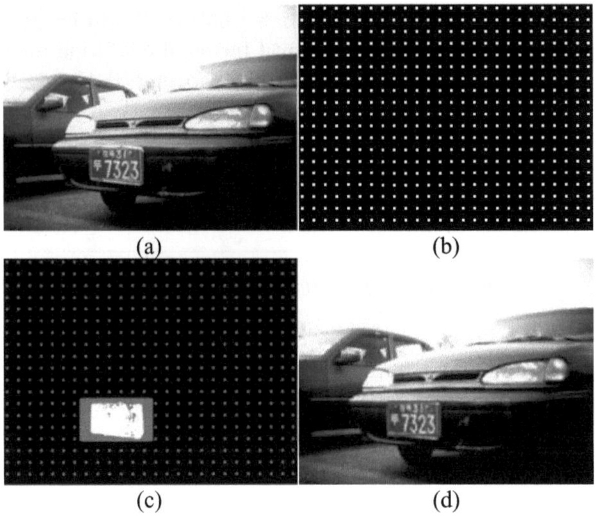

Fig. 9. Example of LP detection using CAMShift: (**a**) input image, (**b**) initial window configuration for CAMShift iteration (5×5-sized windows located at regular interval of (25,25) in horizontal and vertical directions), (**c**) color texture classified region marked as white and gray levels (white: LP region, gray: background region), and (**d**) LP detection result

criterion was 0.1% increase of training error rate. The testing environment was 2.2 GHz CPU with 1.2GB RAM. Table 1 summarizes the performances of various classifiers: the ANN and nonlinear SVM (with polynomial kernel) have shown the best and worst error rates and processing times, respectively. Simply replacing the output layer of the ANN with an SVM did not provide any significant improvement as anticipated in Sect. 3, while the regularization of the ANN has already shown improved classification rate. The combination of the regularized ANN with SVM produced the second best error rate and the processing time which can be regarded as the best overall.

Prior to evaluating the overall performance of the proposed system, the parameters for CAMShift were tuned. The initial locations and sizes of the search windows are dependent on the application. A good selection of initial

Table 1. Performance of different classification methods

Classifier	Error rate (%)	Proc. time (10,000/sec.)
ANN	7.31	0.14
ANN+SVM	6.87	0.14
Regularized ANN	5.10	0.06
Regularized ANN+SVM	4.48	0.08
Nonlinear SVM	4.43	1.74

search windows should be relatively dense and large enough not to miss objects located between the windows, tolerant of noise (classification errors), and yet also moderately sparse and small enough to ensure fast processing. The current study found that 5 × 5-sized windows located at a regular interval of (25, 25) in the horizontal and vertical direction were sufficient to detect the LPs. Variations in the threshold values T_x, T_y, T_w, T_h, and T_θ (for termination condition of CAMShift) did not significantly affect the detection results, except when the threshold values were so large that the search process converged prematurely. Therefore, based on various experiments with the training images, the threshold values were determined as $T_x = T_y = 3$ pixels, $T_w = T_h = 2$ pixels, and $T_\theta = 1°$. The slant angle θ for the finalized search windows was set at 0° if its absolute value was less than 5°, and 90° if it was greater than 85° and less than 95°. This meant that small errors occurring in the orientation estimation process would not significantly affect the detection of horizontally and vertically oriented LPs. Although these parameters were not carefully tuned, the results were acceptable as described below.

The time spent processing an image depended on the image size and number and size of LPs in the image. Most of the time was spent in the classification stage. For the 340 × 240-sized images, an average of 11.2 seconds was taken to classify all the pixels in the image. However, when the classification was restricted to just the pixels located within the search windows identified by CAMShift, the entire detection process only took an average of 1.1 seconds.

To quantitatively evaluate the performance, a criterion was adopted from [23] to decide whether each detection produced automatically by the system is correct. Then, the detection results are summarized based on two metrics as defined by:

$$\text{miss rate}(\%) = \frac{\#\text{ of misses}}{\#\text{ of LPs}} \times 100$$
$$\text{false detection rate}(\%) = \frac{\#\text{ of false detections}}{\#\text{ of LPs}} \times 100\,.$$

The proposed system achieved a miss rate of 2.8% with a false detection rate of 9.9%. Almost all the plates that the system missed were either blurred during the imaging process, stained with dust, or reflecting strong sunshine. In addition, many of the false detections were image patches with green and white textures that looked like parts of LPs. Figures 10 and 11 show examples of the LP detection without and with mistakes, respectively: the system exhibited a certain degree of tolerance with pose variations (Fig. 10-c, h, and l), variations in illumination conditions (Fig. 10-a compared with Fig. 10-d and e), and blurring (Fig. 10-j and k). Although Fig. 10-d and e show a fairly strong reflection of sunshine and thus quite different luminance properties from the surface reflectance, the proposed system was still able to locate the LPs correctly. This clearly shows the advantage of (color) texture-based methods over methods simply based on color.

Fig. 10. LP detection examples

On the other hand, LPs were missed due to bad imaging or illumination conditions (Fig. 11-a and b), a large angle (LP located at upper right part of Fig. 11-c), and excessive blurring Fig. 11-d. While the color texture analysis correctly located a portion of missed LP in Fig. 11-c, the finalized search window was eliminated on account of small size. False detections are present in Fig. 11-e and f where white characters were written on a green background and complex color patterns occurred in glass, respectively. The false detection in Fig. 11-e indicates the limits of a local texture-based system: without information on the holistic shape of the object of interest or other additional problem-specific knowledge, the system has no means of avoiding this kind of false detection.

Accordingly, for a specific problem of LP detection, the system needs to be specialized by incorporating domain knowledge (e.g., width/height ratio for LP detection) in addition to training patterns.

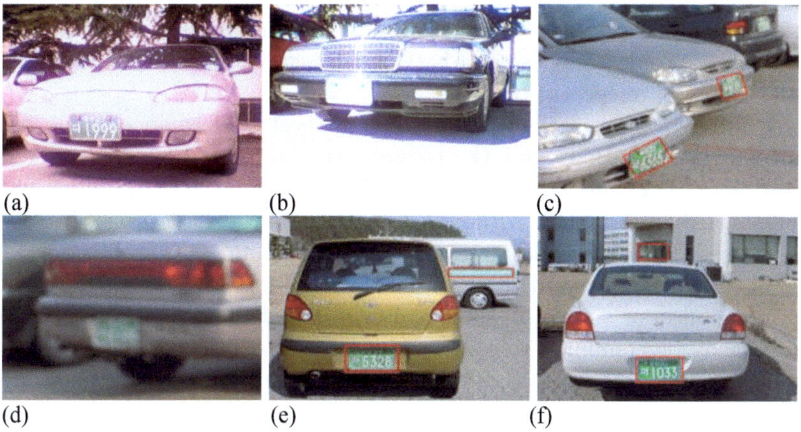

Fig. 11. Examples of LP detection with mistakes

To gain a better understanding of the relevance of the results obtained using the proposed method, benchmark comparisons with other methods were carried out. A set of experiments was performed using a color-based method [18] and color texture-based method [23]. The color-based method segmented an input image into approximately homogeneously colored regions and performed an edge and shape analysis on the green regions. Meanwhile, the color texture-based method adopted an ANN to analyze the color textural properties of horizontal and vertical cross-sections of LPs in an image and performed a projection profile analysis on the classification result to generate LP bounding boxes. Since both methods were developed to detect horizontally aligned LPs without perspective, the comparison was made based on 82 images containing upright frontal LPs.

Table 2 summarizes the performances of the different systems: A (SVM + CAMShift) is the proposed method (ANN feature extractor was used for SVM classification), B (SVM + profile analysis) used SVM for color texture classification and profile analysis [23] for bounding box generation (again ANN was used for feature extraction), C (ANN + CAMShift) used an ANN adopted

Table 2. Performances of various systems

	Miss Rate (%)	False Detection Rate (%)	Avg. Proc. Time Per Image (sec.)
A: SVM + CAMShift	3.9	7.2	0.6
B: SVM + profile analysis	3.9	9.6	1.8
C: ANN + CAMShift	8.5	13.4	0.9
D: ANN + profile analysis [23]	7.3	17.1	2.0
E: Color-based [18]	20.7	26.8	0.5

from [23] for classification and CAMShift for bounding box generation, and D (ANN + profile analysis) and E (Color-based) are the methods described in [23] and [18], respectively.

A and B produced the best and second best performances, plus A was much faster than B. C and D produced similar performances, and C was faster than D. C and D were more sensitive to changes in the illumination conditions than A and B. Although producing the highest processing speed, E produced the highest miss rate, which mainly stemmed from the poor detection of LPs reflecting sunlight or with a strong illumination shadow that often occurs in outdoor scenes. It should also be noted that color-based methods could easily be combined with the proposed method. For example, the color segmentation method in [18] could be used to filter images, then the proposed method applied to only those portions of the image that contain LP color as verification. Accordingly, the use of color segmentation could speed-up the system by reducing the number of calls for SVM classification.

4 Discussion

A generic framework for detecting objects in images was presented. The system analyzes the color and textural properties of objects in images using an SVM and locates their bounding boxes by operating CAMShift on the classification results. The problem of high run-time complexity of SVMs was approached by utilizing a regularized ANN as the feature extractor. In comparison with the standard nonlinear SVM, classification performance of the proposed method was only slightly worse while the run-time is significantly better. Accordingly, it can provide a moderate alternative to the standard kernel SVMs in real-time applications.

As a generic object detection method, the proposed system does not assume the orientation, size, or perspective of objects, is relatively insensitive to variations in illumination conditions, and can also facilitate fast object detection. As regards specific LP detection problems, the proposed system encountered problems when the image was extremely blurred or the LPs were at a fairly large angle yet overall it produced a better performance than various other techniques.

There are a number of directions for future work. While many objects can be effectively located using a bounding box, there are some objects whose location cannot be fully described by only a bounding box. When the precise boundaries of these objects are required, a more delicate boundary location method needs to be utilized. Possible candidates include the deformable template model [30]. Starting with an initial template incorporating priori knowledge of the shape of the object of interest, the deformable template model locates objects by deforming the template to minimize energy, defined as the degree of deformation in conjunction with the edge potential. As such, the object detection problem can be dealt with using a fast and effective ROI selection process (SVM + CAMShift) followed by a delicate boundary location process (deformable template model).

Although the proposed method was applied to the particular problem of LP detection, it is also general enough to be applicable to the detection of an arbitrary class of objects. For LP detection purposes, this implies that the detection performance of the system can be improved by specializing for the task of LP detection. For example, knowledge of the LP size, perspective, and illumination conditions in an image can be utilized, which is often available prior to classification. Accordingly, further work will include the incorporation of problem-specific knowledge into the system as well as the application of the system to detect different types of objects.

Acknowledgement

Kwang In Kim has greatly profited from discussions with M. Hein, A. Gretton, G. Bakir, and J. Kim. A part of this chapter has been published in Proc. International Workshop on Pattern Recognition with Support Vector Machines (2002), pp. 293–309.

References

1. Barroso J, Dagless EL, Rafael A, Bulas-Cruz J (1997) Number plate reading using computer vision. In: Proc. IEEE Int. Symposium on Industrial Electronics, pp 761–766
2. Bradski GR (1998) Real time face and object tracking as a component of a perceptual user interface. In: Proc. IEEE Workshop on Applications of Computer Vision, pp 214–219
3. ter Brugge MH, Stevens JH, Nijhuis JAG, Spaanenburg L (1998) License plate recognition using DTCNNs. In: Proc. IEEE Int. Workshop on Cellular Neural Networks and their Applications, pp 212–217
4. Chapelle O, Vapnik V, Bousquet O, Mukherjee S (2000) Choosing kernel parameters for support vector machines. Machine Learning 46:131–159
5. Cover TM (1965), Geometrical and statistical properties of systems of linear inequalities with applications in pattern recognition. IEEE Trans. Electronic Computers 14: 326–334
6. Cristianini N, Shawe-Taylor J (2000) An introduction to support vector machines and other kernel-based learning methods. Cambridge University Press
7. Cui Y, Huang Q (1998) Extracting characters of license plates from video sequences. Machine Vision and Applications 10:308–320
8. Draghici S (1997) A neural network based artificial vision system for license plate recognition. Int. J. of Neural Systems 8:113–126
9. Duda RO, Hart PE (1973) Pattern classification and scene analysis. A Wiley-interscience publication, New York

10. Gao D-S, Zhou J (2000) Car license plates detection from complex scene. In: Proc. Int. Conf. on Signal Processing, pp 1409–1414
11. Hassibi B, Stork DG (1993) Second order derivatives for network pruning: optimal brain surgeon. In: Hanson SJ, Cowan JD, Giles CL (eds) Advances in neural information processing systems, pp 164–171
12. Haykin S (1998) Neural networks: a comprehensive foundation, 2nd Ed. Prentice Hall
13. Heisele B, Serre T, Prentice S, Poggio T (2003) Hierarchical classification and feature reduction for fast face detection with support vector machines. Pattern Recognition 36:2007–2017
14. Jain AK, Ratha N, Lakshmanan S (1997) Object detection using Gabor filters. Pattern Recognition 30: 295–309
15. Kim HJ, Kim DW, Kim SK, Lee JK (1997) Automatic recognition of a car license plate using color image processing. Engineering Design and Automation Journal 3:217–229
16. Kim KI, Jung K, Park SH, Kim HJ (2002) Support vector machines for texture classification. IEEE Trans. Pattern Analysis and Machine Intelligence 24:1542–1550
17. Kumar VP, Poggio T (2000) Learning-based approach to real time tracking and analysis of faces. In: Proc. IEEE Int. Conf. on Automatic Face and Gesture Recognition, pp 96–101
18. Lee ER, Kim PK, Kim HJ (1994) Automatic recognition of a license plate using color. In: Proc. Int. Conf. on Image Processing, pp 301–305
19. Mohan A, Papageorgiou C, Poggio T (2001) Example-based object detection in images by components. IEEE Trans. Pattern Analysis and Machine Intelligence 23:349–361
20. Naito T, Tsukada T, Yamada K, Kozuka K, Yamamoto S (2000) Robust license-plate recognition method for passing vehicles under outside environment. IEEE Trans. Vehicular Technology 49: 2309–2319
21. Osuna E, Freund R, Girosi F (1997) Training support vector machines: an application to face detection. In: Proc. IEEE Conf. on Computer Vision and Pattern Recognition, pp 130–136
22. Pal NR, Pal SK (1993) A review on image segmentation techniques. Pattern Recognition 29: 1277–1294
23. Park SH, Kim KI, Jung K, Kim HJ (1999) Locating car license plates using neural networks. IEE Electronics Letters 35:1475–1477
24. Platt J (2000) Probabilities for SV machines. In: Smola A, Bartlett P, Schölkopf B, Schuurmans D (eds) Advances in Large Margin Classifiers, MIT Press, Cambridge, pp 61–74
25. Rowley HA, Baluja S, Kanade T (1999) Neural network-based face detection. IEEE Trans. Pattern Analysis and Machine Intelligence 20:23–37
26. Schölkopf B, Smola AJ (2002) Learning with Kernels, MIT Press
27. Sung KK (1996) Learning and example selection for object and pattern detection. Ph.D. thesis, MIT
28. Vapnik V (1995) The nature of statistical learning theory, Springer-Verlag, NY
29. Yang M-H, Kriegman DJ, Ahuja N (2002) Detecting faces in images: a survey. IEEE Trans. Pattern Analysis and Machine Intelligence 24:34–58
30. Zhong Y, Jain AK (2000) Object localization using color, texture and shape. Pattern Recognition 33:671–684

Support Vector Machines for Signal Processing

D. Mattera

Dipartimento di Ingegneria Elettronica e delle Telecomunicazioni,
Università degli Studi di Napoli Federico II,
Via Claudio, 21 I-80125 Napoli (Italy)
mattera@unina.it

Abstract. This chapter deals with the use of the support vector machine (SVM) algorithm as a possible design method in the signal processing applications. It critically discusses the main difficulties related with its application to such a general set of problems. Moreover, the problem of digital channel equalization is also discussed in details since it is an important example of the use of the SVM algorithm in the signal processing.

In the classical problem of learning a function belonging to a certain class of parametric functions (which linearly depend on their parameters), the adoption of the cost function used in the classical SVM method for classification is suggested. Since the adoption of such a cost function (almost peculiar to the basic SVM kernel-based method) is one of the most important achievements of the learning theory, this extension allows one to define new variants of the classical (batch and iterative) minimum-mean-square error (MMSE) procedure. Such variants, which are more suited to the classification problem, are determined by solving a strictly convex optimization problem (not sensitive, therefore, to the presence of local minima). Improvements in terms of the achieved probability of error with respect to the classical MMSE equalization methods are obtained. The use of such a procedure together with a method for subset selection provides an important alternative to the classical SVM algorithm.

Key words: channel equalization, MIMO channel, linear equalizer, sparse filter, cost functions

1 Introduction

The Support Vector Machine (SVM) represents an important method of learning and contains the most important answers (using the results developed in fifty years of research) to the fundamental problems in learning from data. However, the great flourishing of different variations faces us with an important issue in SVM learning: the basic SVM method is given as a specific

D. Mattera: *Support Vector Machines for Signal Processing*, StudFuzz **177**, 321–342 (2005)
www.springerlink.com

algorithm where a small number of minor choices are left to the utilizer: the most important choices have been done by the author of the algorithm [16].

The learning algorithm allows one to implement – on a general purpose computer or in an embedded processing card – a specific processing method. The quality of the overall processing is specified not only by an appropriate performance parameter but also by the computational complexity and by the required storage of the design stage as well as of the resulting processing algorithm.

As it usually happens in engineering practice, each choice is associated with a trade-off among the different parameters specifying the quality of the overall processing method. Such a trade-off needs to be managed by the final utilizer: in fact, only the final user knows the design constraint imposed by the specific environment where the method is applied. For instance, consider the often present trade-off between the performance and the computational complexity: it imposes to the final utilizer to choose the learning algorithm that maximizes a performance parameter but it also imposes the constraint that the chosen method be compatible with the available processing power; moreover, the variations in the available processing power and/or in the real-time processing constraints require to the final user a modification of the learning algorithm in accordance with the new environment. This clearly shows the difficulties of the final utilizer to manage an already defined method where not all the choices are possible.

2 The SVM Method and Its Variants

It is reasonable to assume that the final user, although expert in the considered application, is not so expert of the learning theory; it should be considered an important part of the learning theory the capability to provide to the final utilizer different choices for any possible application environment. Let us first present in the following subsections the basic elements of the considered design method.

2.1 The Basic Processing Choice

The class of systems to be considered for processing the input signal $x(n)$ is linear in the free parameters α_i, i.e., the processing output $y(n)$ can be written as:

$$y(n) = \sum_{i=1}^{M(\ell)} \alpha_i \phi_i^o(x(\cdot), n) \tag{1}$$

where the set of systems $\{\phi_i^o(x(\cdot), n)\ i = 1, \ldots, M(\ell)\}$ cannot be adapted on the basis of the ℓ available examples. Such systems can be very general; for example they could have a recursive nature or they can operate as follows:

they first extract from the sequence $x(n)$ the vector $\mathbf{x}(n) = [x(n-n_1) \ldots x(n-n_m)]^T$ and, successively, obtain $\phi_i^o(x(\cdot), n) \triangleq \phi_i(\mathbf{x}(n))$ where $\phi_i(\mathbf{x})$ is a function of the m-dimensional vector $\mathbf{x}$. The integers $\{n_1, n_2, \ldots, n_m\}$ and the function $\phi_i(\cdot)$ need to be set before starting the design procedure. The number of operators $M(\ell)$ can be equal to a certain number M_0 independent of ℓ but it can also depend on ℓ (e.g., $M(\ell) = M_0 + \ell$) in order to adapt the algorithm to the available set of examples. The choice $M(\ell) = M_0$ is preferable when we are able to select the operators $\phi_i(x(\cdot), n)$ such that in the set of processors (1) there exist values of parameters $\{\alpha_i\}$ that provide a satisfactory performance. The computational complexity of the design stage is obviously dependent on M_0 while the computational complexity of the final algorithm depends on the number of parameters α_i different from zero in the solution determined by the design stage. Note that the linear time-invariant finite-impulse response (FIR) filter belongs to the considered set of functions.

Let us assume that the available samples of the input sequences $x(n)$ and of the desired output $d(n)$ allow us to determine the set of input-output examples $\{(\mathbf{x}(i), d(i)),\ i = 1, \ldots, \ell\}$. The classical choice in SVM learning is the following: $M(\ell) = 1 + \ell$ and $\phi_i(\mathbf{x}) = K(\mathbf{x}, \mathbf{x}(i))$ where $K(\mathbf{x}_1, \mathbf{x}_2)$ is a kernel function whose properties imply that the square matrix $\mathbf{K}$ of size ℓ, with (i,j) element $K(\mathbf{x}(i), \mathbf{x}(j))$, is positive definite. The emphasis given in many presentations of the basic SVM to some properties that follow from the chosen kernel operator should not obscure the obvious fact that the quality of the chosen functions $\{\phi_i(\cdot)\}$ depends on their capability to well-approximate an unknown ideal function, which provides the optimum performance in the considered application; in particular, a choice not based on kernel properties can also outperform a kernel-based choice and two different choices of the functions $\{\phi_i(\cdot)\}$ cannot be compared simply on the basis of their mathematical properties independently of the considered application.

2.2 The Induction Principle

The ideal cost function to be minimized over the free parameter $\boldsymbol{\alpha} = [\alpha_1, \ldots, \alpha_{M_\ell}]^T$ is defined as

$$E[c_I(d(n), y(n))] \tag{2}$$

where $E[\cdot]$ denotes the statistical expectation. Such a cost function cannot be used because the needed joint statistical characterization of $x(n)$ and $d(n)$ is assumed not available; therefore, it is replaced by the cost function

$$\frac{1}{\ell} \sum_{n=1}^{\ell} c_I(d(n), y(n)) + \gamma \boldsymbol{\alpha}^T \mathbf{R} \boldsymbol{\alpha} \tag{3}$$

where γ is a small positive constant and $\mathbf{R}$ is a symmetric positive-definite matrix used for regularization. The use of the approximation (3) of the ideal expression (2) is motivated by several theories. The classical choice of the basic SVM method $\mathbf{R} = \mathbf{K}$ is not a priori more appropriate than other regularization matrices. The additive term $\boldsymbol{\alpha}^T \mathbf{R} \boldsymbol{\alpha}$ may also be independent of some components of $\boldsymbol{\alpha}$.

A different approximation of (2) is also suggested in [16] by using the structural risk minimization. Such an approach provides an improvement of the performance but also an increase of the computational complexity of the design stage, although recently important advances have been introduced [4].

2.3 The Quadratic Cost Function

In order to achieve the best performance, the choice of the cost function should be done with reference to the specific application scenario. For example, when the desired output assumes real values, the classical quadratic function is the optimum choice for the cost function when the environment disturbance is Gaussian while a robust cost function [3] should be used in the presence of disturbances with nonGaussian statistics. Obviously, in order to cope with the difficulties of a considered application, the final user may choose an appropriate cost function. It is important to note that the choice of the cost function does not only affect the achieved performance but also the computational complexity of the design stage. More specifically, the choice of a quadratic cost function allows one to determine the optimum $\boldsymbol{\alpha}_q$ by solving the following square linear system of size $M(\ell)$

$$(\boldsymbol{\Phi}^T \boldsymbol{\Phi} + \gamma \mathbf{R}) \boldsymbol{\alpha}_q = \boldsymbol{\Phi}^T \mathbf{d} \tag{4}$$

where $\boldsymbol{\Phi} \triangleq [\boldsymbol{\phi}(\mathbf{x}_1) \dots \boldsymbol{\phi}(\mathbf{x}_\ell)]^T$, $\boldsymbol{\phi}(\mathbf{x}) \triangleq [\phi_1(\mathbf{x}) \dots \phi_{M(\ell)}(\mathbf{x})]^T$, and $\mathbf{d} \triangleq [d_1 \dots d_\ell]^T$.

When the matrix $\mathbf{R}$ is chosen in the following form $[\mathbf{R}_a, \mathbf{0}; \mathbf{0}, \mathbf{0}]$ where $\mathbf{R}_a$ is the upper-left square subblock of size $M_a < M(\ell)$, i.e. only the first M_a components of $\boldsymbol{\alpha}$ are regularized, then the vector $\boldsymbol{\alpha} = [\boldsymbol{\alpha}_a^T, \boldsymbol{\alpha}_b^T]^T$ can be determined by solving a linear system of smaller size [6].

The choice of the quadratic cost function often (but not always) reduces the computational complexity of the batch design stage since it requires the solution of a simple linear system. However, it determines a computational complexity of the final processing method that is proportional to the chosen $M(\ell)$ since all the components of $\boldsymbol{\alpha}_q$ are nonnull.

Assume first that the final user is able to choose the class of systems in (1) such that the following two conditions are satisfied: (a) $M(\ell)$ small and (b) S_a, defined as the set of vectors $\boldsymbol{\alpha}$ in (1) that guarantee a satisfactorily performance, is not empty. This means that sufficient a priori knowledge about the considered application has been embedded inside the considered learning algorithm. This can be seen equivalently as the result of a semi-automatic learning procedure where the final user has first conceived a large set of possible operators and has successively selected – both on the basis of specific considerations and on the basis of experimental trials – a sufficiently small number of systems included in (1). We have called it a semi-automatic procedure because the $M(\ell)$ systems included in (1) have been selected by means of a human choice from the very large set S_T of systems conceivable by the final user. Starting from S_T a subset S_t is determined on the basis of particular considerations regarding the specific discipline. Finally, from the subset S_t the $M(\ell)$ systems included in (1) are selected by means of experimental data, often not included in the ℓ training examples. Such an achievement is often obtained by a time-consuming procedure involving an human operator; moreover, after a large number of trials, correct generalization is not any more guaranteed by the learning theory also if finally the number of selected systems $M(\ell)$ is much smaller than the number ℓ of learning examples used in (4).

2.4 Forcing a Sparse Solution

An important advance of the learning theory consists in substituting with an automatic procedure the semi-automatic one. Let us therefore assume that a (possibly very large) number $M(\ell)$ of systems (belonging to the above mentioned set S_t) is included in (1). Then, the design stage (4) is not any more the optimum choice from the point of view of computational complexity. Let us further discuss this issue.

Consider the already mentioned set S_a of vectors $\boldsymbol{\alpha}$ in (1) that guarantee a satisfactorily performance. Obviously, the cost function (3) (and perhaps also (2)) achieves its minimum in a specific point $\boldsymbol{\alpha}_q$ (or, perhaps, in a subset of points including $\boldsymbol{\alpha}_q$) but not for any $\boldsymbol{\alpha} \in S_a$; however, the differences in the final performances relative to the vectors $\boldsymbol{\alpha} \in S_a$ are negligible. More correctly, in some application scenarios, the possible performance loss of the worst element of S_a with respect to $\boldsymbol{\alpha}_q$ – although possibly not negligible – may be acceptable when it is compensated by a corresponding reduction of the computational complexity of the design stage and/or of the implemented algorithm. Moreover, the presence of measurement noise may determine a difference between $\boldsymbol{\alpha}_q$ and the true optimum; this is another reason to consider a set S_a of acceptable solutions. We assume that S_a is not empty; this obviously holds provided that the specific application admits solution and the $M(\ell)$ selected systems satisfactorily represent the above-mentioned set S_t.

Let us also introduce the crucial definition of the set $S_{d,n}$ of n-dimensional vectors with a number of null components larger than $n - d$, i.e., the set of vectors with not more than d nonnull components. The problem of automatic learning from examples consists in taking advantage from the fact that the intersection $I_{i,M(\ell)}$ between the set S_a and the set $S_{i,M(\ell)}$ is not empty also for $i \ll M(\ell)$. Let us call d the minimum integer i such that $I_{i,M(\ell)}$ is not empty.

The larger is the ratio $\frac{M(\ell)}{d}$, the more significant the complexity reduction (with respect to (4)) of the design stage and of the processing implementation may be. Exploitation of the property that $d \ll M(\ell)$ is not present in the classical learning algorithms since they were devised for a scenario, resulting from a semi-automatic procedure, where $d \simeq M(\ell)$ and $M(\ell)$ is sufficiently small. Only recently, the learning theory has started to consider methods that are able to take advantage from large values of the ratio $\frac{M(\ell)}{d}$.

There are three basic approaches to exploit the property $d \ll M(\ell)$. The first approach adds to the cost function (3) a term that counts (or approximately counts) the number of nonnull components of $\boldsymbol{\alpha}$. The second approach sets the value of d and tries to optimize the performance in $S_{d,M(\ell)}$ or, at least, to determine an element of $I_{d,M(\ell)}$. The third approach sets the minimum acceptable performance quality and therefore the set S_a and, then, searches the value of d (consequently defined by S_a) and a solution in $S_{d,M(\ell)}$.

The first approach is followed by the basic SVM where a cost function with a "null interval" is chosen; this choice implies the existence of a set S_v of values of $\boldsymbol{\alpha}$ achieving the global minimum of the cost function (3). An appropriate choice of the systems in (1) guarantees that $\boldsymbol{\Phi}$ is positive definite (see previous discussion) and the choice of the regularization matrix $\mathbf{R} = \boldsymbol{\Phi}$ implies that the regularization term is minimized when $\boldsymbol{\alpha}$ has a certain number of null components. The basic SVM can be shown to be equivalent to an alternative method that adopts as cost function a quadratic function of $\boldsymbol{\alpha}$ reaching its minimum in $\boldsymbol{\alpha}_q$ and, as additive term, the sum of the absolute values of the components of $\boldsymbol{\alpha}$, which is one of the best convex approximations of the number of nonnull components of $\boldsymbol{\alpha}$ (see [6] for a detailed discussion).

A new class of methods to force sparsity in the automatic learning can be determined by considering the system $\boldsymbol{\Phi}\boldsymbol{\alpha} \simeq \mathbf{d}$ and by using different methods for its sparse solution, i.e., to determine a sparse vector $\boldsymbol{\alpha}_e$ such that the components of $\mathbf{d} - \boldsymbol{\Phi}\boldsymbol{\alpha}_e$ are sufficiently small. Then, only the systems in (1) corresponding to the nonnull components in $\boldsymbol{\alpha}_e$ are selected and the final vector $\boldsymbol{\alpha}$ is determined from (4). This is a two-stage procedure that can be applied in all the three approaches since the methods for the sparse solution of a linear system exist in all the three settings. According to this two-stage procedure, two methods have been proposed in [6] where a simple example demonstrated that SVM is not always the best way to force the sparsity: in fact, the proposed alternative methods have obtained the same performance with a reduced complexity of both design stage and processing implementation. This was also noted in the first applications of the basic SVM method where the computational complexity of a successive design stage, referred to as reduced-order SVM, has been traded-off with a reduction of the computational complexity of the processing implementation.

Note that many methods have been developed for obtaining a sparse solution of a linear system (e.g., see [12] and references therein) and for obtaining the selection of the systems to include in the expansion (1) (e.g., see [1, 5] and references therein), not always clearly distinguishing the two problems. The important result in [6] consists in achieving a complexity reduction of the processing implementation by using a simpler design stage. The selection methods with a limited computational complexity have particular importance both because they are very suited to cope with real-time constraints and because they are very useful in batch-mode design when very large values of ℓ and $M(\ell)$ are used.

3 Digital Channel Equalization

SVM learning method has been used in several environments as a complete learning tool where the most important choices of the learning procedure have already been done. As previously discussed, this has created problems regarding its use since it is very unlikely that all the choices done in constructing it are adequate to the different scenarios. The attempt to use the SVM method can be limited by the fact that not always the overall set of incorporated choices is well-suited to the application scenarios; also in such a case, however, important advances in the classical design procedures are possible by exploiting some of the principal SVM contributions. This is clearly seen with reference to the problem of digital channel equalization. Consider the discrete-time linear time-invariant noisy communication channel:

$$r(n) = h(n) \otimes x(n) + \eta(n) \tag{5}$$

where $\otimes$ denotes the discrete time convolution, $h(n)$ is the channel impulse response with finite impulse response (FIR), $\eta(n)$ is a zero-mean, independent and identically distributed (IID) noise process with variance σ_n^2 and Gaussian distribution, $x(n) \in \{-1, 1\}$ is an IID sequence of information symbols.

Such a model can be extended to the case of multiple-input/multiple-output (MIMO) digital communication channels where n_o received sequences linearly depends on n_I input sequence:

$$r_j(n) = \sum_{i=1}^{n_I} h_{i,j}(n) \otimes x_i(n) + \eta_j(n) \qquad j \in \{1, \ldots, n_O\} \tag{6}$$

where $h_{i,j}(n)$ represents the dependence on the ith input of the jth output and all input and noise sequences are IID and independent of each other.

The channel models (5) and (6) are often considered in the literature. Channel equalization is the problem of determining a processing system whose input is the received sequence $r(\cdot)$ (or the received sequences $r_j(\cdot)$ in (6)) and whose output is used for deciding about the symbol $x(n) \in \{-1, 1\}$ in (5) (or about the symbols $x_i(n) \in \{-1, 1\}$ in (6)).

The computational complexity of the equalizer that minimizes the probability of error exponentially increases with the length of the channel memory. For such a reason symbol-by-symbol equalizers have been often considered. The linear equalizer is the simplest choice but significant performance advantages are achieved by using a decision-feedback (DF) approach; other nonlinear feedforward (NF) approaches have been proposed in the literature but the performance comparison between NF and DF solutions is often not present.

An equalizer can be based on a direct approach or on a indirect approach. In the former case, the examples are used to learn directly the equalizer. In the latter case, the examples are used to estimate the channel impulse response $h(n)$ and the unknown parameters of the statistics of the input and noise sequences; then, the equalizer is designed on the basis of the assumed model where the ideal values are replaced by the estimated ones. Both approaches have some advantages: the direct approach is more amenable for sample-adaptive equalization and more robust with respect to model mismatches while the indirect approach often presents superior performance in block-adaptive equalization.

The basic SVM method is well-suited to work in a block-adaptive scenario. The computational complexity of the training stage and also the simplification of the software coding has received important contributions in [7] where a very simple iterative algorithm for SVM training has been developed, in [13] where an iterative algorithm requiring at each step the solution of a linear system of size $\ell + 1$ is proposed, and in [15] where an algorithm for multiplicative updates is proposed.

Since the basic SVM method is a block-design method based on a direct approach, a fair comparison with the block-designed linear and DF equalizers operating in an indirect mode should be included. Such a comparison is favorable to the classical methods unless a channel mismatch is considered: in fact, assuming that $h(n) = \delta(n) - z_0\delta(n-1)$, the indirectly block-designed linear and DF equalizers can achieve significant performance advantages provided that the filter memories are sufficiently large. This is especially true when the available number of examples is strongly reduced: in the special case where the no training sequence is available, only the classical indirect approaches based on some methods for blind channel estimation are able to operate. However, also in the presence of a reasonable number of examples, there are problems with the basic SVM method: in fact, the indirect methods estimate the two nonnull values of $h(n)$ by means of the available examples and use them to analytically design the optimum minimum mean square error (MMSE) linear filter. The resulting impulse response $w_{MMSE}(n)$ of the designed filter is very long when $z_0 \to 1$, although the linear equalizer can be easily implemented in a simple recursive form [10]. Such a recursive form, however, cannot be directly learned from the examples and, therefore, a long FIR filter is needed to approximate it. The impulse response of such a long FIR filter cannot be learned from data with the same accuracy and in some cases cannot be learned at all from the data. When the memory of the FIR equalizer is reduced, linear FIR equalizer exhibits unacceptable performances and, therefore, the introduction of nonlinear kernels in the basic SVM is mandatory, so determining a NF equalizer. Note that, when the basic SVM method (or also another NF equalizer) is applied to an input vector with a large number of components practically irrelevant, it may be practically impossible to operate with a radial kernel, which, however, performs well when the most relevant components to use for classification have been selected.

The comparison between the indirectly-designed linear equalizer and directly-designed NF equalizer has been mainly not studied in the literature where the comparison between linear and NF equalizers is often limited to the case of direct design, which is not very fair for the linear equalizer. Moreover, also the comparison within the assumption of direct-design for both the approaches should account for an increase of the memory of the linear filter, which determines a limited increase of the computational complexity and reduces the performance gap. Moreover, in such a scenario, it is well-known that the classical DF equalizer improves the achieved performance (and often reduces the computational complexity).

Although the results of a fair comparison among Linear, NF and DF equalizers are not always clear, we already know that the basic SVM cannot be the best solution among the NF equalizers because it provides a solution where the number of support vectors, straightforwardly related to the computational complexity of the implemented algorithm, is too large. In fact, the large number of support vectors has been shown in all the experiments about the application of the basic SVM method to the equalization of a digital channel, first proposed in [8]. It is fair to note that, in the presence of a nonlinear channel model, the indirect approach cannot use a well-developed method for the design. Therefore, with reference to the nonlinear channel models, it is fair to limit the comparison to the directly-designed linear, DF and NF equalizers. Then, once again the experimental work shows [8] that the performance advantage of the NF equalizer with respect to the classical approach is limited if compared with the significant increase in the computational complexity.

In particular, with reference to the linear channel with impulse response $h(n) = 0.407\delta(n) + 0.815\delta(n-1) + 0.407\delta(n-2)$, followed by the nonlinear zero-memory characteristic $f_n(\zeta) = \zeta$ when $|\zeta| \leq 1$ and $f_n(\zeta) = \text{sign}(\zeta)$ otherwise, where $\text{sign}(x)$ denotes the usual sign function, in [8] the input vector was constructed as $[r(n-2)\ldots r(n+2)]$ and the previous decision $\hat{x}(n-1)$ was also included in the input vector. The classical DF equalizer and the basic SVM method (using the Gaussian kernel $K(\mathbf{x}_1, \mathbf{x}_2) = \exp(-\frac{\|\mathbf{x}_1 - \mathbf{x}_2\|^2}{\sigma^2})$ with $C = 3$ e $\sigma^2 = 4.5$) have been compared in a direct approach by using 500 training examples. The result reported in Fig. 1 shows that 3-4 dB of performance advantage has been achieved with the use of the SVM method. This is, however, paid with a significant increase in the number of support vectors, especially for lower SNR_{nl} defined as $10 \log_{10} \frac{Ez^2(n)}{2E\eta^2(n)}$ where $z(n)$ is the input to the nonlinearity $f_n(\cdot)$. The reported results represent an average over 200 independent trials; in each trial, the estimation P_b of the probability of error has been determined by using 10^6 (or also 10^7 for SNR_{nl} larger than 14 dB) examples not used in the equalizer learning. The coefficients of the two equalizers have also been updated with a decision-directed mode during the test stage by using the LMS algorithm.

Performance advantages very close to that achievable by the basic SVM can also be obtained by using other nonlinear NF equalizers. Such equalizers

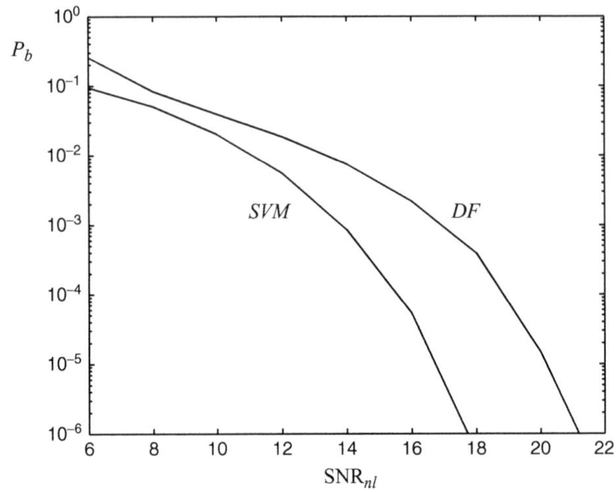

Fig. 1. Performance comparison between the classical DF equalizer and the basic SVM method on a difficult nonlinear channel

force the sparsity of the solutions, without significantly increasing the complexity of the design method, using an approach very similarly to that proposed in [6] with reference to a general learning method: they mainly consist in applying simple methods to determine a sparse approximate solution of the system $\boldsymbol{\Phi\alpha} = \mathbf{d}$.

Therefore, the choices operated in the basic SVM method are well-suited for block-adaptive equalization over nonlinear digital channel when the number of available examples is very small (i.e., a severe nonlinear fast-varying channels), as suggested in [14].

Other applications of the basic SVM method to the problem of digital channel equalization include some contributions where the set of the possible channel states needs to be used during the design stage of the algorithm. This implies that the computational complexity of such design methods exponentially increases with the channel memory and, therefore, such methods cannot be considered acceptable. In fact, when the number of channel states is computationally tractable, the Viterbi algorithm [2], which outperforms all the others, is the obvious choice. Decades of research on channel equalization have been motivated by the need of determining methods exhibiting a weaker dependence on the channel memory. Although the performance of such methods are tested on short memory channels, it should not be forgotten that their principal merit lies in the reasonable computational complexity needed for operating on practical channels with a long memory. When a different equalizer is used and up-dated for each channel state, the large number of channel states determines a very slow convergence of the overall algorithm (i.e., an exponentially large number of training examples is needed to train the state-dependent equalizers).

4 Incorporating SVM Advances in Classical Techniques

The previous discussion has compared the basic SVM approach with the classical methods, looking at SVM as a specific learning method. If we consider it something more than a simple method, important contributions from this approach for a general classification problem can be joined with specific knowledge already incorporated in the classical method.

4.1 The Sparse Channel Estimation

A first issue in existing equalization method is the capability to deal with practical channels in the indirect approach: we can often limit ourselves to consider a linear channel but the channel memory can be very long although many coefficients of the channel impulse response can be neglected; this is the problem of the sparse channel estimation, which derives from the multipath propagation of the wireless communication channels. It is clear from the physics of the field propagation that the significant coefficients of the impulse response are not the closest ones to the time-origin and that for their localization no a priori information is available since the nonnull taps depend on the details of the time-variant electromagnetic channel. Only the assumptions that the channel is linear (and, for a limited time horizon, time invariant) and that the number of nonnull coefficients is small are reasonable.

In this application the systems in (1) are linear systems that produce as output a delayed sequence; the number of such systems to include in (1) in order to guarantee acceptable approximation may also be very large although the number of nonnull coefficients can be very small. The problem fits very well in the general learning scenario considered in the previous section. However, the basic SVM method cannot be applied because the systems, which a priori knowledge suggests to use, in (1) are not compatible with a kernel choice. The linear SVM method is available but unfortunately no sparse solution in the input space can be found by using it. The methods introduced in [6] for sparse solution can be used and in some contributions their use has also been suggested, although the issue needs further investigations. More specifically, in the system design, it would be very useful to know – given a maximum memory length and a small value of the maximum number of nonnull coefficients in the channel impulse response – the minimum number of training examples to use in order to guarantee good performances of the resulting equalizer. The solution of this problem is important to improve the quality of channel estimators used in the indirect approaches and it is crucial to allow the radial-kernel-based NF equalizers to correctly operate by removing the irrelevant components of its input vector. Obviously, the estimation of a sparse channel may also allow significant complexity reduction in the resulting equalizer, although more results are needed about this last issue; such a possible simplified structure of the resulting equalizer may also be learned from the examples in a direct approach by using some methods to force the sparsity in the model.

4.2 Choosing the Cost Function for Classification

A second example of the possible introduction in the equalization methods of the contributions from the SVM approach regards the choice of the cost function to be used in the equalizer design. Such a choice should be not considered peculiar of a specific method but any learning method should be allowed, when possible, to use any cost function. In practice, however, the cost function used in the basic SVM method is not used in any other learning method. Since its introduction is one of the most important advances in fifty years of learning theory, it provides an unfair bias in favor of the kernel-based methods with respect to other parametric methods.

The difficulty in using the same cost function in other learning methods derives from the fact that the derivation of the basic SVM for classification lies on the geometric concept of "maximum margin hyperplane", which, in author's view, does not allow to clearly see the motivation of the choice and does not allow to compare the SVM performances with those achievable by a method relying on a quadratic cost function.

The First Solutions to the Choice of the Cost Function

A simple derivation of the basic SVM for classification has been given in [9] where an useful tool for comparing the SVM approach with all the alternative methods is provided. When the decision $\hat{x}(n)$ about $x(n)$ is taken in accordance with the sign of the equalizer output $y(n)$ (i.e., $\hat{x}(n) = \text{sign}(y(n))$), then the probability of error can be written as

$$P(x(n) \neq \hat{x}(n)) = P(x(n)y(n) < 0) = E[u(-x(n)y(n))] \tag{7}$$

where $P(\cdot)$ denotes the probability, E denotes the statistical expectation, and $u(\cdot)$ denotes the ideal step function (i.e., $u(z) = 1$ for $z > 0$ and null otherwise). By comparing (2) and (7) it is clearly seen that, from the point of view of the performance, the more appropriate cost function for the problem at hand is $c_I(d,y) = c_c(dy) \stackrel{\triangle}{=} u(-dy)$. Although appropriate from the point of view of the performance in (2), the adoption of the cost function $c_c(\cdot)$ implies that the optimization problem (3) exhibits a computational complexity exponentially increasing with ℓ; moreover, such an adoption can also be inappropriate when the number of available examples is small.

Alternative choices of the cost function $c_I(d,y)$ in (3) have been proposed in order to limit the computational complexity of the learning method.

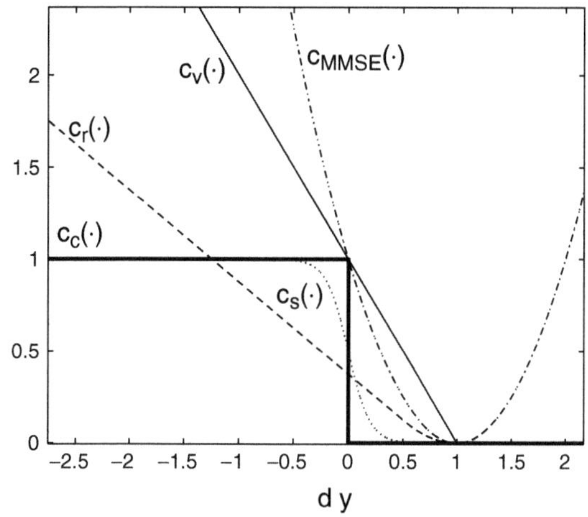

Fig. 2. The different cost functions considered in the Subsect. 4.2: the cost function $c_r(\cdot)$ refers to the choice $(p, \delta) = (2, 0.5)$ and the cost function $c_s(\cdot)$ refers to the choice $A = 0.1$

The first approximation used in the literature proposes the following choice: $c_I(d, y) = c_q(d, y) \stackrel{\triangle}{=} (d - y)^2 = (1 - dy)^2 \stackrel{\triangle}{=} c_{MMSE}(dy)$, where the relation $d = \pm 1$ has been used. Such an approximation has been widely used since the first works in the learning theory because this choice minimizes the computational complexity of the optimization problem (3), as already discussed. The approximation, however, is very crude as shown by Fig. 2 where the functions $c_c(\cdot)$ and $c_{MMSE}(\cdot)$ are represented, together with other cost functions discussed in the sequel.

The Sigmoidal Choice

Performance improvements can be obtained by considering the following approximation: $c_I(d, y) = c_s(dy)$ where $c_s(\cdot)$ is a sigmoidal function such as the logistic function $c_s(z) \stackrel{\triangle}{=} [1 + \exp(\frac{z}{A})]^{-1}$. When the parameter A in the sigmoidal function is set to 0.1, $c_s(\cdot)$ is very close to $c_c(\cdot)$ as shown in Fig. 2; a smaller value of A provides a cost function $c_s(\cdot)$ closer to $c_c(\cdot)$. The cost function $c_s(\cdot)$ has been used in the derivation of the well-known backpropagation algorithm. The main disadvantage lies in the fact that the optimization problem (3) with $c_I(d, y) = c_s(dy)$ exhibits many local minima and, therefore, the result of the learning procedure depends on the initialization procedure. The presence of the local minima restricts the chosen class of functions since an iterative learning algorithm is practically limited to a restricted subset around the local minimum obtained starting from the chosen initialization. This restriction motivates the choice of a very general class of functions such that also the restricted subset is general enough; this can allow one to achieve good performance without solving the problem of the local minima in the global optimization. Consequently, the restricted subset may also have a dimension too large with respect to the available examples; when an iterative algorithm is used for approximating a local minimum, an early stopping procedure is often employed to achieve the regularization. The development of learning methods, which leads to the well-known neural networks, is strongly affected by the initial choice of a nonconvex cost function.

Vapnik's Choice

Basic SVM is based on the choice of an approximating cost function, first proposed forty years ago but very popular only in the last decade [16], that can be written as $c_I(d,y) = c_v(dy) \triangleq (1-dy)\ u(-dy+1)$. It is the simplest way to determine a convex approximation of the ideal cost function $c_c(\cdot)$ as Fig. 2 clearly shows. The choice of the cost function $c_v(\cdot)$ and of the class of functions in (1) guarantees that the former term in (3) is convex. Since the latter term (i.e., $\gamma\boldsymbol{\alpha}^T\mathbf{R}\boldsymbol{\alpha}$) is strictly convex, also the overall cost function (3) is strictly convex and, therefore, every local minimum is also global and the iterative learning methods can achieve the global minimum. This is an important advantage since this can be achieved in a class of functions whose dimension is known and, therefore, it can be controlled. Such a choice still implies an approximation but now it is not as crude as the approximation with $c_{MMSE}(\cdot)$, especially when the input of the cost function is larger than one. This motivates a performance improvement obtained by using $c_v(\cdot)$ instead of $c_{MMSE}(\cdot)$.

Let us now consider the following convex cost function $c_c(\cdot)$: $c_l(z,\delta_1,\delta_2) \triangleq \delta_1 \cdot (1-\frac{z}{\delta_2}) \cdot u(-z+\delta_2)$, with $\delta_1, \delta_2 > 0$. Note that $c_l(z,1,1) = c_v(z)$ and $c_l(z,\delta_1,\delta_2) = \delta_1 c_v(\frac{z}{\delta_2})$. Consequently, the solution obtained with c_l coincides with that obtained with c_v provided that the parameter γ is replaced by $\gamma\frac{\delta_1}{\delta_2^2}$.

The issue of the choice of the cost function is crucial in all the linear and nonlinear equalizers, which can be written in the form (1), operating in both direct and indirect mode. In order to develop a gradient iterative approach or a Newton iterative approach, it is useful to introduce the following convex cost function, which admits derivative everywhere:

$$c_r(z) \triangleq \begin{cases} \delta^{p-1}(1-z) - \left(1-\frac{1}{p}\right)\delta^p & z \leq 1-\delta \\ \dfrac{(1-z)^p}{p} & 1-\delta \leq z \leq 1 \\ 0 & z \geq 1 \end{cases} . \tag{8}$$

When the basic SVM for classification is introduced with the approach followed here, it is possible to extend the approach proposed in [6] for real-valued output, to select a reduced number of $M(\ell)$ systems and, use the same cost function used with reference to nonparametric SVM method also in the resulting parametric optimization.

In such a case, the passage to the dual optimization problem is no longer needed and the learning problem can be solved directly in the input space. The optimization problem (3) is an unconstrained minimization problem. Let us determine the gradient $\mathbf{g}(\boldsymbol{\alpha})$ and the Hessian $\mathbf{H}(\boldsymbol{\alpha})$ of the cost function (3):

$$\mathbf{g}(\boldsymbol{\alpha}) = \frac{1}{\ell}\sum_{n=1}^{\ell} d(n)c^{'}(d(n)y(n))\boldsymbol{\phi}(x(\cdot), n) + 2\gamma\mathbf{R}\boldsymbol{\alpha} \tag{9}$$

$$\mathbf{H}(\boldsymbol{\alpha}) = \frac{1}{\ell}\sum_{n=1}^{\ell} c^{''}(d(n)y(n))\boldsymbol{\phi}(x(\cdot), n)\boldsymbol{\phi}^{T}(x(\cdot), n) + 2\gamma\mathbf{R} \tag{10}$$

where $c^{'}(\cdot)$ and $c^{''}(\cdot)$ denote the first and the second derivative, respectively, of the cost function $c(\cdot)$ chosen to approximate $c_I(d, y)$ in (3): $c_I(d, y) \simeq c(dy)$. In particular, an iterative gradient algorithm can be written as

$$\boldsymbol{\alpha}_{n+1} = \boldsymbol{\alpha}_n - \mu\left(\frac{1}{\ell}\sum_{i=0}^{\ell-1} d(n-i)c^{'}(d_y(n-i))\boldsymbol{\phi}(x(\cdot), n-i) + 2\gamma\mathbf{R}\boldsymbol{\alpha}_n\right) \tag{11}$$

where $d_y(n-i) \triangleq d(n-i)\boldsymbol{\phi}^{T}(x(\cdot), n-i)\boldsymbol{\alpha}_n$, the vector of operators $\boldsymbol{\phi}(x(\cdot), n-i)$ is obviously defined, and the value of ℓ is fixed as a compromise between the computational complexity and the quality of the gradient estimation. Moreover, in a nonstationary environment, a smaller value of ℓ may improve the performances. When $\ell = 1$, the stochastic gradient algorithm is obtained.

When a block-adaptive algorithm is used, it can be useful to improve the convergence of the gradient algorithm by using a Newton method:

$$\boldsymbol{\alpha}_{n+1} = \boldsymbol{\alpha}_n - \mu\mathbf{H}(\boldsymbol{\alpha}_n)^{-1}\mathbf{g}(\boldsymbol{\alpha}_n) \tag{12}$$

Moreover, when the set of functions in (1) is a FIR linear filter, which is an important choice in sample-adaptive equalization, the following adaptive algorithm is obtained

$$\boldsymbol{\alpha}_{n+1} = \boldsymbol{\alpha}_n - \mu\left(\frac{1}{\ell}\sum_{i=0}^{\ell-1} d(n-i)c^{'}(d(n-i)\boldsymbol{\alpha}_n^{T}\mathbf{x}(n-i))\mathbf{x}(n-i) + 2\gamma\mathbf{R}\boldsymbol{\alpha}_n\right) \tag{13}$$

For $\ell = 1$ the stochastic gradient adaptive filter is obtained; it reduces to the classical LMS algorithm for $c(\cdot) = c_q(\cdot)$ and $\gamma = 0$; in fact, $c_q^{'}(dy) = 2(dy-1) = 2d(y-d)$ since $d^2 = 1$. In correspondence of the choice suggested by Vapnik $c(\cdot) = c_v(\cdot)$, then

$$c_v^{'}(z) = \begin{cases} -1 & z \leq 1 \\ 0 & z > 1 \end{cases} \tag{14}$$

In correspondence of the choice proposed in [9] $c(\cdot) = c_r(\cdot)$, then

$$c_r^{'}(z) = \begin{cases} -\delta^{p-1} & z \leq 1-\delta \\ -(1-z)^{p-1} & 1-\delta \leq z \leq 1 \\ 0 & z > 1 \end{cases} \tag{15}$$

$$c_r^{''}(z) = \begin{cases} 0 & z \leq 1-\delta \\ (p-1)(1-z)^{p-2} & 1-\delta \leq z \leq 1 \\ 0 & z > 1 \end{cases} \tag{16}$$

4.3 The Cost Function and the Equalization Algorithms

Although a large number of different cost functions (with reference to the error $d-y$) has been proposed in the literature, mainly with reference to the problem of the robust design in nonGaussian disturbance and of the computational-complexity reduction in the standard LMS, the proposed method is novel for channel equalization and also very recent works [11] do not take into account the method proposed here.

The adoption of the Vapnik cost function may be important also with reference to the problem of the indirect equalizer design. However, also when $h(n)$ is known, the problem of minimizing (2) when $c_I(\cdot) = c_v(\cdot)$ is not as simple as the case of the quadratic cost function. It is important to note that the optimum solution does not depend only on the second-order statistics of the input and output signals but also on their higher-order statistical characterization.

An indirect approach, however, can also be followed by using the methods developed with reference to a direct approach provided that a large number of training examples is artificially generated from the known model and used to train the chosen equalizer. This implies that also the higher-order description of the input sequence and of noise statistics is needed. The assumption of IID input and noise processes is reasonable in many scenarios.

Such an approach can be used with reference to the considered cost function both in the case of linear and nonlinear channels. This may allow us to take advantage of the fact that the linear channel can be estimated well also by using a small number of examples while the learning of a sufficiently long linear equalizer needs a large number of examples.

The design of the algorithm on the basis of the proposed cost function also allows one to generalize it to the case where the desired processing output $d(n)$ belongs to a finite set of N different values. In such a case, unlike the great majority of extensions of the SVM method to multi-class case, we assume that the multiclass decision has to be performed on the basis of the output $y(n)$ of a single equalizer. This is motivated by the need to maintain limited the computational complexity of the processing algorithm also in the presence of large symbol constellations. Then, the use of the considered cost function in order to achieve multiclass extension is straightforward.

5 Simulation Experiments

In this section we report the result of a set of experiments aimed at comparing the classical linear equalizer designed according to the cost functions $c_{MMSE}(\cdot)$ and $c_r(\cdot)$.

When the indirect approach is used, a fair comparison should consider longer filters. We compare, here, the performances in the presence of short FIR filter since we want to use such experiments to investigate the possibility to improve the performance of the classical LMS linear filter by using the proposed iterative algorithm. This possible advantage is clearly present if improved performances are achieved in the block-based indirect training of short linear filters. When the direct approach is mandatory, it may happen that the length of the approximating filter is not sufficiently long and also the choice to distribute the taps between causal and anticausal components cannot be done a priori. Therefore, the same number of causal and anticausal taps should be considered.

The optimum linear MMSE equalizer is designed assuming that the channel (as well as the signal and noise correlations) is perfectly known; such a knowledge of the scenario is also used to generate a large number $\ell = 5000$ of training examples used to train the linear equalizer for classification (C-linear) according to the cost function $c_r(\cdot)$. We also set $\mathbf{R}$ equal to the identity matrix, $\gamma = 10^{-4}$, $p = 2$, $\delta = 1$ and we define the SNR in dB as $10 \log_{10}(\frac{1}{\sigma^2})$ where σ^2 is the noise variance. The performance of each linear equalizer has been determined according to the analytical expression of the probability of error $P(e)$ in each considered scenario; the performance of the C-linear equalizer is obtained by averaging the results over 15 blocks of 5000 examples.

We first consider the simple case where $h(n) = \delta(n) - 0.75\delta(n-1)$ with SNR $= 35\,\mathrm{dB}$ and the following structure for the linear equalizer $w(n) = w_0\delta(n) + w_1\delta(n+1)$ is chosen. Fig. 3 shows the separating line corresponding to the linear MMSE equalizer; the separating lines corresponding to the optimum linear equalizer and to the C-linear equalizer are not reported since they are both very close to the horizontal axis. This is confirmed by the result in Fig. 4 where a significant advantage of the C-linear classifier is shown. We have noticed that the choice $c_v(\cdot)$, instead of $c_r(\cdot)$, does not alter the performance but renders slowly the optimization problem, and, therefore, it has not been considered in the sequel. The considered example shows the capability of the equalizer designed on the basis of the cost function $c_r(\cdot)$ to achieve a performance very close to the ideal cost function.

We repeated the same experiments for a large number of channels and for different choices of the linear structure. We have never found situations where the MMSE-linear equalizer outperforms the C-linear equalizer. We have found instead channels where the C-linear equalizer significantly outperforms the linear MMSE equalizer. We report in the following the most impressive results of our experimental study.

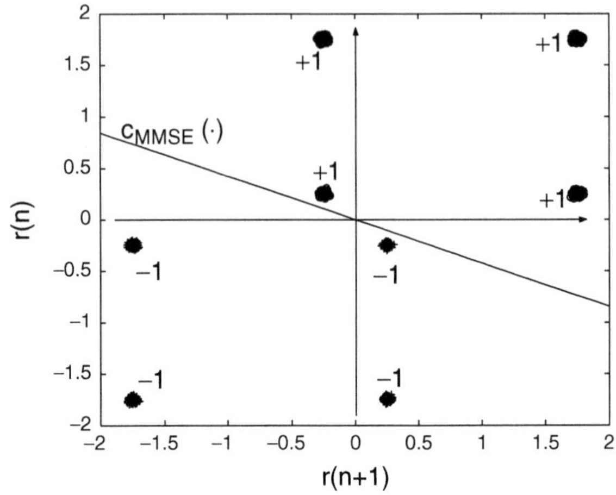

Fig. 3. An example to show the separating line deriving from the cost function $c_{MMSE}(\cdot)$

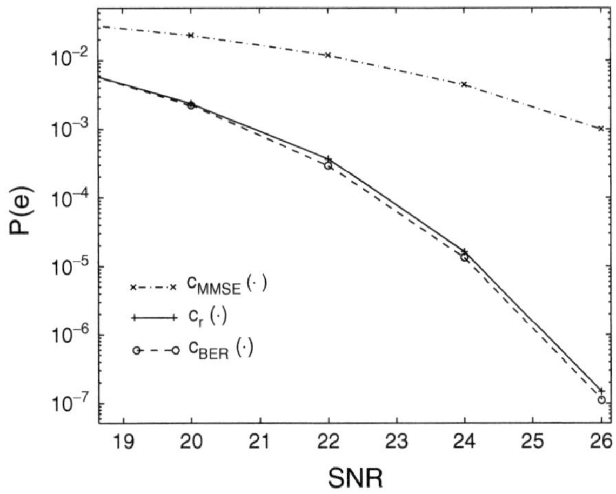

Fig. 4. The performances of the two considered linear equalizers and of the optimum linear equalizer on a simple channel

With reference to the channel $h(n) = \delta(n) + 0.03\delta(n-1) + 0.02\delta(n-2) + 0.9\delta(n-3)$ and to the choice of the linear FIR equalizer with impulse response $w(n) = \sum_{m=-n_b}^{n_f} w_m \delta(n-m)$, we have found impressive differences in the obtained performances. In Fig. 5 we have reported the minimum SNR, say SNR_A, needed to obtain a probability of error $P(e)$ smaller than 10^{-3} versus the number of taps $n_f = n_b$ of the FIR filter. Note that larger values of $n_f = n_b$ do not necessarily improve the equalizer performances.

The most impressive performance improvement has been observed with reference to the 3×3 MIMO channel with impulse responses $h_{1,1}(n) =$

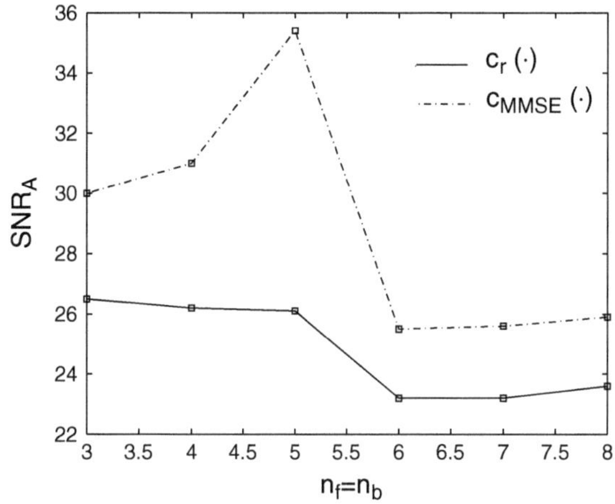

Fig. 5. The performances of the two considered linear equalizers versus the number of equalizer taps

$0.85\delta(n) + 0.03\delta(n-1) + 0.02\delta(n-2)$, $h_{2,1}(n) = 0.7\delta(n-2)$, $h_{1,2}(n) = 0.7\delta(n-1) + 0.02\delta(n-2)$, $h_{2,2}(n) = \delta(n) + 0.03\delta(n-1) + 0.02\delta(n-2) + 0.6\delta(n-3)$, $h_{3,3}(n) = \delta(n) + 0.03\delta(n-1) + 0.02\delta(n-2) + 0.9\delta(n-3)$, $h_{3,1}(n) = h_{1,3}(n) = h_{3,2}(n) = h_{2,3}(n) = 0$. Only the linear equalizer (with two causal and two anticausal taps) for extracting $x_1(n)$ has been considered and the resulting performances, when it is trained according to each of the two different cost functions (including $\ell = 5000$ examples for the case of C-linear equalizer), are reported in Fig. 6.

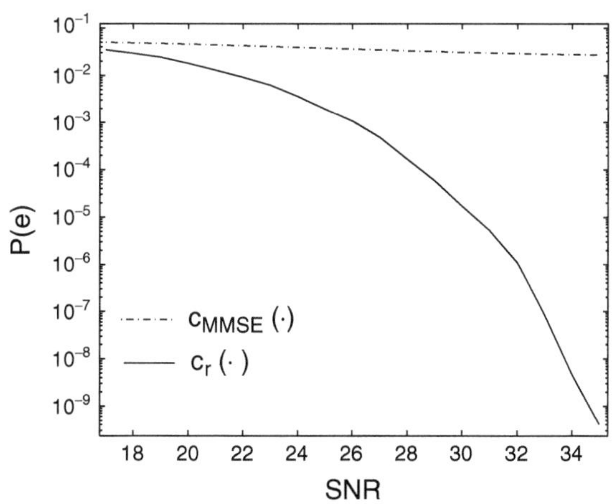

Fig. 6. The performances of the two considered equalizers in the chosen 3×3 MIMO channel

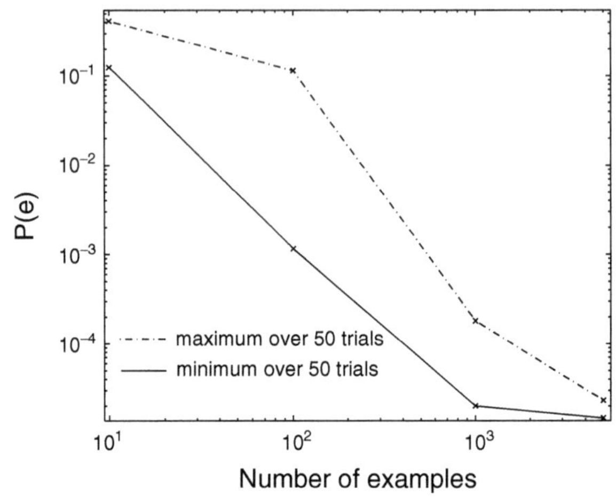

Fig. 7. The dependence on ℓ of the C-linear equalizer in a 3×3 MIMO channel for SNR = 30 dB

Finally, we have considered the effect of a limited number of examples and, with reference to the above mentioned MIMO channel, we have reported in Fig. 7 the probability of error of the C-linear equalizer versus the number ℓ of the training examples. The SNR is set to 30 dB and, for each value of ℓ, the minimum and the maximum values of the probability of error over 50 independent trials are reported. The same experiments have been performed with reference to the first channel ($h(n) = \delta(n) - 0.75\delta(n-1)$); in Fig. 8 it is reported the maximum probability of error achieved over 15 independent

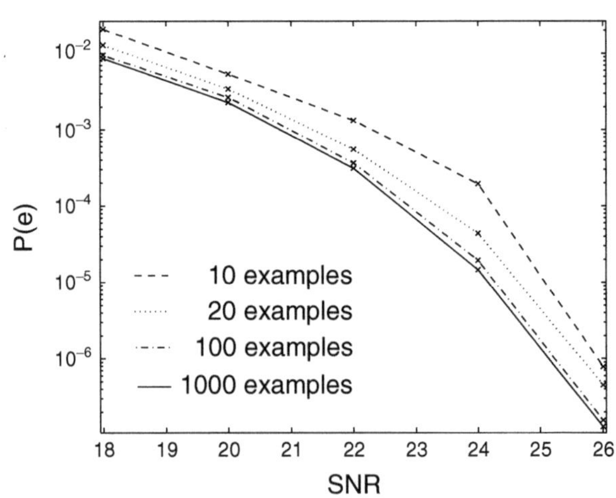

Fig. 8. The dependence on ℓ of the performance of the C-linear equalizer for a simple linear channel

trials. The results show that, in the considered scenario, a small number of examples is sufficient to outperform the linear MMSE equalizer. More experimental studies are needed to compare the performances of the two equalizers in the presence of a limited number of examples.

6 Conclusions

We have discussed the fact that the basic SVM, viewed as a method to force the sparsity of the solution, cannot be the optimum method for any application. Moreover, we have provided an overview and new results with reference to the application of the basic SVM method to the problem of digital channel equalization. We have also provided an unusual derivation of the basic SVM method. This has allowed us to show that the cost function used for classification is a very attractive choice for the final user. We have, therefore, introduced its use in the classical parametric approach, not applied yet in channel equalization. Its application to the classical problem of the linear equalizer design has determined impressive performance advantages with respect to the linear MMSE equalizer.

References

1. Baudat G, Anouar F (2003) Feature vector selection and projection using kernels. Neurocomputing, **55**, 21–38
2. Haykin S (2001) Communication Systems, 4th edn, John Wiley & Sons
3. Huber P J (1981) Robust Statistics. John Wiley and Sons, New York
4. Karacali B, Krim H (2003) Fast minimization of structural risk by nearest neighbor rule. IEEE Trans. on Neural Networks, **14**, 127–137
5. Mao K Z (2004) Feature subset selection for support vector machines through discriminative function pruning analysis. IEEE Trans. on Systems, Man and Cybernetics, **34**, 60–67
6. Mattera D, Palmieri F, Haykin S (1999) Simple and Robust Methods for Support Vector Expansions. IEEE Trans. on Neural Networks, **10**, 1038–1047
7. Mattera D, Palmieri F, Haykin S (1999) An explicit algorithm for training support vector machines. Signal Processing Letters, **6**, 243–245
8. Mattera D Nonlinear Modeling from Empirical Data: Theory and Applications [in italian]. National Italian Libraries of Rome and Florence (Italy), February 1998
9. Mattera D, Palmieri F (1999) Support Vector Machine for nonparametric binary hypothesis testing. In: M. Marinaro e R. Tagliaferri (Eds.), Neural Nets: Wirn Vietri-98, Proceedings of the 10th Italian Workshop on Neural Nets, Vietri sul Mare, Salerno, Italy, 21–23 may 1998, Springer-Verlag, London
10. Mattera D, Palmieri F, Fiore A (2003) Noncausal filters: possible implementations and their complexity. In: Proc. of International Conference on Acoustic, Speech and Signal Processing (ICASSP03), IEEE, 6:365–368

11. Al-Naffouri T Y, Sayed A H (2003) Transient analysis of adaptive filters with error nonlinearities. IEEE Trans. on Signal Processing, **51**, 653–663
12. Natarajan B (1995) Sparse Approximate Solutions to Linear Systems. Siam J. Computing, **24**, 227–234
13. Perez-Cruz F, Navia-Vazquez A, Figueiras-Vidal A R, Artes-Rodriguez A (2003) Empirical risk minimization for support vector classifiers. IEEE Trans. on Neural Networks, **14**, 296–303
14. Perez-Cruz F, Navia-Vazquez A, Alarcon-Diana P L, Artes-Rodriguez A (2003) SVC-based equalizer for burst TDMA transmissions. Signal Processing, **81**, 1681–1693
15. Sha F, Saul L K, Lee D D (2003) Multiplicative updates for nonnegative quadratic programming in support vector machines. In Thrun, Becker and Obermayer, editors, Advances in Neural Information Processing Systems 15, Cambridge, MA, MIT Press
16. Vapnik V (1995) The nature of statistical learning theory. Springer, Berlin Heidelberg New York

Cancer Diagnosis and Protein Secondary Structure Prediction Using Support Vector Machines

F. Chu, G. Jin, and L. Wang

School of Electrical and Electronic Engineering,
Nanyang Technological University,
Block S1, Nanyang Avenue, Singapore, 639798
elpwang@ntu.edu.sg

Abstract. In this chapter, we use support vector machines (SVMs) to deal with two bioinformatics problems, i.e., cancer diagnosis based on gene expression data and protein secondary structure prediction (PSSP). For the problem of cancer diagnosis, the SVMs that we used achieved highly accurate results with fewer genes compared to previously proposed approaches. For the problem of PSSP, the SVMs achieved results comparable to those obtained by other methods.

Key words: support vector machine, cancer diagnosis, gene expression, protein secondary structure prediction

1 Introduction

Support Vector Machines (SVMs) [1, 2, 3] have been widely applied to pattern classification problems [4, 5, 6, 7, 8] and nonlinear regressions [9, 10, 11]. In this chapter, we apply SVMs to two pattern classification problems in bioinformatics. One is cancer diagnosis based on microarray gene expression data; the other is protein secondary structure prediction (PSSP). We note that the meaning of the term *prediction* is different from that in some other disciplines, e.g., in time series prediction where prediction means guessing future trends from past information. In PSSP, "prediction" means supervised classification that involves two steps. In the first step, an SVM is trained as a classifier with a part of the data in a specific protein sequence data set. In the second step (i.e., prediction), we use the classifier trained in the first step to classify the rest of the data in the data set.

In this work, we use the C-Support Vector Classifier (C-SVC) proposed by Cortes and Vapnik [1] available in the LIBSVM library [12]. The C-SVC has radial basis function (RBF) kernels. Much of the computation is spent on

F. Chu, G. Jin, and L. Wang: *Cancer Diagnosis and Protein Secondary Structure Prediction Using Support Vector Machines*, StudFuzz **177**, 343–363 (2005)
www.springerlink.com

tuning two important parameters, i.e., γ and C. γ is the parameter related to the span of an RBF kernel: the smaller the value is, the wider the kernel spans. C controls the tradeoff between the complexity of the SVM and the number of nonseparable samples. A larger C usually leads to higher training accuracy. To achieve a good performance, various combinations of the pair (C, γ) have to be tested, ideally, to find the optimal combination.

This chapter is organized as follows. In Sect. 2, we apply SVMs to cancer diagnosis with microarray data. In Sect. 3, we review the PSSP problem and its biological background. In Sect. 4, we apply SVMs to the PSSP problem. In the last section, we draw our conclusions.

2 SVMs for Cancer Type Prediction

Microarrays [15, 16] are also called gene chips or DNA chips. On a microarray chip, there are thousands of spots. Each spot contains the clone of a gene from one specific tissue. At the same time, some mRNA samples are labelled with two different kinds of dyes, for example, Cy5 (red) and Cy3 (blue). After that, the mRNA samples are put on the chip and interact with the genes on the chip. This process is called *hybridization*. The color of each spot on the chip changes after hybridization. The image of the chip is then scanned out and reflects the characteristics of the tissue at the molecular level. Using microarrays for different tissues, biological and biomedical researchers are able to compare the difference of those tissues at the molecular level. Figure 1 summarizes the process of making microarrays.

In recent years, cancer type/subtype prediction has drawn a lot of attention in the context of the microarray technology that is able to overcome some limitations of traditional methods. Traditional methods for diagnosis of different types of cancers are mainly based on morphological appearances

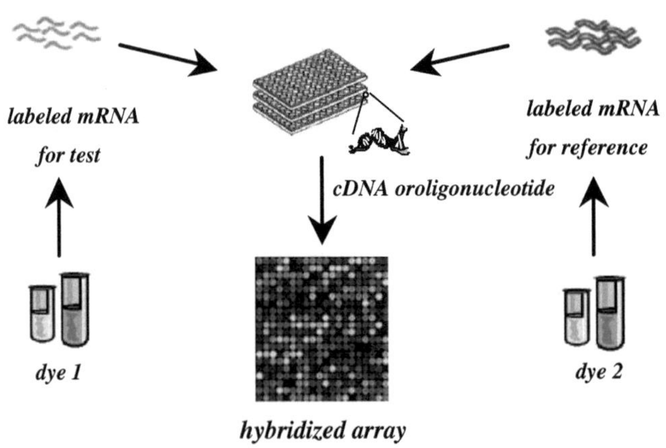

Fig. 1. The process of making microarrays

of cancers. However, sometimes it is extremely difficult to find clear distinctions between some types of cancers according to their appearances. Thus, the newly appeared microarray technology is naturally applied to this muddy problem. In fact, gene-expression-based cancer classifiers have achieved good results in classifying lymphoma [17], leukemia [18], breast cancer [19], liver cancer [20], and so on.

Gene-expression-based cancer classification is challenging due to the following two properties of gene expression data. Firstly, gene expression data are usually very high dimensional. The dimensionality usually ranges from several thousands to over ten thousands. Secondly, gene expression data sets usually contain relatively small numbers of samples, e.g., a few tens. If we treat this pattern recognition problem with supervised machine learning approaches, we need to deal with the shortage of training samples and high dimensional input features.

Recent approaches to this problem include artificial neural networks [21], an evolutionary algorithm [22], nearest shrunken centroids [23], and a graphical method [24]. Here, we use SVMs to solve this problem.

2.1 Gene Expression Data Sets

In the following parts of this section, we describe three data sets to be used in this chapter. One is the small round blue cell tumors (SRBCTs) data set [21]. Another is the lymphoma data set [17]. The last one is the leukemia data set [18].

The SRBCT Data Set

The SRBCT data set (http://research.nhgri.nih.gov/microarray/Supplement/) [21] includes the expression data of 2308 genes. Khan et al. provided totally 63 training samples and 25 testing samples, five of the testing samples being not SRBCTs. The 63 training samples contain 23 Ewing family of tumors (EWS), 20 rhabdomyosarcoma (RMS), 12 neuroblastoma (NB), and 8 Burkitt lymphomas (BL). And the 20 SRBCTs testing samples contain 6 EWS, 5 RMS, 6 NB, and 3 BL.

The Lymphoma Data Set

The lymphoma data set (http://llmpp.nih.gov/lymphoma) [17] has 62 samples in total. Among them, 42 samples are derived from diffuse large B-cell lymphoma (DLBCL), 9 samples from follicular lymphoma (FL), and 11 samples from chronic lymphocytic lymphoma (CLL). The entire data set includes the expression data of 4026 genes. We randomly divided the 62 samples into two parts, 31 for training and the other 31 for testing. In this data set, a small part of data is missing. We applied a k-nearest neighbor algorithm [25] to fill those missing values.

The Leukemia Data Set

The leukemia data set (www-genome.wi.mit.edu/MPR/data_set_ALL_AML.html) [18] contains two types of samples, i.e. the acute myeloid leukemia (AML) and the acute lymphoblastic leukemia (ALL). Golub et al. provided 38 training samples and 34 testing samples. The entire leukemia data set contains the expression data of 7129 genes.

Ordinarily, raw gene expression data should be normalized to reduce the systemic bias introduced during experiments. For the SRBCT and the lymphoma data sets, normalized data can be found on the web. However, for the leukemia data set, such normalized data are not available. Thereafter, we need to do normalization ourselves.

We followed the normalization procedure used in [26]. Three steps were taken, i.e., (a) setting threshold with a floor of 100 and a ceiling of 16000, that is, if a value is greater (smaller) than the ceiling (floor), this value is replaced by the ceiling (floor); (b) filtering, leaving out the genes with $\max / \min \leq 5$ or $(\max - \min) \leq 500$ (max and min refer to the maximum and minimum of the expression values of a gene, respectively); (c) carrying out logarithmic transformation with 10 as the base to all the expression values. 3571 genes survived after these three steps. Furthermore, the data were standardized across experiments, i.e., subtracted by the mean and divided by the standard deviation of each experiment.

2.2 A T-Test-Based Gene Selection Approach

The t-test is a statistical method proposed by Welch [27] to measure how large the difference is between the distributions of two groups of samples. If a gene shows large distinctions between 2 groups, the gene is important for classification of the two groups. To find the genes that contribute most to classification, t-test has been used in gene selection [28] in recent years.

Selecting important genes using t-test involves several steps. In the first step, a score based on the t-test (named t-score or TS) is calculated for each gene. In the second step, all the genes are rearranged according to their TSs. The gene with the largest TS is put in the first place of the ranking list, followed by the gene with the second largest TS, and so on.

Finally, only some top genes in the list are used for classification. The standard t-test is applicable to measure the difference between only two groups. Therefore, when the number of classes is more than two, we need to modify the standard t-test. In this case, we use the t-test to measure the difference between one specific class and the centroid of all the classes. Hence, the definition of the TS for gene i can be described as follows:

$$TS_i = \max\left\{\left|\frac{\overline{x}_{ik} - \overline{x}_i}{m_k s_i}\right|, \quad k = 1, 2, \ldots K\right\} \tag{1}$$

$$\overline{x}_{ik} = \sum_{j \in C_k} \overline{x}_{ij}/n_k \tag{2}$$

$$\overline{x}_i = \sum_{j=1}^{n} x_{ij}/n \tag{3}$$

$$s_i^2 = \frac{1}{n-K}\sum_{k}\sum_{j \in C_k}(x_{ij} - \overline{x}_{ik})^2 \tag{4}$$

$$m_k = \sqrt{1/n_k + 1/n} \tag{5}$$

There are K classes. $\max\{y_k, k = 1, 2, \ldots K\}$ is the maximum of all y_k. C_k refers to class k that includes n_k samples. x_{ij} is the expression value of gene i in sample j. $\overline{x}_{ik}$ is the mean expression value in class k for gene i. n is the total number of samples. $\overline{x}_i$ is the general mean expression value for gene i. s_i is the pooled within-class standard deviation for gene i.

2.3 Experimental Results

We applied the above gene selection approach and the C-SVC to process the SRBCT, the lymphoma, and the leukemia data sets.

Results for the SRBCT Data Set

In the SRBCT data set, we firstly ranked the importance of all the genes with TSs. We picked out 60 of the genes with the largest TSs to do classification. The top 30 genes are listed in Table 1. We input these genes one by one to the SVM classifier according to their ranks. That is, we first input the gene ranked No.1 in Table 1. Then, we trained the SVM classifier with the training data and tested the SVM classifier with the testing data. After that, we repeated the whole process with the top 2 genes in Table 1, and then the top 3 genes, and so on. Figure 2 shows the training and the testing accuracies with respect to the number of genes used.

In this data set, we used SVMs with RBF kernels. C and γ were set as 80 and 0.005, respectively. This classifier obtained 100% training accuracy and 100% testing accuracy using the top 7 genes. In fact, the values of C and γ have great impact on the classification accuracy. Figure 3 shows the classification results with different values of γ. We also applied SVMs with linear kernels (with kernel function $K(\mathbf{X}, \mathbf{X}_i) = \mathbf{X}^T\mathbf{X}_i$) and SVMs with polynomial kernels (with kernel function $K(\mathbf{X}, \mathbf{X}_i) = (\mathbf{X}^T\mathbf{X}_i + 1)^p$ and order $p = 2$) to the SRBCT data set. The results are shown in Fig. 4 and Fig. 5. The SVMs with linear kernels and the SVMs with polynomial kernels obtained 100% accuracy with 7 and 6 genes, respectively. The similarity of these results indicates that the SRBCT data set is separable for all the three kinds of SVMs.

Table 1. The 30 top genes selected by the t-test in the SRBCT data set

Rank	Gene ID	Gene Description
1	810057	cold shock domain protein A
2	784224	fibroblast growth factor receptor 4
3	296448	insulin-like growth factor 2 (somatomedin A)
4	770394	Fc fragment of IgG, receptor, transporter, alpha
5	207274	Human DNA for insulin-like growth factor II (IGF-2); exon 7 and additional ORF
6	244618	ESTs
7	234468	ESTs
8	325182	cadherin 2, N-cadherin (neuronal)
9	212542	Homo sapiens mRNA; cDNA DKFZp586J2118 (from clone DKFZp586J2118)
10	377461	caveolin 1, caveolae protein, 22 kD
11	41591	meningioma (disrupted in balanced translocation) 1
12	898073	transmembrane protein
13	796258	sarcoglycan, alpha (50kD dystrophin-associated glycoprotein)
14	204545	ESTs
15	563673	antiquitin 1
16	44563	growth associated protein 43
17	866702	protein tyrosine phosphatase, non-receptor type 13 (APO-1/CD95 (Fas)-associated phosphatase)
18	21652	catenin (cadherin-associated protein), alpha 1 (102 kD)
19	814260	follicular lymphoma variant translocation 1
20	298062	troponin T2, cardiac
21	629896	microtubule-associated protein 1 B
22	43733	glycogenin 2
23	504791	glutathione S-transferase A4
24	365826	growth arrest-specific 1
25	1409509	troponin T1, skeletal, slow
26	1456900	Nil
27	1435003	tumor necrosis factor, alpha-induced protein 6
28	308231	Homo sapiens incomplete cDNA for a mutated allele of a myosin class I, myh-1c
29	241412	E74-like factor 1 (ets domain transcription factor)
30	1435862	antigen identified by monoclonal antibodies 12E7, F21 and O13

For the SRBCT data set, Khan et al. [21] 100% accurately classified the 4 types of cancers with a linear artificial neural network by using 96 genes. Their results and our results of the linear SVMs both proved that the classes in the SRBCT data set are linearly separable. In 2002, Tibshirani et al. [23] also correctly classified the SRBCT data set with 43 genes by using a method named nearest shrunken centroids. Deutsch [22] further reduced the number of genes required for reliable classification to 12 with an evolutionary algorithm. Compared with these previous results, the SVMs that we used can achieve

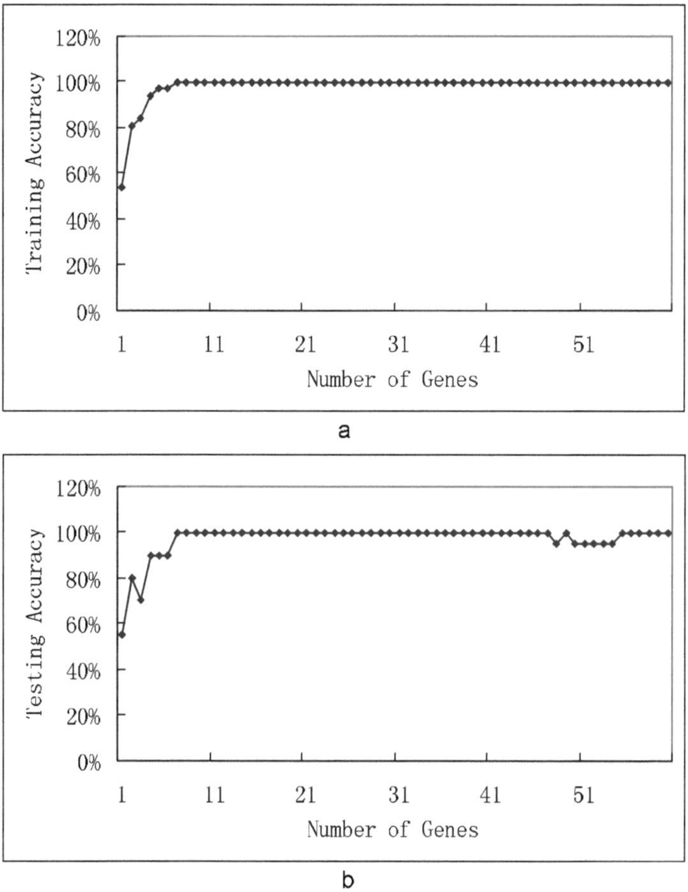

Fig. 2. The classification results vs. the number of genes used for the SRBCT data set: (**a**) the training accuracy; (**b**) the testing accuracy

100% accuracy with only 6 genes (for the polynomial kernel function version, $p = 2$) or 7 genes (for the linear and the RBF kernel function versions). Table 2 summarizes this comparison.

Results for the Lymphoma Data Set

In the lymphoma data set, we selected the top 70 genes. The training and testing accuracies with the 70 top genes are shown in Fig. 6. The classifiers used here are also SVMs with RBF kernels. The best C and γ obtained are equal to 20 and 0.1, respectively. The SVMs obtained 100% accuracy for both the training and testing data with only 5 genes.

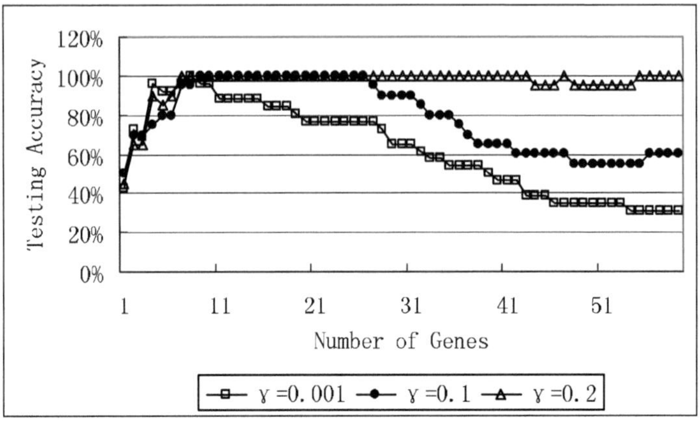

Fig. 3. The testing results of SVMs with RBF kernels and different values of γ for the SRBCT data

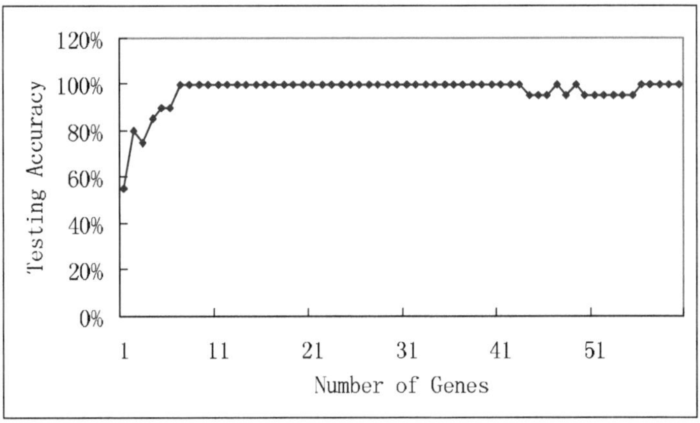

Fig. 4. The testing results of the SVMs with linear kernels for the SRBCT data

For the lymphoma data set, nearest shrunken centroids [29] used 48 genes to give a 100% accurate classification. In comparison with this, the SVMs that we used greatly reduced the number of genes required.

Table 2. Comparison of the numbers of genes required by different methods to achieve 100% classification accuracy

Method	Number of Genes Required
Linear MLP neural network [21]	96
Nearest shrunken centroids [23]	43
Evolutionary algorithm [22]	12
SVM (linear or RBF kernel function)	7
SVM (polynomial kernel function, $p = 2$)	6

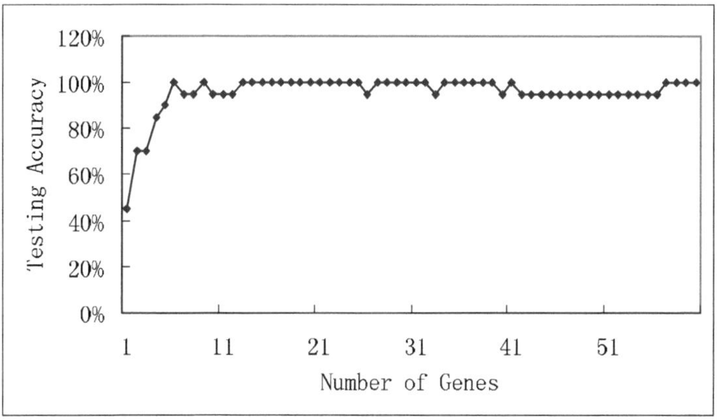

Fig. 5. The testing result of the SVMs with polynomial kernels ($p = 2$) for the SRBCT data

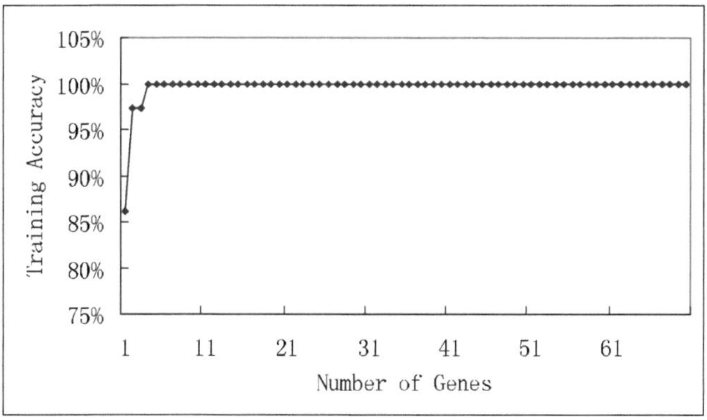

a

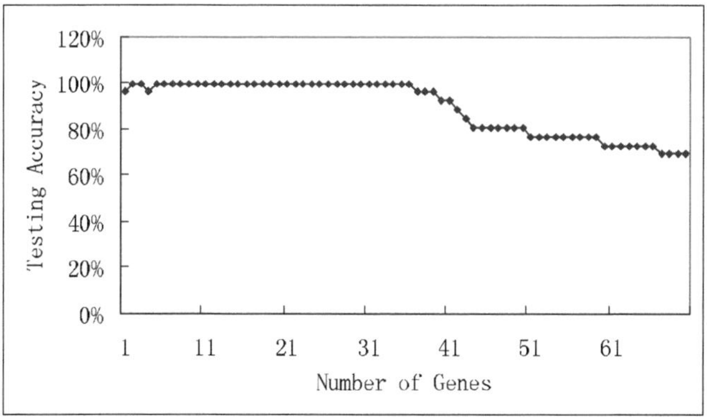

b

Fig. 6. The classification results vs. the number of genes used for the lymphoma data set: (**a**) the training accuracy; (**b**) the testing accuracy

Results for the Leukemia Data Set

Alizadeh et al. [17] built a 50-gene classifier that made 1 error in the 34 testing samples; and in addition, it cannot give strong prediction to another 3 samples. Nearest shrunken centroids made 2 errors among the 34 testing samples with 21 genes [23]. As shown in Fig. 7, we used the SVMs with RBF kernels with 2 errors for the testing data but with only 20 genes.

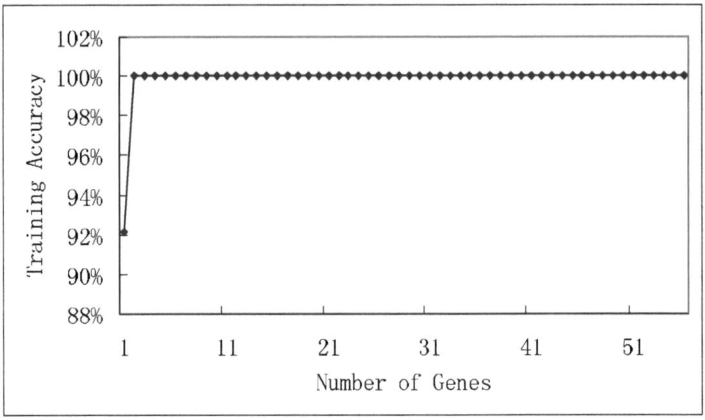

a

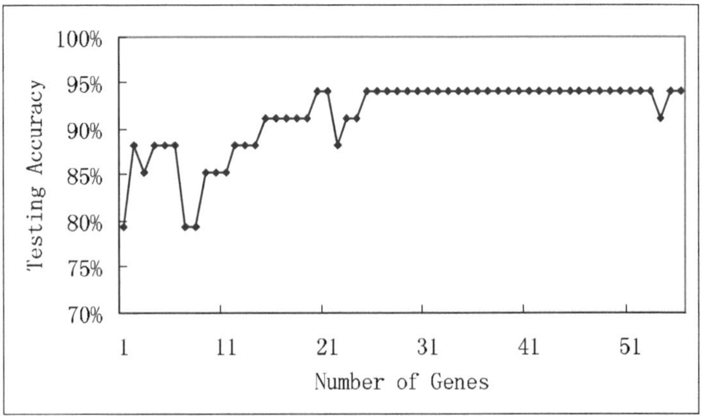

b

Fig. 7. The classification results vs. the number of genes used for the leukemia data set: (**a**) the training accuracy; (**b**) the testing accuracy

Name: Complex Of Troponin C With A 47 Residue (1-47) Fragment Of Troponin I
PDB ID: 1A2X:B
Sequence:

1) GDEEKRNRAI	TARRQHLKSV	MLQIAATELE	KEEGRREAEK	QNYLAEH
2) GDEEKRNRAI	TARRQHLK _ _	MLQIAATELE	KEEGRREAEK	QNYLAEH
3) GDEEKRNRAI	TARRQHLKSV	MLQIAATELEFFE	KEEGRREAEK	QNYLAEH
4) GDEEKGFRAI	TARRQHLKSV	MLQIAATELE	KEEGRREAEK	QNYLAEH

Note
(1) Original Protein Sequence (47 Residues)
(2) Deletion: several amino acids deleted from the chain
(3) Insertion: amino acids FFE was inserted into the original sequence
(4) Substitution: the replacement of amino acids segment by GF

Fig. 8. An example of alphabetical representations of protein sequences and protein mutations. PDB stands for protein data bank [30]

3 Protein Secondary Structure Prediction

3.1 The Biological Background of the PSSP

A protein sequence is a linear array of amino acids. Each amino acid consists of 3 consecutively ordered DNA bases (A, T, C, or G). An amino acid carries various kinds of information determined by its DNA combination. An amino acid is a basic unit of a protein sequence and is called a *residue*. There are altogether 20 types of amino acids and each type of amino acids is denoted by an English character. For example, the character "A" is used to represent the type of amino acid named Alanine. Thus, a protein sequence in the alphabetical representation is a long sequence of characters, as shown in Fig. 8. Given a protein sequence, various evolutionary environments may induce mutations, including insertions, deletions, or substitutions, to the original protein, thereby producing diversified yet biologically similar organisms.

3.2 Types of Protein Secondary Structures

Secondary structures are formed by hydrogen bonds between relatively small segments of protein sequences. There are three common secondary structures in proteins, namely *α-helix, β-sheet (strand)* and *coil.*

Figure 9 visualizes protein secondary structures. In Fig. 9, the dark ribbons represent helices and the gray ribbons are sheets. And the strings in between are coils that bind helices and sheets.

3.3 The Task of PSSP

In the context of PSSP, "prediction" carries similar meaning as that of classification: given a residue of a protein sequence, the predictor should classify the residue into one of the three secondary structure states according to the residue's characteristics. PSSP is usually conducted in two stages: sequence-structure (Q2T) prediction and structure-structure (T2T) prediction.

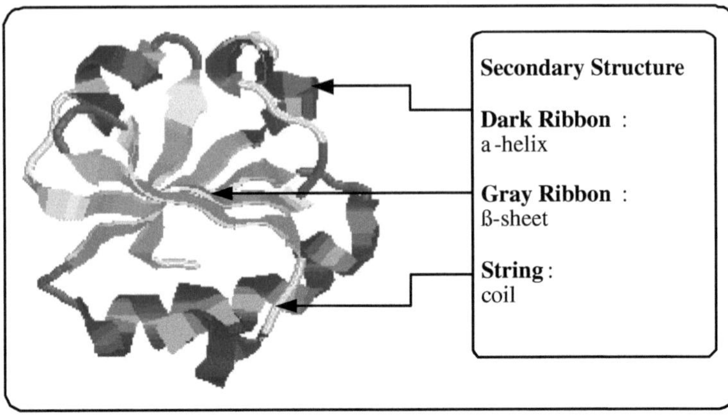

Fig. 9. Three types of protein secondary structures: α-helix, β-strand, and coil

Sequence-Structure (Q2T) Prediction

Q2T prediction predicts the protein secondary structure from protein sequences. Given a protein sequence, a Q2T predictor maps each residue of the sequence to a relevant secondary structure state by inspecting the distinct characteristics of the residue, e.g., the type of the amino acid, the sequence context (that is, what are the neighboring residues), and evolutionary information. The sequence-structure prediction plays the most important role in PSSP.

Structure-Structure (T2T) Prediction

For common pattern classification problems, it would be the end of the task once each data point (each residue in our case) has been assigned a class label. Classification usually do not continue to a second phase. However, the problem we are dealing with is different from most pattern recognition problems. In a typical pattern classification problem, the data points are assumed to be independent. But this is not true for the PSSP problem because the neighboring sequence positions usually provide some meaningful information. For example, an α-helix usually consists of at least 3 consecutive residues of the same secondary structure state (e.g., $\ldots\alpha\alpha\alpha\ldots$). Therefore, if an alternative occurrence of the α-helix and the β-strand (e.g., $\ldots\alpha\beta\alpha\beta\ldots$) is predicted, it would be incorrect. Thus, T2T prediction based on the Q2T results is usually carried out. This step helps to correct errors incurred in Q2T prediction and hence enhances the overall prediction accuracy. Figure 10 illustrates PSSP with the two stages.

Note that amino acids of the same type do not always have the same secondary structure state. For instance, in Fig. 10, the *12*-th and the *20*-th amino residues counted from the left side are both F. However, they are assigned to two different secondary structure states, i.e., α and β.

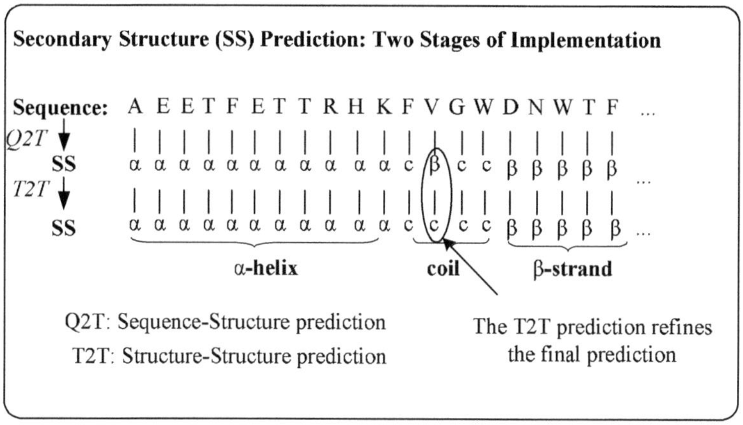

Fig. 10. Protein secondary structure prediction: the two-stage approach

Prediction of the secondary structure state at each sequence position should not solely rely on the residue at that position. A window expanding towards both directions of the residue should be used to include the sequence context.

3.4 Methods for PSSP

PSSP was stimulated by research on protein 3D structures in the 1960s [31, 32], which attempted to find the correlations between protein sequences and secondary structures. This was the first generation of PSSP, where most methods carried out prediction based on single residue statistics [33, 34, 35, 36, 37]. Since only particular types of amino acids from protein sequences were extracted and used in experiments, the accuracies of these methods were more or less over-estimated [38].

With growth of knowledge on protein structures, the second generation PSSP made use of segment statistics. A segment of residues was studied to find out how likely the central residue of the segment belonged to a secondary structure state. Algorithms of this generation include statistical information [36, 40], sequence patterns [41, 42], multi-layer networks [43, 44, 45, 49], multivariate statistics [46], nearest-neighbor algorithms [47], etc.

Unfortunately, the methods in both the first and the second generations could not reach an accuracy higher than 70%.

The earliest application of artificial neural networks to PSSP was carried out by Qian and Sejnowski in 1988 [48]. They used a three-layered back-propagation network whose input data was encoded with a scheme called BIN21. Under BIN21, each input data was a sliding window of 13 residues obtained by extending 6 sequence positions from the central residue. The focus of each observation was only on the central residue, i.e., only the central residue was assigned to one of the three possible secondary structure states

(α-helix, β-strand, and coil). Modifications to the BIN21 scheme were introduced in two later studies. Kneller et al. [49] added one additional input unit to present the hydrophobicity scale of each amino acid residue and showed a slightly higher accuracy. Sasagawa and Tajima [51] used the BIN24 scheme to encode three additional amino acid alphabets, B, X, and Z. The above early work had an accuracy ceiling of 65%. In 1995, Vivarelli et al. [52] used a hybrid system that combined a Local Genetic Algorithm (LGA) and neural networks for PSSP. Although LGA was able to select network topologies efficiently, it still could not break through the accuracy ceiling, regardless of the network architectures applied.

A significant improvement of the 3-state secondary structure prediction came from Rost and Sander's method (PHD) [53, 54], which was based on a multi-layer back-propagation network. Different from the BIN21 coding scheme, PHD took into account evolutionary information in the form of *multiple sequence alignments* to represent the input data. This inclusion of the protein family information improved the prediction accuracy by around six percentages. Moreover, another cascaded neural network conducted structure-structure prediction. Using the 126 protein sequences (RS126) developed by themselves, Rost and Sander achieved the overall accuracy as high as 72%.

In 1999, Jones [56] used a Position-Specific Scoring Matrix (PSSM) [57, 58] obtained from the online alignment searching tool PSI-Blast (http://www.ncbi.nlm.nih.gov/BLAST/) to numerically represent the protein sequence. A PSSM was constructed automatically from a multiple alignment of the highest scoring hits in an initial BLAST search. The PSSM was generated by calculating position-specific scores for each position in the alignment. Highly conserved positions of protein sequence received high scores and weakly conserved positions received scores near zero. Due to its high accuracy in finding the biologically similar protein sequences, the evolutionary information carried by the PSSM is more sensitive than the profiles obtained by other multiple sequence alignment approaches. With a neural network similar to that of Rost and Sander's, Jones' PSIPRED method achieved an accuracy as high as 76.5% using a much larger data set than RS126.

In 2001, Hua and Sun [6] proposed an SVM approach. This was an early application of the SVM to the PSSP problem. In their work, they first constructed 3 one-versus-one and 3 one-versus-all binary classifiers. Three tertiary classifiers were designed based on these binary classifiers through the use of the largest response, the decision tree and votes for the final decision. By making use of the Rost's data encoding scheme, they achieved the accuracy of 71.6% and the segment overlap accuracy of 74.6% for the RS126 data set.

4 SVMs for the PSSP Problem

In this section, we use the LIBSVM, or more specially, the C-SVC, to solve the PSSP problem.

The data set used here was originally developed and used by Jones [56]. This data set can be obtained from the website (http://bioinf.cs.ucl.ac.uk/psipred/). The data set contains a total of 2235 protein sequences for training and 187 sequences for testing. All the sequences in this data set have been processed by the online alignment searching tool PSI-Blast (http://www.ncbi.nlm.nih.gov/BLAST/).

As mentioned above, we will conduct PSSP in two stages, i.e., Q2T prediction and T2T prediction.

4.1 Q2T Prediction

Parameter Tuning Strategy

For PSSP, there are three parameters, i.e., the window size N, and SVM parameters (C,γ), to be tuned. N determines the span of the sliding window, i.e., how many neighbors to be included in the window. Here, we test four different values for N, i.e., 11, 13, 15, and 17.

Searching for the optimal (C,γ) pair is also difficult because the data set used here is extremely large. In [50], Lin and Lin found an optimal pair, $(C,\gamma) = (2, 0.125)$, for the PSSP problem with a much smaller data set (about 10 times smaller compared to the data set used here). Despite the difference of data sizes, we find that their optimal pair also benefits our search as a proper starting point. During our search, we change only one parameter at a time. If the change (increase/decrease) leads to a higher accuracy, we continue to do a similar change (increase/decrease) next time; otherwise, we reverse the change (decrease/increase). Both C and γ are tuned with this scheme.

Results

Tables 3, 4, 5, and 6 show the experimental results for various (C,γ) pairs with the window size $N \in \{11, 13, 15, 17\}$, respectively. Here, Q_3 stands for

Table 3. Q2T prediction accuracies of the C-SVC with different (C,γ) values: window size $N = 11$

		Accuracy			
C	γ	Q_3(%)	Q_α(%)	Q_β(%)	Q_c(%)
1	0.02	*73.8*	71.7	54.0	85.5
1	0.04	*73.8*	72.4	53.9	85.1
1.5	0.03	*73.9*	72.6	54.2	84.9
2	0.04	*73.7*	73.1	54.4	84.0
2	0.045	*73.7*	73.3	54.5	83.8
2.5	0.04	*73.6*	73.3	54.8	83.4
2.5	0.045	*73.7*	73.3	55.2	83.4
4	0.04	*73.3*	73.4	55.9	82.0

Table 4. Q2T prediction accuracies of the C-SVC with different (C, γ) values: window size $N = 13$

		Accuracy			
C	γ	$Q_3(\%)$	$Q_\alpha(\%)$	$Q_\beta(\%)$	$Q_c(\%)$
1	0.02	*73.9*	72.3	54.8	84.9
1.5	0.008	*73.6*	71.4	54.3	85.0
1.5	0.02	*73.9*	72.6	54.7	84.8
1.7	0.04	*74.1*	73.6	54.8	83.4
2	0.025	*74.0*	73.0	55.1	84.3
2	0.04	*74.1*	73.9	55.0	83.9
2	0.045	*74.2*	74.1	55.9	83.5
4	0.04	*73.2*	73.9	55.5	81.7

Table 5. Q2T prediction accuracies of the C-SVC with different (C, γ) values: window size $N = 15$

		Accuracy			
C	γ	$Q_3(\%)$	$Q_\alpha(\%)$	$Q_\beta(\%)$	$Q_c(\%)$
2	0.006	*73.4*	70.8	54.2	85.2
2	0.03	*74.1*	73.6	55.6	84.0
2	0.04	*74.2*	73.9	55.7	83.7
2	0.045	*74.0*	73.7	55.4	83.7
2	0.05	*74.0*	73.7	55.4	83.6
2	0.15	*69.0*	63.3	32.7	91.9
2.5	0.02	*74.0*	73.0	55.6	84.0
2.5	0.03	*74.1*	74.0	55.9	83.5
4	0.025	*74.0*	73.8	55.8	83.4

Table 6. Q2T prediction accuracies of the C-SVC with different (C, γ) values: window size $N = 17$

		Accuracy			
C	γ	$Q_3(\%)$	$Q_\alpha(\%)$	$Q_\beta(\%)$	$Q_c(\%)$
1	0.125	*70.0*	63.6	36.0	91.3
2	0.03	*74.1*	73.5	56.2	83.7
2.5	0.001	*71.3*	68.1	52.4	83.5
2.5	0.02	*74.0*	68.1	52.4	83.5
2.5	0.04	*74.0*	75.0	55.8	83.1

the overall accuracy; Q_α, Q_β, and Q_c are the accuracies for α-helix, β-strand, and coil, respectively.

From these tables, we could see that the optimal (C, γ) values for window size $N \in \{11, 13, 15, 17\}$ are $(1.5, 0.03)$, $(2, 0.045)$, $(2, 0.04)$, and $(2, 0.03)$,

Table 7. Q2T prediction accuracies of the multi-class classifier of BSVM with different (C, γ) values: window size $N = 15$

		Accuracy			
C	γ	$Q_3(\%)$	$Q_\alpha(\%)$	$Q_\beta(\%)$	$Q_c(\%)$
2	0.04	*74.18*	73.90	56.39	84.18
2	0.05	*74.02*	73.68	56.09	83.39
2.5	0.03	*74.20*	73.95	56.85	83.22
2.5	0.035	*74.06*	73.93	56.70	82.99
3.0	0.35	*73.77*	73.88	56.55	82.44

respectively. The corresponding Q_3 accuracies achieved are 73.9%, 74.2%, 74.2%, and 74.1%, respectively. A window size of 13 or 15 seems to be the optimal window size that could most efficiently capture the information hidden in the neighboring residues. The best accuracy achieved is 74.2%, with $N = 13$ and $(C, \gamma) = (2, 0.045)$, or $N = 15$ and $(C, \gamma) = (2, 0.04)$.

The original model of SVMs was designed to do binary classification. To deal with multi-class problems, one usually needs to decompose a large classification problem into a number of binary classification problems. The LIBSVM that we used does such a decomposition with the "one-against-one" scheme [59].

In 2001, Crammer and Singer proposed a direct method to build multi-class SVMs [60]. We also applied such a multi-class SVMs to PSSP with the BSVM (http://www.csie.ntu.edu.tw/ cjlin/bsvm/). The results are shown in Table 7. Through comparing Table 5 and Table 7, we found that the multi-class SVMs using Crammer and Singer's scheme [60] and the group of binary SVMs using "one-against-one" scheme [59] obtained similar results.

4.2 T2T Prediction

The T2T prediction uses the output of the Q2T prediction as its input. In T2T prediction, we use the same SVMs as the ones we use in the Q2T prediction. Therefore, we also adopt the same parameter tuning strategy as in the Q2T prediction.

Results

Table 8 shows the best accuracies reached for window size $N \in \{15, 17, 19\}$ with the corresponding C and γ values. From Table 8, it is unexpectedly observed that the structure-structure prediction has actually degraded the prediction performance. A close look at the accuracies for each secondary structure class reveals that the prediction for the coils becomes much less accurate. In comparison to the early results (Tables 3, 4, 5 and 6) in the first

Table 8. The T2T prediction accuracies for window size $N = 15, 17$, and 19

Window Size (N)	C	γ	Accuracy $Q_3(\%)$	$Q_\alpha(\%)$	$Q_\beta(\%)$	$Q_c(\%)$
15	1	2^{-5}	*72.6*	77.9	60.8	74.3
17	1	2^{-4}	*72.6*	78.0	60.4	74.5
19	1	2^{-6}	*72.8*	78.2	60.1	74.9

stage, the Q_c accuracy dropped from 84% to 75%. By sacrificing the accuracy for coils, the predictions for the other two secondary structures improved. However, because coils have a much larger population than the other two kinds of secondary structures, the overall 3-state accuracy Q_3 decreased.

5 Conclusions

To sum up, SVMs performs well in both bioinformatics problems that we discussed in this chapter. For the problem of cancer diagnosis based on microarray data, the SVMs that we used outperformed most of the previously proposed methods in terms of the number of genes required and the accuracy. Therefore, we conclude that the SVMs can not only make highly reliable prediction, but also can reduce redundant genes. For the PSSP problem, the SVMs also obtained results comparable with those obtained by other approaches.

References

1. Cortes C, Vapnik VN (1995) Support vector networks. Machine Learning 20:273–297
2. Vapnik VN (1995) The nature of statistical learning theory. Springer-Verlag, New York
3. Vapnik VN (1998) Statistical learning theory. Wiley, New York
4. Drucker N, Donghui W, Vapnik VN (1999) Support vector machines for spam categorization. IEEE Transaction on Neural Networks 10:1048–1054
5. Chapelle O, Haffner P, Vapnik VN (1999) Support vector machines for histogram-based image classification. IEEE Transaction on Neural Networks 10:1055–1064
6. Hua S, Sun Z (2001) A novel method of protein secondary structure prediction with high segment overlap measure: support vector machine approach. Journal of molecular Biology 308:397–407
7. Strauss DJ, Steidl G (2002) Hybrid wavelet-support vector classification of waveforms. J Comput and Appl 148:375–400
8. Kumar R, Kulkarni A, Jayaraman VK, Kulkarni BD (2004) Symbolization assisted SVM classifier for noisy data. Pattern Recognition Letters 25:495–504

9. Mukkamala S, Sung AH, Abraham A (2004) Intrusion detection using an ensemble of intelligent paradigms. Journal of Network and Computer Applications, In Press
10. Norinder U (2003) Support vector machine models in drug design: applications to drug transport processes and QSAR using simplex optimisations and variable selection. Neurocomputing 55:337–346
11. Van GT, Suykens JAK, Baestaens DE, Lambrechts A, Lanckriet G, Vandaele B, De Moor B, Vandewalle J (2001) Financial time series prediction using least squares support vector machines within the evidence framework. IEEE Transactions on Neural Networks 12:809–821
12. Chang CC, Lin CJ LIBSVM: A library for support vector machines. available at http://www.csie.ntu.edu.tw/˜cjlin/libsvm
13. Schölkopf B, Smolar A, Williamson RC, Bartlett PL (2000) New support vector algorithms. Neural Computation 12:1207–1245
14. Schöklopf B, Platt JC, Shawe-Taylor J, Smola AJ, Williamson RC (2001) Estimating the support of a high-dimensional distribution. Neural Computation 13:443–1471
15. Slonim DK (2002) From patterns to pathways: gene expression data analysis comes of age. Nature Genetics Suppl. 32:502–508
16. Russo G, Zegar C, Giordano A (2003) Advantages and limitations of microarray technology in human cancer. Oncogene 22:6497–6507
17. Alizadeh AA, Eisen MB, Davis RE, Ma C, Lossos IS, Rosenwald A, Boldrick JC, Sabet H, Tran T, Yu X, et al. (2000) Distinct types of diffuse large B-cell lymphoma identified by gene expression profiling. Nature 403:503–511
18. Golub TR, Slonim DK, Tamayo P, Huard C, Gaasenbeek M, Mesirov JP, Coller H, Loh ML, Downing JR, Caligiuri MA et al. (1999) Molecular classification of cancer: class discovery and class prediction by gene expression monitoring. Science 286:531–537
19. Ma X, Salunga R, Tuggle JT, Gaudet J, Enright E, McQuary P, Payette T, Pistone M, Stecker K, Zhang BM et al. (2003) Gene expression profiles of human breast cancer progression. Proc Natl Acad Sci USA 100:5974–5979
20. Chen X, Cheung ST, So S, Fan ST, Barry C (2002) Gene expression patterns in human liver cancers. Molecular Biology of Cell 13:1929–1939
21. Khan J, Wei JS, Ringner M, Saal LH, Ladanyi M et al. (2001) Classification and diagnostic prediction of cancers using gene expression profiling and artificial neural networks. Nature Medicine 7:673–679
22. Deutsch JM (2003) Evolutionary algorithms for finding optimal gene sets in microarray prediction. Bioinformatics 19:45–52
23. Tibshirani R, Hastie T, Narashiman B, Chu G (2002) Diagnosis of multiple cancer types by shrunken centroids of gene expression. Proc Natl Acad Sci USA 99:6567–6572
24. Bura E, Pfeiffer RM (2003) Graphical methods for class prediction using dimension reduction techniques on DNA microarray data. Bioinformatics 19:1252–1258
25. Troyanskaya O, Cantor M, Sherlock, G et al. (2001) Missing value estimation methods for DNA microarrays. Bioinformatics 17:520–525
26. Dudoit S, Fridlyand J, Speed T (2002) Comparison of discrimination methods for the classification of tumors using gene expression data. J Am Stat Assoc 97:77–87

27. Welch BL (1947) The generalization of student's problem when several different population are involved. Biomethika 34:28–35
28. Tusher, VG, Tibshirani R, Chu G (2001) Significance analysis of microarrays applied to the ionizing radiation response. Proc Natl Acad Sci USA 98:5116–5121
29. Tibshirani R, Hastie T, Narasimhan B, Chu G (2003) Class prediction by nearest shrunken centroids with applications to DNA microarrays. Statistical Science 18:104–117
30. Berman HM, Westbrook J, Feng Z, Gilliland G, Bhat TN, Weissig H, Shindyalov IN, Bourne PE (2000) The protein data bank. Nucleic Acids Research 28:235–242
31. Kendrew JC, Dickerson RE, Strandberg BE, Hart RJ, Davies DR et al. (1960) Structure of myoglobin: a three-dimensional fourier synthesis at 2 Å resolution. Nature 185:422–427
32. Perutz MF, Rossmann MG, Cullis AF, Muirhead G, Will G et al. (1960) Structure of haemoglobin: a three-dimensional fourier synthesis at 5.5 Å resolution. Nature 185:416–422
33. Scheraga HA (1960) Structural studies of ribonuclease III. A model for the secondary and tertiary struture. J Am Chem Soc 82:3847–3852
34. Davids DR (1964) A correlation between amino acid composition and protein structure. Journal of Molecular Biology 9:605–609
35. Robson B, Pain RH (1971) Analysis of the code relating sequence to conformation in proteins: possible implications for the mechanism of formation of helical regions. Journal of Molecular Biology 58:237–259
36. Chou PY, Fasma UD (1974) Prediction of protein conformation. Biochem 13:211–215
37. Lim VI (1974) Structural principles of the globular organization of protein chains. A stereochemical theory of globular protein secondary structure. Journal of Molecular Biology 88:857–872
38. Rost B, Sander C (1994) Combining evolutionary information and neural networks to predict protein secondary structure. Proteins 19:55–72
39. Robson B (1976) Conformational properties of amino acid residues in globular proteins. Journal of Molecular Biology 107:327–56
40. Nagano K (1977) Triplet information in helix prediction applied to the analysis of super-secondary structures. Journal of Molecular Biology 109:251–274
41. Taylor WR, Thornton JM (1983) Prediction of super-secondary structure in proteins. Nature 301:540–542
42. Rooman MJ, Kocher JP, Wodak SJ (1991) Prediction of protein backbone conformation based on seven structure assignments: influence of local interactions. Journal of Molecular Biology 221:961–979
43. Bohr H, Bohr J, Brunak S, Cotterill RMJ, Lautrup B et al (1988) Protein secondary structure and homology by neural networks. FEBS Lett 241:223–228
44. Holley HL, Karplus M (1989) Protein secondary structure prediction with a neural network. Proc Natl Acad Sci USA 86:152–156
45. Stolorz P, Lapedes A, Xia Y (1992) Predicting protein secondary structure using neural net and statistical methods. Journal of Molecular Biology 225:363–377
46. Muggleton S, King RD, Sternberg MJE (1992) Protein secondary structure predictions using logic-based machine learning. Prot Engin 5:647–657

47. Salamov AA, Solovyev VV (1995) Prediction of protein secondary structure by combining nearest-neighbor algorithms and multiple sequence alignment. Journal of Molecular Biology 247:11–15
48. Qian N, Sejnowski TJ (1988) Predicting the secondary structure of globular proteins using neural network models. Journal of Molecular Biology 202:865–84
49. Kneller DG, Cohen FE, Langridge R (1990) Improvements in protein secondary structure prediction by an enhanced neural network. Journal of Molecular Biology 214:171–82
50. Lin KM, Lin CJ (2003) A study on reduced support vector machines. IEEE Transactions on Neural Networks 12:1449–1559
51. Sasagawa F, Tajima K (1993) Prediction of protein secondary structures by a neural network. Computer Applications in the Biosciences 9:147–152
52. Vivarelli F, Giusti G, Villani M, Campanini R, Fraiselli P, Compiani M, Casadio R (1995) LGANN: a parallel system combining a local genetic algorithm and neural networks for the prediction of secondary structure of proteins. Computer Application in the Biosciences 11:763–9
53. Rost B, Sander C (1993) Prediction of protein secondary structure at better than 70% accuracy. Journal of Molecular Biology 232:584–599
54. Rost B (1996) PHD: predicting one-dimensional protein secondary structure by profile-based neural network. Methods in Enzymology 266:525–539
55. Riis SK, Krogh A (1995) Improving prediction of protein secondary structure using structured neural networks and multiple sequence alignments. Journal of Computational Biology 3:163–183
56. Jones DT (1999) Protein secondary structure prediction based on position-specific scoring matrices. Journal of Molecular Biology 292:195–202
57. Stephen FA, Warren G, Webb M, Engene WM, David JL (1990) Basic local alignment search tool. Journal of Molecular Biology 215:403–410
58. Altschul SF, Schaffer AA, Zhang J, Zhang Z, Miller W, Lipman FJ (1997) Gapped BLAST and PSI-BLAST: a new generation of protein database search programs. Nucleic Acids Research 25:3389–3402
59. Hsu CW, Lin CJ (2002) A comparison of methods for multiclass support vector machines. IEEE Transactions on Neural Networks 13:415–425
60. Crammer K, Singer Y (2001) On the algorithmic implementation of multiclass kernel-based vector machines. Journal of Machine Learning Research 2:265–292

Gas Sensing Using Support Vector Machines

J. Brezmes[1], E. Llobet[1], S. Al-Khalifa[2], S. Maldonado[3], and J.W. Gardner[2]

[1] Departament d'Enginyeria Electrònica, Elèctrica i Automàtica, Universitat Rovira i Virgili, Av. Paisos Catalans 26, 43007 Tarragona, Spain
[2] School of engineering, University of Warwick, Coventry, CV4 7 AL, UK
[3] Dpto. de Teoría de la Señal y Comunicaciones, Universidad de Alcalá, 28871 Alcalá de Henares, Madrid, Spain

Abstract. In this chapter we deal with the use of Support Vector Machines in gas sensing. After a brief introduction to the inner workings of multisensor systems, the potential benefits of SVMs in this type of instruments are discussed. Examples on how SVMs are being evaluated in the gas sensor community are described in detail, including studies in their generalisation ability, their role as a valid variable selection technique and their regression performance. These studies have been carried out measuring different blends of coffee, different types of vapours (CO, O_2, acetone, hexanal, etc.) and even discriminating between different types of nerve agents.

Key words: electronic nose, gas sensors, odour recognition, multisensor systems, variable selection

1 Introduction

The use of support vector machines in gas sensing applications is discussed in this chapter. Although traditional gas sensing instruments do not use pattern recognition algorithms, recent developments based on the Electronic Nose concept employ multivariate pattern recognition paradigms. That is why, in the second section, a brief introduction to electronic nose systems is provided and their operating principles discussed. The third section identifies some of the drawbacks that have prevented electronic noses from becoming a widely used instrument. The section also discusses the potential benefits that could be derived from using SVM algorithms to optimise the performance of an Electronic Nose. Section 4 describes some recent work that has been carried out. Different reports from research groups around the world are presented and discussed. Finally, Sect. 5 summarises our main conclusions about support vector machines.

J. Brezmes et al.: *Gas Sensing Using Support Vector Machines*, StudFuzz **177**, 365–386 (2005)
www.springerlink.com

2 Electronic Nose Systems

Gas sensing has been a very active field of research for the past thirty years. Nowadays issues such as environmental pollution, food poisoning and fast medical diagnosis, are driving the need for faster, simpler and more affordable instruments capable of characterising chemical headspaces, so that appropriate action can be taken as soon as possible. The so-called "Electronic Nose" appeared in the late eighties to address the growing needs in these fields and others such as the cosmetic, and chemical industries [1].

The Electronic Nose, also known sometimes as "electronic olfactory system", borrowed its name from its analogue counterpart for two main reasons:

1. It mimics the way biological olfaction systems work [1, 2].
2. It is devised to perform tasks traditionally carried out by human noses.

A more complete and formal definition would describe these systems as "instruments comprising an array of chemical sensors with overlapping sensitivities and appropriate pattern-recognition software devised to recognize or characterize simple or complex odors" [3].

In order to understand how an electronic olfactory system works it is important to note its differences from conventional analytical instruments. While traditional instruments (e.g. gas chromatography and mass spectrometry) analyse each sample by separating out its components (so each one of them can be identified and quantified), electronic noses (ENs) evaluate a vapour sample (simple or complex) as a whole, trying to differentiate or characterise the mixture without necessarily determining its basic chemical constituents. This is especially true when working with complex odours such as food aroma, where hundreds of chemicals can coexist in a headspace sample and it is very difficult to identify every single contributor to the final aroma [4]. This approach is achieved exploiting the concept of overlapping sensitivities in the sensor array.

Figure 1 shows this concept in a very simple graphical manner. From the plot it can be seen that each sensor from the chemical array is sensitive to a range of aromas, although with different sensitivities. At the same time, each single odorant (or chemical) is sensed by more than one sensor since their sensitivity curves overlap. The resulting multivariate data can be plotted in a radar plot. Ideally, the same aroma will always have the same radar pattern (see Fig. 2), and an increase in its concentration would only retain the same shape but scaled larger, whereas a different aroma would have a different shape.

This approach has two fundamental advantages. First, solid-state chemical sensors tend to be non-selective, and a conventional analytical instrument solves that problem by separating out the constituents of the mixture by chromatography, a rather expensive, tedious and slow method. In contrast, processing the information generated by an array of non-specific sensors with overlapping sensitivities should improve the resolution of the unit, be much

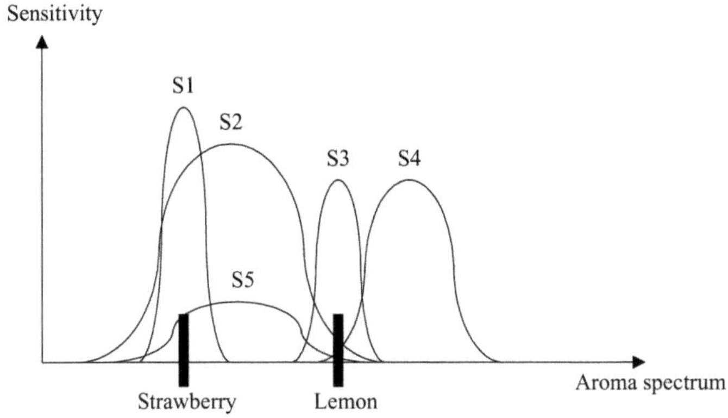

Fig. 1. The concept of overlapping sensitivities in a sensor array

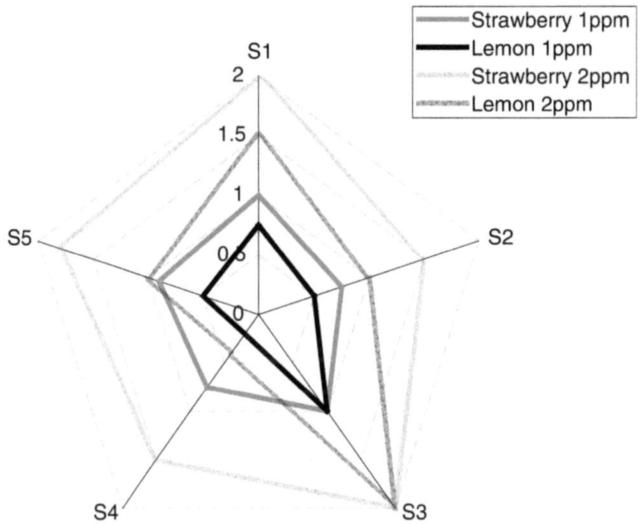

Fig. 2. Radar plot of 2 aromas in 2 different concentrations

more economical and easier to build. In this manner, if the selectivity is enhanced sufficiently, no separation stage is necessary, rendering much faster results.

Secondly, since the sensors from the array are non-selective, they can sense a wider range of odours. In conventional gas sensing, if no separation stage is used, and the sensors used are specific then the system cannot detect a wide range of aromas. For specific sensors, as many sensors as species to be sensed would be needed, whereas in the innovative approach with a few sensors hundreds of different aromas may be sensed.

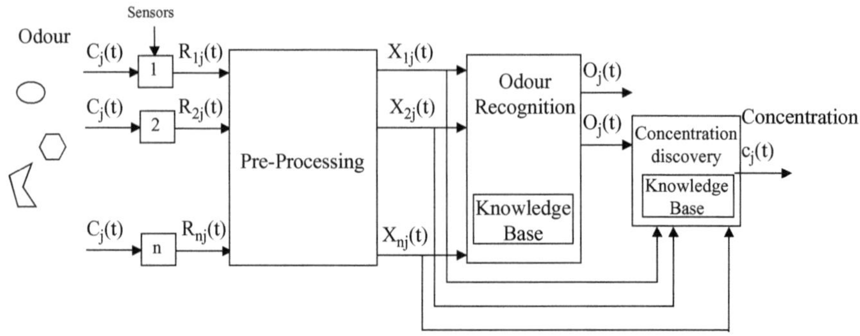

Fig. 3. Mathematical modeling of a general olfactory system. Adapted from [5] with permission

Figure 3 shows a general arrangement of an electronic nose system, viewed from a mathematical/system point of view [5]. A complex odor in a given concentration *Cj(t)* is sensed by n sensors from a chemical array.

Many different types of sensors have been used in electronic olfactory systems [6, 7]. Commonly, the parameter that is affected by the interaction between the odour and the sensing layer is the conductivity of the active material. This is the case with metal oxide semiconductor gas sensors and conductive polymers, in which $R_{ij}(t)$ is a resistance measurement. Figure 4 shows a typical response of a semiconductor gas sensor to a step change in the concentration of a chemical vapour in air.

The curve represents conductance and typical pre-processing parameters (or features) that can be extracted from such a response are the final conductance (Gf), conductance change (ΔG = Gf − Gi) or normalised conductance, ΔG/Gi. Other parameters, such as the conductance rise time (Tr1090) of the sensors have also been used [8, 9].

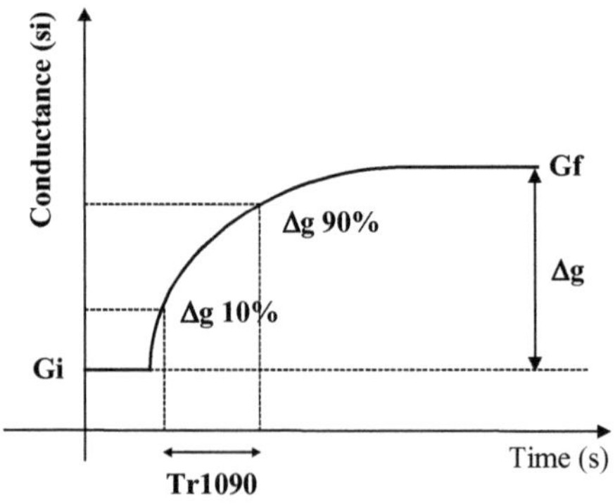

Fig. 4. Main parameters extracted from a conductance transient in a semiconductor gas sensor

Other sensing mechanisms are based on mass absorption which is often related to a frequency shift (QMB devices) or time delays (SAW sensors). Other sensors change their work (potential) function under the presence of oxidising or reducing species, such as Pd-gate MOSFETS.

The physical response of each sensor to odour j is then converted to an electrical parameter that can be measured ($R_{ij}(t)$, $i = 1$ to n). This electrical response is then pre-processed to extract relevant features from each sensor ($x_{ij}(t)$), so that the entire response to odour j is described by the n-dimensional vector $x_j(x_j = [x_{1j}, x_{2j}, \ldots x_{3j}, x_{nj}])$.

One of the most important parts of an electronic nose, the pattern recognition engine, compares this odour vector to a knowledge base to identify or/and quantify the vapour sample. In many cases, identification comes first: the vapour is compared to the training patterns acquired during the calibration phase and the class predicted.

Many algorithms have been developed for different applications envisaged for electronic noses [10, 11]. Some of these algorithms are mainly for qualitative identification/classification purposes, while others are used for quantification tasks. Most of them require a calibration (training) phase to generate the knowledge base used to classify/identify or even quantify vapour samples. Table 1 outlines a few of the most common algorithms with their characteristics, advantages and drawbacks.

Table 1. Some of the most popular algorithms for electronic noses

Method	Learning	Application	Type	Advantages/drawbacks
PCA	Unsupervised	Classification	Linear	Simple and graphical/bad performance in non-linear problems and noisy data
DFA	Supervised	Classification	Linear	Powerful/Easily overfits data
PLS	Supervised	Quantification	Linear	Fast/Fails in non-linear problems
Back-propagation	Supervised	Classification/ Quantification	Neural	Universal approximator/ slow training, linear boundaries
Fuzzy Art	Unsupervised	Classification	Neural	Fast and simple/ unsupervised
Fuzzy Artmap	Supervised	Classification	Neural	Fast and simple/does not quantify
SOM	Unsupervised	Classification	Neural	They tend to adapt to drift/Only unsupervised and for classification
Radial Basis Functions	Supervised	Classification/ Quantification	Neural	Fast training and good performance/defines boundary tightly so drift sensitive

Electronic nose instruments have been tested in many applications. Some of them involve measurements with complex odours (food aromas [12, 13], cosmetic perfumes, breath analysis in patients [14], etc.) and some measure simpler vapours (single, binary or tertiary mixtures [15, 16]). In the first case most applications require attribute classification while in the second case a numerical variable classification is sought.

3 Electronic Nose Optimisation Using SVMs

Although the Electronic Nose concept seems to hold a great potential for a high number of applications, the truth is that after a decade of research and development very few systems have been commercialised successfully. Commercial products have been around for some years [17, 18], and used to evaluate specialised application fields, such as the food or chemical industry. The outcome of most of these studies is that the Electronic Nose seems to work well at the laboratory level (in a highly controlled environment) but its practical implementation in the field under variable ambient conditions is problematic. Many reasons might be behind this issue. Most of them are associated to the sensing technologies used. This is where new pattern recognition algorithms, such as SVMs, may improve the performance of the instruments.

The major drawback that has prevented the use of electronic olfactory systems in the industry is the calibration/training process, which is usually a lengthy and costly task. Before using the olfactory system it is necessary to calibrate the instrument with all odours that it is likely to experience. The problem is that obtaining measurements is a costly, time consuming and complicated process for most applications. For supervised pattern recognition algorithms this training set has to be statistically representative of the measurement scenario in which the system will regularly operate, which means obtaining a similar number of samples for each category that has to be identified and large data-sets to ensure a good degree of generalisation. SVM networks are specifically trained to optimise generalisation with a reduced training set, and they do not need to have the same number of measurements for each category. Therefore, using SVM the training procedure can be reduced or the generalisation ability of the network optimised, whichever is the priority. This single reason is sufficient to encourage the inclusion of the SVM paradigm in the processing engine of electronic noses, since a reduction in the training time/effort may make a unit practically viable.

On the other hand, one of the most complex problems to solve in gas sensing is drift. Drift can be defined as an erratic behaviour that causes different responses to the same stimulus. In metal oxide sensors, the most common sensing technology used in olfactory systems, drift is associated with sensor poisoning or ageing. In the first case, the active layer changes its behaviour due to the species absorbed in previous measurements. This effect can be reversible if the species are desorbed in a short period of time (in that situation it is usually called a memory effect). Ageing occurs either because the microstructure of the sensing layers changes during the 4operating life of the sensor or because some surface reactions are irreversible and change the device sensitivity. This is major problem in electronic nose systems. A straightforward solution to the problem is to retrain the instrument periodically, but this is rarely done given the cost and efforts necessary in each recalibration process.

Many authors have tried to easy the burden associated with recalibration procedures with different approaches. One of the first approaches proposed [19] uses a single gas as a reference value to compute a correction that will be applied to subsequent readings. This approach, although may work in simple applications, is of little use when very different species are measured. Component correction [20], a linear approach based on PCA and PLS is a new method that tries to solve this problem using more reference gases. Other groups have proposed successful pattern recognition algorithms [21, 22] that modify their inner parameters while regular measurements are executed, without the need to include additional calibration procedures. These algorithms adapt to drift as long as sufficient regular measurements are executed to adapt to the changing situation. SVM algorithms could easily be adapted to this approach by changing the support vectors that define category regions each time a new regular measurement is performed. Since SVMs are not computationally intensive, the system could be retrained each time a support vector is added or replaced.

Environmental drift can be defined as the lack of selectivity against changes in the environmental conditions, such as humidity and temperature. Most of the sensing technologies used in electronic noses need a high working temperature that, if it is not controlled, can lead to very strong fluctuations in the sensor response under the same odour input. The temperature dependence of sensor response is non-linear and in many situations a temperature control feedback loop is needed to ensure that the sensor active layer is working at the proper temperature range. Even with this mechanism it is very difficult to isolate ambient temperature fluctuations which indeed change sample characteristics and the sensor operating temperature. This situation results in the low reproducibility of results with scattered measurements belonging to identical samples. SVMs can reduce this effect thanks to the way the algorithm chooses the separating hyperplane, maximising the distance to the support vectors (the most representative training measurements) and, therefore, giving ample tolerances in order to obtain a robust classification against temperature fluctuations.

Humidity has a very strong influence on sensor response. The baseline signal is altered by this environmental parameter in a non-linear manner that also depends on temperature. In most situations, fast humidity changes can be considered as an additional interfering species like any other chemical, raising the complexity of recognition. Since humidity may be monitored in most applications, a possible solution is to measure its value and use it as additional information, provided that the pattern recognition engine can cope with the additional dimension. SVMs might be well suited to these situations thanks to the fact that they optimise generalisation, and therefore, the effects of training and operating at different humidity levels will be minimised.

All these influences, along with poor sampling procedures in many applications, generate a high number of erroneous measurements, also known as outliers. Using these erroneous measurements during the training process can lead to a bad calibration of the instrument and, therefore, result in poor performance during real operation. Since SVMs reduce the training measurements to a few training vectors (the so-called support vectors) to define the separation hyperplane, the chance that one of the outliers will be in this small group is minimised. Moreover, since the algorithm allows for a compromise between separation distance between groups and erroneous classification during training, giving a chance to learn with some mistaken measurements can help ignore outliers, maximising performance during evaluation.

4 Progress on Using SVMs on Electronic Noses

Although SVMs where first introduced in 1995, it was not until the end of the last century that they became popular among scientists working in different fields where pattern recognition was necessary. By 2002 this paradigm started to be studied by the electronic nose community, and that is why relatively few papers have been published in this field. In the following subsections the literature is reviewed.

4.1 Comparison of SVM to Feed-Forward Neural Networks

Most of the studies published to date deal with simple systems that try to classify single or binary mixtures of common vapours. The majority of these works compare the performance of SVM against more traditional paradigms such as the feed forward back propagation network or the more efficient Radial Basis Function.

For example, in [5], Distante et al. evaluate the performance of a SVM paradigm on an electronic nose based on sol-gel doped SnO_2 thin film sensors. They measured seven different types of samples (water, acetone, hexanal, pentanone and binary mixtures between the last three of them) and used the raw signals from the sensors to classify samples, comparing the performance of SVMs against other well-known methods.

Since the most common SVMs solve two-class problems, they built 7 different machines to differentiate each species from the rest. To validate the network they use the leave-one-out approach which was also used to determine the best regularisation parameter C. Since the problem was not linearly separable, a second degree polynomial kernel function was used to translate the non-linear problem into a higher dimension linearly separable problem.

Table 2. Confusion matrix using SVMs. Adapted from [5], with permission

	Water	Acetone	M1	Hexanal	M2	M3	Pentanone
Water	28	0	0	0	0	0	0
Acetone	0	28	0	0	0	0	0
M1	0	0	33	0	3	0	0
Hexanal	0	0	0	34	0	0	1
M2	1	0	4	1	32	0	0
M3	0	0	0	0	1	50	0
Pentanone	0	0	0	0	0	0	24

Table 3. Confusion matrix using RBF. Adapted from [5], with permission

	Water	Acetone	M1	Hexanal	M2	M3	Pentanone
Water	19	0	3	0	6	0	0
Acetone	0	26	0	1	0	0	1
Ml	0	0	27	1	8	0	0
Hexanal	0	0	0	33	0	0	2
M2	0	0	7	0	31	0	0
M3	0	0	4	0	3	44	0
Pentanone	0	0	0	1	0	0	23

Comparison with back propagation networks helped to highlight the superior performance of this type of networks. Using a classical back-propagation training algorithm, a classification error of 40% was the minimum achievable, while the more elaborated RBF network reduced this quantity to 15%. The SVM classifier outperformed both networks giving a 4.5% classification error. Tables 2 and 3 compare the confusion matrixes obtained for SVM and RBF methods respectively, where rows show the true class and the columns show the predicted class. It can be seen that while SVM has a small number of errors concentrated on mixtures, RBF shows more prediction errors, some of them even on single vapours.

4.2 A Study on Over/Under-Fitting Data Using a PCA Projection Coupled to SVMs

Other studies that evaluate the performance of SVMs use data acquired with a commercial e-nose. That is the case in [23] where Pardo and Sberveglieri measured different coffee blends using the Pico-1 Electronic Nose. This electronic nose comprises five thin film semiconductor gas sensors. The goal of the study was to evaluate the generalisation ability of SVMs with two different kernel functions (polynomial and Gaussian) and their corresponding kernel values. They had a total of 36 measurements for each of the 7 different blends of coffee analysed. To fit a binary problem, they artificially converted the seven-class problem to a two-category measurement set based on PCA projections. In this study, the regularisation parameter was fixed to a standard value of 1. Then, using 4-fold validation, they evaluated the performance of each network against two parameters: the number of principal components retained from the PCA projection and a kernel related figure (the polynomial order in the polynomial kernel and the variance value for the Gaussian kernel).

Their study showed that for RBF kernel SVMs, the minimum error is found for a small variance value (higher values result in over fitting) and that more than 2 PC components have to be used in order to avoid under fitting. In the case of the polynomial kernel, the minimum error is obtained for a second degree, while a polynomial of order one slightly under fits the data. Again, more than two principal components are necessary to avoid under fitting.

4.3 Exploring the Regression Capabilities of SVMs

Other works explore the use of SVM as regression machines. Ridge regression [24] is considered a linear kernel regression method that can be used to quantify gas mixtures. In [25] two pairs of TiO_2 sensors were paired to detect both CO and O_2 in a combustion chamber.

$$K(\boldsymbol{x}, \boldsymbol{x}') = e^{|\boldsymbol{x}, \boldsymbol{x}'|^2/\sigma^2} \tag{1}$$

Two different kernels were used to compare the generalisation abilities of the regression machines, namely Gaussian and reciprocal kernels. Equation (1) shows the Gaussian kernel formula. A value for the spread constant σ of 2 was used for all regressions done with this kernel. Figures 5 (a) and (b) compare the regression surfaces with the calibration points on them. As it can be seen, the Gaussian kernel performs the regression in a sinusoidal like manner, while the reciprocal approximation has a more monotonically increasing behaviour. In this problem, where the sensor behaviour is monotonically increasing, the reciprocal regression works much better.

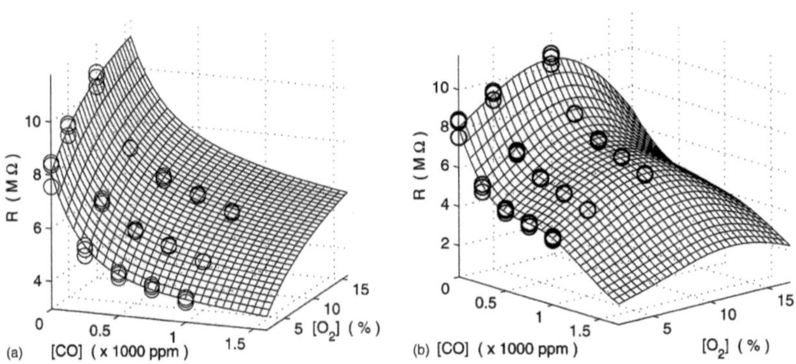

Fig. 5. Regression surfaces for a Reciprocal kernel (**a**) and a Gaussian Kernel (**b**). Reproduced from [25], with permission

Table 4. Comparison between predicted and real values for CO and O_2 using Ridge Regression. Adapted from [25], with permission

Actual % O_2	Predicted % O_2	Actual [CO] (ppm)	Predicted [CO] (ppm)
3	2.59	250	259.18
3	1.81	350	228.55
3	2.87	600	524.21
3	1.86	800	494.94
3	2.36	900	701.7
4	2.78	250	215.95
4	2.38	350	294.95
4	2.73	600	404.26
4	2.03	800	407.02
4	2.04	900	580.68
8	8.57	250	254.64
8	11.44	700	1270.39
8	10.24	900	1005.11
9	8.79	250	241.56
9	3.36	700	405.15
9	16.88	900	1775.03

The regression works well, as long as there is sufficient orthogonality between both sensors. Experimentally that was achieved between 200–400 ppm of CO. Table 4 compares actual and predicted values using two differently doped TiO_2 sensors with the reciprocate kernel. The authors concluded that Ridge regression coupled to two differently doped TiO_2 sensors was able to predict concentration of CO and O_2 in high temperature samples, an encouraging result that can be applied to optimise high temperature combustion processes.

Another interesting application of the regression abilities using kernel functions can be found in [26]. Blind source separation (BSS) consists of recovering a set of original source signals from their instantaneous linear mixtures [27]. When the mixture is non-linear (a common problem in gas sensing), the method cannot be applied because infinite solutions exist. The authors propose a kernel-based solution that takes the non-linear problem to a higher dimension where it can be linearised using the Stone's linear BBS algorithm.

Stone's algorithm explodes the time dependence statistics of a signal, considering each source signal as a time series. Although, in theory, a simple non-linear mapping into a higher dimension can convert the problem into a linear separable classification, the explicit mapping can lead to an impractical calculation problem, with too many dimensions to be computed. To avoid working with mapped data directly, a kernel method is proposed in the same way they are used in SVMs. This method embeds the data into a very high dimension (or even infinite) feature vectors and allows the linear BSS algorithm to be carried out in feature space without this space ever being explicitly represented or computed.

This approach seems well-suited to the gas sensing problem since, in an electronic olfactory system, gas sensors tend to be highly non-selective and they usually respond to a wide variety of gases. When measuring, the objective of the instrument is to separate the different gases present in a mixture using the sensor array. However, due to the sensor nonlinearities, the mixtures are also non-linear.

The experiment consisted on 105 measurements done on tertiary gas mixtures comprising carbon monoxide (0, 200, 400, 1000 and 2000 ppm), methane (0, 500, 1000, 2000, 5000, 7000 and 10000 ppm) and water vapour (25, 50, 90% relative humidity). The electronic nose used comprised a sensor array with 24 commercially available tin oxide gas sensors.

Since the sample set was too small and had no temporal structure, an imaginative process was followed: first, a feed-forward back-propagation trained neural network was implemented with the 102 samples to build a model relating gas concentrations (model input) to sensor response (model output). Then, thousands of measurements were artificially created using the trained network. Figure 6 shows the time dependence of the simulated gas signals, which are sinusoids of different frequencies, assuring in this way the time independence between them.

Prior to the KBSS algorithm, 104 support vectors were extracted. An RBF kernel was used with a standard deviation $\sigma = 2$. Figure 7 shows the recovered signals when the algorithm was applied. It is interesting to note that the solution obtained with a linear kernel fails to recover the signals in a significant way, as shown in Fig. 8.

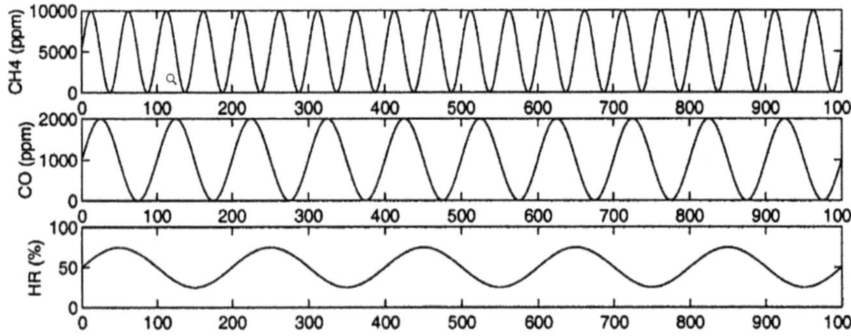

Fig. 6. Initial simulated vapor concentrations. Reproduced from [26], with permission

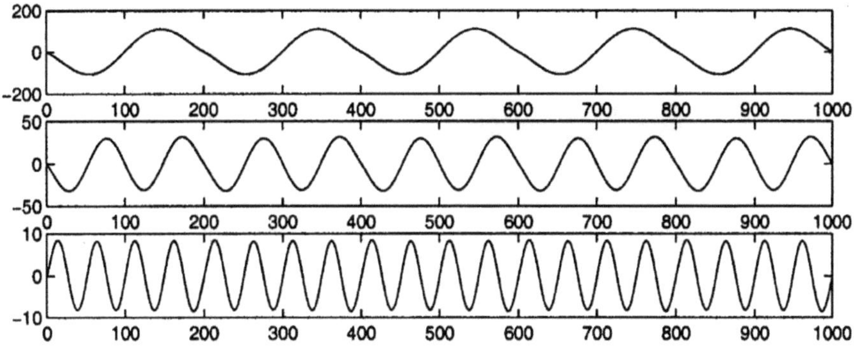

Fig. 7. Recovered KBBS signal. Reproduced from [26], with permission

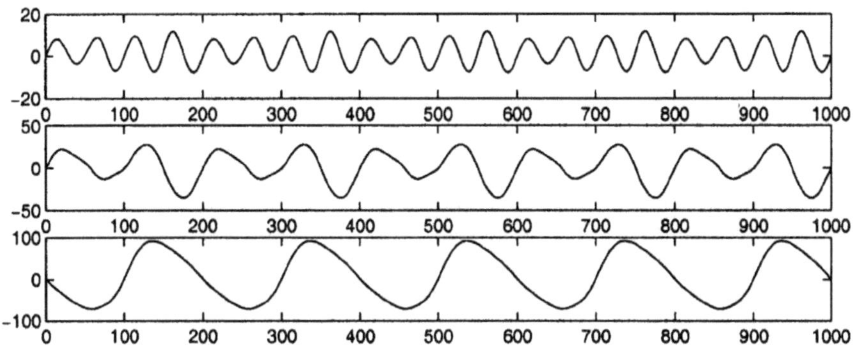

Fig. 8. Recovered signals using a linear kernel. Reproduced from [26], with permission

4.4 The Use of Support Vector Machines for Binary Gas Detection

As was mentioned at the start of the chapter, electronic noses have many potential applications. One of them is the detection and/or identification of hazardous vapour leaks. In this context, Al-Khalifa presented in [28, 29] a complete description of a single sensor gas analyser designed to discriminate between CO, NO_2 and their binary mixtures. The goal of the study was to minimise power and system requirements to pave the way to truly portable electronic olfactory instruments.

The sensor they used was deposited on a micro-machined substrate with a very low thermal inertia that allows the device to be modulated in temperature. The architecture of the sensor includes a heating resistance (Rh) that heats the active layer up to 500°C. This resistor is a thermistor that allows an accurate monitoring of the sensor temperature. Figure 9 shows a schematic diagram of how such a sensor is built and electrically connected.

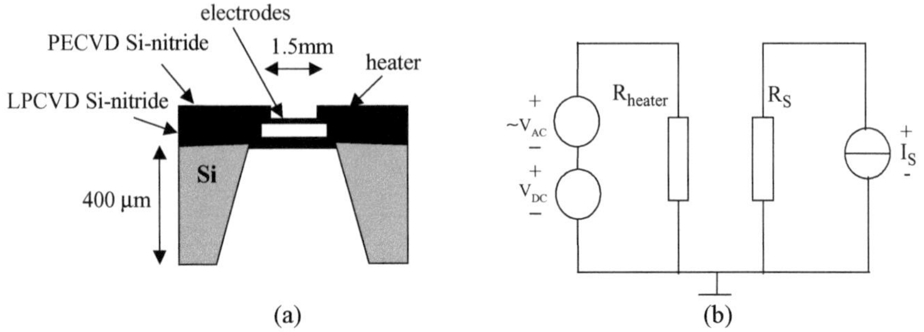

Fig. 9. Sensor schematics: (**a**) construction, (**b**) electrical connections. Adapted from [28], with permission

Temperature modulation of semiconductor gas sensors has been a classical technique used to offset some of the drawbacks generally associated to electronic nose semiconductor gas sensors. This approach enhances selectivity, reproducibility and even minimises power consumption if the sensors used are micro-machined devices.

Figure 10 shows a general diagram of the approach. In any sort of modulation (temperature, flux or concentration), sensor kinetics are monitored and characterised in order to obtain more information of each sensing element. Generally speaking, the time-domain information obtained has to be expressed in a more compact, pre-processed manner. A typical approach is to use the Fourier transform as the pre-processing step, but lately wavelets are becoming more popular due to their superior performance in non-stationary processes [30]. In this study, temperature modulation was performed with sinusoidal signals, and the sensor response was sampled at 50 mHz. The resulting time-domain signal was then transformed onto the wavelet domain using 8 tap Daubechies filters. In all, each measurement performed was described by 102 wavelet coefficients. In a first experiment the goal was to classify the samples according to their nature (CO, NO_2 and binary mixtures). Since binary classification SVMs were used, a two step approach was used; first, a SVM was trained to discriminate CO samples from the remaining ones (NO_2 and binary mixture samples). Then, a second SVM was included to differentiate between the later ones.

Since the study was conducted to advance in the design of portable, low-cost instruments, SVMs are the perfect choice since the computation

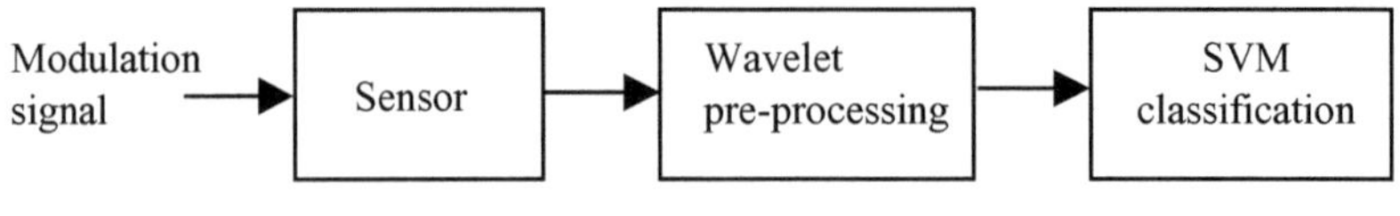

Fig. 10. Schematic diagram of the approach

requirements of the algorithm are not very demanding. Using 109 point vectors for each measurement actually drives up the computational requirements. That is why support SVMs were used in an innovative manner to reduce the number of descriptors for each measurement, since they were used as a variable selection algorithm.

According to the basic theory of SVMs, during training, support vector machines generate a hyperplane (defined by its perpendicular vector w) in order to maximise the distance between any support vector to such a threshold surface. The well known basic (2) determines, once the hyperplane has been defined during training, a dot product between the new measurement x and the hyperplane vector w to find the category for measurement x.

$$f(x) = w\mathbf{\Phi}(x) + b \qquad (2)$$

In a two-dimensional world (that can be generalised to any n-dimension problem), x-vector components (or features) parallel to the hyperplane are of no relevance to the binary classification problem, while perpendicular features are of great value when classifying samples into possible categories. In other words, in the dot product, x components multiplied by higher w components are the most important values to classify the sample. Therefore, in order to reduce the number of descriptors (variables) for each measurement, a valid criterion would be to keep those components that give higher values in the hyperplane vector w.

In Fig. 11 the magnitude of the w vector, once trained to discriminate between CO and NO_2/mixtures is presented. As it can be seen, there is a very high correlation between adjacent values. In order to reduce dimensionality, the maximum values of each peak were retained and the ten highest were selected. Table 5 shows the results obtained for the first and second SVMs.

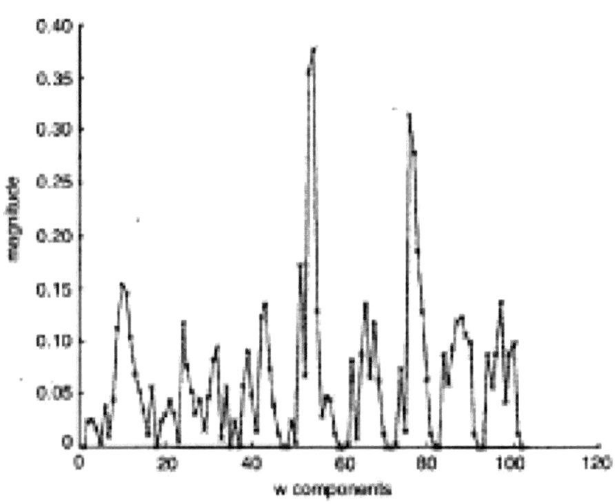

Fig. 11. Hyperplane vector (w) components. Reproduced from [29], with permission

Table 5. SVM binary classification results. Adapted from [29], with permission

	CO	NOT CO
1ST SVM	100%	100%
	NO_2	$CO + NO_2$
2ND SVM	100%	94%

As it can be seen, the overall process gave a 94% success rate when classifying samples in three different categories (CO, NO_2 and mixtures).

In a second experiment, SVMs were used for quantification purposes. In ε-SVM regression the goal is to find a function $f(x)$ that has, at most a ε deviation from the actual value during training with the restriction of being as flat as possible to optimise generalisation. As usual in any supervised regression algorithm, a number of training pairs (x_i, y_i), where x_i is the sample vector and y_i the real quantification value for such a measurement, are used during training. The regularisation parameter C determines how many training measurements can have a greater error than ε, and therefore it is a trade-off between the flatness of the fitting function and the number of measures that do no lie within the ε error area.

In this work, the target values were gas concentrations. Four different regression SVMs were trained and evaluated for each gas: CO, NO_2, CO in the binary mixture and NO_2 in the same sample. An exponential kernel was used in all four regressions.

Initially, 102 coefficients from the wavelet transform were used for the regressions. Again, the authors wanted to reduce the dimensionality of the data to lower the computational requirements if used in a handheld unit. The strategy followed this time required an iterative process where the final square error was evaluated with and without each wavelet coefficient.

Figure 12 shows the normalised relative error obtained for each variable in two cases. When the removal of a variable resulted in a higher error rate, this meant that the removed variable was an important feature that should be retained. Those removals that decreased the error rate were indicating that the variables removed added noise rather than useful information. Table 6 shows the characteristics of each of the four SV regression machines implemented. It can be seen that with a reduced parameter set (a maximum of 47 variables were used from the initial 102 coefficients), a highly accurate predictive model could be obtained, with a relative error lower than 8%.

In summary, this work shows how SVMs can be used in innovative and imaginative ways such as a variable selection method. The study illustrates how SVMs can reduce computation requirements in classification and quantification problems, proving the feasibility of enhancing the selectivity of a single sensor using temperature modulation techniques coupled to signal processing algorithms such as wavelets and SVMs.

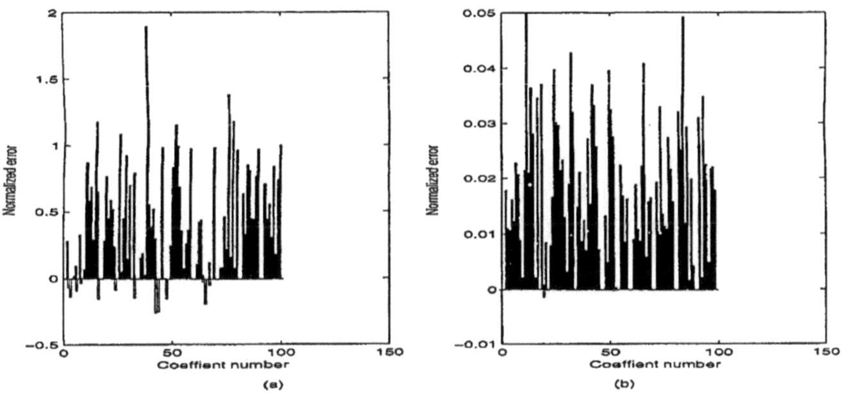

Fig. 12. Relative error (**a**) CO, (**b**) CO from the mixtures. Reproduced from [29], with permission

Table 6. SV regression results for the four mixtures. Adapted from [29], with permission

Gas	CO	NO_2	CO From Mix	NO_2 From Mix
No Vector components	15	12	22	47
No of Support Vectors	56	22	48	29
Relative error	6.37%	4.57%	5.70%	7.55%

4.5 Multisensor System Based on Support Vector Machines to Detect Organophosphate Nerve Agents

Under the increasing worries about terrorist threats in the form of chemical and biological agents, many government agencies are funding research aiming at developing early detection instruments that overcome the main drawbacks of the existing ones. Actual systems fall in four different categories:

- DNA sequence detectors
- immune-detection systems that use antibodies
- tissue based systems
- mass spectrometry systems

These systems are either complex to operate, have a big size that makes them unsuitable to perform in-field measurements or their accuracy is limited.

In [31] a new system that cannot be included in any of the mentioned categories is presented. It combines a commercial electronic nose with SVM pattern recognition algorithms to discriminate between different organophosphate nerve agents such as Parathion, Paraoxon, Dichlorvos and Trichlorfon.

The commercial electronic nose used in the study is based on a polypyrrole sensor array with 32 different sensing layers (AromaScan Ltd.). In this system, different sensitivities are achieved fabricating different sized polypyrrole membranes with multiple pore shapes. The system comes with a standard signal processing package that uses feed forward neural networks trained using a back-propagation algorithm based on the stochastic gradient descent method. The authors replaced the signal processing software with their own based on the Structural Risk Minimisation principle.

Although the authors tested the system with different samples, they concentrated at discriminating Paraoxon from Parathion, since both molecules have almost an identical structure. As it can be seen in Fig. 13, the only difference lies in the $p = 0$ bond that is replaced by the $p = s$ bond in parathion.

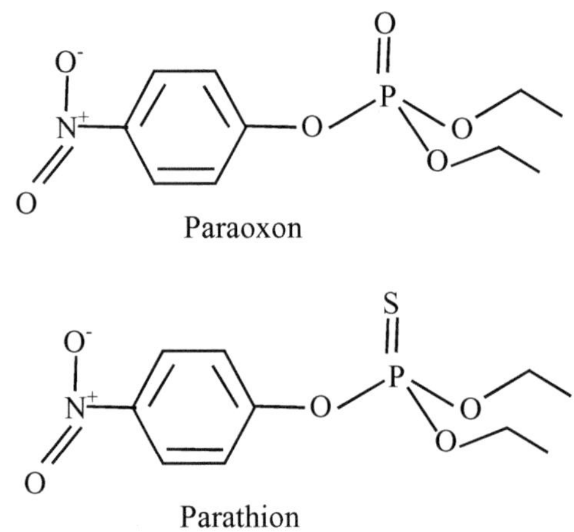

Fig. 13. Differences between Paraoxon and Parathion. Adapted from [31], with permission

To test the system they used 250 measurements performed with the Aromascan system. They evaluated their processing algorithm using a five-fold validation procedure in which they iterated five times, training with 200 measurements and evaluating with the remaining 50. They compared the results obtained with three different kernels (Polynomial, RBF, and s2000). To understand the benchmark results they obtained a few definitions have to be made:

- The ROC curve is a plot of the True Positive Ratio (TPR) as a function of the False Positive Ratio (FPR). The area under this curve, known as the AZ index, represents an overall performance over all possible (TPR, FPR) operating points. In other words, the AZ index makes a performance average through different threshold values.
- Sensitivity is defined as the ratio of TP/(TP + FN) which represents the likelihood that an event will be detected if that event is present. (TP = true positive event, FN = false negative event)

- Specificity is defined as the ratio of TN/(TN + FP) which represents the likelihood that the absence of an event will be detected given that that event is absent. (TN = true negative events, FP = false positives)
- Positive predictive value (PPV) is defined as the ratio of TP/(TP + FP) which represents the likelihood that a signal is related to the true event that actually occurred. (FP = false positives)

All this figures are associated to the threshold value applied. Higher thresholds will minimise false positives (increasing specificity) at the expense of increasing false negatives (reducing sensitivity). On the other hand, lowering threshold values will reduce specificity and increase sensitivity.

Table 7 shows the results obtained for the three different kernels. Two different sensitivity values (100% and 98%) were used to evaluate specificity and PPV. This was done controlling the threshold value. As it can be seen, the polynomial kernel seems to obtain superior performances compared to the other 2. These are excellent results considering there is a single atom difference between both nerve agents. Although the study was centred around the distinction between Paraoxon and Parathion, additional results on the discrimination of other binary pairs were reported. For example, when classifying parathion samples from dichlorvos, a 23% improvement on the overall ROC AZ index was obtained when using the s2000 kernel compared to the standard back-propagation algorithm supplied with the electronic nose. Moreover, specificity was improved a surprising 173% while making no false positive errors. Similar improvements where seen using the Gaussian and polynomial (degree 2) kernels. In the Dichlorvos vs. Trichlorfon classification problem all SVM classifiers performed flawlessly.

Table 7. Results obtained discriminating Parathion from Paraxon with 3 kernels. Adapted from [31], with permission

Kernel	Az	Az90	Spec at 100%	PPV at 100%	Spec at 98%	PPV at 98%
RBF	0.9275	0.7881	0.7633	0.7304	0.7633	0.7304
S2000	0.9844	0.9002	0.8701	0.8359	0.8701	0.8359
POLY	0.9916	0.9344	0.8739	0.8366	0.8739	0.8366

5 Conclusions

SVM algorithms have certain characteristics that make them very attractive for use in artificial odour sensing systems. Their well-founded statistical behaviour, their generalisation ability and their low computational requirements are the main reasons for the recent interest in these new types of paradigms. Moreover, their regression capabilities add an additional dimension to their possible use in gas sensing instruments.

From all of the possible advantages that SVMs can offer to the electronic nose community, perhaps the generalisation ability and the robustness in front of outliers can be considered as the most interesting ones. Both advantages address important drawbacks of conventional electronic noses, namely the lengthy calibration process and often poor reproducibility of results.

The literature presented has shown that the SVM paradigm compares favourably to other methods in simple and complex vapour analysis. Moreover, SVM algorithms have been used in classification, quantification and variable selection, giving good results in all cases.

Although SVMs hold a great potential as the pattern recognition paradigm of choice in many multisensor gas systems, no commercial system yet offers them. In fact, only a few research studies have explored their possibilities in this type of applications with very promising results. Therefore, a lot of work remains to be done in this interesting field of application.

Studies on how the SVMs can cope with sensor drift, how they compare with other algorithms and how different kernel functions perform under similar problems should be performed in a systematic manner. The objective would be to determine the optimal way and in which applications they should be used. Moreover, since they are well founded, mathematically speaking, algorithm modifications can be proposed and explored. Despite the fact that in these initial results the original (unmodified) SVM algorithms have been used, results compare favourably to other type of algorithms. Therefore, it can be anticipated that optimised paradigms can give even better results than those reported in the studies reported to date.

References

1. Wilkens, W.F., Hatman, A.D. (1964) An electronic analog for the olfactory processes, Ann. NY Acad. Sci. 116, 608–620.
2. Persaud K.C., Dodd G.H. (1982) Analysis of discrimination mechanisms of the mammalian olfactory system using a model nose, Nature, 299, 352–355.
3. Gardner J.W., Barlett P.N. (1994) A brief history of electronic noses, Sensors and Actuators B, 18, 211–220.
4. Brezmes J., Llobet E., Vilanova X., Correig X. (1997) Neural-network based electronic nose for the classification of aromatic species, Anal. Chim. Acta, 348, 503–509.
5. Distante C., Ancona N., Siciliano P. (2003) Support vector Machines for olfactory signals recognition, Sensors and Actuators B, 88, 30–39.
6. Göpel W. (1991), Sensors: A Comprehensive Survey, Vol. 2/3: Chemical and Biochemical Sensors, VCH, Weinheim.
7. Gardner J.W., Hines E.L. (1997) Pattern analysis techniques, Handbook of Biosensors and Electronic Noses, Medicine, Food, and the Environment, AEG Frankfurt, Germany, 633–652.
8. Di Natale C., Davide F., D'Amico A. (1995) Pattern recognition in gas sensing; well-stated techniques and advances, Sensors and Actuators B, 23, 111–118.

9. Llobet E., Brezmes J., Vilanova X., Fondevila L., Correig X. (1997) Quantitative vapor analysis using the transient response of non-selective thick-film tin oxide gas sensors, Proceedings Transducers'97, 971–974.
10. Hines E.L., Llobet E., Gardner J.W. (1999) Electronic Noses: a review of signal processing techniques, IEE Proc.-Circuits Devices Syst., 146, 297–310.
11. Dinatale C. et al. (1995) Pattern recognition in gas sensing: well-stated techniques and advances, Sensors and Actuators B, 23, 111–118.
12. Brezmes J., Llobet E., Vilanova X., Sáiz G., Correig X. (2000) Fruit ripeness monitoring using an Electronic Nose, Sensors and Actuators B 69, 223–29.
13. Shin H.W., Llobet E., Gardner J.W., Hines E.L., Dow C.S. (2000) The classification of the strain and growth phase of cyanobacteria in potable water using an electronic nose system, IEE Proc. Sci. Meas. Technology, 147, 158–164.
14. Fleischer M. et al. (2002) Detection of volatile compounds correlated to human diseases through breath analysis with chemical sensors, Sensors and Actuators B, 83, 245–249.
15. Llobet E., Ionescu R. et al. (2001) Multicomponent gas mixture analysis using a single tin oxide sensor and dynamic pattern recognition, IEEE Sensors Journal, 1, 207–213.
16. Llobet E., Brezmes J., Ionescu R. et al. (2002) Wavelet Transform and Fuzzy ARTMAP Based Pattern Recognition for Fast Gas Identification Using a Micro-Hotplate Gas Sensor, Sensors & Actuators B Vol, 83, 238–244.
17. www.Alpha-mos.com
18. www.Cyranonose.com
19. Fryder M., Holmberg M., Winquist F., Lundstrom I. (1995) A calibration technique for an electronic nose, Proceedings of the Transducers' 95 and Eurosensors IX, 683–686.
20. Artursson T., Eklov T., Lundstrom I., Marteusson P., Sjostrom M., Holmberg M. (2000) Drift correction for gas sensors using multivariate methods, J. Chemomet. 14, 1–13.
21. Holmberg M., Winquist F., Lundstrom I., Davide F., Di Natale C., d'Amico A. (1996) Drift counteraction for an electronic nose, Sensors & Actuators B, 35/36, 528–535.
22. Holmberg M., Winquist F., Lundstrom I., Davide F., Di Natale C., D'Amico A. (1997), Drift counteraction in odour recognition application: lifelong calibration method, Sensors & Actuators B, 42, 185–194.
23. Pardo M., Sberveglieri G., (2002) Classification of electronic nose data with Support Vector Machines, Proc. ISOEN'02, 192–196.
24. Cristianini N., Shawe-Taylor J. (2000) An introduction of Support Vector Machines and other Kernel-based learning methods, Cambridge University Press, Cambridge, UK.
25. Frank M.L., Fulkerson M.D. et al., (2002) TiO_2-based sensor arrays modeled with nonlinear regression analysis for simultaneously determining CO and O_2 concentrations at high temperatures, Sensors & Actuators B, 87, 471–479.
26. Martinez D., Bray A. (2003) Nonlinear blind source separation using kernels, IEEE Transaction on Neural Networks, 14, 1, 228–235.
27. Stone J.V. (2001) "Blind source separation using temporal predictability", Neural Comput., 13, 1559–1574.
28. Al-Khalifa S., Maldonado-Bascon S., Gardner J.W. (2003) Identification of CO and NO_2 using a thermally resistive microsensor and support vector machine, IEE Proceedings Measurement and Technology, 150, 11–14.

29. MaldonadoS, Al-Khalifa S., López-Ferreras F. (2003) Feature reduction using support vector machies for binary gas detection, Lecture notes in computer science, LNCS 2687 "Artificial neural nets.roblem solving methods", IWANN 2003 798–805 Proceedings vol. 2 ISSN 0302-9743.
30. Mallat, S (1997) A Wavelet Tour of Signal Processing, Academic Press.
31. Land W.H. et al. (2003) New results using multiarray sensors and support vector machines for the detection and classification of organophosphate nerve agents, proc. IEEE International Conference on Systems, Man and Cybernetics, Washington DC, 2883–2888.

Application of Support Vector Machines in Inverse Problems in Ocean Color Remote Sensing

H. Zhan

LED, South China Sea Institute of Oceanology, Chinese Academy of Sciences, Guangzhou 510301, China

Abstract. Neural networks are widely used as transfer functions in inverse problems in remote sensing. However, this method still suffers from some problems such as the danger of over-fitting and may easily be trapped in a local minimum. This paper investigates the possibility of using a new universal approximator, support vector machine (SVM), as the nonlinear transfer function in inverse problem in ocean color remote sensing. A field data set is used to evaluate the performance of the proposed approach. Experimental results show that the SVM performs as well as the optimal multi-layer perceptron (MLP) and can be a promising alternative to the conventional MLPs for the retrieval of oceanic chlorophyll concentration from marine reflectance.

Key words: transfer function, ocean color remote sensing, chlorophyll, support vector machine

1 Introduction

In remote sensing, retrieval of geophysical parameters from remote sensing observations usually requires a data transfer function to convert satellite measurements into geophysical parameters [1, 2]. Neural networks have gained popularity for modeling such a transfer function in the last almost twenty years. They have been applied successfully to derive parameters of the oceans, atmosphere, and land surface from remote sensing data. The advantages of this approach are mainly due to its ability to approximate any nonlinear continuous function without a priori assumptions about the data. It is also more noise tolerant, having the ability to learn complex systems with incomplete and corrupted data. Different models of NNs have been proposed, among which, multi-layer perceptrons (MLPs) with the backpropagation training algorithm are the most widely used [3, 4].

However, MLPs still suffers from some problems. First, the training algorithm may be trapped in a local minimum. The objective function of MLPs

H. Zhan: *Application of Support Vector Machines in Inverse Problems in Ocean Color Remote Sensing*, StudFuzz **177**, 387–397 (2005)
www.springerlink.com

is very often extremely complex. The conventional training algorithms can easily be trapped in a local minimum and never converge to an acceptable error. In that case even the training data set cannot be fit properly. Second, it is generally a difficult task to determine the best architecture of MLPs, such as the selection of the number of hidden layers and the number of nodes therein. Third, over-fitting of the training data set may also pose a problem. MLP training is based on the so-called empirical risk minimization (ERM) principle, which minimizes the error on a given training data set. A drawback of this principle is the fact that it can lead to over-fitting and thereby poor generalization [5, 6].

These problems can be avoided by using a promising new universal approximator, i.e., support vector machines (SVMs). SVMs have been developed by Vapnik within the area of statistical learning theory and structural risk minimization (SRM) [6]. SVM training leads to a convex quadratic programming (QP) problem, rather than a non-convex, unconstrained minimization problem as in MLP training, hence it always converges to the global solution for a given data set, regardless of initial conditions. SVMs use the principle of structural risk minimization to simultaneously control generalization and performance on the training data set, which provides it with a greater ability to generalize. Furthermore, there are few free parameters to adjust and architecture of the SVM does not need to be found by experimentations.

The objective of this paper is to illustrate the possibility of using support vector machines as the nonlinear transfer function to retrieve geophysical parameters from satellite measurements. As an example of their use in a remote sensing problem with real data we consider the nonlinear inversion of oceanic chlorophyll concentration from ocean color remote sensing data. The results show that SVMs perform as well as the optimal multi-layer perceptron (MLP) and can be a promising alternative to the conventional MLPs for modeling the transfer function in remote sensing.

2 Inverse Problem in Ocean Color Remote Sensing

2.1 Ocean Color Remote Sensing

In oceanography, the term ocean color is used to indicate the visible spectrum of upwelling radiance as seem at the sea surface or from space. This radiance contains significant information on water constituents, such as the concentration of phytoplankton pigments (can be regarded as the chlorophyll concentration), suspended particulate matter (SPM) and colored dissolved organic matter (CDOM, the so-called yellow substance) in surface waters. Ocean color is the result of the process of scattering and absorption by the water itself and by these constituents. Variations of these constituents modify the spectral and geometrical distribution of the underwater light field, and thereby alter the color of the sea. For example, biologically rich and productive waters are characterized by green water, and the relatively depauperate open ocean regions are blue. Information on these constituents can be used to investigate biological productivity in the oceans, marine optical properties, the interaction of winds and currents with ocean biology, and how human activities influence the oceanic environment [7, 8, 9].

Since the Coastal Zone Color Scanner (CZCS) aboard the Nimbus 7 satellite was launched in 1978, it has became apparent that ocean color remote sensing is a powerful means in synoptic measurements of the optical properties and oceanic constituents over large areas and over long time periods. More than a decade after the end of the pioneer CZCS emission, a series of increasingly sophisticated sensors, such as SeaWiFS (the Sea-Viewing Wide Field-of-View Sensor) has emerged [7]. The concentration of optically active water constituents can be derived from ocean color remote sensing data by the interpretation of the received radiance at the sensor at different wavelength.

Figure 1 illustrates different origins of light received by satellite sensor. The signal received by the sensor is determined by following contributors: (1) scattering of sunlight by the atmosphere, (2) reflection of direct sunlight at the sea surface, (3) reflection of sunlight at sea surface, and (4) light reflected within the water body [10, 11]. Only the portion of the signal originating from

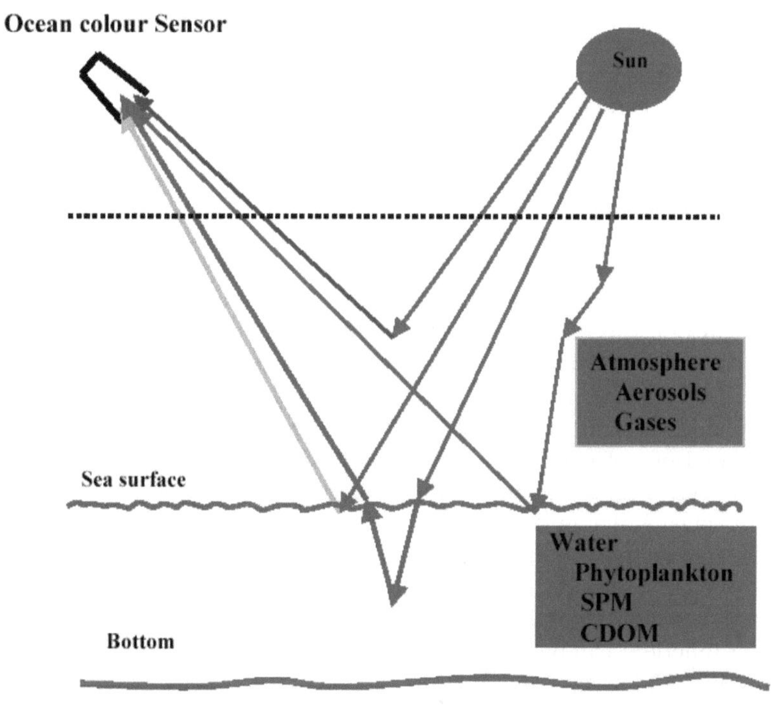

Fig. 1. Graphical depiction of different origins of light received by ocean color sensor

the water body contains information on the water constituents; the remaining portion of the signal, which takes up more than 80% of the total signal, has to be assessed precisely to extract the contribution from the water body. Therefore, there exit two strategies to derive water constituents from the signal received by ocean color sensor. One is that the water leaving radiance (or reflectance) is firstly derived from the signal received by the sensor (this procedure is called "atmospheric correction"), and then oceanic constituents are retrieved from water leaving radiance (or reflectance). Another is that oceanic constituents are directly derived from the signal received by the satellite sensor.

In the remote sensing of ocean color, two major water types, referred to as "case 1" and "case 2" waters, can be identified [9]. Case 1 waters are ones where the optical signature is due to the presence of phytoplankton and their by-products. Case 2 waters are ones where the optical properties may also be influenced by the presence of SPM and CDOM. In general, case 1 waters are those of the open ocean, while case 2 waters are those of the coastal seas (represent less than 1% of the total ocean surface). Therefore, estimation of water constituents from case 1 waters and case 2 waters can be identified as a one-variable and a multivariate problem respectively, and interpretation of an optical signal from case 2 waters can therefore be rather difficult [9].

2.2 Inverse Algorithms in Ocean Color Remote Sensing

Ocean color inverse algorithms, like in most other geophysical inverse problems, can be classified into two categories: implicit and explicit [10]. In implicit inversion, water constituents are estimated simultaneously by matching the measured with the calculated spectrum. The match is quantified with an objective function, which expresses a measure of "goodness of fit". Water constituents associated with the calculated spectrum that most closely match the measured spectrum are then taken to be the solution of the problem. As a model-based approach, successes of implicit algorithm rely on accuracy of the forward optical models and search ability of the optimization algorithms. This type of algorithms have been employed mostly for in case 2 waters, in which they outperform traditional explicit approaches because information of all available spectral bands can be flexible involve in the objective function and extractions of constituent concentrations are carried out pixel-by-pixel [9]. However, computing time may be a limitation to implicit algorithms, especially when global optimization algorithms, such as simulated annealing or genetic algorithm were used [12].

In explicit inversion, concentrations of water constituents are expressed as inverse transfer functions of measured radiance spectrum. These inverse transfer functions can be obtained by empirical, semi-analytical and analytical approaches. Empirical equations derived by statistical regression of radiance versus water constituent concentrations are the most popular algorithms for estimation of water constituent concentrations. They do not require a full understanding of the relationship between radiance (or reflectance) and the water constituent concentrations. The advantages of empirical approaches are the simplicity and rapidity in data processing, which are important for the retrieval of information from large data sets such as satellite images. The semi-analytical and analytical approaches are based on solutions to the radiative transfer equation and attempt to model the physics of ocean color. They have advantages over empirical algorithms in that they can be applied to retrieve multiple water properties simultaneously from a single radiance spectrum [9].

In recent years, neural networks have been increasingly applied as inverse transfer functions to retrieval of water properties from radiance (or reflectance) in both case 1 and case 2 waters [13, 14, 15, 16, 17, 18, 19]. The two most important steps in applying NNs to inverse problems in ocean color remote sensing are the selection (including the selection of network configuration and training data) and learning stages, since these directly influence the performance of the inverse models. Inputs of NNs may be (1) radiance or reflectance at the top of the atmosphere; (2) rayleigh-corrected radiance or reflectance; (3) direct water-leaving remote-sensing reflectance, after atmospheric correction; and (4) normalized water-leaving remote-sensing reflectance, after atmospheric correction. Optional outputs are the concentrations of water constituents or optical properties used as intermediate variables, which can be converted to concentrations using regional conversion factors [9]. Since the number of inputs and outputs is fixed, main attention have been paid to the number of hidden layers and the number of neurons therein. Data of different origin may be used to construct NNs: field data collected from in situ measurements, and synthetic data obtained from forward simulations of optical models. The two most frequent problems related to the selection of training data are unrepresentative and the so-called over-fitting. In the first case, too many bad examples are selected, as a result, the trained models are not appropriate to waters with dissimilar characteristics. To circumvent this problem, the training dataset may be fit to some statistical distribution. In the latter case, NNs will be able to model the training data very well but it may be very inaccurate for other data that were not part of the training data. To guard against this possibility, some methods for estimating generalization error based on "resampling", such as cross validation and bootstrapping have been used to control over-fitting [13].

3 Examples of the Use of Support Vector Machine in Ocean Color Remote Sensing

In this section we present an example of the use of SVMs as nonlinear transfer function, and compare their performance to other explicit methods. Further details are in Zhan et al. [21].

3.1 Data Description and Preprocessing

To carry out an experimental analysis to validate the performance of SVM, we considered an in situ data set that was archived by the NASA SeaWiFS Project as the SeaBAM data set [22]. This data set consists of coincident remote sensing reflectance (Rrs) at the SeaWiFS wavelengths (412, 443, 490, 510, and 555 nm) and surface chlorophyll concentration measurements at 919 stations around the United States and Europe. It is encompassed a wide range of chlorophyll concentration between 0.019 and 32.79 μgL^{-1} with a geometric mean of 0.27 μgL^{-1}. Most of the data are from Case 1 nonpolar waters, and about 20 data collected from the North Sea and Chesapeake Bay should be considered as case 2 waters. Log-transferred Rrs values and chlorophyll concentrations are used as the inputs and output respectively. The advantage of this transformation is that the distribution of transformed data will become more symmetrical and closer to normal. To facilitate training of SVMs, the values of each input and output were scaled into the range of $[-1\ 1]$.

3.2 Training of the SVM

The training software used in our experiments is LIBSVM [23]. LIBSVM is an integrated software package for support vector classification, regression and distribution estimation. It uses a modified sequential minimal optimization (SMO) algorithm to perform training of SVMs. SMO algorithm breaks the large QP problem into a series of smallest possible QP problems. These small QP problems are solved analytically, which avoids using a time-consuming numerical QP optimization as an inner loop [24]. The RBF kennel function was chosen because it is much more flexible than the two-layer perceptron and the polynomial kennel function. Consequently, it tends to perform best over a range of applications, regardless of the particulars of the data [25]. This kennel function always satisfies Mercer's condition [26].

There are three free parameters, namely C, ε and σ should be determined to find the optimal solution. In our experiments, we split the SeaBAM data set into two subsets and used the split-sample validation approach to tune these free parameters. This split-sample validation approach estimates the free parameters by using one subset (training set) to train various candidate models and the other subset (validation set) to validate their performance [13, 14]. In order to ensure the representative of the data sets, the SeaBAM data set was first arranged in increasing order of the chlorophyll concentrations and then, starting from the top, the odd order samples ($n = 460$) were picked up as the training set and the remaining samples ($n = 459$) were used as the validation set. We set $C = 15, \varepsilon = 0.03$ and $\sigma = 0.7$, because these values were found to produce the best possible results on the validation set by slit-sample validation approach. After these parameters are fixed, the SVM automatically determines the number (how many SVs) and locations (the SVs) of RBF centers during its training.

3.3 Experimental Results

The performance of the SVM was evaluated using the same criteria as [22], namely, root mean square error (RMSE), coefficient of determination (R^2), and scatterplot of derived versus in situ chlorophyll concentrations. The RMSE index is defined as

$$\mathrm{RMSE} = \sqrt{\frac{1}{N}\sum_{k=1}^{N}\left(\log_{10}\left(c_k^d\right) - \log_{10}\left(c_k^m\right)\right)^2}$$

where N is the number of examples, c is the chlorophyll concentration, and the superscripts d and m indicate derived and measured values. Coefficient of determination and scatterplot are also based on log-transformed data.

Figure 2 displays the scatterplots of the SVM-derived versus the in situ chlorophyll (Chl) concentration on the training and the validation set. The RMSE for the training set is 0.122 and its R^2 is 0.958. The RMSE for the validation set is 0.138 and its R^2 is 0.946. The number of support vectors (SVs) is 288, which close to 60 percent of the training data. These SVs contains all the information necessary to model the nonlinear transfer function.

The performance of the SVM was compared with those of MLPs and SeaWiFS empirical algorithms. Two SeaWiFS empirical algorithms are

$$\begin{aligned} \text{OC2:} \quad & C = 10^{(0.341-3.001*R+2.811*R*R-2.041*R*R*R)} - 0.04\,, \\ & R = \log(\mathrm{Rrs490}/\mathrm{Rrs555}) \end{aligned}$$

$$\begin{aligned} \text{OC4:} \quad & C = 10^{(0.4708-3.8469*R+4.5338*R*R-2.4434*R*R*R)} - 0.0414 \\ & R = \log(\max(\mathrm{Rrs443}, \mathrm{Rrs490}, \mathrm{Rrs510})/\mathrm{Rrs555}) \end{aligned}$$

In order to allow for this comparison, the training and validation data were preprocessed in similar manner for SVM and MLP uses, and results of MLPs

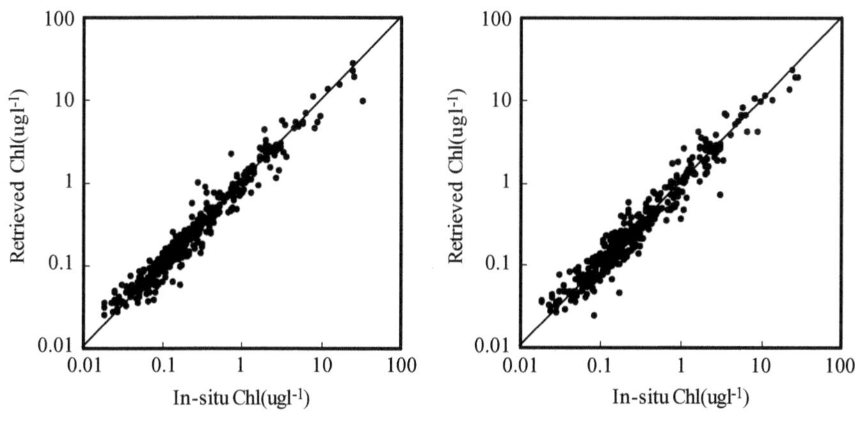

Fig. 2. Comparison of the SVM-derived versus in situ chlorophyll concentrations on training (*left*) and validation data set (*right*)

Table 1. Statistical results of MLPs, SVM, and empirical algorithms on the validation set

			RMSE				
MLP			Number of Hidden Nodes				
Trial	4	5	6	7	8	9	10
1	0.177	0.143	0.157	0.155	0.188	0.143	0.156
2	0.139	0.152	0.139	0.149	0.140	2.563	0.157
3	0.144	0.140	0.149	0.157	0.152	0.548	0.206
4	0.138	0.596	0.336	0.143	0.197	0.159	0.229
5	0.140	0.141	0.160	0.241	1.841	0.155	0.220
6	0.137	0.140	0.140	0.156	0.158	0.282	0.180
7	0.195	0.142	0.305	0.176	5.169	0.150	2.878
8	0.154	0.219	0.143	0.135	0.151	0.137	0.155
9	0.144	0.140	0.144	0.140	0.301	0.471	0.236
10	0.162	0.146	0.146	0.151	0.145	0.171	0.205
SVM				0.138			
OC2				0.172			
OC4				0.161			
			R^2				
1	0.912	0.942	0.931	0.933	0.908	0.943	0.932
2	0.946	0.936	0.945	0.938	0.945	0.014	0.931
3	0.942	0.945	0.938	0.931	0.935	0.516	0.891
4	0.946	0.427	0.724	0.942	0.823	0.929	0.867
5	0.944	0.943	0.930	0.848	0.158	0.931	0.878
6	0.947	0.944	0.945	0.932	0.930	0.806	0.910
7	0.893	0.943	0.763	0.915	0.002	0.937	0.002
8	0.933	0.878	0.942	0.949	0.935	0.947	0.931
9	0.942	0.944	0.941	0.944	0.770	0.560	0.854
10	0.928	0.940	0.939	0.936	0.941	0.916	0.886
SVM				0.946			
OC2				0.919			
OC4				0.929			

and SeaWiFS empirical algorithms were based on the same validation set as was used for the SVM. There are a large number of factors control the performance of MLPs, such as the number of hidden layers, the number of hidden nodes, activation functions, epochs, weights initialization methods and parameters of the training algorithm. It is a difficult task to obtain an optimal combination of these factors that produces the best retrieval performance. We used MLPs with one hidden layer and tan-sigmoid activation, and trained them using Matlab Neural Network Toolbox 4.0 with Levenberg-Marquardt algorithm. The epoch was set to 500 and other training parameters were set to the default values of the software. The training process was run 10 times with different random seeds for the number of hidden nodes from 4 to 10. The statistical results of the MLPs, the SVM and the SeaWiFS algorithm OC2 and OC4 on the validation set are reported in Tables 1. Several results can be found from this table. First, the performance of the SVM is as good as the optimal MLP solution. There are only two trials in which the RMSE of the best MLP is slight smaller than that of the SVM. Second, the optimal number of hidden nodes is difficult to determine because it varies with different weights initialization. Third, large errors occurred in some trials due to the training algorithm was trapped in a local minimum. Finally, the SVM and the best MLP with different weights initialization outperform the SeaWiFS empirical algorithms.

4 Conclusions

The use of SVMs as a transfer function in inverse problem in ocean color remote sensing was demonstrated in this paper. Experiments on a field data set indicated that the performance of SVMs were comparable in accuracy to the best MLP. Advantages of SVMs over MLPs include the existence of fewer parameters to be chosen, a unique, global minimum solution, and high generalization ability. The proposed method seems to be a promising alternative to the conventional MLPs for modeling the nonlinear transfer function between chlorophyll concentration and marine reflectance.

It is worthy to note that SVM generalization performance, unlike those of conventional neural networks such as MLPs, does not depend on the dimensionality of the input space. The SVM can have good performance even in problems with a large number of inputs and thus provide a way to avoid the curse of dimensionality. This makes it attractive for case 2 waters, since more spectral channels are needed for retrieval of some parameters in such waters. Further research will be carried out to validate the performance of SVMs in inverse problem in case 2 waters.

Acknowledgements

The author acknowledges the SeaWiFS Bio-optical Algorithm Mini-Workshop (SeaBAM) for their SeaBAM data and Chih-Chung and Chih-Jen Lin for their software package LIBSVM. This work is supported by the National Natural Science Foundation of China (40306028), the Funds of Knowledge Innovation Program of South China Sea Institute of Oceanology (LYQY200308), and the Guangdong Natural Science Foundation (32616).

References

1. Krasnopolsky, V. M., Schiller, H. (2003) Some Neural Network Applications in Environmental Sciences. Part I: Forward and Inverse Problems in Geophysical Remote Measurements. Neural Networks, 16, 321–334
2. Krasnopolsky, V. M., Chevallier, F. (2003) Some Neural Network Applications in Environmental Sciences. Part II: Advancing Computational Efficiency of Environmental Numerical Models. Neural Networks, 16, 335–348
3. Atkinson, P. M., Tatnall, A. R. L. (1997) Neural networks in remote sensing. Int. J. Remote Sens., 18, 699–709
4. Kimes, D. S., Nelson, R. F., Manry, M. T., Fung, A. K. (1998) Attributes of neural networks for extracting continuous vegetation variables from optical and radar measurements. Int. J. Remote Sens., 19, 2639–2663
5. Vapnik, V. N. (1999) An Overview of Statistical Learning Theory. IEEE Trans. Neural Networks, 10, 988–1000
6. Vapnik, V. N. (2000) The Nature of Statistical Learning Theory (2nd Edition). New York, Springer-Verlag
7. IOCCG. (1997) Minimum Requirements for an Operational Ocean-Colour Sensor for the Open Ocean. Reports of the International Ocean-Colour Coordinating Group, No. 1, IOCCG, Dartmouth, Nova Scotia, Canada
8. IOCCG. (1998) Status and Plans for Satellite Ocean-Color Missions: Considerations for Complementary Missions. J. A. Yoder (ed.). Reports of the International Ocean-Colour Coordinating Group, No. 2, IOCCG, Dartmouth, Nova Scotia, Canada, 1998
9. IOCCG. (2000) Remote Sensing of Ocean Colour in Coastal, and Other Optically-Complex, Waters. S. Sathyendranath (ed.). Reports of the International Ocean-Colour Coordinating Group, No. 3, IOCCG, Dartmouth, Nova Scotia, Canada
10. Mobley, C. D. (1994) Light and Water: Radiative Transfer in Natura Waters. New York, Academic
11. Bukata, R. P., Jerome, J. H. K., Konrdatyev Ya., Pozdnyakov, D. V. (1995) Optical Properties and Remote Sensing of Inland and Coastal Waters. Boka Raton, CRC
12. Zhan, H. G., Lee, Z. P., Shi, P., Chen, C. Q., Carder, K. L. (2003) Retrieval of Water Optical Properties for Optically Deep Waters Using Genetic Algorithms. IEEE Trans. Geosci. Remote Sensing, 41, 1123–1128
13. Keiner, L. E., Yan, X. H. (1998) A neural network model for estimating sea surface chlorophyll and sediments from Thematic Mapper imagery. Remote Sens. Environ., 66, 153–165
14. Keiner, L. E., Brown, C. W. (1999) Estimating oceanic chlorophyll concentrations with neural networks. Int. J. Remote Sens., 20, 189–194
15. Schiller H., Doerffer, R. (1999) Neural network for estimating of an inverse model-operational derivation of Case II water properties from MERIS data. Int. J. Remote Sens., 20, 1735–1746
16. Buckton, D., Mongain, E. (1999) The use of neural networks for the estimation of oceanic constituents based on MERIS instrument. Int. J. Remote Sens., 20, 1841–1851
17. Lee, Z. P., Zhang, M. R., Carder, K. L., Hall, L. O. (1998) A neural network approach to deriving optical properties and depths of shallow waters. In S. G.

Ackleson, J. Campell (eds) Proceedings, Ocean Optics XIV, Office of Navel Research, Washington, DC
18. Tanaka, A., Oishi, T., Kishino, M., Doerffer, R. (1998) "Application of the neural network to OCTS data," in in S. G. Ackleson, J. Campell (eds) Proceedings, Ocean Optics XIV, Office of Navel Research, Washington, DC
19. Gross, L., Thiria, S., Frouin, R., Mitchell, B. G. (2000) Artificial neural networks for modeling the transfer function between marine reflectance and phytoplankton pigment concentration. J. Geophys. Res., 105, 3483–3495
20. Kwiatkowska E. J., Fargion, G. S. (2002) Merger of Ocean Color Data from Multiple satellite Missions within the SIMBIOS Project. SPIE Proceedings, 4892, 168–182
21. Zhan, H. G., Shi, P., Chen, C. Q. (2003) Retrieval of Oceanic Chlorophyll Concentration using Support Vector Machines. IEEE Trans. Geosci. Remote Sensing, 41, 2947–2951
22. O'Reilly, J. E., Maritorena, S., Mitchell, B. G., Siegel, D. A., Carder, K. L., Garver, S. A., Kahru, M., McClain, C. (1998) Ocean color chlorophyll algorithms for SeaWiFS. J. Geophys. Res., 103, 24937–24953
23. Chang C. C., Lin, C. J. (2001) LIBSVM: a library for support vector machines. Software available at http://www.csie.ntu.edu.tw/~cjlin/libsvm
24. Platt, J. (1999) Fast training of SVMs using sequential minimal optimization. in B. SchPolkopf, C. Burges, A. Smola, (Eds), Advances in Kernel Methods: Support Vector Learning, Cambridge, MIT press
25. Smola, A. J., (1998) Learning with Kernels. PhD Thesis, GMD, Birlinghoven, Germany
26. Haykin, S. (1999) Neural Networks: A Comprehensive Foundation. (2nd edition), New Jersey, Prentice-Hall

Application of Support Vector Machine to the Detection of Delayed Gastric Emptying from Electrogastrograms

H. Liang

School of Health Information Sciences, University of Texas Health Science Center at Houston, Houston, TX 77030, USA
hualou.liang@uth.tmc.edu

Abstract. The radioscintigraphy is currently the gold standard for gastric emptying test, but it involves radiation exposure and considerable expenses. Recent studies reported neural network approaches for the non-invasive diagnosis of delayed gastric emptying from the cutaneous electrogastrograms (EGGs). Using support vector machines, we show that this relatively new technique can be used for detection of delayed gastric emptying and is in fact able to improve the performance of the conventional neural networks.

Key words: support vector machine, genetic neural networks, spectral analysis, electrogastrogram, gastric emptying

1 Introduction

Delayed gastric emptying, also called gastroparesis, is a disorder in which the stomach takes too long to empty its contents. It often occurs in people with type 1 diabetes or type 2 diabetes. If food lingers too long in the stomach, it can cause problems like bacterial overgrowth from the fermentation of food. Also, the food can harden into solid masses called bezoars that may cause nausea, vomiting, and obstruction in the stomach. Bezoars can be dangerous if they block the passage of food into the small intestine. Therefore, accurate detection of delayed gastric emptying is important for diagnosis and treatment of patients with motility disorders of the stomach.

The currently standard gastric emptying test, known as radioscintigraphy [1], is performed by instructing a patient to digest a meal with radioactive materials, and then to stay under a gamma camera for acquiring abdominal images for 2 to 4 hours. Although radioscintigraphy is the gold standard for gastric emptying test, the application of this technique involves radiation exposure and more, it is considerably expensive, and usually limited to very

H. Liang: *Application of Support Vector Machine to the Detection of Delayed Gastric Emptying from Electrogastrograms*, StudFuzz **177**, 399–412 (2005)
www.springerlink.com

sick patients. It is, therefore, imperative to develop non-invasive and low-cost methods for the diagnosis of delayed gastric emptying.

Gastric myoelectric activity is known to be the most fundamental activity of the stomach and it modulates gastric motor activity [2, 3]. Gastric myoelectric activity consists of two components, slow wave and spikes. The slow wave is omnipresent and its normal frequency in humans ranges from 2 to 4 cycles per minute (cpm). Both the frequency and propagation direction of gastric contractions are controlled by the gastric slow wave. Spikes, bursts of rapid changes in GMA are directly associated with antral contractions. The antral muscles contract when slow waves are superimposed with spike potentials [4, 5]. Abnormalities in the frequency of the gastric slow wave have been linked with gastric motor disorders and gastrointestinal symptoms. The abnormal frequencies of gastric slow wave ranges from slow activity termed bradygastria (0.5–2 cpm) to fast activity termed tachygastria (4–9 cpm). The electrogastrogram (EGG) is a cutaneous recording of the gastric slow wave from abdominal surface electrodes. It is attractive due to its non-invasive nature and the minimal disturbance of ongoing activity of the stomach. Many studies have shown that the EGG is an accurate measure of the gastric slow wave [6, 7, 8]. It was recently shown [2] that there were significant differences in a number of EGG parameters between the patients with actual delayed gastric emptying and those with normal gastric emptying.

Neural network techniques have been recently used for the non-invasive diagnosis of delayed gastric emptying based on cutaneous electrogastrograms (EGGs) [3, 9, 10]. In spite of its success, they still suffer from some problems such as over-fitting which results in low generalization ability. Specially, the irrelevant variables will hurt the performance of neural network.

Support Vector Machine (SVM) is a new promising pattern classification technique proposed recently by Vapnik and co-workers [11, 12]. It is based on the idea of structural risk minimization [11], which shows that the generalization error is bounded by the sum of the training set error and a term depending on the Vapnik-Chervonenkis (VC) dimension [11] of the learning machine. Unlike traditional neural networks, which minimize the empirical training error, SVM aims to minimize the upper bound of the generalization error, so that higher generalization performance can be achieved. Moreover, SVM generalization error is related not to the input dimensionality of the problem, but to the margin with which it separates the data. This explains why SVM can have good performance even in problems with a large number of inputs. We use the diagnosis of delayed gastric emptying from EGGs as an example to illustrate the potent performance of SVM. The materials presented here have been reported previously [10, 13].

The remainder of this chapter is organized as follows. In Sect. 2, we first provide some background about the measurements of the EGG and the procedure of the gastric emptying test. We then introduce a variant of neural network based on genetic algorithms. We next review briefly the SVM, followed by the performance criteria used for the algorithm comparisons. In Sect. 3, we present the application of SVM for detection of delayed gastric emptying from the EGGs and contrast its performance with the genetic neural network, an improved version of conventional neural network. Section 4 presents a set of discussion and conclusions.

2 Methods

2.1 Measurements of the EGG and Gastric Emptying

The EGG data used in this study were obtained from 152 patients with suspected gastric motility disorders who underwent clinical tests for gastric emptying. A 30-min baseline EGG recording was made in a supine position before the ingestion of a standard test meal in each patient. Then, the patient set up and consumed a standard test meal within 10 minutes. After eating, the patient resumed supine position and simultaneous recordings of the EGG and scintigraphic gastric emptying were made continuously for 2 hours. Abdominal images were acquired every 15 min. The EGG signal was amplified using a portable EGG recorder with low and high cutoff frequencies of 1 and 18 cpm, respectively. On-line digitization with a sampling frequency of 1 Hz was performed and digitized samples were stored on the recorder (Synectics Medical Inc., Irving, TX, USA). All recordings were made in a quiet room and the patient was asked not to talk and to remain as still as possible during the recording to avoid motion artifacts.

The technique for gastric emptying test was previously described [2]. Briefly, the standard test meal for determining gastric emptying of solids consisted of 7.5 oz of commercial beef stew mixed with 30 g of chicken livers. The chicken livers were microwaved to a firm consistency and cut into 1-cm cubes. The cubes were then evenly injected with 18.5 MBq of ^{99m}Tc sulfur colloid. The liver cubes were mixed into beef stew, which was heated in a microwave oven. After the intake of this isotope-labeled solid meal, the subject was asked to lie supine under the gamma camera for 2 hours. The percentage of gastric retention after 2 hours and $T^{1/2}$ for gastric emptying were calculated. Delayed gastric emptying was defined as the percentage of gastric retention in 2 hours equal to or greater than 70% or the $T^{1/2}$ equal to or greater than 150 min, or both. The interpretation of gastric emptying results was made by the nuclear medicine physicians.

2.2 EGG Data Preprocessing

Previous studies have shown that spectral parameters of the EGG provide useful information regarding gastrointestinal motility and symptoms [14] whereas the waveform of the EGG is unpredictable and does not provide reliable information. Therefore, all EGG data were subjected to computerized spectral

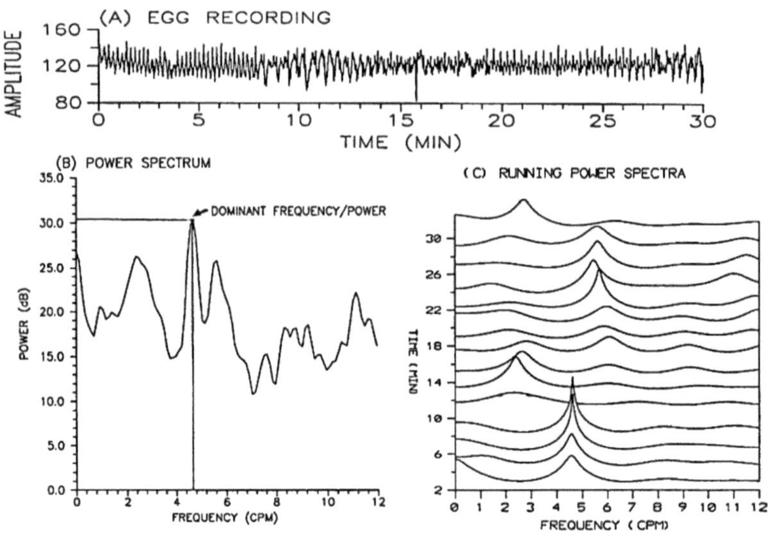

Fig. 1. Computation of EGG power spectrum. (**A**) A 30-min EGG recording, (**B**) its power spectrum, and (**C**) running power spectra showing the calculation of the dominant frequency/power of the EGG, the percentage of normal 2–4 cpm gastric slow waves and tachygastria (4–9 cpm) [16]

analysis using the programs previously described [15]. The following EGG parameters were extracted from the spectral domain of the EGG data in each patient and were used as candidate for the input to the classifiers.

(1) *EGG dominant frequency and power:* The frequency at which the EGG power spectrum has a peak power in the range of 0.5–9.0 cpm was defined as the EGG dominant frequency. The power at the corresponding dominant frequency was defined as EGG dominant power. Decibel (dB) units were used to represent the power of the EGG. Figure 1 illustrates the computation of the EGG dominant frequency and power. An example of a 30-min EGG recording in the fasting state obtained in one patient is shown in Fig. 1(A). The power spectrum of this 30-min EGG recording is illustrated in the Fig. 1(B). Based on this spectrum, the dominant frequency of the 30-min EGG shown in Fig. 1 (A) is 4.67 cpm and the dominant power is 30.4 dB. The smoothed power spectral analysis method [15] was used to compute an averaged power spectrum of the EGG during each recording, including the 30 min fasting EGG and 120 minutes postprandial EGG. These two parameters represent the mean frequency and amplitude of the gastric slow wave.
(2) *Postprandial change of EGG dominant power:* The postprandial increase of EGG dominant power was defined as the difference between the EGG dominant powers after and before the test meal, i.e., the EGG dominant power during the recording period B minus that during the recording

period A. The reason for the use of the relative power of the EGG as a feature is that the absolute value of the EGG power is associated with several factors unrelated to gastric motility or emptying, such as the thickness of the abdominal wall and the placement of the electrodes. The relative change of EGG power is related to the regularity and amplitude of the gastric slow wave, and has been reported to be associated with gastric contractility.

(3) *Percentages of normal gastric slow waves and gastric dysrhythmias:* The percentage of the normal gastric slow wave is a quantitative assessment of the regularity of the gastric slow wave measured from the EGG. It was defined as the percentage of time during which normal 2–4 cpm slow waves were observed in the EGG. It was calculated using the running power spectral analysis method [15]. In this method, each EGG recording was divided into two blocks of 2 min without overlapping. The power spectrum of each 2-min EGG data was calculated and examined to see if the peak power was within the range of 2–4 cpm. The 2-min EGG was called normal if the dominant power was within the 2–4 cpm range. Otherwise, it was called gastric dysrhythmia.

Gastric dysrhythmia includes tachygastria, bradygastria and arrhythmia. Tachygastria has been shown to be associated with gastric hypomotility [14], though the correlation between bradygastria and gastric motility is not completely understood. The percentage of tachygastria, thus, was calculated and used as a feature to be input into the SVM or genetic neural network. It was defined as the percentage of time during which 4–9 cpm slow waves were dominant in the EGG recording. It was computed in the same way as for the calculation of the percentage of the normal gastric slow wave. Details can be found in Liang et al. (2000) for an example of the EGG recording, its running power spectra for the calculation of the percentage of normal 2–4 cpm waves and tachygastria (4–9 cpm).

In summary, we ended up with five EGG spectral parameters which included the dominant frequency in the fasting state, the dominant frequency in the fed state, the postprandial increase of the EGG dominant power, the percentage of normal 2–4 cpm slow waves in the fed state and the percentage of tachygastria in the fed state. We used these five parameters extracted from the spectral domain as the inputs for both the SVM and the genetic neural network described in the following sections.

In order to preclude the possibility of some features dominating the classification process, the value of each parameter was normalized to the range of zero to one. Experiments were performed using all or part of the above parameters as the input to the classifier to derive an optimal performance.

2.3 Genetic Neural Network

The genetic neural network classifier was designed by using genetic algorithm [17] in conjunction with the cascade correlation algorithm architecture [18], hence termed the genetic cascade correlation algorithm (GCCA) [19]. The GCCA is an improved version of cascade correlation learning architectures, in which the genetic algorithm is used to select the neural network structure. The main advantage of this technique over the conventional back propagation (BP) for supervised learning is that it can automatically grow the architecture of neural networks to give a suitable network size for a specific problem.

The basic idea of the GCCA is first to apply the genetic algorithm over all the possible sets of weights in the cascade correlation learning architecture and then to apply the gradient descent technique (for instance, Quickprop [20]) to converge on a solution. This approach can automatically grow the architecture of the neural network to give a suitable network size for a specific problem. The GCCA algorithm is outlined in the following five-step procedure [19]:

(1) Initialize the network: Set up initial cascade-correlation architecture.
(2) Train output layer: The output layer weights are optimized by the genetic algorithm using populations of chromosomes on which the weights of the output layer are encoded. If the error could not be reduced significantly in a patience number of generations or the timeout (i.e., the maximum number of generations allowed) has been reached, then use Quickprop to adjust weights of output layer. If the learning is complete, then stop; else if error could not be reduced significantly in a patience number of consecutive epochs or the timeout has been reached, then go to the next step.
(3) Initialize candidate units: Create and initialize a random population sized to fit the problem. Each string in the population represents weights linking a candidate unit to input units, all pre-existing hidden units and bias unit.
(4) Train candidate units: Perform genetic search in weight space of candidate unit and use Quickprop to adjust weights of candidate unit so as to maximize the correlation between activation of the candidate unit and the error of the network. If the correlation could not be improved significantly in a patience number of consecutive epochs or the timeout has been reached, then go to the next step.
(5) Install a new hidden unit: Select the candidate unit with the highest correlation value and install it in the network as a new hidden unit. Now freeze its incoming weights and initialize the newly established outgoing weights. Go to step 2.

In essence, experiments have demonstrated that the network obtained with this technique is of small size and is superior to the standard BPNN classifier [19].

2.4 Support Vector Machine

In this section we briefly sketch the ideas behind SVM for classification and refer readers to [11, 12, 21] as well as the first chapter in the book for a full description of the technique.

Given the training data $\{(x_i, y_i)\}_{i=1}^{N}, x_i \in \Re^m, y_i \in \{\pm 1\}$ for the case of two-class pattern recognition, the SVM first maps x from input data x into a high dimensional feature space by using a nonlinear mapping ϕ, $z = \phi(x)$. In case of linearly separable data, the SVM then searches for a hyperplane $\mathbf{w}^{\mathbf{T}}\mathbf{z} + b$ in the feature space for which the separation between the positive and negative examples is maximized. The $\mathbf{w}$ for this optimal hyperplane can be written as $\mathbf{w} = \sum_{i=1}^{N} \alpha_i y_i \mathbf{z}_i$ where $\alpha = (\alpha_1, \ldots, \alpha_N)$ can be found by solving the following quadratic programming (QP) problem: maximize

$$\alpha^T - \frac{1}{2}\alpha^T Q \alpha$$

subject to

$$\alpha \geq 0\,, \qquad \alpha^T \mathbf{Y} = 0$$

where $\mathbf{Y}^{\mathbf{T}} = (y_1, \ldots, y_N)$ and Q is a symmetric $N \times N$ matrix with elements $Q_{ij} = y_i y_j \mathbf{z}_i^T \mathbf{z}_j$. Notice that Q is always positive semidefinite and so there is no local optimum for the QP problem. For those α_i that are nonzero, the corresponding training examples must lie closest to the margins of decision boundary (by the Kuhn-Tucker theorem [22], and these examples are called the *support vectors* (*SVs*).

To obtain Q_{ij}, one does not need to use the mapping ϕ to explicitly get $\mathbf{z}_i$ and $\mathbf{z}_j$. Instead, under certain conditions, these expensive calculations can be reduced significantly by using a suitable *kernel* function K such that $K(x_i, x_j) = z_i^T z_j$, Q_{ij} is then computed as $Q_{ij} = y_i y_j K(x_i, x_j)$. By using different kernel functions, the SVM can construct a variety of classifiers, some of which as special cases coincide with classical architectures:

Polynomial classifiers of degree p:

$$K(x_i, x_j) = \left(x_i^T x_j + 1\right)^p$$

Radial basis function (RBF) classifier:

$$K(x_i, x_j) = e^{-\|x_i - x_j\|^2/\sigma}$$

Neural networks (NN):

$$K(x_i, x_j) = \tanh(\kappa x_i x_j + \theta)$$

In the RBF case, the SVM *automatically* determines the number (how many SVs) and locations (the SVs) of RBF centers and gives excellent result compared to classical RBF [23]. In case of the neural networks, the SVM gives a particular kind of two-layer sigmoidal neural network. In such a case, the first layer consists of N_s (the number of SVs) sets of weights, each set consisting of d (the dimension of the data) weights, and the second layer consists of N_s weights (α_i). The architecture (the number of weights) is determined by SVM training.

During testing, for a test vector $x \in R^m$, we first compute

$$a(x, \mathbf{w}) = \mathbf{w^T z} + b = \sum_i \alpha_i y_i K(x, x_i) + b$$

and then its class label $o(\mathbf{x}, \mathbf{w})$ is 1 if $a(x, \mathbf{w}) > 0$, otherwise, it is -1.

The above algorithm for separable data can be generalized to the non-separable data by introducing nonnegative slack variables $\xi_i, i = 1, \ldots, N$ [12]. The resultant problem becomes minimizing

$$\frac{1}{2}\|\mathbf{w}\|^2 + C \sum_{i=1}^{N} \xi_i$$

subject to

$$\xi_i \geq 0\,, \qquad y_i a(x_i, \mathbf{w}) \geq 1 - \xi_i\,, \qquad i = 1, \ldots, N$$

Thus, once an error occurs, the corresponding ξ_i which measures the (absolute) difference between $a(x_i, \mathbf{w})$ and y_i must exceed unity, so $\sum_i \xi_i$ is an upper bound on the number of training error. C is a constant controlling the tradeoff between training error and model complexity. Again, minimization of the above equation can be transformed to a QP problem: maximize (1) subject to the constraints $0 \leq \alpha \leq C$ and $\alpha^T \mathbf{Y} = \mathbf{0}$. Its convergence behavior is nothing but solving a linearly constrained convex quadratic programming problem.

To provide some insight into how SVM behaves in input space, we show a simple binary toy problem solved by a SVM with a polynomial kernel of degree 3 (see Fig. 2). The support vectors, indicated by extra circles, define the margin of largest separation between the two classes (circles and disks). It was shown [12] that the classification with optimal decision boundary (the solid line) generalizes well as opposed to the other boundaries found by the conventional neural networks.

2.5 Evaluation of the Performance

We evaluated the performance of the classifier by computing the percentages of correct classification (CC), sensitivity (SE) and specificity (SP), which are defined as follows:

$$\begin{aligned} \mathrm{CC} &= 100^*(\mathrm{TP} + \mathrm{TN})/\mathrm{N} \\ \mathrm{SE} &= 100^*\mathrm{TP}/(\mathrm{TP} + \mathrm{FN}) \\ \mathrm{SP} &= 100^*\mathrm{TN}/(\mathrm{TN} + \mathrm{FP}) \end{aligned}$$

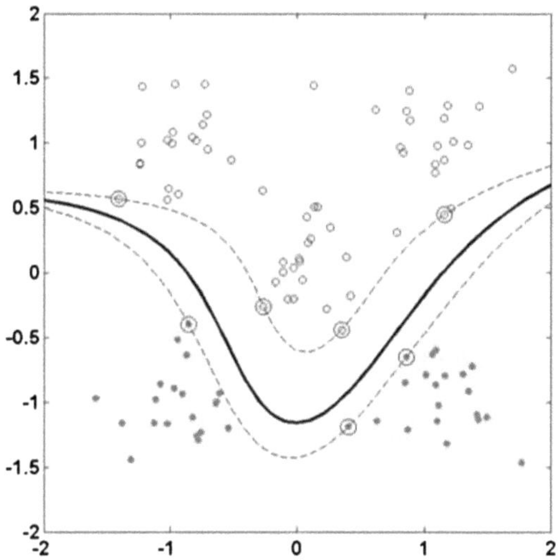

Fig. 2. A simulation of a two-dimensional classification problem solved by a SVM with a cubic polynomial kernel [12]. *Circles* and *disks* are two classes of training examples. The *solid line* is the decision boundary; the two *dashed lines* are the margins of the decision boundary. The support vectors found by the algorithm (marked by *extra circles*) are examples which are critical for the given classification task

where N was the total number of the patients studied, TP was the number of true positives, TN was the number of the true negatives, FN was the number of false negatives, and FP was the number of false positives [24].

3 Results

Based on the result of the established radioscintigraphic gastric emptying tests, the EGG data obtained from 152 patients were split into two groups: 76 patients with delayed gastric emptying and 76 patients with normal gastric emptying. Half of the EGG data from each group were selected at random as the training set, and the remaining data were used as the testing set. Ten-fold cross-validation was also employed.

The feature selection of surface EGGs was based on the statistical analysis of the EGG parameters between the patients with normal and delayed gastric emptying [3, 10]. Among the five parameters used as the input, statistical differences existed between the two groups of the patients in the percentages of the regular 2–4 cpm wave ($90.0 \pm 1.0\%$ vs. $77.8 \pm 2.2\%$, $p < 0.001$ for patients with normal and delayed gastric emptying in the fed state, for example), the percentage of tachygastria ($4.1 \pm 0.6\%$ vs. $13.9 \pm 1.8\%$, $p < 0.001$, patients with delayed gastric emptying in the fed state had significantly higher level) and the postprandial increase in EGG dominant power (4.6 ± 0.5 dB vs. $1.2 \pm 0.6\,\mathrm{dB}$, $p < 0.001$, the increase was significantly lower in patients with delayed gastric emptying). The size of the training set in this study is equal to that of the testing set. We use the balanced training and testing set so as to conveniently compare with the previous result obtained by BP algorithm [9].

Table 1 shows the experimental results for test set using networks with 2, 3, and 7 of hidden units developed by the GCCA. It can be seen from this table that the network with 3 hidden units seems to be a good choice for this specific application, which exhibits a correct diagnosis of 83% cases with a sensitivity of 84% and a specificity of 82%. The result achieved with 3 hidden units is comparable with that previously obtained by the BPNN [9], and with more recent result [3]. However, the GCCA provides an automatic model selection procedure without guessing the size and connectivity pattern of the network in advance for a given task.

Table 1. Results of tests for genetic neural networks with different hidden units developed by GCCA. The CC, SE and SP are respectively the percentages of correct classification, sensitivity and specificity [16]

# of Hidden Units	CC (%)	SE (%)	SP (%)
2	80	79	82
3	83	84	82
7	76	71	82

Results for the three different kernels (polynomial kernels, radial basis function kernels, and sigmoid kernels are summarized in Table 2. In all experiments, we used the support vector algorithm with standard quadratic programming techniques and $C = 5$. Note changing C while keeping the same number of outliers provides an alternative for the merit measure of the SVM. We use the above criteria which allow us to make direct comparison with previous result. It can be seen from Table 2 that the SVM with the radial basis kernels performs best (89.5%) among the three classifiers.

Table 2. Testing results for the three different kernels of the SVM with the parameters in parentheses. The CC, SE and SP are respectively the percentages of correct classification, sensitivity and specificity. The numbers of SVs found by different classifiers are also shown in the last column (© 2001 IEEE)

	CC (%)	SE (%)	SP (%)	# of SVs
Polynomial ($\rho = 5$)	88.2	81.6	94.7	30
RBF ($\sigma = 0.6$)	89.5	84.2	94.7	48
NN ($\kappa = 0.6$, $\theta = 0.9$)	85.5	79.0	92.1	42

In all three cases, the SVMs exhibit higher generalization ability compared to the best performance achieved (83%) on the same data set with the genetic neural network [10]. The low sensitivity and the high specificity observed in Table 2 are consistent with the results in [10]. Table 2 (last column) also shows the numbers of SVs found by different types of support vector classifiers which contains all the information necessary to solve a given classification task.

4 Discussion and Conclusions

We have reviewed that the SVM approach can be used for the non-invasive diagnosis of delayed gastric emptying from the cutaneous EGGs. We have shown that, compared to the neural network techniques, the SVM exhibits higher prediction accuracy of delayed gastric emptying.

Radioscintigraphy is currently the gold standard for quantifying gastric emptying. The application of this technique is radioactive and considerably expensive, and usually limited to very sick patients. This motivates ones to develop the low-cost and non-invasive methods based on the EGG. EGG is attractive because of its non-invasiveness (no radiation and no intubation). Once the technique is learned, studies are relatively easy to perform. Unlike radiositigraphy, EGG provides information about gastric myoelectrical activity in both the fasting and postprandial periods [1]. Numerous studies have been performed on the correlation of the EGG and gastric emptying [2, 3, 14, 25, 26, 27, 28]. Although some of the results are still controversial, it is generally accepted that an abnormal EGG usually predicts delayed gastric emptying [1, 2]. This is because gastric myoelectrical activity modulates gastric motor activity. Abnormalities in this activity may cause gastric hypomotility and/or uncoordinated gastric contractions, yielding delayed gastric emptying. Moreover, the accuracy of the prediction is associated with the selection of EGG parameters and methods of prediction. Previous studies [2, 3, 14] have shown that spectral parameters of the EGG provide useful information regarding gastrointestinal motility and symptoms, whereas the waveform of the EGG is unpredictable and does not provide reliable information. This leaded us to use the spectral parameters of the EGG as the inputs of the SVM. The feature selection of surface EGGs was based on the statistical analysis of the EGG parameters between the patients with normal and delayed gastric emptying [1, 2, 3, 10].

Although the diagnosis result for the genetic neural network approach is comparable with that obtained by the BPNN, the main advantage of the GCCA over the BP algorithm is that it can automatically grow the architecture of neural networks to give a suitable network size for a specific problem. This feature makes the GCCA very attractive for real world applications. In addition to no need to guess the size and connectivity pattern of the network in advance, speedup of the GCCA over the BP is another benefit. This is because in the BP algorithm each training case requires a forward and a backward pass through all the connections in the network; the GCCA requires only a forward pass and genetic search of limited generations, many training epochs are run while the network is much smaller than its final size.

Based on the foregoing discussion, it is evident that the genetic neural network shows several advantages over the standard neural network. It, nevertheless, is still inferior by comparison with the SVM, at least for our specific example discussed here. It is important to stress the essential differences between the SVM and neural networks. First, the SVM always finds a global solution which is in contrast to the neural networks, where many local minima usually exist [21]. Second, the SVM does not minimize the empirical training error alone which neural networks usually aim at. Instead it minimizes the sum of an upper bound on the empirical training error and a penalty term that depends on the complexity of the classifier used. Despite the high generalization ability of the SVM the optimal choice of kernel for a given problem is still a research issue.

All in all, the SVM seems to be a potentially useful tool for the automated diagnosis of delayed gastric emptying. Further research in this field will include adding more EGG parameters as inputs to the SVM to improve the performance.

Acknowledgements

The author would like to thank Zhiyue Lin for providing the EGG data in our previous work reviewed here.

References

1. Parkman, H.P., Arthur, A.D., Krevsky, B., Urbain, J.-L. C., Maurer, A.H., Fisher, R.S. (1995) Gastroduodenal motility and dysmotility: An update on techniques available for evaluation. Am. J. Gastroenterol., **90**, 869–892.
2. Chen, J.D.Z., Lin, Z., McCallum, R.W. (1996) Abnormal gastric myoelectrical activity and delayed gastric emptying in patients with symptoms suggestive of gastroparesis. Dig. Dis. Sci., **41**, 1538–1545.
3. Chen, J.D.Z., Lin, Z., McCallum, R.W. (2000) Non-invasive feature-based detection of delayed gastric emptying in humans using neural networks. IEEE Trans. Biomed Eng., **47**, 409–412.
4. Sarna, S.K. (1975) Gastrointestinal electrical activity: Terminology. Gastroenterology, **68**, 1631–1635.
5. Hinder, R.A., Kelly, K.A. (1978) Human gastric pacemaker potential: Site of origin, spread and response to gastric transection and proximal gastric vagotomy. Amer. J. Surg., **133**, 29–33.
6. Smout, A.J.P.M., van der Schee, E.J., Grashuis, J.L. (1980) What is measured in electrogastrography? Dig. Dis. Sci., **25**, 179–187.
7. Familoni, B.O., Bowes, K.L., Kingma, Y.J., Cote, K.R. (1991) Can transcutaneous recordings detect gastric electrical abnormalities? Gut, **32**, 141–146.

8. Chen, J., Schirmer, B.D., McCallum, R.W. (1994) Serosal and cutaneous recordings of gastric myoelectrical activity in patients with gastroparesis. Am. J. Physiol., **266**, G90–G98.
9. Lin, Z., Chen, J.D.Z., McCallum, R.W. (1997) Noninvasive diagnosis of delayed gastric emptying from cutaneous electrogastrograms using multilayer feedforward neural networks. Gastroenterology, **112**(4): A777 (abstract).
10. Liang, H.L. Lin, Z.Y., McCallum, R.W. (2000) Application of combined genetic algorithms with cascade correlation to diagnosis of delayed gastric emptying from electrogastricgrams. Med. Eng. & Phys., **22**, 229–234.
11. Vapnik, V. (1995) The Nature of Statistical Learning Theory. Berlin, Germany: Springer-Verlag.
12. Cortes, C., Vapnik, V. (1995) Support Vector Networks. Machine Learning, **20**, 273–297.
13. Liang, H.L., Lin, Z.Y. (2001) Detection of delayed gastric emptying from electrograstrograms with support vector machine. IEEE Trans. Biomed Eng., **48**, 601–604.
14. Chen, J.D.Z., McCallum, R.W. (1994) EGG parameters and their clinical significance. 45–73, Electrogastrography: Principles and Applications, New York: Raven Press.
15. Chen, J. (1992) A computerized data analysis system for electrogastrogram. Comput. Bio. Med., **22**, 45–58.
16. Reprinted from Medical Engineering & Physics, V22: 229–234, Liang H. et al, © 2000, with permission from The Institute of Engineering and Physics in Medicine.
17. Goldberg, D.E. (1989) Genetic Algorithm in Search, Optimizing, and Machine Learning. New York, Addison-Wesley, NY.
18. Fahlman, S.E., Lebiere, C. (1990) The Cascade Correlation Learning Architecture. Technical Report CMU-CS-90-100, School of Computer Science, Carnegie Mellon University.
19. Liang, H.L., Dai, G.L. (1998) Improvement of cascade correlation learning algorithm with an evolutionary initialization. Information Sciences, **112**, 1–6.
20. Fahlman, S.E., Lebiere, C. (1990) An Empirical study of learning speed in backpropagation networks. Technical Report CMU-CS-88-162, School of Computer Science, Carnegie Mellon University.
21. Burges, C.J.C. (1998) A tutorial on support vector machines for pattern recognition. Data Mining and Knowledge Discovery, **2**, 955–974.
22. Fletcher, R. (1987) Practical Methods of Optimization, John Wiley and Sons, Inc., 2nd edition.
23. Scholkopf, B., Sung, K., Burges, C., Girosi, F., Niyogi, P., Poggio, T., Vapnik, V. (1997) Comparing support vector machines with gaussian kernels to radial basis function classifiers. IEEE Trans. Sign. Processing, **45**, 2758–2765.
24. Eberhart, R.C., Dobbins, R.W. (1990) Neural Network PC Tools, San Diego: Academic Press, Inc.
25. Dubois, A., Mizrahi, M. (1994) Electrogastrography, gastric emptying, and gastric motility. 247–256, Electrogastrography: Principles and Applications, New York: Raven Press.
26. Hongo, M. Okuno, Y. Nishimura, N. Toyota, T., Okuyama, S. (1994) Electrogastrography for prediction of gastric emptying asate. 257–269, Electrogastrography: Principles and Applications, New York: Raven Press.

27. Abell, T.L., Camilleri, M., Hench, V.S., Malagelada, J.-R. (1991) Gastric electromechanical function and gastric emptying in diabetic gastroparesis. Eur. J. Gastroenterol. Hepatol., **3**, 163–167.
28. Koch, K.L., Stern, R.M., Stewart, W.R., Vasey, M.W. (1989) Gastric emptying and gastric myoelectrical activity in patients with diabetic gastroparesis: Effect of long-term domperidone treatment. Am. J. Gastroenterol., **84**, 1069–1075.

Tachycardia Discrimination in Implantable Cardioverter Defibrillators Using Support Vector Machines and Bootstrap Resampling

J.L. Rojo-Álvarez[1], A. García-Alberola[2], A. Artés-Rodríguez[1], and Á. Arenal-Maíz[3]

[1] Universidad Carlos III de Madrid (Leganés-Madrid, Spain)
[2] Hospital Universitario Virgen de la Arrixaca (Murcia, Spain)
[3] Hospital General Universitario Gregorio Marañón (Madrid, Spain)

Abstract. Accurate automatic discrimination between supraventricular (SV) and ventricular (V) tachycardia (T) in implantable cardioverter defibrillators (ICD) is still today a challenging problem. An interesting approach to this issue can be Support Vector Machine (SVM) classifiers, but their application in this scenario can exhibit limitations of technical (reduced data sets are only available) and clinical (the solution is inside a hard-to-interpret black box) nature. We first show that the use of bootstrap resampling can be helpful for training SVM with few available observations. Then, we perform a principal component analysis of the support vectors that leads to simple discrimination rules that are as effective as the black box SVM. Therefore, a low computational burden method can be stated for discriminating between SVT and VT in ICD.

Key words: tachycardia, defibrillator, support vector, bootstrap resampling, principal component analysis

1 Introduction

Automatic and semiautomatic medical decisions is a widely scrutinized framework. On the one hand, the lack of detailed physiological models is habitual; on the other hand, the relationships and interactions among the predictor variables involved can often be nonlinear, given that biological processes are usually driven by complex nature dynamics. These are important reasons encouraging the use of nonlinear machine learning approaches in medical diagnosis.

A wide range of methods have been used to the date in this framework, including neural networks, genetic programming, Markov models, and many

J.L. Rojo-Álvarez et al.: *Tachycardia Discrimination in Implantable Cardioverter Defibrillators Using Support Vector Machines and Bootstrap Resampling*, StudFuzz **177**, 413–431 (2005)
www.springerlink.com

others [1]. During the last years, the Support Vector Machines (SVM) have strongly emerged in the statistical learning community, and they have been applied to an impressive range of knowledge fields [13]. The interest of SVM for medical decision problems arises from the following properties:

1. The optimized functional has a single minimum, which avoids convergence problems due to local minima that can appear when using other methods.
2. The cost function and the maximum margin requirement are appropriate mathematical conditions for the usual case when the underlying statistical properties of the data are not well known.
3. SVM classifiers work well when few observations are available. This is a very common requirement in medical problems, where data are often expensive because they are obtained from patients, and hence, we will often be dealing with just about one or some few hundred data sets.

However, one of the drawbacks for nonlinear SVM classifiers is that the solution still remains inside a complex, difficult to interpret equation, i.e., we obtain a black-box model. Two main implications for the use of SVM in medical problems arise: first, we will not be able to learn anything about the underlying dynamics of the modeled biological process (despite we have properly *captured* it in our model!); and second, a health care professional is likely neither going to rely on a black box (whose working principles are unknown), nor assuming the responsibility for the automatic decisions of this obscure diagnostic tool. Thus, if we want to gain benefit from the SVM properties in medical decision problems, the following question arises: can we use a nonlinear classifier, and still be able to learn something about the underlying complex dynamics?

The answer could be waiting for us in another promising property of SVM classifiers: the solution is build by using a subset of the observations, which are called the *support vectors*, and all the remaining samples are then clearly and correctly classified. This suggests the possibility of exploring the properties of those critical-for-classification samples.

Another desirable requirement for using SVM in medical environments is the availability of a statistical significance test for the machine performance, for instance, confidence intervals on the error probability. The use of nonparametric bootstrap resampling [2] will make possible this featuring in a conceptually easy, yet effective way. We can afford the computational cost of bootstrap resampling for SVM classifiers in low-sized data sets, which is the case of many medical diagnosis problems. But bootstrap resampling can provide not only nonparametric confidence intervals, but also bias-corrected means, and thus, it also represents an efficient way to obtain the best free parameters (margin-losses trade-off and nonlinearity) of the SVM classifier without splitting the available samples into training and validation subsets.

Here, we present an application example of SVM to automatic discrimination of cardiac arrythmia. First, the clinical problem is introduced, the clinical hypothesis to be tested is formulated, and the patient data-bases are described. Then, the general methods (SVM classifiers and SVM bootstrap resampling) are briefly presented. A black box approach is first used to determine the most convenient preprocessing. We suggest two simple analysis (principal component and linear SVM) of the support vectors of the previously obtained machine to study their properties, leading to propose three simple parameters that can be clinically interpreted and with similar diagnostic performance to the black box SVM. Descriptive contents in this chapter have been mainly condensed from [5, 6, 7], and detailed descriptions of the methods and results can be found therein.

2 Changes in Ventricular EGM Onset

Automatic Implantable Cardioverter Defibrillators (ICD) have supposed a great advance in arrhythmia treatment during the last two decades [12]. The automatic tasks carried out by these devices are: (1) continuous monitoring of the heart rate; (2) detection and automatic classification of cardiac abnormal rhythms; and (3) delivering of appropriate and severity-increasing therapy (cardiac pacing, cardioversion, and defibrillation) [11].

The *sinus rhythm* (SR) is the cardiac activation pattern under non-pathological conditions (Fig. 1), and cardiac arrhythmias are originated by any alteration on it. According to their anatomical origin, tachyarrhythmias (more than 100 beats per minute) can be classified into two groups:

1. Supraventricular tachycardias (SVT), which are originated in a given location at the atria.

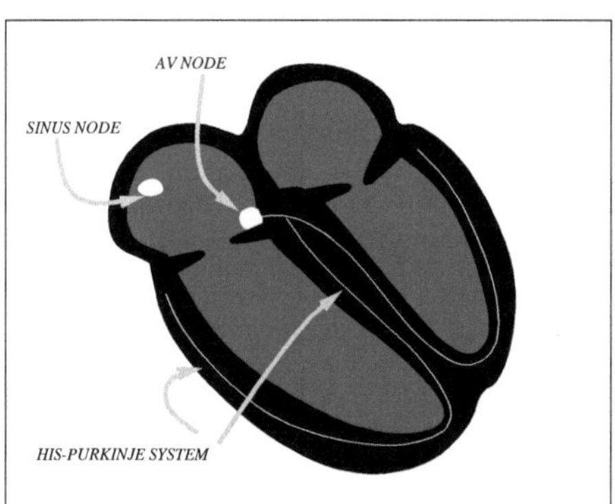

Fig. 1. The heart consists of four chambers: two atria and two ventricles. The electrical activation is driven by the rhythmic discharges of the sinus node. The depolarizing stimulus propagates along the atria (producing atrial contraction), reaches the atrio-ventricular (AV) node, rapidly propagates through the specific conduction tissue His-Purkinje system, and gets simultaneously to near all the ventricular myocardial fibers (producing ventricular contraction)

2. Ventricular tachyarrhythmias, which are originated at a given location in the ventricles, and distinguishing between ventricular tachycardia (VT), (electric pattern that is repetitive and has a well defined shape), and ventricular fibrillation (VF), which is an aleatory and ineffective electric activation (electric pattern that is almost random).

The most dangerous tachyarrhythmias are VF and fast VT, as there is no effective blood ejection when they happen, and sudden cardiac death follows in very few minutes. Hence, these rhythms require an immediate therapy (electrical shock!) delivered. On the contrary, SVT rarely imply acute hemodynamic damage, as about 80% of blood continues to flow by the circulatory system, so that they do not require an immediate shock, though they often need pharmacological interventions.

Due to the limited lifetime of the batteries, arrhythmia discrimination algorithms in ICD must demand a low computational burden. The most commonly implemented algorithms are the Heart Rate Criterion [11], the QRS Width Criterion [4] and the Correlation Waveform Analysis [3]. While VF detection cycle rank is commonly accepted as an appropriate criterion, there is a strong overlap between SVT and VT cycle ranks, so that the estimated number of inappropriate shocks (i.e., shocks delivered to SVT) is estimated between 10 and 30% [9]. Inappropriate shocks shorten the battery lifetime, deteriorate the patient's quality of life, and can even originate a new VT or VF episode.

Hypothesis

The analysis of the initial changes in the intracardiac ventricular electrograms (EGM) has been proposed as an alternative arrhythmia discrimination criterion, as it does not suffer from the drawbacks of the Heart Rate, the QRS Width, or the Correlation Waveform Analysis [5, 6, 7]. The clinical hypothesis underlying this criterion is as follows:

During any supraventricular originated rhythm, both ventricles are depolarized through the His-Purkinje system, whose conduction speed is high (4 m/s); however, the electric impulse for a ventricular originated depolarization travels initially through the myocardial cells, whose conduction speed is slow (1 m/s). Then, we hypothesize that changes in the ventricular EGM onset can discriminate between SVT and VT.

Waveform changes can be observed in the EGM first derivative. Figure 1 shows the anatomical elements involved in the hypothesis. Figure 2 depicts examples of SR, SVT, and VT episodes recorded in ICD, together with their first derivatives. There, the noisy activity preceding the beat onset has been previously removed. Note the EGM being a sudden activation in both SR and SVT beats, but an initially less energetic activation in VT beats.

Once the clinical hypothesis has been stated, the next issue is how can this criterion be implemented into an efficient algorithm. As there is no statistical

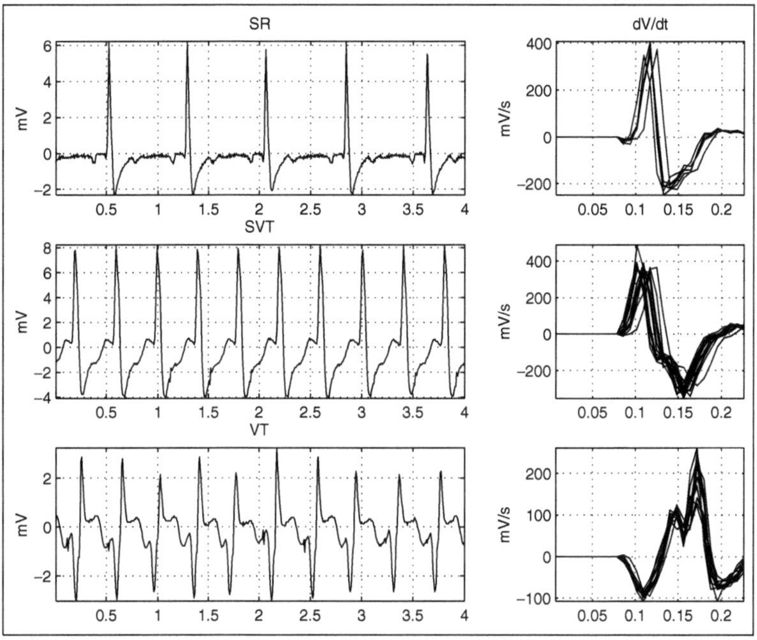

Fig. 2. Examples of SR, SVT and VT EGM recorded in ICD (*left*), and their corresponding first derivative (*right*). Changes are initially less strong (smaller derivative modulus) during the early stage of the ventricular activation. *Horizontal axis* are in seconds

model for the cardiac impulse propagation being detailed enough to allow a detailed analytical or simulation research, statistical learning from samples can be a valuable approach.

The next step is to get a representative data base of ICD stored EGM.

Patients Data Base

Assembling an ICD stored EGM data base is a troublesome task, due to the need of their exact correct labelling. Two different data bases were assembled for this analysis, one of them (Base C) for control-training-purposes, and the other (Base D) for final test purposes.

- **Base C** (control episodes). A number of 26 patients, with a third generation ICD (Micro-Jewel 7221 and 7223, Medtronic), were included in this study. In these patients, monomorphic VT EGM were obtained during an electrophysiologic study performed three days after the implant. The EGM source between the subpectoral can and the defibrillation coil in the left ventricle was programmed, as it was previously shown to be the most appropriate electrode configuration for the criterion. The ICD pacing capabilities were used to induce monomorphic VT. The EGM were stored in the ICD during induced sustained monomorphic VT and during its preceding SR. In order to obtain a group of SVT, a treadmill test, modified Bruce protocol, was performed in the post-absorptive state, at least 4 days after the implantation procedure, if no contraindication was present. The EGM recorded during VT, sinus tachycardia and SR were downloaded in a computer system (A/D conversion: 128 Hz, 8 bits per sample, range ± 7.5 mV). In this group, spontaneous tachycardia stored in the device during the follow-up were included if the EGM morphology of the recurrence was identical to either the induced VT morphology or the exercise induced SVT morphology. A total of 38 SVT episodes (cycle 493 ± 54 ms) and 68 VT episodes (cycle 314 ± 49 ms) were assembled.

- **Base D** (spontaneous episodes). An independent group of spontaneous tachycardias from 54 patients with a double chamber ICD (Micro-Jewel 7271, Medtronic) was assembled. Only data from this type of device were admitted in order to reduce the diagnostic error during arrhythmia classification. Let V(A) denote the time for the ventricular (atrial) EGM maximum peak; a VT was diagnosed when there was: (a) V-A dissociation; (b) irregular atrial rhythm, even if this was faster than a regular ventricular rhythm; or (c) V-A association with a V-A $<$ A-V. This allowed to label 299 SVT (cycle 498 ± 61 ms) and 1088 VT (cycle 390 ± 81 ms) episodes.

The number of available episodes is high enough to be considered as significant in a clinical study. However, learning machines are usually trained with a far greater number of data in order to avoid the overfitting, so that robust procedures are recommendable in order to adequately extract the information from the available observations.

3 The Support Vector Machines

The SVM was first proposed to obtain maximum margin separating hyperplanes in classification problems, but in a short time it has grown to a more general learning theory, and it has been applied to a number of real data problems. A comprehensive description of the method can be found in Chap. 1 of this book.

Be $\mathbf{V}$ a set of N observed and labeled data

$$\mathbf{V} = \{(\mathbf{x}_1, y_1), \ldots, (\mathbf{x}_N, y_N)\} \tag{1}$$

where $\mathbf{x}_i \in \mathbb{R}^n$ and $y_i \in \{-1, +1\}$. Be a nonlinear transformation $\phi(\mathbf{x}_i)$ to a generally unknown, higher dimensional space $\mathbb{R}^{n}$, where a separating hyperplane is given by $(\phi(\mathbf{x}_i) \cdot \mathbf{w}) + b = 0$. The problem is to minimize

$$\frac{1}{2}\|\mathbf{w}\|^2 + C \sum_{i=1}^{N} \xi_i \tag{2}$$

with respect to $\mathbf{w}, b$, and ξ_i, and constrained to

$$y_i\{(\phi(\mathbf{x}_i) \cdot \mathbf{w}) + b\} - 1 + \xi_i \geq 0 \tag{3}$$

$$\xi_i \geq 0 \tag{4}$$

for $i = 1, \dots, N$, where ξ_i represent the losses; C represents a trade-off between margin $\|\mathbf{w}\|^2$ and losses; and $(\cdot)$ expresses the dot product. By using the Lagrange Theorem, (2) can be rewritten into its dual form, and then, the problem consists of maximizing

$$\sum_{i=1}^{N} \alpha_i - \frac{1}{2} \sum_{i,j=1}^{N} \alpha_i y_i \alpha_j y_j K(\mathbf{x}_i, \mathbf{x}_j) \tag{5}$$

constrained to $0 \leq \alpha_i \leq C$ and $\sum_{i=1}^{N} \alpha_i y_i = 0$, where α_i are the Lagrange multipliers corresponding to constrains (3), and $K(\mathbf{x}_i, \mathbf{x}_j) = (\phi(\mathbf{x}_i) \cdot \phi(\mathbf{x}_j))$ is a Mercer's kernel that allows us to calculate the dot product in high-dimensional space without explicitly knowing the nonlinear mapping. The two kernels used here are the linear, $K(\mathbf{x}_i, \mathbf{x}_j) = (\mathbf{x}_i \cdot \mathbf{x}_i)$, and the Gaussian,

$$K(\mathbf{x}_i, \mathbf{x}_j) = \exp\left\{-\frac{\|\mathbf{x}_i - \mathbf{x}_j\|^2}{2\sigma^2}\right\}$$

As it can be seen, width σ is a free parameter that has to be previously settled when using the gaussian kernel, and the knowledge of the trade-off parameter C is always mandatory (except in separable problems). The search of the free parameters is usually done by using cross validation, but when working with low sized data sets, a dramatic reduction in the size of the training set arises.

4 Bootstrap Resampling and SVM Tuning

We propose to adjust the free parameters in SVM basing on the bootstrap resampling techniques [2]. A dependence estimation process between pairs of data in a classification problem, where the data are drawn from a joint distribution

$$p(\mathbf{x}, y) \rightarrow \mathbf{V} \tag{6}$$

can be solved by using a SVM. The estimated SVM coefficients with the whole data set are

$$\alpha = [\alpha_1, \dots, \alpha_N] = s\left(\mathbf{V}, C, \sigma\right) \tag{7}$$

where $s()$ is the operator that account for the SVM optimization, and it depends on the data and on the values of the free parameters. The empirical risk for the current coefficients is defined as the training error fraction of the machine,

$$R_{\text{emp}} = t(\alpha, \mathbf{V}) \tag{8}$$

where $t()$ is the operator that represents the empirical risk estimation.

A *bootstrap resample* is a data set drawn from the training set according to the empirical distribution, i.e., it consists of sampling with replacement the observed pairs of data:

$$\hat{p}(\mathbf{x}, y) \rightarrow \mathbf{V}^* = \{(\mathbf{x}_1^*, y_1^*), \ldots, (\mathbf{x}_N^*, y_N^*)\} \tag{9}$$

Therefore, $\mathbf{V}^*$ contains elements of $\mathbf{V}$ that appear zero, one, or several times. The resampling process is repeated $b = 1, \ldots, B$ times. A partition of $\mathbf{V}$ in terms of resample $\mathbf{V}(b)^*$ is

$$\mathbf{V} = (\mathbf{V}_{\text{in}}^*(b), \mathbf{V}_{\text{out}}^*(b)) \tag{10}$$

where $\mathbf{V}_{\text{in}}^*(b)$ is the subset of samples included in resample b, and $\mathbf{V}_{\text{out}}^*(b)$ is the subset of non-included samples. SVM coefficients for each resample are given by:

$$\alpha^* = s(\mathbf{V}_{\text{in}}^*(b), C, \sigma) \tag{11}$$

The empirical risk estimation for the resample is known as its *bootstrap replicate*,

$$R_{\text{emp}}^*(b) = t(\alpha^*, \mathbf{V}_{\text{in}}^*(b)) \tag{12}$$

and its normalized histogram for the B resamples approximates the empirical risk density function. However, a better estimation can be obtained by taking

$$R_{\text{act}}^*(b) = t(\alpha^*, \mathbf{V}_{\text{out}}^*(b)) \tag{13}$$

which in fact represents an approximation to the actual (i.e., not only empirical) risk. The bias due to overfitting by a non convenient choice of the free parameters will be detected (and in part corrected) by analyzing (13). A proper choice for B is typically from 150 to 300 resamples.

Example 1: Linear SVM Weights

Let us consider a biclassic classification problems with bidimensional input space, given by Gaussian distributions

$$N\left(\left[\frac{3}{2}, 0\right], \mathbf{I}_2\right) \rightarrow y_i = -1 \quad \text{and} \quad N\left(\left[\frac{-3}{2}, 0\right], \mathbf{I}_2\right) \rightarrow y_i = +1 \tag{14}$$

where $\mathbf{I}_2$ denotes the 2×2 identity matrix. The SVM solution for a given N samples set is

$$y = f(\mathbf{x}) = \sum_{i=1}^{N} \alpha_i (\mathbf{x} \cdot \mathbf{x}_i) + b = w_1 x_1 + w_2 x_2 + b \tag{15}$$

where x_1, x_2 are the components of input space vector $\mathbf{x}$. Classifier parameters w_1, w_2, b, are random variables, and the optimum solution for the maximum margin criterion will trend to

$$w_1 < 0 \; ; \quad w_2 = 0 \; ; \quad b = 0 \tag{16}$$

The constrain of w_1 being strictly negative is due to the maximum margin criterion, and its value will depend on the observed samples.

The approximate parameters for $N = 50$ samples (using $C = 10$) are

$$w_1^{\mathrm{SVM}} = -1.1 \; ; \quad w_2^{\mathrm{SVM}} = -0.8 \; ; \quad b^{\mathrm{SVM}} = 0.12 \tag{17}$$

By bootstrap resampling (B = 2000), the following parameter replicates are obtained:

$$w_1^* \sim -2.4 \pm 1.0 \; ; \quad w_2^* \sim -0.6 \pm 0.8 \; ; \quad b^* \sim -0.2 \pm 0.6 \tag{18}$$

It can be tested that both w_2^{SVM} and b^{SVM} are noisy and null coefficients, which can in fact be suppressed from the model.

We can estimate the error probability of this machine by using its bootstrap replicates, as shown in Fig. 3. The empirical error of a SVM trained with the complete training set is estimated in the same set as low as $P_e = 6\%$, but the bootstrap estimator of the distribution is $P_e^* \sim 8.9\% \pm 2\%$, $P_e^* \in (4, 14)$. Bayes' error is $P_e^{\mathrm{true}} = 13.6\%$. It can be seen that the empirical estimate is biassed toward very optimistic error probability values. Though the bias-corrected bootstrap estimate of error is not close to Bayes' error, it still represents a better approximation, thus allowing to detect that the free parameter value $C = 10$ produces overfitting.

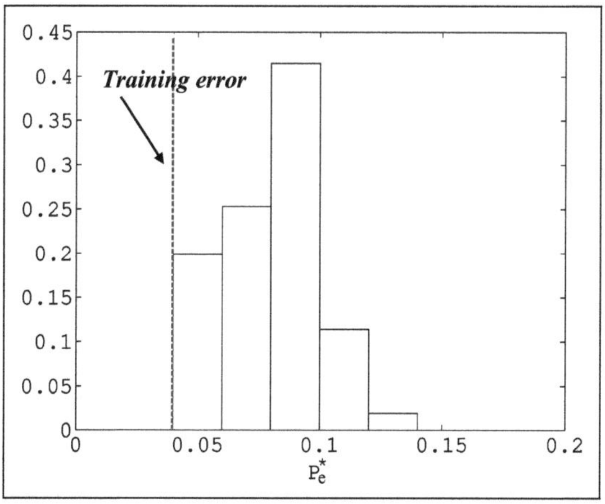

Fig. 3. Example 1. Bootstrap resampling estimation of the density function of the error probability

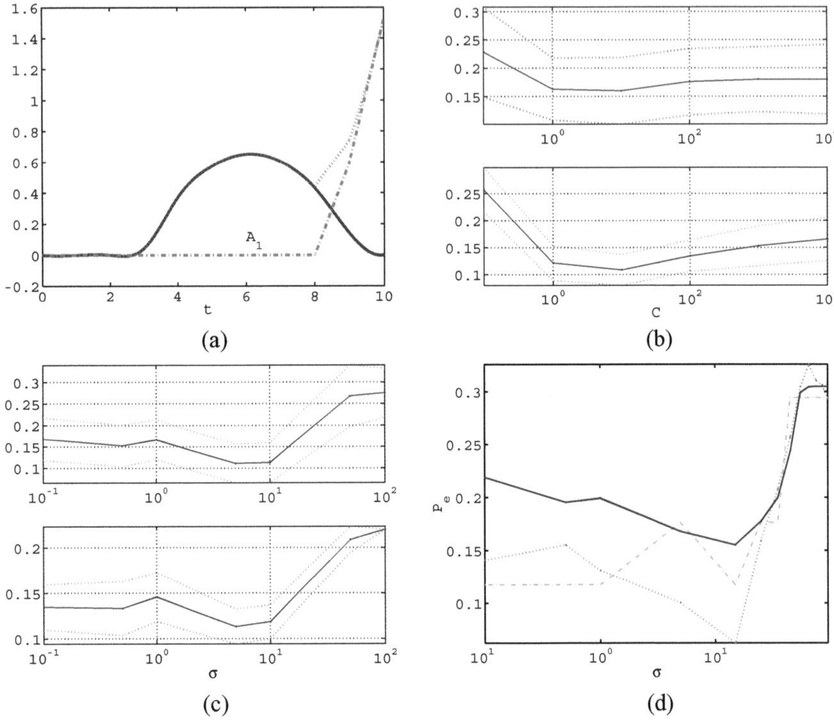

Fig. 4. Example 2. (**a**) Rectified slow component (continuous), rectified fast component (discontinuous), and their sum (*dotted*). The area of the slow component A_1 is the criterion for classifying the generated vectors. (**b**) Error in parabolas example as a function of C for a linear SVM, with bootstrap (*up*) and test (*down*). (**c**) The same for σ in a Gaussian kernel SVM. (**d**) Error in parabolas example as a function of σ in a Gaussian kernel SVM, using cross validation (*dotted*), bootstrap resampling (*dashed*) and test set (continuous)

Example 2: Parabolas

A simulation problem is proposed which qualitatively (and roughly) emulates the electrophysiological principle of the initial ventricular activation criterion. Input vectors $\mathbf{v} \in \mathbf{R}^{11}$ consisted of two summed convex, half-wave rectified parabolas (a slow and a fast component), between 0 and 10 seconds, sampled at $f_s = 1\,\text{Hz}$ (Fig. 4.a), and according to

$$v(t) = (t - t_s)^2 + v_s + (t - t_f)^2 + v_f \tag{19}$$

$$\mathbf{v} = v(t)_{|t=0,\ldots,10} \tag{20}$$

where t_s, v_s (t_f, v_f) are the slow (fast) component parameters. These parameters are generated by following the rules given in Table 1. The class for each vector is assigned according to area of the slow parabola A_1 being minor $(y_i = +1)$ or greater $(y_i = -1)$ than a threshold level 3. We generated

Table 1. Parabolas model parameters. For each component, center and interception were generated according these rules. $U[a, b]$ denotes uniform distribution in (a, b)

	t_{cen}	v_{cen}	t_{inter}	v_{inter}
slow	U[2,8]	U[0,1]	U[1,7]	0
fast	10	U[1,3]	U[6,9]	0

200 training vectors and 10 000 test vectors. In order to model errors in class labeling, about 3% of randomly selected training vectors are changed their label.

Bootstrap error probability in the training set and averaged error probability in the test set are calculated as a function of (1) C parameter using a linear kernel SVM, and (2) width σ using a Gaussian kernel SVM, as shown in Fig. 4.b,c. In both cases, there is a close agreement between the optimum value of the free parameter estimated with bootstrap from the training set, and the actual best value for the free parameter given by the test set.

Cross validation is often used to determine the optimum free parameter values in SVM classifiers, but when low sized training data are split, the information extracted by the machine can be dramatically reduced. As an example, we compare the choice of σ for a Gaussian kernel SVM (using $C = 10$). A number of 30 training vectors are generated, and error probability is obtained by using boostrap and by using cross validation (50% samples for training and 50% for validation). Figure 4.d shows that, in this situation, cross validation becomes a misleading criterion, and the optimum width is not accurately determined, but bootstrap selection still indicates clearly the optimum value to be used.

5 Black Box Tachycardia Discriminant

The analysis in this and next sections is a brief description that has been mainly condensed from [5, 6, 7]. More detailed results can be found there, and we only draw the main ideas therein.

A previous study [6] showed that the EGM onset always happens at most as early as 80 ms before the maximum peak of the ventricular EGM. If we call this maximum peak *R wave* (by notation similarity with surface ECG), and assign it a relative time origin (i.e., $t = 0$ at R wave for each beat), we can limit the EGM waveform of interest to the $(-80, 0)$ ms time intervals. This EGM interval will contain all the information related to initial changes in the EGM onset for all (SR, SVT, and VT) episodes. For each episode, we use: (1) a SR template, obtained by alignment of consecutive SR beats previous to the arrhythmia episode, that provides an intra-patient reference measurement; and (b) a T template, obtained in the same way from the arrhythmia episode beats.

However, EGM preprocessing can be a determining issue for the final performance of the classifier, as it can either deteriorate or improve the separability between classes. We will focus here on the following aspects:

- The clinical hypothesis suggests the observation of changes through the EGM first derivative, which corresponds to a rough high-pass filtering. But it must be previously shown that the attenuation on the low frequency components involved by the derivative does not degrade the algorithm performance.
- A previous discriminant analysis upon the electrophysiological features had revealed the onset energies as relevant variables [6], so that rectification could benefit the classification.
- Although beat alignment using R wave reference is habitual, beat synchronization with respect to the maximum of the first derivative is also possible (and often used in other ECG applications). This maximum will be denoted as M_d wave. The best synchronization point is to be tested.
- Inter-patient variability could be reduced by amplitude normalization with respect to the R wave of the patient SR.

Using episodes in Base C, the averaged samples in the 80 ms previous to the synchronization wave were used as the feature space of a Gaussian kernel SVM. Starting from a basic preprocessing scheme, where the EGM first derivative was obtained and R wave synchronization was used, a single preprocessing block was changed each time; rectification incorporation, first derivative removal, M_d synchronization, and SR normalization, led in each case to a different feature space and to a different SVM classifier. Optimum values of the SVM free parameters were found by bootstrap resampling.

Table 2 shows sensitivity, specificity, and complexity (percentage of support vectors) for each classifier. Neither rectification nor the first derivative preprocessing step gives any performance improvement to the classifier. Also, M_d synchronization worsens all the classification rates, probably due to the higher instability of this fiducial point when compared to R wave. Finally, the SR normalization increases the complexity of the related SVM without improving the performance, and hence, it should be suppressed and EGM amplitude should be taken into account.

Table 2. Bootstrap average ± standard deviation for the tested preprocessing schemes

	Sensitivity	Specificity	% of Support Vectors
Basic	90 ± 7	91 ± 10	56 ± 4
Rectifying	91 ± 7	90 ± 9	52 ± 47
M_d aligned	84 ± 7	76 ± 11	78 ± 4
No derivative	91 ± 6	93 ± 8	41 ± 4
SR normalized	89 ± 7	93 ± 8	99 ± 7

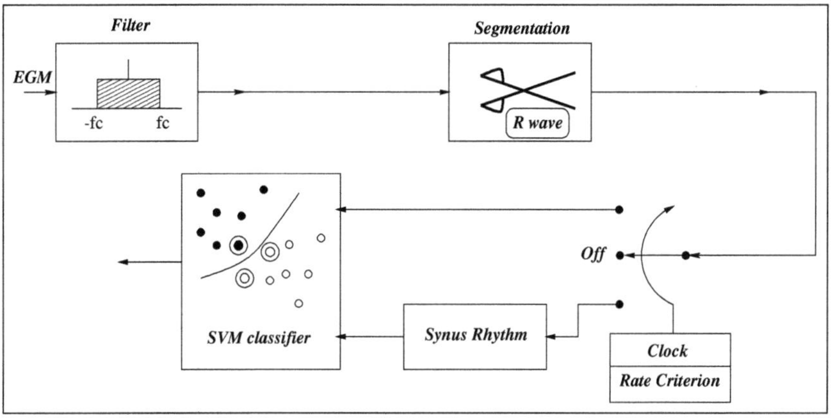

Fig. 5. Black-box discrimination scheme

Therefore, we can conclude that non linear SVM classifiers are robust with respect to preprocessing enhancements that affect the feature space. Information distortion (like unstable synchronization) can deteriorate the classifier performance.

Final Scheme

The black box algorithmic implementation is depicted in Fig. 5. The EGM goes through the next consecutive stages:

1. Noise filtering: cascade of a low pass (45 Hz) and a notch (50 Hz) FIR 32.
2. Segmentation: includes a conventional beat detector, and extracts the 80 ms previous to the R wave to be used as feature vector.
3. SR recorder: periodically stores the SR feature vector.
4. Commuter: allows to switch off the system (when the Rate Criterion for presence of tachycardia is not fulfilled), and to switch on the periodic SR storing or the transmission of the arrhythmic beat.
5. Trained SVM classifier.

The optimum pair of free parameters minimizing the error rate was iteratively adjusted by bootstrap resampling search ($C = 10, \sigma = 5$). For this pair, empirical and bootstrap sensitivity, specificity and complexity were obtained for both Base C (training) and Base D (test set, completely independent of the classifier design process). The result was a performance of 91 ± 6 sensitivity and 94 ± 7 specificity in Base C, and 76 sensitivity and 94 specificity in Base D. For the training set, both high sensitivity and high specificity are provided by the SVM, and all the SVT are correctly classified. For the independent data set, the output of the previously trained classifier agreed with the results in the training set in terms of specificity, but not in sensitivity. This was later observed to be due, in general, to new VT episodes showing a very different morphology from Base C observations. Therefore, the learning procedure has correctly extracted the features in Base C, but it cannot generalize wisely when facing to not previously observed VT episodes. Significant improvement can be obtained by considering an intra-patient algorithm, as proposed in [6].

6 Support Vector Interpretation

In the preceding section, discrimination between SVT and VT from Base C episodes (38 SVT and 68 VT from 26 patients) following the ventricular EGM onset criterion was achieved. The samples contained in the 80 ms preceding the R wave in the SR and tachycardia templates were used as a single input feature vector for each episode. A Gaussian kernel SVM was trained, and the free parameters (kernel width and margin-losses trade-off) were fixed with the bootstrap resampling method in order to avoid the overfitting to the training set. The resulting non linear SVM classifier had 35 support vectors (from 106 total feature vectors), 22 corresponding to saturated coefficients (8 SVT and 14 VT) and 13 corresponding to non saturated coefficients (5 SVT and 8 VT).

Principal Component Analysis

Not surprisingly, the morphological similarity among the support vectors of one and another class is considerable (see [7]), due to the SVM is built in term of the most difficult-to-classify (the most similar) observations. We propose a simple geometrical analysis of the whole set input space and its comparison to the same analysis when performed in the support vectors.

For this purpose, covariance matrices of SR, SVT, and VT vectors are separately obtained, i.e., one matrix per rhythm. Also, covariance matrix only for the SR, SVT, and VT support vectors are calculated. Covariances are factorized by conventional principal component analysis, and eigenvectors are sorted and denoted by $v_1, \ldots, v_{11}$, with decreasing associated eigenvalue.

Figures 6 and 7 show the eigenvectors for the whole data and for the support vectors, respectively. Several considerations can be drawn:

- In the whole set, differences in tachycardia eigenvector morphology appear specially in the most significant (greater eigenvalues) vectors (v_5 to v_{11}), whereas in the support vector set they appear mainly in the least significant (minor eigenvalues) vectors (v_1, v_2). The most significant direction, v_1, differs greatly in each set; however, in both cases it points to important differences between atrial rhythms (SR and SVT) and ventricular rhythms during the first 40 ms.
- In the support vector set, v_1 and v_2 seem to point at differences in two complementary regions, early and late. Hence, critical differences seem to be appearing in two time intervals, early and late activation. This suggests grouping the tachycardia according to their distances to the SR.

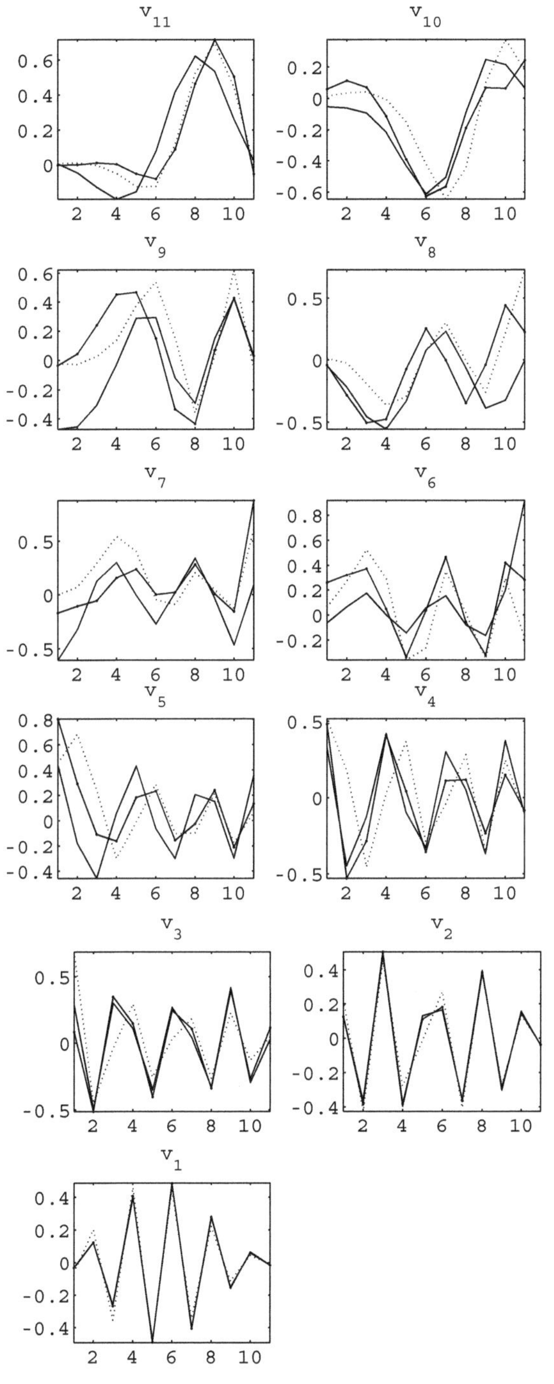

Fig. 6. Eigenvectors for SR (*dotted*), SVT (continuous *dotted*) and VT (continuous) for the covariance matrices of SR, SVT, and VT vectors in Base C

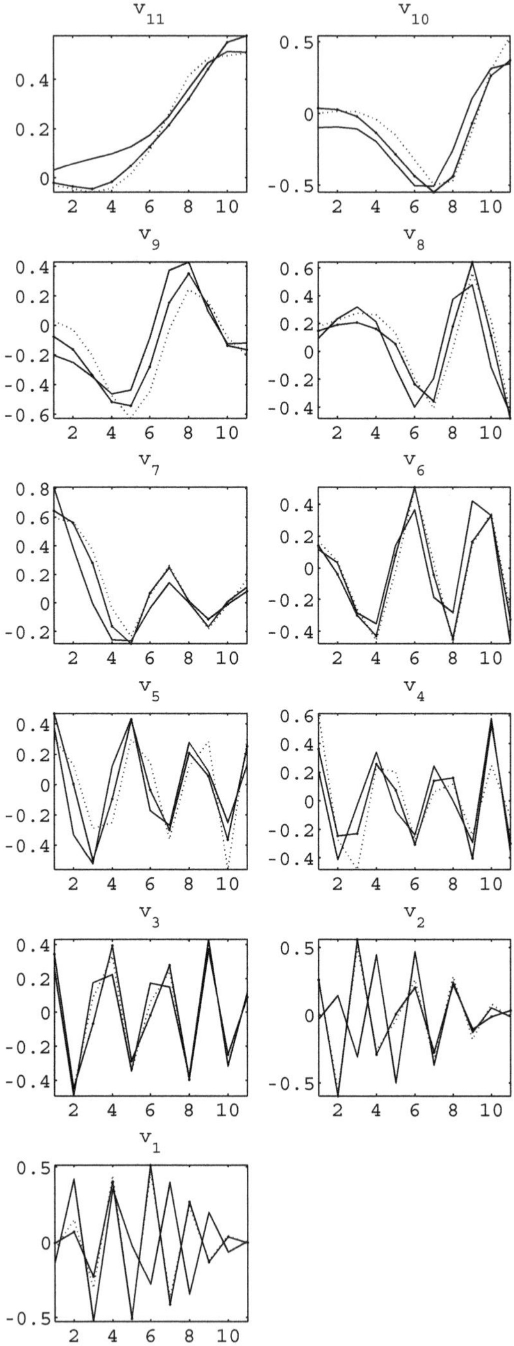

Fig. 7. Eigenvectors for SR (*dotted*), SVT (continuous *dotted*) and VT (continuous) for the covariance matrices of SR, SVT, and VT support vectors only in Base C

- The differences at the principal directions in the support vector set are higher in SVT with respect to SR, while VT differences are still much greater than in the rest of the cases. So, a reasonable approach is to center the search of SVT using their distance to the SR, excluding as VT those vectors appearing far away from this SR center.

Finally, it seems convenient to cluster the SVT vectors, excluding as VT vectors those ones with features being far from SR in any direction. By normalizing the data by SVT mean vector and covariance matrix, a radial geometry can be obtained, and a single parameter (the modulus of the vector) could be used to classify a vector as close or far from the SVT center[4].

Linear SVM Analysis

A more accurate analysis of the relative importance of each time interval can be performed with the aid of the SVM. A non linear machine would lead to a better classification function, but it should be hardly useful for interpretation purposes. However, despite of a poorer performance, a linear machine could provide a clearer information about the relevance of each time sample according to the corresponding weight.

Figure 8(a) represents the input feature space for each episode, consisting on the beat averaged, derived and rectified EGM onset (80 ms previous to the R wave) for both the tachycardia and the preceding SR. Figure 8(b) depicts the weights for the SVM classifier, comparing SR to T weights. Three different activation regions should be considered, which are early, transient or middle, and late.

Given that these three different time regions can be observed in the analysis, the following statistics (see Fig. (5.c)) are obtained for each episode, according to:

$$V_1 = \int_{t=-80\text{ ms}}^{t=-60\text{ ms}} f(t)\,dt \tag{21}$$

$$V_2 = \int_{t=-60}^{t=-20} f(t)\,dt \tag{22}$$

$$V_3 = \int_{t=-20}^{t=0\text{ ms}} f(t)\,dt\ . \tag{23}$$

where $f(t) = |\frac{dEGM(t)^{SR}}{dt}| - |\frac{dEGM(t)^{T}}{dt}|$. In this case, the normalization with respect to the SVT average vector and covariance matrix clearly enhances the detection [7].

The area under the curve was obtained for the black box classifier (0.99 for Base C, 0.92 for Base D), and for the simple rules classifier (0.96 for

[4]Not included here, it can also be shown that, in this case, taking the first derivative and rectifying it enhances the classifier capabilities [7].

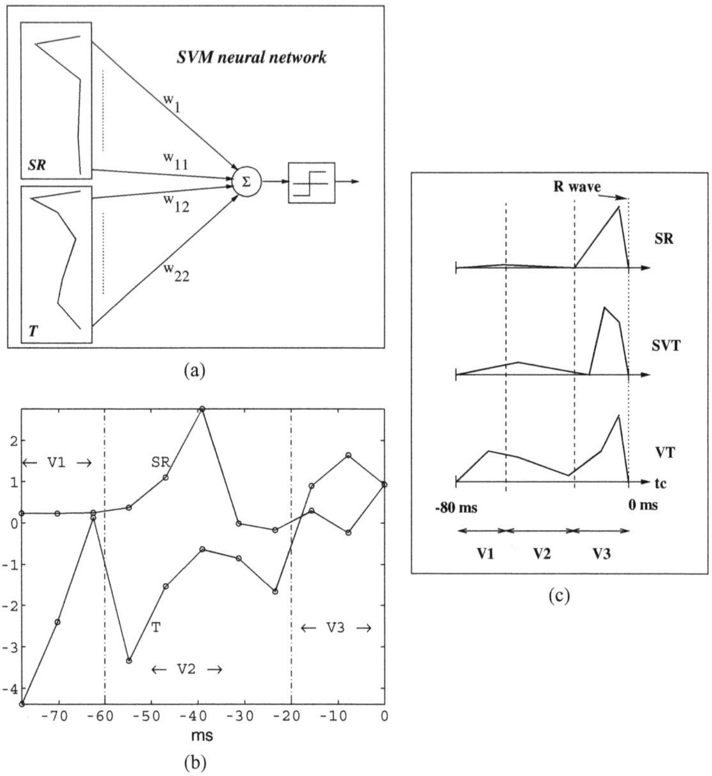

Fig. 8. (**a**) Linear SVM scheme. (**b**) Linear SVM coefficients, comparing SR with T coefficients. (**c**) Intervals for featuring changes in QRS onset: *early* (V_1), *middle* (V_2), and *late* (V_3) activations

Base C, 0.98 for Base D). The performance of the simple rules scheme is not significantly different from the black-box model.

7 Conclusions

The analysis of voltage changes during the initial ventricular activation process is feasible using the detected EGM and the computational capabilities of an ICD system, and may be useful to discriminate SVT and VT. The proposed algorithm yields high sensitivity and specificity for arrhythmia discrimination in spontaneous episodes. The next analysis should be the specific testing of the proposed algorithms with data bases containing a significant number of bundle-branch-block cases [3], as this is still the most challenging problem for most of the SVT-VT discrimination algorithms.

The SVM can provide not only high-quality medical diagnosis machines, but also interpretable black-box model, which is an interesting promise for clinical applications. In this sense, the analysis presented here is mainly heuristical, but statistically detailed and systematic analysis of the support vectors could be developed in order to take profit of the information lying in these critical samples.

Finally, it is remarkable that, in the absence of a statistical featuring for the method, the bootstrap resampling can be used as a tool for complementing the SVM analysis. Also, it can be useful for selecting the SVM free parameters when analyzing low-sized data sets. The usefulness in other SVM algorithms, such as SV regression, kernel-based principal/independent component analysis, or SVM-ARMA modeling [8, 10, 13]

References

1. Bronzino, J.D. (1995) The biomedical engineering handbook. CRC Press and IEEE Press, Boca Raton, FL.
2. Efron, B., Tibshirani, R.J. (1998) An introduction to the bootstrap. Chapman & Hall.
3. Jenkins, J.M., Caswell, S.A. (1996) Detection algorithms in implantable cardioverter defibrillators. Proc. IEEE, 84:428–45.
4. Klingenheben, T., Sticherling, C., Skupin, M., Hohnloser, S.H. (1998) Intracardiac QRS electrogram width. An arrhhythmia detection feature for implantable cardioverter defibrillator. exercise induced variation as a base for device programming. PACE, 8:1609–17.
5. Rojo-Álvarez, J.L., Arenal-Maíz, A., García-Alberola, A., Ortiz, M., Valdés, M., Artés-Rodríguez, A. (2003) A new algorithm for rhythm discrimination in cardioverter defibrillators based on the initial voltage changes of the ventricular electrogram. Europace, 5:77–82.
6. Rojo-Álvarez, J.L., Arenal-Maíz, A., Artés-Rodríguez, A. (2002) Discriminating between supraventricular and ventricular tachycardias from egm onset analysis. IEEE Eng. Med. Biol., 21:16–26.
7. Rojo-Álvarez, J.L., Arenal-Maíz, A., Artés-Rodríguez, A. (2002) Support vector black-box interpretation in ventricular arrhythmia discrimination. IEEE Eng. Med. Biol., 21:27–35.
8. Rojo-Álvarez, J.L., Martínez-Ramón, M., Figueiras-Vidal, A.R., dePrado-Cumplido, M., Artés-Rodríguez, A. (2004) Support Vector Method for ARMA system identification. IEEE Trans. Sig. Proc., 1:155–64.
9. Schaumann, A., von zur Muhlen, F., Gonska, B.D., Kreuzer, H. (1996) Enhanced detection criteria in implantable cardioverter defibrillators to avoid inappropiate therapy. Am. J. Cardiol., 78:42–50.
10. Schölkopf, B. (1997) Support Vector Learning. R. Oldenbourg Verlag.
11. Singer, I. (1994) Implantable cardioverter defibrillator. Futura Publishing Inc.
12. Singh, B.N. (1997) Controlling cardiac arrhythmias: an overview with a historical perspective. Am. J. Cardiol., 80:4G–15G.
13. Vapnik, V. (1995) The nature of statistical learning theory. Springer–Verlag, New York.

MIX
Papier aus verantwortungsvollen Quellen
Paper from responsible sources
FSC® C105338

If you have any concerns about our products,
you can contact us on
ProductSafety@springernature.com

In case Publisher is established outside the EU,
the EU authorized representative is:
Springer Nature Customer Service Center GmbH
Europaplatz 3, 69115 Heidelberg, Germany

Printed by Libri Plureos GmbH
in Hamburg, Germany